O.F.X. Almeida   T.S. Shippenberg (Eds.)

Neurobiology of Opioids

with 57 Figures and 23 Tables

Foreword by A. Goldstein

Springer-Verlag
Berlin Heidelberg New York
London Paris Tokyo
Hong Kong Barcelona
Budapest

Dr. Osborne F.X. Almeida
Department of Neuroendocrinology
Max Planck Institute for Psychiatry
Kraepelinstraße 2
D-8000 München 40
FRG

Dr. Toni S. Shippenberg
Department of Neuropharmacology
Max Planck Institute for Psychiatry
D-8033 Martinsried
FRG

Cover photograph:
Immunocytochemical demonstration of opioid receptors on neuroblastoma × glioma cells (NG 108-15 cell line), using an anti-idiotypic antibody against opioid receptors (courtesy of A.H.S. Hassan and O.F.X. Almeida). See *Neuroscience*, 1989, 32: 269-278, for details.

ISBN-13: 978-3-642-46662-5 e-ISBN-13: 978-3-642-46660-1
DOI: 10.1007/978-3-642-46660-1

Library of Congress Cataloging-in-Publication Data
Neurobiology of opioids/O. Almeida, T. Shippenberg, (eds.). p. cm. Includes bibliographical references. Includes index.
ISBN-13: 978-3-642-46662-5 1. Endorphins-Congresses.
I. Almeida, O. (Osborne), 1954- . II. Shippenberg, T. (Toni), 1957- . [DNLM: 1. Endorphins-physiology. 2. Neurobiology. 3. Receptors, Endorphin-physiology. QU 68 N4935 1989]
QP552.E53N484 1990 612.8'22-dc20 DNLM/DLC

Typesetting: International Typesetters, Inc., Manila, Philippines
31/3145-543210 Printed on acid-free paper

Foreword

It is a singular honor to have been asked to write the Foreword to this volume, which largely consists of the papers presented at the symposium held 2–5 July 1989 at Ringberg, Federal Republic of Germany, to honor the retirement of our colleague Albert Herz. I have known Professor Herz for 20 years in his several roles – leading investigator in the field of opioids, distinguished department head at the basic science unit of the Max-Planck Institute for Psychiatry at Martinsried, one-time host and current Secretary of the International Narcotics Research Conference, member of the Deutsche Akademie der Naturforscher Leopoldina, recipient (1988) of the prestigious Nathan B. Eddy Memorial Award of the Committee on Problems of Drug Dependence. Our relations have been sometimes those of friendly collaborators and sometimes those of friendly competitors. The field of opioid research has been an example – in contrast to some other fields – of how, even in a fast-moving and highly competitive area, friendly collegial relations and mutual respect can be maintained. Albert Herz exemplifies this quality.

One who writes a Foreword has usually had the opportunity to read the book, so that the prefatory remarks can be germane and, at least, not contradictory to some major theme expounded in the main text. Not having seen anything but a Table of Contents puts me at a distinct disadvantage. But there is this offsetting advantage – I can stand back and comment on the state of the field and the contributions of the Herz group to it without the complication of having critically to dissect the individual chapters. I have watched the field of opioid research develop, and have been fortunate enough to be a part of that development over a long enough time to have gained some perspetive. I present these remarks, therefore, not as a scientific review, but simply as the personal opinions that they are.

First, let me explain a matter of style. I hold the view that the scientific leader of a research group is responsible (i.e., gets both the credit and the blame) for the scientific accomplishments of the group. Therefore – and especially since this volume is a testimonial to that leader – I mention no one else by name, even though a large number of younger scientists have been associated with (and often have been the key figures in) the research described here. A number of them are authors of chapters in this book, and the names of some others appear in the selected citations I have provided to aid readers who are unfamiliar with the field. The comprehensive textbook on the opioids (Herz 1978), written as a collective effort of the group, offers an excellent view of the state of the field and the composition of the Martinsried department a dozen or so years ago.

I served as chairman of the international Scientific Advisory Board (Fachbeirat) that site-visited the basic science departments of the Max-Planck Institute for Psychiatry in 1986 and 1988. That experience confirmed what I knew already from the literature, namely, the scientific stature of Professor Herz's Department of Neuropharmacology on the international scene. After each visit, and based on our study of the scientific progress reports, we agreed unanimously – and so informed the president of the Max-Planck Society – that this department's work in the field of opioids was not only of world-class quality but actually ranked among the very few most productive groups anywhere. Let me begin, then, with a straightforward quantitative analysis of this productivity and then furnish a brief qualitative assessment.

Figure 1 shows the number of papers published by the Herz group in each of the past 20 years. There are two reasons why I chose this period of time for analysis. First, two decades is a convenient, round number, approximately the second half of the scientific life of someone who has just reached retirement age. Second, the year 1969 was just about when Herz shifted his interests seriously from other neuropharmacologic problems to the opioids.

A comment is in order concerning the impressive numbers here. The total of 361 publications – the majority, incidentally, in refereed journals of recognized high quality – does not in itself speak for the significance of the work. Indeed, we tend nowadays to look with some suspicion on "too big" a list of publications. This list, however, represents the efforts of a large and extremely talented group, operating (despite turnover) at a consistently high level of scientific activity. An uninformed reader might suspect that the old European system (not unknown in some American institutions) was operating here, whereby the professor routinely appends his name to every paper, without being truly responsible for the work. But every knowledgeable scientist in our field knows of Albert Herz's intimate involvement in the planning and interpretation of experiments, his talent for recruiting the outstanding younger scientists who produced the data, and his ability to expound and defend all aspects of the work of his group.

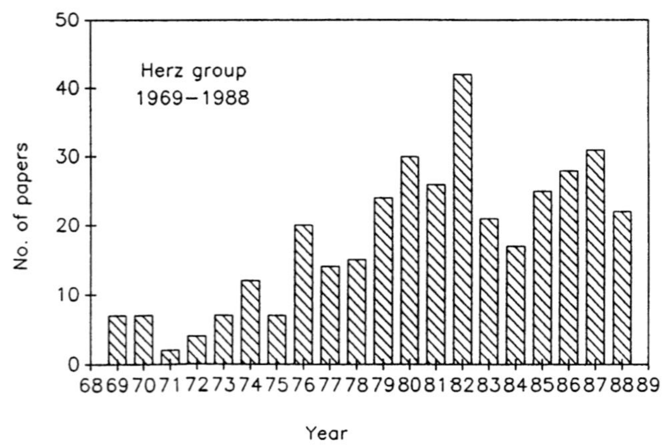

Fig. 1. Number of papers published by the Herz group between 1969–1988

Figure 1 is also an eloquent commentary on the practice of forced retirement at any age, but especially at the early age of 67, when many biological scientists are in the prime of productive life. The figure speaks so well for itself in this regard that no more need be said about this mindlessly rigid and foolish policy!

In Table 1 I break down the scientific productivity by subject area. Naturally, this involved some arbitrary judgments on my part as to how a particular paper was to be classified, especially when it pertained to more than a single category. Nevertheless, one can see a kind of ontogenesis of a career in opioid research here. It begins with studies on the sites of action of morphine and other opiates, especially in relation to antinociception, and with a heavy emphasis on catecholamines and on tolerance and physical dependence. Then comes the end of 1975 and beginning of 1976, and the discovery of the enkephalins and β-endorphin. The Herz group jumped in at once, and already in 1976 published several papers on the endogenous opioid peptides (Blasig and Herz 1976; Schulz and Herz 1976; Zieglgänsberger et al. 1976).

Dynorphin comes onto the scene in 1979, and the Herz group – well positioned now for peptide research – immediately (beginning with Wüster et al. 1980) becomes a foremost contributor. Then, beginning in 1981 (see below) the group gives increasing attention to two areas of wider scope: interrelationships of opioid peptides with other endocrine systems and motivational properties of opioids. It would be interesting to know more about what factors determined their decision to undertake these new and very important research problems.

Enough about the numbers. What has always struck me about the work of the Herz group is the originality and broad impact of their discoveries. Obviously, every scientist would choose differently in assessing which were the most important contributions; these are my own subjective judgments.

Twenty years ago it was unclear where morphine and other opiates acted to produce their antinociceptive effects. The differentiation between spinal and supraspinal sites was especially unclear. As the receptors had not yet been identified, sites of action had to be localized by direct pharmacologic means – injecting at local sites and assaying the effects. Herz devised ingenious ways to confine microinjections of morphine to discrete regions of the central core of the brainstem (Herz et al. 1968, 1970). These early studies focused attention on the ventricular system, especially the floor of the fourth ventricle and the periaqueductal gray (PAG) as key sites (though by no means the only ones) of the antinociceptive actions of morphine. This work may have stimulated Liebeskind's experiments showing that analgesia could be produced by electrical stimulation in the PAG. Those, in turn, led to Akil's discovery (in Liebeskind's laboratory) that the effect of such stimulation could be blocked by naloxone, a result that foreshadowed the discovery of the endogenous opioids.

Before the opioid receptor types (μ, δ, κ, ε) had been well characterized, and without sufficiently type-selective opioid ligands available, most investigators regarded tolerance as a generalized reaction to chronic administration of any opioid. It was therefore not regarded as surprising that opioid peptides should produce tolerance and dependence themselves, as well as cross-tolerance and cross-dependence to morphine; indeed, the Herz group was among the first to demonstrate that (Blasig and Herz 1976; Schulz and Herz 1976). The subsequent discovery of

Table 1. Herz group: analysis of productivity

Year:	1969	1970	1971	1972	1973	1974	1975	1976	1977	1978	1979	1980	1981	1982	1983	1984	1985	1986	1987	1988	Total
Category[a]																					
A	2	4	1	2	5	1	4	2	2	3	1	1									28
B	1	1	1					1				2		3	2	1	2	1	4		19
C				1	2	9	3	14	4	2	7	6	4	5	1		2		2		62
D									2	1	1	2	3	3	4	5		2		1	24
E									1	4	3	7	2	6	3	5	6	5	5	6	53
F											1			2			2	4	3	2	14
G								1		1	3	3	6	9	2	2	2	2	2	2	35
H													2	2		2	2	6	7	3	24
I								1	1	2	4	5	6	4	3		1	2		1	30
J													1	5	2	1	3	3	3	2	20
K														1	2					1	4
L						2		1	2	2	3	3	2	2	2	1	1	1	2		24
M	4	2		1					2		1	1					4	2	3	4	24
Total	7	7	2	4	7	12	7	20	14	15	24	30	26	42	21	17	25	28	31	22	361

[a] A Opioid sites of ction, relation to biogenic amines; B antinociception; C opioid tolerance, dependence, addiction; D localization of opioid peptides; E expression and release of opioid peptides; F opioid receptors, properties, and signal transduction; G opioid receptors, regulation, distribution, binding; H opioids and other hormones, endocrine interrelatioships; I opioid peptide functions, pathophysiology; J motivational properties of opioids; K antibodies to opioids and opioid receptors; L opioids: general, theoretical, reviews; M other neuropharmacology, not opioid.

selective tolerance was therefore unexpected. It was found (Schulz et al. 1980) that a tissue preparation like the mouse vas deferens could be treated chronically (in vivo by means of an osmotic minipump) with a type-selective opioid agonist and thereby made selectively tolerant to agonists of that receptor type.

The phenomenon of selective tolerance was found to apply quite generally, both in brain and in peripheral tissues, and more dramatically as more selective ligands came to hand. It was employed to confirm that there is a receptor type, ε, that is highly selective for β-endorphin. The presence of such a receptor in rat vas deferens, previously reported, was confirmed; and it was demonstrated that a tissue made tolerant to other opioid agonists remained sensitive to β-endorphin (Schulz et al. 1981).

The Herz group raised the first monoclonal anti-idiotypic antibodies that would recognize opioid receptors (Gramsch et al. 1988). An antibody was first raised against an opioid pharmacophore. In this case the known structure-activity relations among the various opioid peptides led to the rational decisions to use β-endorphin as hapten, and then to select a monoclonal antibody that would recognize the essential N-terminal YGGF sequence, shared by all three families of opioid peptides. Then a monoclonal anti-idiotypic antibody was raised against this primary (idiotypic) antibody. Consistent with the way immunoglobin hypervariable regions are thought to fit the three-dimensional shapes of antigens, the anti-idiotypic antibody did, indeed, behave as though it were the "internal image" of the idiotypic binding site; it mimicked opioid ligands in interacting with the binding site of opioid receptors.

The studies on the motivational properties of opioids have truly opened new prospects for understanding "reward systems" in the brain. I have long been puzzled by a curious fact concerning the historical development of opioid research. From early times a strong impetus was the desire to understand opiate addiction. The central point about drug addiction is that people – some people, at least – seek out certain drugs and self-administer them. Tolerance and physical dependence, although they are fascinating neuropharmacological phenomena, are consequences, not causes, of opiate addiction. But nevertheless, studies on self-administration and on other ways of directly studying opioid primary reinforcement (reward) lagged far behind research on tolerance and physical dependence. Until relatively recently they remained entirely in the domain of experimental behavioral psychology, unconnected with neuroanatomy or neurochemistry. This is not the place to analyze the scientific, sociologic, and psychologic reasons for this strange imbalance in opioid research. Suffice it to say that – belated though it may have been – when the Herz group turned their attention to this area, they struck gold.

One of the central questions about the opioid receptor types is whether their functions are highly differentiated, or whether they serve largely redundant purposes. Herz found that whereas μ-receptors mediate positive reinforcement, activation of κ-receptors is clearly aversive. This has been established not only in the hard data of animal experiments using the conditioned place preference technique (Mucha and Herz 1985), but even in human experiments in which Albert Herz himself was one of the volunteer subjects (Pfeiffer et al. 1986). Here, the demonstrated psychotomimetic effects of a κ-agonist were shown to be stereospecific and to be abolished by naloxone, two important properties essential to the definition of an opioid effect.

The group's studies on a possible role of endogenous opioid dysfunction in psychosis, through therapeutic trials with naloxone (Emrich et al. 1977), suggested – in accord with many similar studies by others – that sometimes, in some patients, to some degree, some endogenous opioids may play some role. Not a very illuminating conclusion, that, and deja vu for those of us old enough to have lived through the biogenic amine hypotheses. It is a reminder of how complex the chemical pathology is, of such multifaceted "diseases" as schizophrenia or bipolar depression. There must be hundreds of different genetic-biochemical ways to go crazy! However, the evidence of "aversion systems" that counteract "reward systems" (both systems, perhaps, tonically active to maintain a delicate balance) strikes me as of seminal importance in modifying our thinking about the chemical regulation of behavior.

The group's work over the years on the endogenous opioid peptides and their receptors has been extraordinarily productive. They discovered amidorphin (Liebisch et al. 1985), a previously unknown product of the enkephalin gene, which is amidated at its C-terminus. They discovered opioid activity in fragments of cytochrome (Brantl et al. 1985) and hemoglobin (Brantl et al. 1986); these are interesting as examples of sequences other than YGGF (here, YPWT and YPFT) with affinity for opioid receptors. Analogous discoveries in other laboratories include the casomorphins (and morphiceptin), α-gliadin, dermorphin, and Erspamer's recently discovered deltorphins.

At this writing none of the opioid receptors have yet been cloned and sequenced, but the Herz group has been in the forefront of the attempt to understand their physical properties. Those investigations employed target size techniques to estimate molecular size (Ott et al. 1986) and, over many years, have sought to clarify the signal transduction mechanisms that operate through G-protein and adenylate cyclase interactions (Costa et al. 1988).

Sometimes advances in our knowledge come not from startlingly novel discoveries but from years of patient effort to understand complexity. In this category I would put the numerous contributions of the Herz group to a field in which I have a special fatherly interest – the dynorphin peptides, their distribution, regulation, and physiologic functions (e.g., Przewłocki et al. 1983; Höllt et al. 1987). This research was part of a broad effort to understand how the various opioid peptides regulate the transcription and secretion of other hormones, and reciprocally, how other hormones regulate the opioid peptides (Pfeiffer and Herz 1984).

Let me say, in conclusion, something that is suggested by the undiminished vigorous activity depicted in Fig. 1. I am sure I express the feelings of all opioid researchers in hoping that Albert Herz's influence on the field of opioid research, to which many of the papers in this volume bear witness, will in some manner be able to continue for many years to come.

Stanford, California, USA Avram Goldstein

References

Blasig J, Herz A (1976) Tolerance and dependence induced by morphine-like pituitary peptides in rats. Naunyn-Schmiedeberg's Arch Pharmacol 294:297–300

Brantl V, Gramsch C, Lottspeich F, Henschen A, Jaeger KH, Herz A (1985) Novel opioid peptides derived from mitochondrial cytochrome b: cytochrophins. Eur J Pharmacol 111:293–294

Brantl V, Gramsch C, Lottspeich F, Mertz R, Jaeger KH, Herz A (1986) Novel opioid peptides derived from hemoglobin: hemorphins. Eur J Pharmacol 125:309–310

Costa T, Klinz FJ, Vachon L, Herz A (1988) Opioid receptors are coupled tightly to G proteins but loosely to adenylate cyclase in NG-108–15 cell membranes. Mol Pharmacol 34:744–754

Emrich HM, Cording C, Piree S, Kolling A, von Zerssen D, Herz A (1977) Indication of an antipsychotic action of the opiate antagonist naloxone. Pharmakopsychiatr Neuropsychopharmakol 10:265–270

Gramsch C, Schulz R, Kosin S, Herz A (1988) Monoclonal anti-idiotypic antibodies to opioid receptors. J Biol Chem 263:5853–5859

Herz A (ed) (1978) Developments in opiate research. Marcel Dekker, New York

Herz A, Metys J, Schondorf N, Hoppe S (1968) Über den Angriffspunkt der analgetischen Wirkung von Morphin. Naunyn-Schmiedeberg's Arch Pharmacol 260:143

Herz A, Albus K, Metys J, Schubert P, Teschemacher H (1970) On the central sites for the antinociceptive action of morphine and fentanyl. Neuropharmacology 9:539–551

Höllt V, Haarman I, Millan MJ, Herz A (1987) Prodynorhin gene expression is enhanced in the spinal cord of chronic arthritic rats. Neurosci Lett 73:90–94

Liebisch DC, Seizinger BR, Michael G, Herz A (1985) Novel opioid peptide amidorphin: characterization and distribution of amidorphin-like immunoreactivity in bovine, ovine, and porcine brain, pituitary, and adrenal medulla. J Neurochem 45:1495–1503

Mucha RF, Herz A (1985) Motivational properties of kappa and mu opioid receptor agonists studied with place and taste preference conditioning. Psychopharmacology 86:281–285

Ott S, Costa T, Wüster M, Hietel B, Herz A (1986) Target size analysis of opioid receptors. No difference between receptor types, but discrimination between two receptor states. Eur J Biochem 155:621–630

Pfeiffer A, Herz A (1984) Endocrine actions of opioids. Horm Metab Res 16:386–397

Pfeiffer A, Brantl V, Herz A, Emrich HM (1986) Psychotomimesis mediated by kappa opiate receptors. Science 233:774–776

Przewłocki R, Lason W, Konecka AM, Gramsch C, Herz A (1983) The opioid peptide dynorphin, circadian rhythms, and starvation. Science 219:71–73

Schulz R, Herz A (1976) Dependence liability of enkephalin in the myenteric plexus of the guinea pig. Eur J Pharmacol 39:429–432

Schulz R, Wüster M, Krenss H, Herz A (1980) Selective development of tolerance without dependence in multiple opiate receptors of mouse vas deferens. Nature 285:242–243

Schulz R, Wüster M, Herz A (1981) Pharmacological characterization of the epsilon-opiate receptor. J Pharmacol Exp Ther 216:604–606

Wüster M, Schulz R, Herz A (1980) Highly specific opiate receptors for dynorphin-(1–13) in the mouse vas deferens. Eur J Pharmacol 62:235–236

Zieglgänsberger W, Fry JP, Herz A, Moroder L, Wunsch E (1976) Enkephalin-induced inhibition of cortical neurones and the lack of this effect in morphine tolerant/dependent rats. Brain Res 115:160–164

Preface

Professor Albert Herz

This book is dedicated to Professor Albert Herz whose leadership and contributions to opioid research are described in the Preface. It is comprised of articles written by friends and colleagues with whom he has cooperated over the years.

We have attempted to portray the many facets of opioid research, with particular emphasis given to those aspects of the field in which Professor Herz has been involved. We necessarily had to omit contributions from many eminent scientists for which we hope to be excused. Nevertheless, their contributions to the area are mirrored in the lists of citations.

We wish to thank the authors for their patience and cooperation, and Dr. Jutta Lindenborn of Springer-Verlag for her kind support. Lastly, on behalf of past members of the Department of Neuropharmacology at the Max Planck Institute for Psychiatry, we would like to express our appreciation to Professor Herz for providing a stimulating environment within which researchers from different disciplines could interact. Although Professor Herz has officially retired, he continues to persue an active interest in opioids, as reflected in his contribution to this book (final chapter). We wish him the best for the future.

Munich, Martinsried
September 1990

O.F.X. Almeida
T.S. Shippenberg

Contents

Section I
General Aspects

Chapter 1
Opioid Receptor Subtypes: Past, Present and Future 3
H.W. Kosterlitz

Chapter 2
Opioid Peptide Genes: Structure and Regulation 11
V. Höllt

Chapter 3
Distribution of Opioids in Brain and Peripheral Tissues 53
M.K.-H. Schafer, R. Day, S.J. Watson, and H. Akil

Section II
Opioid Receptor Biochemistry and Signal Transduction

Chapter 4
Multiple Opioid Receptors 75
E.J. Simon

Chapter 5
Molecular Approaches to Isolating Opioid Receptors 89
A.P. Smith and H.H. Loh

Chapter 6
Interactions Between Opioid Receptors and Guanine Nucleotide Regulatory Proteins in Intact Membranes: Studies in a Neurotumor Cell Line 101
T. Costa, S. Ott, L. Vachon, and F.-J. Klinz

Chapter 7
Guanine Nucleotide Binding Proteins and Their Coupling to Opioid Receptors 121
J. Lang

Chapter 8
Opioid Receptors and Ion Channels 141
R.A. North

Chapter 9
Cellular Signaling Mechanisms Regulating Opioid Peptide Gene Expression 151
N.A. Kley, C.-J. Farin, and J.P. Loeffler

Section III
CNS Opioidergic Systems: Distribution and Modulation

Chapter 10
Central Distribution of Opioid Receptors: A Cross-Species Comparison of the Multiple Opioid Systems of the Basal Ganglia 169
A. Mansour, M.K.-H. Schaffer, S.W. Newman, and S.J. Watson

Chapter 11
Modulation of Central Opioid Receptors 185
B.J. Morris

Chapter 12
The Control of Hypothalamic Opioid Peptide Release 199
G. Burns and K.E. Nikolarakis

Chapter 13
Modulation of Catecholamine Release in the Central Nervous System by Multiple Opioid Receptors 213
P. Illes and R. Jackisch

Section IV
Functional Aspects

Chapter 14
Adaptation of Opioid Systems to Stress 229
R. Przewłocki, B. Przewłocka and W. Lasón

Chapter 15
Endogenous Opioid Systems in the Control of Pain 245
M.J. Millan, E. Weihe, and A.C. Czlonkowski

Chapter 16
Opioid Interactions with Other Neuropeptides in the Spinal Cord: Relevance to Nociception 261
M. Satoh and Y. Kuraishi

Chapter 17
Opioid Analgesia at Peripheral Sites 273
C. Stein

Chapter 18
Regulation of Intestinal Motility by Peripheral Opioids: Facts and Hypotheses 287
W. Kromer

Chapter 19
A Neuroendocrine Perspective of Sex and Drugs 301
O.F. X. Almeida and D.G. Pfeiffer

Chapter 20
Motivational Aspects of Opioids: Neurochemical and Neuroanatomical Substrates 331
T.S. Shippenberg and R. Bals-Kubik

Chapter 21
Psychotomimetic Effects of Opioids 351
A. Pfeiffer

Chapter 22
Psychiatric Aspects of Opioid Research 363
H. Emrich and C. Schmauss

Chapter 23
Functional Correlates of Opioid Receptor Activation as Measured by Local Cerebral Glucose Utilization 369
A. Ableitner

Section V
Opioid Tolerance and Dependence

Chapter 24
Tolerance and Physical Dependence: Physiological Manifestations of Chronic Exposure to Opioids 387
F.J. Ayesta

Chapter 25
Aspects of Opioid Tolerance and Dependence in Peripheral Nervous Tissues 407
R. Schulz

Chapter 26
Overview of Opioid Tolerance and Physical Dependence 417
A.K. Louie and E.L. Way

Section VI
Epilogue

Chapter 27
Some Thoughts on the Present Status and Future of Opioid Research 439
A. Herz

Contributors

Ableitner Annemarie*, Institute for Pharmacology, Toxicology and Pharmacy, Ludwig Maximilian University, Königinstraße 16, D-8000 Munich 22, FRG

Akil Huda, Mental Health Research Institute, University of Michigan, 205 Washtenaw Place, Ann Arbor, Michigan 48109–0720, USA

Almeida Osborne F.X.*, Institute for Pharmacology, Toxicology and Pharmacy, Ludwig Maximilian University, Königinstraße 16, D-8000 Munich 22, FRG

Ayesta F. Javier*, Department of Physiology and Pharmacology, Division of Pharmacology, University of Cantabria, 39011 Santander, Spain

Bals-Kubik Regine*, Department of Neuropharmacology, Max Planck Institute for Psychiatry, D-8033 Martinsried, FRG

Burns Geoffrey*, Department of Neuropharmacology, Max Planck Institute for Psychiatry, D-8033 Martinsried, FRG

Costa Tommaso*, Laboratory for Theoretical Biology, National Institute of Child Health and Human Development, National Institutes of Health, Bethesda, Maryland 20982, USA

Czlonkowski Andrez*, Department of Pharmacology, Institute of Physiology, Medical Academy of Warsaw, Poland

Day Robert, Mental Health Research Institute, University of Michigan 205 Washtenaw Place, Ann Arbor, Michigan 48109–0720, USA

Emrich Hinderk, Max Planck Institute for Psychiatry, Kraepelinstraße 10, D-8000 Munich 40, FRG

Farin-Claus-Jürgen*, Department of Neuropharmacology, Max Plank Institute for Psychiatry, D-8033 Martinsried, FRG

Goldstein Avram, 735 Dolores, Stanford, California 94305, USA

Herz Albert*, Department of Neuropharmacology, Max Planck Institute for Psychiatry, D-8033 Martinsried, FRG

Höllt Volker*, Institute of Physiology, Ludwig Maximilian University, Pettenkofferstraße 12, D-8000 Munich 2, FRG

Illes Peter*, Institute of Pharmacology and Toxicology, Albert-Ludwig University, Hermann-Herder Straße 5, D-7800 Freiburg, FRG

Jackisch Rolf, Institute of Pharmacology and Toxicology, Albert-Ludwig University, Hermann-Herder Straße 5, D-7800 Freiburg, FRG

Kley Nikolai A.*, Neurogenetics Laboratory, Harvard Medical School, Massachusetts General Hospital, Cambridge, Massachusetts 02114, USA

Klinz Franz-Josef*, Institute of Pharmacology, Free University of Berlin, Thielallee 69–73, D-1000 Berlin 33, FRG

Kosterlitz Hans W, Unit for Research on Addictive Drugs, Marischal College, University of Aberdeen, Aberdeen AB9 1AS, UK

Kromer Wolfgang, Department of Pharmacology, Byk Gulden Pharmaceuticals, Byk-Gulden-Straße 2, D-7750 Constance, FRG

Kuraishi Yashushi, Department of Pharmacology, Faculty of Pharmaceutical Sciences, Kyoto University, Sakyo-ku, Kyoto 606, Japan

Lang Jochen*, Department of Medicine, Division of Endocrinology, Thyroid Laboratory, 25 Rue Micheli-du-Crest, CH-1211 Geneva, Switzerland

Lason Wladysław, Department of Neuropeptide Research, Institute of Pharmacology, Polish Academy of Sciences, 12 Smetna Street, 31–343 Krakow, Poland

Loeffler Jean-Phillipe*, Institute of Biochemical Physiology, Louis Pasteur University, Rue Rene des Cartes, Strasbourg, France

Loh Horace H., Department of Pharmacology, University of Minnesota Medical School, 435 Delaware Street S.E., Minneapolis, Minnesota 55455, USA

Louie Alan K., Departments of Psychiatry, Pharmacology and Pharmaceutical Chemistry, University of California, San Francisco, California 94143, USA

Mansour Alfred, Mental Health Research Institute, University of Michigan, 205 Washtenaw Place, Ann Arbor, Michigan 48109–0720, USA

Millan Mark J.*, FONDAX-Groupe de Recherche SERVIER, 7 Rue Ampere, 92800 Puteaux, Paris, France

Morris Brain J.*, MRC Molecular Neurobiology Unit, MRC Centre, Hills Road, Cambridge CB2 2QH, UK

Newman Sarah W., Mental Health Research Institute, University of Michigan, 205 Washtenaw Place, Ann Arbor, Michigan 48109–0720, USA

Nikolarakis Konstantinos E.*, Department of Neuropharmacology, Max Planck Institute for Psychiatry, D-8033 Martinsried, FRG

North R. Alan, Institute for Advanced Biomedical Research, Oregon University of the Health Sciences, 3181 S. W. Sam Jackson Park Road, Portland, Oregon 97201, USA

Ott Susanna*, Department of Biopharmacy, ETH, Institute for Pharmacy, University of Zurich, Claudstraße 25, CH-8092 Zurich, Zwitzerland

Pfeiffer Andreas*, Medizinische Klinik und Poliklinik, Klinikum Bergmannsheil, Universität Bochum, Gilsingstraße 14, D-4630 Bochum 1, FRG

Pfeiffer Doris*, Frauenklinik, Klinikum Großhadern, Ludwig Maximilian University, Marchionistraße 15, D-8000 Munich 70, FRG

Przewłocka Barbara, Department of Neuropeptide Research, Institute of Pharmacology, Polish Academy of Sciences, 12 Smetna Street, 31–343 Krakow, Poland

Przewłocki Rysard*, Department of Neuropeptide Research, Institute of Pharmacology, Polish Academy of Sciences, 12 Smetna Street, 31–343 Krakow, Poland

Satoh Masamichi, Department of Pharmacology, Faculty of Pharmaceutical Sciences, Kyoto University, Sakyo-ku, Kyoto 606, Japan

Schafer Martin K.-H., Mental Health Research Institute, University of Michigan, 205 Washtenaw Place, Ann Arbor, Michigan 48109–0720, USA

Schmauss Claudia, Max Planck Institute for Psychiatry, Kraepellinstraße 10, D-8000 Munich 40, FRG

Schulz Rüdiger*, Institute of Pharmacology, Toxicology and Pharmacy, Ludwig Maximilian University, Königinstraße 16, D-8000 Munich 22, FRG

Shippenberg Toni S.*, Department of Neuropharmacology, Max Planck Institute for Psychiatry, D-8033 Martinsried, FRG

Simon Eric J., Department of Psychiatry and Pharmacology, New York University Medical Center, New York, New York 10016, USA

Smith Andrew P., Department of Pharmacology, University of Minnesota Medical School, Minneapolis, Minnesota 55455, USA

Stein Christoph*, Department of Anaesthesiology, Klinkum Großhadern, Ludwig Maximilian University, Marchionistraße 15, D-8000 Munich 70, FRG

Watson Stanley J., Mental Health Research Institute, University of Michigan, 205 Washtenaw Place, Ann Arbor, Michigan 48109-0720, USA

Way E. Leong, Departments of Psychiatry, Pharmacology and Pharmaceutical Sciences, University of California, San Francisco, California 94143, USA

Vachon Luc*, Sandoz, Montreal, Canada

Weihe Eberhardt, Institute of Anatomy, Johannes-Gutenberg University, Saarstraße, D-6500 Mainz, FRG

*Current or previous associates of Professor Albert Herz.

Section I

General Aspects

CHAPTER 1

Opioid Receptor Subtypes: Past, Present and Future

H.W. Kosterlitz

1 Introduction

As this presentation is in honour of Albert Herz, I wish to give as introduction the occasion when I first met him at the International Pharmacological Congress in Basle in July 1969. Harry Collier and I arranged to have a meeting of pharmacologists interested in problems of drug dependence. It was then that I had the good fortune of being asked by Albert Herz to participate in a Symposium on Pain to be held about a month later. This was the beginning of getting to know one of my best friends in the area of the problems of opioid drugs and their tendency to drug dependence. When the paper on the discovery of endogenous [Met]enkephalin and [Leu] enkephalin was published, he sent me a touching letter: "Fortuna verteilt ihre Gaben doch nicht immer blind".

It is not necessary to consider his work and that of his colleagues in detail as it is well known in the scientific world. Because of its outstanding quality, it is an important addition to our knowledge of opioid physiology and pharmacology.

In the investigation of the actions of neuroactive compounds in the brain, progress has been facilitated by the use of peripheral tissues. A classical example is the work by P. Trendelenburg in 1917. He found that the peristaltic reflex caused by distension of the lumen of the in vitro guinea-pig ileum is blocked by concentrations of morphine similar to those present in tissues after administration of therapeutic doses given to whole animals. A number of narcotic analgesic drugs had similar effects which were correctly graded by their potency to inhibit the peristaltic effects in similar tests (Gyang et al. 1964). For systematic analysis electrical stimulation was much more successful than distension of the lumen of the intestine (Paton 1957).

There has been a period during which progress of our understanding of the role of opioids has been rather slow. Probably the main reason was the lack of highly selective antagonists. When naloxone became available, the circumstances changed fundamentally. Two receptor systems became available, one depending on bioassay in guinea-pig intestine (Kosterlitz and Watt 1968; Kosterlitz et al. 1972; Lees et al. 1972) and the other on their effects in binding assays in mammalian brain and guinea-pig intestine (Pert and Snyder 1973). In this context, a review and a book may be of interest: *The best laid schemes o'mice an'men gang aft agley* (Kosterlitz 1979) and *Brainstorming* (Snyder 1989).

One of the main problems of our understanding of the mode of action of the endogenous opioid peptides is the multiplicity of the fragments of proopiocortin, proenkephalin and prodynorphin. Furthermore, none of the fragments interact with only one of the μ-, δ- and κ-sites of the receptors (Kosterlitz and Paterson 1985). The complexity of the system has recently been aggravated by the finding that the non-peptide morphine is present in low concentrations in animal tissue where it can be biosynthesized from its non-morphinan precursor reticuline (Donnerer et al. 1986, 1987; Weitz et al. 1986, 1987). The possible physiological significance of morphine in animal tissue is still uncertain (Kosterlitz 1987).

A difficulty of the interpretation of the physiological effects of the endogenous opioid peptides is due to the fact that almost all are degraded by peptidases. It is therefore necessary either to use antipeptidases or synthetic peptidase-resistant analogues, the action of which may be different from the endogenous peptides.

2 Binding Assays of μ-, δ- and κ-Ligands in Membranes of Guinea-Pig Brain

It is important to note that the binding interaction of a molecule may or may not lead to a biological response of either excitatory or inhibitory activity. Pharmacologically, such activity would indicate that the response is that of an agonist. An antagonist compound would also bind to the receptor but block its excitatory or inhibitory response.

The affinity of a ligand is given as its binding affinity constant, $(K_i,\text{nM})^{-1}$, which is the reciprocal of its inhibition constant (Tables 1–3). In addition, it is useful to determine the relative binding affinities by the ratio K_i^{-1} for μ, δ or $\kappa/(K_i^{-1}$ for $\mu + K_i^{-1}$ for $\delta + K_i^{-1}$ for $\kappa)$, the maximum being 1.00. The temperature of the binding assays was 25°C when peptidase-resistant compounds were used and 0°C when peptides sensitive to enzyme activity were used.

Table 1. Binding affinities of μ-selective opioids and their relative binding affinities for the μ-, δ- and κ-sites in homogenates of guinea-pig brain[a]

	μ-Affinity $(K_i, \text{nM})^{-1}$	Relative affinity[b] μ	δ	κ
[Met]enkephalyl-Arg-Arg-Val-NH_2	16.7	0.77	0.03	0.20
[Met]enkephalyl-Arg-Arg-Val-Gly-Arg-Pro-Glu-Trp Trp-Met-Asp-Tyr-Gln (BAM 18)	3.4	0.68	0.06	0.26
[Met]enkephalyl-Arg-Phe	0.29	0.60	0.36	0.04
β-Endorphin	0.49	0.52	0.45	0.03
Morphine	0.56	0.97	0.02	0.01
[D-Ala^2, MePhe^4, Gly-ol^5]enkephalin	0.54	0.99	0.01	0
Tyr-D-Arg-Phe-Lys-NH_2	0.28	0.998	0	0.002

[a] Hurlbut et al. (1987); Kosterlitz (1985); Kosterlitz and Paterson (1985); Paterson et al. (1984); Schiller et al. (1989).

[b] Relative binding affinities at the μ-, δ- and κ-sites are: K_i^{-1} for μ, δ or $\kappa/(K_i^{-1}$ for $\mu + K_i^{-1}$ for $\delta + K_i^{-1}$ for $\kappa)$

Table 2. Binding affinities of δ-selective opioids and their relative binding affinities for the μ-, δ- and κ-sites in homogenates of guinea-pig brain[a]

	δ-Affinity $(K_i, nM)^{-1}$	Relative affinity μ	δ	κ
[Leu]enkephalin	0.85	0.06	0.94	0
[Met]enkephalin	1.10	0.09	0.91	0
[D-Ala^2, D-Leu^5]enkephalin	0.74	0.10	0.90	0
[D-Pen^2, D-Pen^5]enkephalin	0.37	0.004	0.996	0

[a] Kosterlitz and Paterson (1985); Paterson et al. (1984).

Table 3. Binding affinities of κ-selective opioids and their relative binding affinities for the μ-, δ- and κ-sites in homogenates of guinea-pig brain[a]

	κ-Affinity $(K_i, nM)^{-1}$	Relative affinity μ	δ	κ
Dynorphin A	8.7	0.13	0.04	0.83
Dynorphin B	8.5	0.14	0.03	0.83
α-Neo-endorphin	5.1	0.10	0.23	0.67
Dynorphin A (1–8)	0.75	0.22	0.16	0.62
[D-Pro^{10}]dynorphin A (1–11)[b]	34.7	0.05	0.01	0.94
U-69,593	0.74	0.001	0	0.999
PD 117302[c]	1.72	0.003	0	0.997

[a] Kosterlitz (1985); Kosterlitz and Paterson (1985); Paterson et al. (1984).
[b] Gairin et al. (1985).
[c] (±)-trans-N-methyl-N-[2-(1-pyrrolidinyl)-cyclohexyl] benzo[b]thiophene-4-acetamide (Birchmore et al. 1987).

In Table 1 four peptides are shown that, at the μ-site, have relative affinities varying between 0.52 and 0.77. Two of these are [Met]enkephalins extended at the terminus with -Arg-Arg in positions 6 and 7. [Met]enkephalyl-Arg-Arg-Val-NH_2 is of particular interest since it has the very high affinity of 16.7 nM^{-1}. The relative affinities of this peptide and its possible precursor, BAM 18 (Hurlbut et al. 1987) are 0.77 and 0.68 at the μ-site, 0.20 and 0.26 at the κ-site and very low at the δ-site. In contrast, [Met]enkephalyl-Arg-Phe and β-endorphin have much lower binding affinities of 0.29 nM^{-1} and 0.49 nM^{-1}. Their relative affinities are 0.60 and 0.52 at the μ-site, 0.36 and 0.45 at the δ-site and very low at the κ-site.

The binding pattern of morphine is very different. As discussed in Section 1, morphine is an endogenous compound, but is present only in low concentrations. Its binding affinity is of an average order but it is of particular interest that this plant opioid has the very high relative μ-affinity of 0.97 which is much higher than the relative affinities of any of the known endogenous μ-opioids in animal tissue.

Since none of the endogenous peptides is sufficiently selective for the μ-site, it was important to obtain synthetic compounds. Tyr-D-Arg-Phe-Lys-NH_2 and [D-Ala^2,MePhe^4,Gly-ol^5]enkephalin fulfil this purpose (Table 1).

The endogenous δ-ligands [Met]enkephalin and [Leu]enkephalin have relative affinities of 0.91 and 0.94 but low values of 0.09 and 0.06 at the μ-site (Table 2). These high values at the δ-site should be compared with the corresponding maximum values for the peptide μ-ligands of 0.77 (Table 1) and for the κ-ligand of 0.83 (Table 3). The availability of an even more selective δ-ligand, [D-Pen^2,D-Pen^5]enkephalin, has become important as also the δ-antagonist naltrindole (Table 5).

With regard to κ-ligands (Table 3), there are no endogenous peptidase-resistant compounds of high potency, with the possible exception of dynorphine A (1–17). Dynorphin A (1–17) and dynorphin B (1–13) and α-neo-endorphin have high affinities at the κ-sites (5.1–8.7 nM^{-1}). In contrast, the fragment dynorphin A (1–8), present in many areas of the nervous system in higher concentrations than dynorphin A (1–17), has a much lower affinity at the κ-site (0.75 nM^{-1}) and is also much less selective. The physiological significance of this difference between the two peptides is still not understood.

Non-peptide ligands with a very high degree of selectivity for the κ-binding site are U-69,593 and PD 117,302 with relative affinities at the κ-site of more than 0.99. Another interesting compound is [D-Pro^{10}]dynorphin A (1–11) which has the highest affinity to the κ-site so far observed but its relative binding affinity at the κ-site is somewhat lower than the values obtained with the two non-peptide compounds.

3 Bioassays of Opioid μ-, δ- and κ-Receptors in Isolated Tissue Preparations

While a final understanding of the mode of action of opioid peptides or non-peptide opioids has to await investigation in vivo, the pharmacological effects of opioid compounds and their physiological properties are often more readily analyzed in excised tissues. By such an approach, it is possible to decide whether a synthetic opioid has agonist or antagonist properties or a combination of both. Such a differentiation is not possible in binding assays on homogenized brain membranes.

In the bioassays used in this section, the opioid peptides or non-peptides have high selectivity for one of the μ-, δ- or κ-receptors. The assay tissues are from five different species which have opioid receptors of one, two or three subtypes (Table 4). The opioid receptors of the rabbit vas deferens interact with κ-ligands, e.g. dynorphin A and U-69,593, those of the hamster vas deferens with δ-ligans, e.g. [D-Pen^2,D-Pen^5]enkephalin, those of the rat vas deferens mainly, but not solely, with μ-ligands, e.g. morphine and [D-Ala^2,MePhe^4,Gly-ol^5]enkephalin and β-endorphin, those of the guinea-pig ileum with μ- and κ-ligands, e.g. [D-Ala^2,MePhe^4,Gly-ol^5]enkephalin, dynorphin A and U-69,593, and, finally, those of the mouse vas deferens interact with all three μ-, δ- and κ-ligands (Kosterlitz and Paterson 1985).

There are unsolved problems concerning the physiology and pharmacology of opioid compounds. Firstly, almost all endogenous opioid peptides are liable to degradation by peptidases and, secondly, potent selective agonists and antagonists have become available only recently.

Table 4. Opioid receptor activity (37°C) in rat vas deferens (RVD), hamster vas deferens (HVD), rabbit vas deferens (LDV), guinea-pig ileum myenteric plexus (GPI) and mouse vas deferens (MVD)[a]

	LVD (κ)	HDV (δ)	RVD (μ)	GPI ($\mu+\kappa$)	MVD ($\mu+\delta+\kappa$)
Morphine	0	0	A	A	A
[D-Ala^2, MePhe^4, Gly-ol^5]enkephalin	0	0	A	A	A
β-Endorphin	0	A	A	A	A
[Met]- and [Leu]enkephalin	0	A	A	A	A
[D-Pen^2, D-Pen^5]enkephalin	0	A	0	(A)	A
Dynorphin A	A	(A)	0	A	A
U-69,593[b]	A	(A)	0	A	A
Naloxone	ANT	ANT	ANT	ANT	ANT
(−)-Bremazocine	A	ANT	ANT	A	A
(−)-Ethylketazocine	A	ANT	ANT	A	A

[a]0 = no effect, A = agonist, ANT = antagonist. Antipeptidases were present for assay of [Met]- and [Leu]enkephalin and dynorphin A (Kosterlitz 1985; Kosterlitz and Paterson 1985; McKnight et al. 1985; Paterson et al. 1984).

[b](5α, 7α, 8β)-(+)-N-methyl-N(7-(1-pyrrolidinyl)-1-oxaspiro-[4,5]dec-8-yl)-benzeneacetamide (Lahti et al. 1985).

The new antagonist for the κ-site, norbinaltorphimine (Portoghese et al. 1987), has considerable selectivity for the κ-receptor but appears to be degraded in brain tissue (Birch et al. 1987). One of the selective δ-antagonists, ICI 174,864 is highly selective but is of low potency (Cotton et al. 1984). It is now superceded by the potent and selective δ-antagonist naltrindole (Portoghese et al. 1988). Recently, a selective μ-antagonist, CTOP, has become available (Hawkins et al. 1989) but its antagonist potency (K_e = 16 nM) is lower than that of the δ-antagonist naltrindole (K_e = 0.11 nM) or that of the κ-antagonist norbinaltorphimine (K_e = 0.14 nM) (Table 5).

It is important to be aware of the fact that ligands may be agonists at opioid receptors in some tissues but antagonists in others. Ethylketazocine and bremazo-

Table 5. Binding affinities of μ-, δ- and κ-opioid antagonists and their relative binding affinities in homogenates of guinea-pig brain. Antagonist activity (K_e)

	Affinity $(K_i, nM)^{-1}$	Relative affinity μ	δ	κ	Antagonist[d] (K_e, nM)
CTOP[a]	0.52(μ)	>0.998	–	–	16.1
Naltrindole[b]	8.3 (δ)	0.01	0.98	0.01	0.11
Norbinaltor-phimine[c]	2.9 (κ)	0.02	0.03	0.95	0.14

[a]Hawkins et al. (1989).

[b]Portoghese et al. (1988).

[c]Portoghese et al. (1987).

[d]Selective agonist for μ-site, [D-Ala^2, MePhe^4, Gly-ol^5]enkephalin (GPI), for the δ-site, DPDPE (MVD), and for the κ-site, U-69,593 (GPI).

cine are examples of this behaviour, as they are κ-agonists in the guinea-pig ileum and the vasa deferentia of the mouse and rabbit but μ-antagonists in the vas deferens of the rat and δ-antagonists in the vas deferens of the hamster (Table 4).

4 Conclusions

The investigation of the present state of our knowledge of the opioid receptors and their endogenous ligands confirms that the analysis of their mode of action is increasingly complex. Two recent independent events are likely to be of importance.

More than 20 years ago progress was made by the availability of the opioid antagonist naloxone which has negligible agonist activity but interacts with three μ-, δ- and κ-sites. The synthesis by Portoghese and his group of naltrindole, a selective δ-antagonist, and of norbinaltorphimine, an almost selective κ-antagonist, is of high significance. Furthermore, the structure of the cyclic octapeptide CTOP, a selective μ-antagonist, has been published quite recently (Hawkins et al. 1989). We may expect that our understanding of the mode of action of endogenous opioids will be much facilitated.

The second event is the discovery by the groups of Goldstein and of Spector that morphine is an endogenous opioid. Although present only in low concentrations, it may not only be of pharmacological but also of physiological significance. Whatever the results of future experiments may indicate, it is interesting that morphine is the only known member of the morphinan group that is such a highly selective μ-opioid ligand. The known endogenous opioid peptides are much less selective in this respect.

Acknowledgements. Supported by grants from the Medical Research Council and the National Institute on Drug Abuse (DA-0062).

References

Birch PJ, Hayes AG, Sheehand MJ, Tyres MB (1987) Norbinaltorphimine: antagonist profile at κ opioid receptors. Eur J Pharmacol 144:405–408

Birchmore B, Clark CR, Hill DC, Horwell JC, Hunter J, Hughes J, Sharif N (1987) PD 117302: a selective agonist at the κ opioid receptor. Br J Pharmacol 91:299P

Cotton R, Giles MG, Miller L, Shaw JS, Timms D (1984) ICI 174864: a highly selective antagonist of the δ-opioid receptor. Eur J Pharmacol 97:331–332

Donnerer J, Oka K, Brossi A, Rice KC, Spector S (1986) Presence and formation of codeine and morphine in the rat. Proc Natl Acad Sci USA 83:4566–4567

Donnerer J, Cardinale G, Coffey J, Lisak CA, Jardine I, Spector S (1987) Chemical characterization and regulation of endogenous morphine and codeine in the rat. J Pharmacol Exp Ther 242:583–587

Gairin JE, Gouarderes C, Mazarguil H, Alvinerie P, Cros I (1985) [D-Pro^{10}]dynorphin-(1–11) is a highly potent and selective ligand for κ-opioid receptor. Eur J Pharmacol 106:457–458

Gyang EA, Kosterlitz HW, Lees GM (1964) The inhibition of autonomic neuro-effector transmission by morphine-like drugs and its use as a screening test for narcotic analgesic drugs. Naunyn-Schmiedebergs Arch Pharmacol 248:231–246

Hawkins KN, Knapp RJ, Lui GK, Gulya K, Kazmierski W, Wan Y-P, Pelton JT, Hruby VJ, Yamamura HI (1989) [^{3}H]-[H-D-Phe-Cys-Tyr-D-Trp-Orn-Thr-Pen-Thr-NH_2] ([^{3}H]CTOP), a potent and highly selective peptide for Mu opioid receptors in rat brain. J Pharmacol Exp Ther 248:73–80

Hurlbut DA, Evans CJ, Barchas JD, Leslie FM (1987) Pharmacological properties of a proenkephalin-derived opioid peptide: BAM 18. Eur J Pharmacol 138:359–366
Kosterlitz HW (1979) The best laid schemes o'mice an'men gang aft agley. Annu Rev Pharmacol Toxicol 19:1–12
Kosterlitz HW (1985) Opioid peptides and their receptors. The Wellcome Foundation Lecture, 1982. Proc R Soc London Ser B 225:27–40
Kosterlitz HW (1987) Biosynthesis of morphine in the animal kingdom. Nature (London) 330:606
Kosterlitz HW, Paterson SJ (1985) Types of opioid receptors: relation to antinociception. Philos Trans R Soc London Ser B 308:291–297
Kosterlitz HW, Watt AJ (1968) Kinetic parameters of narcotic agonists and antagonists, with particular reference to N-allylnoroxymorphone (naltrexone). Br J Pharmacol 33:266–267
Kosterlitz HW, Lord JAH, Watt AJ (1972) Morphine receptor in the myenteric plexus of the guinea-pig ileum. In: Kosterlitz HW, Collier HOJ, Villarreal JE (eds) Agonist and antagonist actions of narcotic analgesic drugs. Macmillan, London Basingstoke, pp 45–61
Lahti RA, Mickelson MM, McCall JM, von Voigtlander PF (1985) [^{3}H]-U-69,593 a highly selective ligand for the opioid κ-receptor. Eur J Pharmacol 109:281–284
Lees GM, Kosterlitz HW, Waterfield AA (1972) Characteristics of morphine-sensitive release of neuro-transmitter substances. In: Kosterlitz HW, Collier HOJ, Villarreal JE (eds) Agonist and antagonist actions of narcotic analgesic drugs. Macmillan, London Basingstoke, pp 142–152
McKnight AT, Corbett AD, Marcoli M, Kosterlitz HW (1985) The opioid receptors in the hamster vas deferens are of the δ-type. Neuropharmacology 24:1011–1017
Paterson SJ, Robson LE, Kosterlitz HW (1984) Opioid receptors. In: Udenfriend S, Meienhofer J (eds) The peptides, vol 6: The peptides, analysis, synthesis and biology. Academic Press, New York London, pp 147–180
Paton WDM (1957) The action of morphine and related substances on contraction and on acetylcholine output of coaxially stimulated guinea-pig ileum. Br J Pharmacol 12:119–127
Pert CB, Snyder SH (1973) Opiate receptor: demonstration in nervous tissue. Science 179:1011–1014
Portoghese PS, Libkowski AW, Takemori AE (1987) Binaltorphimine and norbinaltorphimine, potent and selective κ-opioid receptor antagonists. Life Sci 40:1287–1292
Portoghese PS, Sultana M, Nagase H, Takemori AE (1988) Application of the message-address concept in the design of highly potent and selective non-peptide δ-opioid receptor antagonists. J Med Chem 31:281–282
Schiller PW, Nguyen TM-D, Chung NN, Lemieux C (1989) Dermorphin analogues carrying an increased positive net charge in their "message" domain display extremely high μ-opioid receptor selectivity. J Med Chem 32:698–702
Snyder SH (1989) Brainstorming. Harvard University Press, Cambridge, Mass London
Trendelenburg P (1917) Physiologische und pharmakologische Versuche über die Dünndarmperistaltik. Nauyn-Schmiedebergs Arch Pharmacol 81:55–129
Weitz CJ, Lowney LI, Faull KF, Feistner G, Goldstein A (1986) Morphine and codeine from mammalian brain. Proc Natl Acad Sci USA 83:9784–9788
Weitz CJ, Faull KF, Goldstein A (1987) Synthesis of the skeleton of the morphine molecule by mammalian liver. Nature (London) 330:674–677

CHAPTER 2

Opioid Peptide Genes: Structure and Regulation

V. Höllt

1 Introduction

All mammalian opioid peptides belong to one of three peptide families, each deriving from distinct precursors: proopiomelanocortin (POMC), proenkephalin (PENK), and prodynorphin (PDYN), respectively. These precursor molecules are translation products from separate genes. Their structures have been determined using recombinant DNA techniques and have been extensively described in excellent reviews (e.g., Numa 1984). In addition, the proteolytic processing of the precursor proteins and the receptor selectivity of the opioid peptides have been reviewed in detail (Höllt 1986). The recent development of mRNA hybridization techniques has provided new data regarding the distribution and regulation of the opioid peptide genes. In this review an attempt is made to summarize these recent findings concerning the structures and regulation of the three opioid peptide genes.

2 Proopiomelanocortin (POMC)

POMC is the precursor of several biologically active peptides, such as β-endorphin, ACTH, and various peptides with melanocyte stimulating activity (α-MSH, β-MSH, γ-MSH). The POMC gene is predominantly expressed in the pituitary. In addition, cells expressing POMC have also been found in the brain, particularly in the arcuate nucleus, and in a variety of peripheral tissues such as the gonads, placenta, spleen macrophages (Lolait et al. 1986), and adrenal medulla (DeBold et al. 1988a), and in several pituitary and nonpituitary tumors (Steenbergh et al. 1984; de Keyzer et al, 1985, 1989a).

2.1 *Structure*

The structure of the POMC gene has been determined for several species: human (Takahashi et al. 1981; Cochet et al. 1982; Whitfeld et al. 1982), bovine (Nakanishi et al. 1981), rat (Drouin et al. 1985) and mouse (Notake et al. 1983a; Uhler and Herbert 1983). In addition, cDNA sequences have been obtained for porcine mRNA (Boileau et al. 1983; Oates and Herbert 1984), mink (Khlebodarova et al. 1988), chum salmon (Kitahara et al. 1988), and *Xenopus laevis* (Martens 1986). The complete sequence of the human gene (Takahashi et al. 1983) is 7665 base pairs long and is comprised as follows: exon 1 (86 bp), exon 2 (152 bp), exon 3 (833 bp); intron A (3708 bp); intron B (2886 bp). The 5′-flanking region and the introns of the human genes contain middle repetitive sequences which belong to the Alu family (Whitfeld et al. 1982; Takahashi et al. 1983). Although this general structure of the POMC gene is similar among the various species (three exons separated by two introns; Fig. 1), there are minor differences in the coding regions. Thus, the mouse gene lacks about 40 nucleotides coding for the amino-terminal region of γ-lipotropin (Uhler and Herbert 1983), whereas in the chum salmon the sequence coding for γ-melanocyte-stimulating hormone is missing (Kitahara et al. 1988). In addition, with the exception of two structurally different genes in *Xenopus* (Martens 1986), there appears to exist only one functional gene in the other species investigated. The mouse has an additional pseudogene which does not produce a functional mRNA (Uhler et al. 1983; Notake et al. 1983a). It is also known that the human POMC gene is localized on chromosome 2 (Owerbach et al. 1981) and that of the mouse on chromosome 19 (Uhler et al. 1983).

The size of POMC mRNA in the pituitary is 1100–1200 nucleotides. An mRNA species with a longer poly(A) tail has been observed in the hypothalamus (Jeannotte et al. 1987). In some peripheral tissues shorter POMC transcripts have been found. In the cow, these smaller species have been found in the adrenal medulla, thyroid, thymus, duodenum, and lung, in addition to mRNA of the size found in the hypothalamus (Jingami et al. 1984). These smaller mRNA species are derived from aberrant transcription initiations next to the 5′-end of exon 3 and, thus, do not contain any exon 1 or exon 2 sequences (Jingami et al. 1984; Jeannotte et al. 1987; Kilpatrick et al. 1987). Diversity of POMC transcription also occurs by alternate splicing. Thus, in the intermediate pituitary of rats, a different splicing occurs between exon 1 and exon 2 giving rise to two POMC mRNAs which differ in 30 nucleotides within the 5′-untranslated portion of the POMC mRNA (Oates and Herbert 1984). Larger, 5′-extended POMC transcripts of about 1450 nucleotides have been observed in ectopic tumors. In the human POMC gene, these are derived from initiations at promoter sites located 100–350 nucleotides upstream (de Keyzer et al. 1989b). Similar upstream initiation sites have been found in pheochromocy-

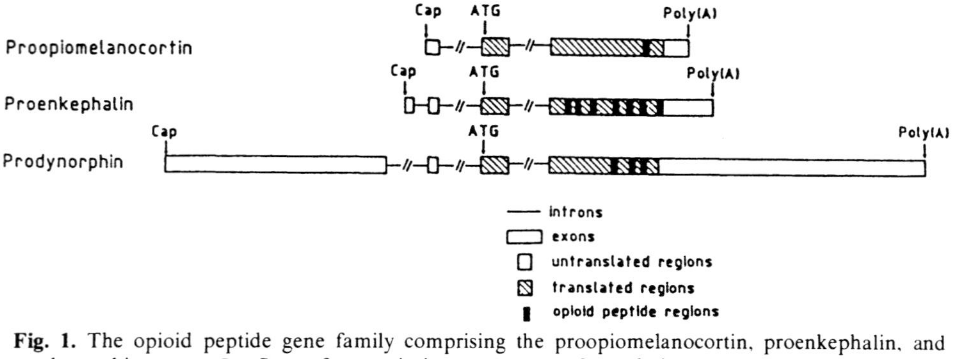

Fig. 1. The opioid peptide gene family comprising the proopiomelanocortin, proenkephalin, and prodynorphin genes. *Cap* Start of transcription; *ATG* start of translation; *Poly(A)* start of polyadenylation; opioid peptide regions refer to met-enkephalin and/or leu-enkephalin (modified according to Horikawa et al. 1983)

toma cells (DeBold et al. 1988c). Although such a variation in human POMC gene expression has also been found in normal tissues, the activity of the upstream promotors appears to be enhanced in nonpituitary tumors (de Keyzer et al. 1989b).

2.2 Regulation

The regulation of POMC gene expression has been investigated in the anterior and intermediate lobe of the pituitary, in the arcuate nucleus of the hypothalamus, in testis, ovary, and in human mononuclear cells.

Adenohypophysis

In the anterior pituitary, POMC gene expression is under the negative feedback control of adrenal steroids. Adrenalectomy results in a marked increase of POMC mRNA levels in the adenohypophysis (Nakanishi et al. 1979; Birnberg et al. 1983). This appears to be due to an increased transcription rate per cell and to an increased number of POMC synthesizing cells, as revealed by in-situ hybridization experiments using a probe specific for the first intron of the POMC gene (Fremeau et al. 1986). Marked differences in the POMC gene expression of individual corticotropes (and also melanotrophes) has recently been detected by in situ hybridization (Hatfield et al. 1989).

Injection of dexamethasone into adrenalectomized rats causes a rapid fall in POMC transcription (Gagner and Drouin 1985) and a more protracted decline in POMC mRNA levels (Birnberg et al. 1983). Glucocorticoids have also been shown to inhibit POMC gene expression in primary cultures of anterior pituitaries and in mouse AtT-20 tumor cells (Nakamura et al. 1978; Roberts et al. 1979; Gagner and Drouin 1985; Eberwine et al. 1987).

In contrast to the marked effect of adrenalectomy, impairment of thyroid function by chronic treatment with propylthiouracil did not affect POMC mRNA levels in the anterior pituitary, indicating that thyroid hormones do not alter POMC gene expression (Samuels et al. 1989).

POMC gene expression in the anterior pituitary is under positive control by corticotropin releasing hormone (CRH). Chronic administration of exogenous CRH for 3–7 days results in a marked increase in the POMC mRNA levels in the anterior pituitary of rats. This is associated with elevated glucocorticoid levels in the adrenals and in plasma (Bruhn et al. 1984; Höllt and Haarmann 1984). This finding indicates that the inductive effect of CRH on POMC gene expression overrides the inhibitory action of glucocorticoids.

An increase in POMC mRNA levels in the anterior pituitary is also seen after various stress treatments, such as chronic foot shock (Höllt et al. 1986; Shiomi et al. 1986) or hypoglycemic shock induced by insulin (Tozawa et al. 1988). Since these treatments cause a release of CRH, it is likely that their actions on POMC gene expression are mediated by CRH.

Repeated administration of morphine has also been shown to increase POMC mRNA levels in the anterior pituitary of rats, as measured by RNA blot analysis (Höllt and Haarmann 1985). Since morphine releases CRH from rat hypothalamic slices (Buckingham 1982), it is reasonable to assume that CRH also mediates the effect of morphine in elevating POMC mRNA levels in the adenohypophysis. Recently, in situ hybridization techniques revealed no change in the levels of POMC mRNA in the anterior pituitary after either acute or chronic morphine treatment in rats (Lightman and Young 1988). The reasons for this discrepancy are unknown. It is possible, however, that only high doses of morphine induce POMC gene expression in the anterior pituitary. On the other hand, naloxone-precipitated withdrawal in morphine-tolerant rats increased POMC mRNA levels, indicating that withdrawal stress activates POMC gene expression in the anterior pituitary (Lightman and Young 1988).

Chronic treatment of rats with ethanol in a vapor chamber decreased levels of POMC mRNA in the anterior pituitary (Dave et al. 1986). However, other experiments, in which ethanol was chronically administered by liquid diet, revealed an increase in the biosynthesis of POMC in the anterior pituitary (Seizinger et al. 1984a). The different findings might be related to the different modes of ethanol administration. Dehydration, commonly associated with ethanol administration in the drinking water, is also likely to affect POMC mRNA levels in the pituitary. In fact, salt loading in mice increases POMC mRNA levels in the anterior pituitary more than two fold (Elkabes and Loh 1988).

The ontogeny of POMC gene expression in the anterior pituitary parallels that of hypothalamic CRH. Both start at embryonic day 17 and progressively increase until day 21 (Grino et al. 1989). POMC mRNA levels in the anterior pituitary decrease transiently in the first week after birth and increase steadily thereafter. During the so-called stress nonresponsive period in the first week after birth, the glucocorticoid regulation of the hypothalamus is immature, since adrenalectomy does not change hypothalamic CRH mRNA in the 7-day-old rat as it does in adults (Grino et al. 1989). The levels of POMC mRNA in the sheep anterior pituitary increase before birth in line with the reported increase in fetal plasma levels of ACTH at this time (McMillen et al. 1988). There appears to be no age-dependent gene expression in the mouse anterior pituitary, since POMC mRNA levels remain constant (Nelson et al. 1988).

In primary cultures of rat anterior pituitary cells and in AtT-20 mouse tumor cells, CRH has been shown to increase POMC mRNA levels (Affolter and Reisine 1985; Loeffler et al. 1985; Knight et al. 1987; von Dreden et al. 1988) as a result of increased gene transcription (Gagner and Drouin 1985; Eberwine et al. 1987). The effect of CRH appears to be mediated via cAMP followed by activation of protein kinase A. In fact, insertion of a protein kinase A inhibitor in AtT 20/D16-16 tumor cells blocked the ability of CRH and 8-Br-cAMP to increase POMC mRNA levels (Reisine et al. 1985). Moreover, in primary cultures of rat anterior pituitary cells 8-Br-cAMP, forskolin (a biterpene derivative which activates the adenylate cyclase), cholera toxin, and drugs which elevate cellular cAMP levels by inhibiting phosphodiesterase (e.g., Ro 20-1724), increase POMC mRNA levels (Affolter and Reisine 1985; Loeffler et al. 1986b; Simard et al. 1986; Dave et al. 1987; Stalla et al. 1988; Suda et al. 1988a). In addition, imidazole derivatives, such as ketoconazole or isoconazole, which inhibit adenylate cyclase, block the CRH- or forskolin-induced increase in POMC message levels in rat anterior pituitary cultures (Stalla et al. 1988;

Stalla et al. 1989b). CRH also enhances POMC gene expression in cultured human corticotrophic tumor cells (Stalla et al. 1989a). The inducing effect of CRH on POMC mRNA levels in cultured anterior lobe cells was partially inhibited by voltage-dependent Ca^{2+} channel blockers, such as verapamil and nifedipine (Loeffler et al. 1986b; Dave et al. 1987). Similar findings were reported with mouse AtT-20/D-16v cell lines (von Dreden et al. 1988). This indicates that CRH exerts is effect on POMC gene expression via entry of Ca^{2+} ions. It is possible that this effect of CRH is, at least partially, mediated by cAMP, since cAMP analogs can elevate intracellular Ca^{2+} levels (Luini et al. 1985).

Although arginine vasopressin (AVP) releases POMC peptides from anterior pituitary cells and potentiate the secretory effect of CRH in vitro, it does not have a major effect on POMC biosynthesis (von Dreden et al. 1988; Stalla et al. 1989a; Suda et al. 1989). Moreover, AVP does not potentiate the stimulation of POMC gene expression by CRH. The mechanism whereby AVP exerts its secretory effect on corticotrophic cells appears to be the activation of phospholipase C which causes an increase in intracellular diacylglycerol (which activates protein kinase C) and phosphatidyl-inositol-phosphates (which release Ca^{2+} from intracellular stores). However, other secretagogues which activate phospholipase C, such as angiotensin II and bombesin (Höllt and Sincini 1988) failed to increase POMC mRNA levels in primary cultures of rat anterior pituitary cells and in AtT-20/D-16v tumor cells (Kessler et al. 1989). Tumor-promoting phorbol esters, such as phorbol 12-myristate 13-acetate (TPA), which activate protein kinase C, increase POMC mRNA levels in AtT-20/D16-16 tumor cells (Affolter and Reisine 1985). However, in primary cultures of rat anterior pituitary and in the AtT-20/D16-v cell line (another tumor cell line), TPA had no effect on POMC mRNA levels (Suda et al. 1989; J.P. Loeffler and V. Höllt, unpubl.). This indicates that the protein kinase C-dependent signal transduction system contributes to the release, but not the biosynthesis, of POMC peptides. Similarly, interleukin-1 (alpha and beta) releases POMC peptides by a direct action on anterior pituitary cells; however, its effect on POMC gene expression is minimal, if any (Suda et al. 1988b).

Intermediate Pituitary

In general, POMC gene expression in the melanotrophic cells of the intermediate pituitary is regulated differently from that in the corticotrophic cells of the anterior pituitary.

Adrenalectomy affects POMC mRNA levels in the intermediate pituitary only slightly (Schachter et al. 1982). Normally, melanotrophic cells do not contain any glucocorticoid receptors (Antakly and Eisen 1984). However, after denervation in vivo, or prolonged culture in vitro, they express glucocorticoid receptors (Antakly et al. 1985; Seger et al. 1988). Interestingly, glucocorticoids cause an elevation of POMC mRNA levels in the intermediate pituitary of rats after denervation of the intermediate pituitary by hypothalamic lesions (Seger et al. 1988) or after chronic haloperidol treatment (Autelitano et al. 1987, 1989). These findings indicate that the direction in which POMC gene expression is altered is not only dependent on the structure of the POMC gene, but also on the presence of glucocorticoid receptors. In primary cultures of the rat intermediate pituitary, however, glucocorticoids have been shown to slightly inhibit basal and/or CRH-induced POMC transcription (Eberwine and Roberts 1984) or to have no effect (Gagner and Drouin 1985).

POMC gene expression in the intermediate pituitary is also influenced by CRH. Thus, CRH administration (Loeffler et al, 1985, 1988) and 8-Br-cAMP, forskolin, cholera toxin, and the phosphodiesterase inhibitor Ro 20–1724 increased POMC mRNA levels in intermediate lobe cultures (Loeffler et al. 1986b). The effect of forskolin was partially blocked by the Ca^{2+} channel antagonist nifedipine and potentiated by Bay K 8644, a dihydropyridine which increases the opening probability of voltage-dependent calcium channels (Loeffler et al. 1986b). This indicates that Ca^{2+} entry triggered by cAMP is also involved in the induction of POMC gene expression in melanotrophic cells. In cultured porcine intermediate lobe cells, the calmodulin antagonists W7 and W13 decrease POMC mRNA levels. In addition, the phorbol esters phorbol 12-myristate 13-acetate and phorbol 12,13-dibutyrate decrease POMC gene expression in this preparation, indicating that activation of the Ca^{2+}/calmodulin- and the protein kinase C- dependent signal transduction pathways can modulate POMC gene expression in these cells (Loeffler et al. 1989).

There is increasing evidence that long-term administration of CRH to rats in vivo decreases POMC mRNA levels (Höllt and Haarmann 1984; Lundblad and Roberts 1988). The differential effect of CRH on intermediate pituitary POMC gene expression in vivo, versus in vitro, indicates that CRH affects POMC gene expression by both direct and indirect actions.

Various stressors, such as insulin shock or chronic foot shock for up to 7 days, increase POMC mRNA levels in the anterior pituitary, but do not affect the levels of POMC mRNA in the intermediate pituitary (Höllt et al. 1986; Tozawa et al. 1988). However, prolonged electrical foot-shock stress (administered for 2 weeks) has been reported to also cause an increase in POMC mRNA levels in the intermediate lobe (Shiomi et al. 1986).

The intermediate lobe of the rat pituitary is innervated by dopaminergic (DAergic), GABAergic, and serotonergic fibers arising from the hypothalamus (Oertel et al. 1982; Friedman et al. 1983). The release of POMC-derived peptides from the intermediate pituitary is tonically inhibited by DA via D_2-receptors (Munemura et al. 1980). Blockade of these receptors by haloperidol causes an increase in the levels of POMC mRNA (Höllt et al. 1982; Chen et al. 1983) and POMC gene transcription (Pritchett and Roberts 1987) in the intermediate pituitary lobe of rats. In contrast, chronic injection of bromocryptine, a DA receptor agonist, substantially decreased POMC mRNA levels in the rat intermediate pituitary (Chen et al. 1983; Levy and Lightman 1988). A combined morphometric analysis by light and electron microscopy, as well as by in-situ hybridization, revealed differences in the biosynthetic activity of individual melanotrophic cells. Haloperidol treatment increased the number of dark melanotrophic cells and the amount of POMC mRNA in each cell (Chronwall et al. 1988). DA and bromocryptine inhibited POMC gene expression in cultured intermediate lobes (Cote et al. 1986) and intermediate lobe cells (Loeffler et al. 1988) in vitro. The inhibitory effect of the DAergic compounds on POMC mRNA levels was abolished by pretreatment with pertussis toxin. Conversely, compounds that activate the cAMP pathway (8-Br-cAMP, cholera toxin, forskolin) counteracted the DAergic inhibition of POMC mRNA biosynthesis (Loeffler et al. 1986b,c). These findings suggest that the inhibitory guanyl nucleotide binding protein G_i and, possibly, adenylate cyclase mediate the DAergic inhibition of POMC biosynthesis.

In addition to DA, GABA has also been shown to inhibit the release of POMC-derived peptides from the intermediate pituitary (Tomiko et al. 1983). When endogenous GABA levels in the hypothalamus and pituitary of rats were elevated by chronic administration of inhibitors of the GABA metabolizing enzyme GABA-transaminase, a time-dependent decrease in the levels of POMC mRNA was found (Loeffler et al. 1986a). GABA caused a marked reduction of POMC mRNA levels when applied to primary cultures of intermediate lobe cells (Loeffler et al. 1986a).

Chronic application of morphine decreases POMC mRNA levels in the intermediate pituitary lobe, as revealed by RNA blot analysis (Höllt and Haarmann 1985). In contrast, in-situ hybridization techniques revealed no effect of morphine upon POMC levels in this tissue (Lightman and Young 1988). It is possible, however, that high doses of morphine are required to cause a decrease of POMC in the intermediate pituitary. In addition, chronic treatment of rats with NAL, or NAL-induced withdrawal in morphine-tolerant rats, failed to alter POMC mRNA levels in the intermediate pituitary (Lightman and Young 1988).

Chronic ethanol treatment decreases POMC biosynthesis in the intermediate pituitary lobe of rats (Seizinger et al. 1984b; Dave et al. 1986). As discussed previously, this effect may result from the dehydrating action of ethanol, since salt loading has been shown to decrease POMC mRNA levels in the intermediate pituitary of mice (Elkabes and Low 1988). Moreover, ethanol administered by a controlled drinking schedule (liquid diet) increases POMC biosynthesis in the intermediate lobe of rats (Seizinger et al. 1984a; V. Höllt, unpubl. data).

POMC- expressing cells are first evident in the intermediate pituitary at embryonic day 14.5 (Elkabes et al. 1989). In contrast to POMC gene expression in the anterior pituitary which decreases during the first postnatal week, POMC mRNA levels in the intermediate lobe increase steadily during this period (Grino et al. 1989). In adult rats POMC gene expression exhibits circadian rhythmicity. Thus, diurnal content and secretion of POMC peptides are associated with parallel changes in POMC mRNA concentrations and are preceded by similar changes in POMC gene transcription (Millington et al. 1986). The pars intermedia of the amphibian *Xenopus laevis* is known to adapt to light and darkness, and POMC mRNA levels in tissues of black-background adapted *Xenopus* are much higher than those in white-background adapted animals (Martens et al. 1987).

Hypothalamus

Within the brain, the cells synthesizing POMC are almost exclusively localized in the periarcuate region of the hypothalamus as revealed by in-situ hybridization (Gee et al. 1983).

Recently, transcription of the POMC gene in individual neurons of the arcuate nucleus has been detected by using in-situ hybridization with a probe complementary to nonrepetitive sequences to the first intron of the rat POMC gene

(Fremeau et al. 1989). Although POMC mRNA has also been found in other brain areas, such as the cortex or the amygdala by Northern blot or solution hybridization (Civelli et al. 1982), the individual cells producing the POMC mRNA have not yet been identified by in-situ hybridization techniques.

It was initially reported that POMC mRNA levels in the hypothalamus were unchanged 2 weeks following adrenalectomy (Birnberg et al. 1983). However, a recent paper reported that POMC mRNA levels in the hypothalamus increase after adrenalectomy and that glucocorticoids reverse this response (Beaulieu et al. 1988). Such findings indicate that glucocorticoid regulation of POMC gene expression is similar in the anterior pituitary and hypothalamus.

There is also increasing evidence that hypothalamic POMC mRNA levels are controlled by gonadal steroids. Thus, estrogen treatment in ovarectomized rats decreases hypothalamic levels of POMC mRNA (Wilcox and Roberts 1985), and POMC mRNA levels are decreased in the rat arcuate nucleus following castration, an effect that is testosterone-reversible (Chowen-Breed et al. 1989a,b). The effect of castration and testosterone treatment was confined to the most rostral area of the arcuate nucleus, indicating that there is a heterogeneous population of POMC neurons in this nucleus and that testosterone regulates POMC gene expression in a select group of cells. These observations indicate that testosterone may regulate gonadotropin releasing hormone (GnRH) secretion by increasing the synthesis of POMC in the arcuate nucleus. Other studies, however, showed that castration of male rats results in an increase in POMC mRNA levels in the medial basal hypothalamus (MBH) and that testosterone replacement reverses this effect, indicating that androgens have an inhibitory effect on POMC gene expression in the MBH (Blum et al. 1989). The reason for these discrepant results is not known. Evidence for an involvement of POMC in the control of GnRH came from studies showing that a marked increase in the levels of hypothalamic POMC mRNA occurs contemporaneously with the onset of puberty and that this increase was confined to the rostral portion of the arcuate nucleus (Wiemann et al. 1989). Thus, the increase in POMC mRNA during the onset of puberty may be important for pulsatile GnRH secretion.

Chronic morphine has been shown to decrease POMC mRNA levels in the rat hypothalamus (Mocchetti et al. 1989). In our hands, however, POMC mRNA levels in the hypothalamus were unchanged following the chronic application of morphine according to various administration schedules (Höllt et al. 1989b). Prolonged exposure of rats to low levels of lead caused a dramatic increase in the POMC mRNA levels in the hypothalamus (Rosen and Polakiewicz 1989a).

Recently, electrical stimulation of the periaqueductal gray, a terminal field of arcuate nucleus POMC-producing cells, has been reported to alter hypothalamic POMC mRNA levels, indicating that the POMC arcuate neurons might be activated by retrograde electrical activation in vivo (Bronstein and Akil 1990).

Peripheral Tissues

POMC mRNA has been localized in many peripheral tissues, apart from the anterior pituitary. Northern blot analysis of bovine tissues revealed that the adrenal medulla, thyroid gland, duodenum, and lung contain a POMC mRNA species which is

200–300 bases smaller than the species found in the pituitary and hypothalamus (Jingami et al. 1984). This smaller mRNA of the adrenal medulla has been sequenced and found to be shortened at the 5′-end of the transcript. It appears unlikely that this smaller transcript is effectively translated and/or processed, since the bovine adrenal medulla does not contain measurable levels of immunoreactive β-endorphin (β-END). Similarly, a short POMC-like mRNA of about 900 bases has been found in many nonpituitary human tissues, such as the adrenal gland, testis, spleen, kidney, ovary, lung, thyroid, liver, colon, and duodenum (DeBold et al. 1988b). In addition, POMC-like mRNA species of 1200 to 1300 bp have been found in some human tissues. These are larger than that those seen in the pituitary (1150 bases). In all these nonpituitary human tissues, significant amounts of POMC-derived peptides (ACTH, β-END) have been measured. POMC-derived immunoreactive peptides and a small POMC mRNA species of about 800 bases were also found in many nonpituitary rat tissues, such as the testis, duodenum, kidney, colon, liver, lung, and stomach. They were not, however, present in muscle (Lacaze-Masmonteil et al. 1987; DeBold et al. 1988a). Primer extension and S1 nuclease mapping studies showed that this small RNA lacked exon 1 and exon 2 of the gene and that it corresponded to a set of molecules 41 to 162 nucleotides downstream from the 5′-end of exon 3. The ratio of POMC-like mRNA to POMC-derived peptide concentrations was 1000 times greater in nonpituitary tissues than in pituitary, indicating that the POMC mRNA is much less efficiently translated and/or that the POMC-derived peptides are more rapidly released from nonpituitary tissues. In the rat pineal gland, a POMC mRNA species of the same size as that in the pituitary has recently been detected (Aloyo et al. 1990).

The localization and regulation of a smaller POMC mRNA in gonadal tissues have been extensively investigated. In the rat testis, POMC-like mRNA was predominantly found in Leydig cells (Chen et al. 1984; Pintar et al. 1984; Gizang-Ginsberg and Wolgemuth 1985). In addition, POMC mRNA was shown to be present in testicular germ cells (Gizang-Ginsberg and Wolgemuth 1987; Kilpatrick et al. 1987). Recent studies, however, revealed that administration of ethane dimethane sulfonate (EDS), a drug which selectively destroys Leydig cells, did not alter POMC mRNA levels in the testis of rats (Li et al. 1989). These data suggest that the predominant site of rat POMC gene expression is in testicular interstitial cells rather than Leydig cells.

A POMC-like transcript smaller than that found in the testis was also found in the rat ovary and placenta (Chen et al. 1986). Gonadotropins markedly increase ovarian levels of POMC mRNA in immature and adult rats (Chen et al. 1986; Melner et al. 1986; Chen and Madigan 1987). Similarly, androgens have a potent stimulatory effect on ovarian POMC mRNA levels (Melner et al. 1986). In addition, the levels of ovarian POMC mRNA are higher in pregnant than in nonpregnant rats (Chen and Madigan 1987; Jin et al. 1988).

Tumors

POMC mRNA has been detected in many tumorous tissues or cell lines. POMC mRNA was first detected in metastases of a medullary carcinoma of the thyroid, but not in the primary tumor (Steenbergh et al. 1984). Northern blot analysis revealed that this POMC mRNA was larger than that found in normal pituitary tissue. Such abnormal POMC transcripts have frequently been found in nonpituitary tumors, e.g., a larger, 1450-kb POMC mRNA species has been found in a thymic carcinoid tumor (de Keyzer et al. 1985), in human pheochromocytomas (DeBold et al. 1988b), and in other nonpituitary tumors (de Keyzer et al. 1989a). S1 mapping studies revealed that these longer POMC mRNA species result from a variable mode of transcription induced by promoters located at upstream start sites of transcription located between -400 and -100 (DeBold et al. 1988c; de Keyzer et al. 1989b). The relative activities of these promoters appear to be increased in some nonpituitary tumors. However, another study, in which a variety of human tumors was analyzed, indicated that most POMC transcripts in tumors originated from the conventional promoter. The size heterogeneities observed was due to longer poly-(A) tails (Clark et al. 1989). In addition, some tumors expressed a short POMC mRNA (800 bp) which may lack the first two exons.

Immune System

There is increasing evidence that POMC-derived peptides play a functional role in the immune system. Leukocytes and splenic macrophages synthesize immunoreactive ACTH and β-END (Smith et al. 1986). Furthermore, as in the pituitary, the synthesis of these peptides has been shown to be increased by CRH and to be inhibited by glucocorticoids as in the pituitary (Smith et al. 1986). The presence of a specific POMC mRNA in normal human peripheral mononuclear cells was recently demonstrated (Buzzetti et al. 1989). Low levels of POMC mRNA are constitutively expressed in the thymus and the spleen of pathogen free mice. These levels are suppressed by infection of animals with high amounts of murine hepatitis virus (Linner et al. 1989).

2.3 *Regulatory Elements*

The sequence requirements for transcription of the human POMC gene have been studied using transfection experiments. Sequences of the human POMC gene were joined with an SV 40 vector and introduced into COS monkey cells. Transfection with 5′-deletion mutants of the fusion gene revealed that sequences located 53–59 bp upstream of the capping site enhance transcription about three fold (Mishina et al. 1982). This indicates that the upstream region contains a silencer element which exerts a suppressive effect on POMC gene transcription. Similar results were obtained in experiments in which the transcription of the human POMC gene was studied in cell-free systems derived from extracts of HeLa and AtT-20 mouse pituitary cells in vitro (Notake et al. 1983b). In the AtT-20 system, deletion of the sequence lying between 53 and 59 bp upstream from the capping site increased transcriptional efficiency. Since this effect was markedly less in the HeLa cell system, it has been suggested that the interaction of the silencer sequence with some factors in the AtT-20 cell extract is responsible for the negative modulation of transcription of the human POMC gene.

The specificity of POMC promoter utilization was also assessed in gene transfer studies in which a rat POMC-pRSV neo-fusion gene was expressed in pituitary (AtT-20) or fibroblastic cells (L) (Jeannotte et al. 1987). The rat POMC promoter was only efficiently utilized and correctly transcribed in AtT-20 cells. Sequences conferring tissue specificity have been localized 480 to 34 bp upstream of the capping site. The tissue specificity of the rat POMC promoter was also shown in transgenic mice in which a chimeric rat POMC neo-gene was introduced into the germ line. In these mice, high levels of the fusion gene transcripts were detected in the intermediate and the anterior pituitary. They were not detected in any other tissue except for very low levels in the testes (Tremblay et al. 1988).

Transfection of a human POMC-thymidine kinase fusion gene into mouse fibroblasts revealed that the sequences responsible for the negative regulation by glucocorticoids are located within a DNA segment that extends 670 bp upstream from the cap site for POMC mRNA (Israel and Cohen 1985). Similarly, gene transfer studies with the rat POMC gene indicated that DNA sequences responsible for the direct, negative regulation of transcription reside within 706 nucleotides upstream of the transcription start site (Charron and Drouin 1986). Moreover, expression of a rat POMC-neomycin fusion gene in transgenic mice revealed that no more than 769 bp of the rat POMC gene promoter sequences are required for specific glucocorticoid inhibition of the POMC gene in the anterior pituitary (Tremblay et al. 1988). The mechanism whereby glucocorticoids inhibit transcription of the POMC gene has still to be resolved. Glucocorticoid recognition elements (GREs), through which glucocorticoid receptors activate the transcription of several genes, have been determined; however, they have thus far not been found on human or rat POMC genes. It is possible that glucocorticoid receptors interact with other (unknown) sequences to inhibit transcription of POMC or other negatively regulated genes, e.g., prolactin. Inhibition of transcription of the POMC gene by glucocorticoids in cell culture has been shown to be essentially unaffected by the protein synthesis inhibitor cycloheximide (Gagner and Drouin 1985). Similarly, earlier data have shown that the decrease in POMC mRNA caused by dexamethasone in AtT-20 cells cannot be prevented by cycloheximide (Johnson et al. 1980). These findings indicate that glucocorticoid receptors directly inhibit POMC transcription without any involvement of a protein-like intermediate factor (Gagner and Drouin 1985). Recent data, however, suggest that the DNA binding domain of the glucocorticoid receptor is not required for inhibition of the prolactin gene (Adler et al. 1988). Interactions between the glucocorticoid receptor and other proteins might therefore be important for transcriptional inhibition by glucocorticoids (Burnstein and Cidlowski 1989).

The regulatory elements on the rat POMC gene which underlie the inducing effects of CRH and forskolin were studied by transfection experiments with chimeric genes, in which sequences of the 5′-flanking region of the rat POMC gene were fused with the gene for bacterial chloramphenicol transferase (CAT) under control of the herpes simplex virus thymidine kinase (tk) promoter. A fragment of the rat POMC gene containing DNA sequences 794 to 38 nucleotides upstream of the capping site confers both elevated basal activity and CRH inducibility on the tk-CAT fusion gene when transiently expressed in AtT-20 cells. A fragment extending from 478 to 320 bp upstream of the capping site, which confers elevated basal activity and a moderate inducibility by CRH, was identified. Another fragment, extending from 320 to 133 nucleotides upstream of the start of transcription, was shown to be required for the strong CRH inducibility. This fragment did not confer elevated basal activity (Roberts et al. 1987). Since this DNA fragment does not have significant homology with the cAMP-responsive elements of other cAMP-regulated genes, it appears that there are other elements which are responsible for mediating the cAMP responsiveness of the rat POMC gene. The precise structures of these elements and the factors which interact with them have still to be elucidated.

2.4 Conclusions

The POMC gene is expressed in pituitary and many nonpituitary tissues. A POMC mRNA species of about 1100–1200 bases appears to be translated into the various POMC peptides (ACTH, β-END, etc.) in the pituitary, brain, and pineal gland. In nonpituitary tissues a shorter POMC species of 800–900 bases is the predominant mRNA species. This shorter POMC mRNA species lacks exon 1 and exon 2 sequences and results from an aberrant transcription. The translation of this transcript does not appear to be very effective. In some human nonpituitary tumors, larger POMC mRNA species of about 1400 to 1500 bases have been found. This larger POMC mRNA results from transcription of the POMC gene at upstream promoter sites.

The regulation of the POMC gene is tissue specifically regulated by steroid hormones and by neuropeptides/neurotransmitters which modulate intracellular signal transduction pathways involving cAMP and Ca^{2+}. The DNA sequences conferring tissue specificity and regulation by glucocorticoids and cAMP reside within 700 bp upstream of the transcription start site of the POMC gene.

3 Proenkephalin (PENK)

Bovine, human, and rat PENK (also termed PENK A by Noda et al. 1982b) contain four copies of met-enkephalin (ENK) and one copy each of leu-ENK, the heptapeptide met-ENK-arg^6-phe^7 and the octapeptide met-ENK-arg^6-gly^7-leu^8 (Comb et al. 1982; Gubler et al. 1982; Legon et al. 1982; Noda et al. 1982b). In the mouse, a different octapeptide (met-ENK-arg^6-ser^7-leu^8) has been found (Zurawski et al. 1986). The PENK sequence in the frog contains an additional met-ENK and no leu-ENK (Martens and Herbert 1984).

All these peptides are bounded by pairs of basic amino acids which serve as signals for proteolytic processing. PENK can also be processed into larger ENK-containing peptides, such as peptide F, peptide E, BAM-12P, BAM-18P, BAM-20P, BAM-22P, metorphamide (or adrenorphin), and amidorphin (for a review, see Höllt 1986). The PENK gene is predominantly expressed in various regions of the CNS (striatum, hypothalamus, cortex, hippocampus, and spinal cord). A high degree of PENK expression has also been found in the chromaffin cells of the adrenal medulla. In addition, cells containing PENK mRNA have been localized in the pituitary, pineal gland, gonads, placenta, blood cells, and in several tumors, and cell lines.

3.1 Structure

The sequence of the PENK mRNA has been determined for several species: bovine (Gubler et al. 1982; Noda et al. 1982b), human (Comb et al. 1982; Legon et al. 1982; Noda et al. 1982a), rat (Howells et al. 1984; Rosen et al. 1984; Yoshikawa et al. 1984), mouse (Zurawski et al. 1986), and *Xenopus laevis* (Martens and Herbert 1984). In addition, the structure of the genes of human (Noda et al. 1982a) and rat (Rosen et al. 1984) have been elucidated. The human PENK gene contains four exons separated by two large and one short intron (Fig. 1; Noda et al. 1982a). The gene is about 5.2 kb long and consists of exon 1 (70 bp), intron A (87 bp), exon 2 (56 bp), intron B (469 bp), exon 3 (141 bp), intron C (about 3400 bp), and exon 4 (980 bp). In contrast, the rat PENK gene contains three exons only: the portion of PENK mRNA that is contained between exons 1 and 2 in the human PENK gene is located in a single exon (exon 1) in the rat gene. All of the species investigated appear to have only a single PENK gene. In the human, the PENK gene has been localized to chromosome 8 (Litt and Buder 1988).

The size of the PENK mRNA in the brain and adrenal medulla is about 1450 nucleotides (Jingami et al. 1984; Pittius et al. 1985). A second, slightly larger mRNA species, which results from alternate splicing and contains the 87 nucleotides of intron A, have been detected as a minor component in bovine hypothalamus using S1 mapping (Noda et al. 1982a). In addition to these mRNA species, larger forms of PENK mRNA have been found in some peripheral tissues. Thus, rat testis expresses a PENK mRNA species with a molecular size of 1700 to 1900 nucleotides (Kilpatrick et al. 1985; Kilpatrick and Millette 1986b; Kew and Kilpatrick 1989). This larger transcript is found in spermatogenic cells, whereas the 1450-nucleotide form is of somatic cell origin (Kilpatrick et al. 1987). The larger PENK mRNA in the haploid sperm cells has recently been cloned and found to contain an additional 5′-flanking sequence which is derived from an alternate splicing of the PENK gene within intron A (Yoshikawa et al. 1989a). There is also evidence for the existence of a shorter PENK mRNA species which occurs together with the 1450-nucleotide form in the human caudate nucleus and in pheochromocytoma tissues (Monstein and Geijer 1988). Whether these different mRNA species result from alternative splicing or from aberrant transcription of the gene has not been investigated.

3.2 Regulation

The regulation of the PENK gene has been investigated in the brain, adrenal medulla, and testis as well as in several cell lines.

Striatum

In the brain, the highest levels of PENK mRNA have been found in the striatum where more than half of the neurons express the gene. The majority of these cells appear to be projection neurons which innervate the globus pallidus (Pittius et al. 1985; Shivers et al. 1986; Morris et al. 1988b, 1989). PENK gene expression in these neurons appears to be under negative control of DA. Thus, chronic treatment of rats with haloperidol increases the levels of PENK mRNA in the striatum (Tang et al. 1983; Blanc et al. 1985; Sivam et al. 1986a,b; Angulo et al. 1987; Romano et al. 1987; Morris et al. 1988b, 1989). Similarly, the 6-hydroxy-DA lesion of the substantia nigra and/or the ventral tegmental area of rats caused a marked ipsilateral increase in the levels of PENK mRNA (Tang et al. 1983; Normand et al. 1988; Vernier et al. 1988; Morris et al. 1989). Time-course studies indicated that a maximal ipsilateral increase was observed 3 to 4 weeks after 6-hydroxy-DA lesion of the nigra with only slightly elevated levels of PENK mRNA 4 weeks postlesioning (Vernier et al. 1988). The 6-hydroxy-DA lesions caused a similar increase in the PENK mRNA in the striatum of neonatally and lesioned adult rats (Sivam et al. 1987).

There appears to be a different regulation of the PENK mRNA via D_1 and D_2 receptors. However, there is some discrepancy in the reports. Whereas Mocchetti et al. (Mocchetti et al. 1987) reported that chronic treatment of rats with the selective D_1 antagonist SCH 23390 increases PENK mRNA levels in the striatum, a clear decrease in striatal PENK mRNA was found by Morris et al. (1988b).

In monkeys, bilateral or unilateral destruction of the DAergic neurons in the substantia nigra by systemic or intracarotid 1-methyl-4-phenyl-1,2,3,6-tetrahydropyridine (MPTP) injection induces Parkinsonism associated with increased PENK mRNA levels in the denervated striatal neurons and increased PENK-producing striatal neurons (Augood et al. 1989). It is still an unresolved question as to whether PENK gene expression is also increased in patients with Parkinson's disease. Levels of PENK-derived peptides are decreased in the basal ganglia of Parkinson patients, a finding which suggests a decreased production and/or an increased release of PENK-derived peptides (Agid and Javoy-Agid 1985).

DA also causes a slight decrease in PENK mRNA levels in primary cultures of rat striatal cells (Kowalski et al. 1989), indicating that DA directly inhibits PENK gene expression. In these cultured striatal cells, cAMP analogs and the phorbol ester TPA increase PENK mRNA levels, suggesting that cAMP- and protein kinase C-dependent pathways are involved in the induction of PENK gene expression.

An increase in PENK mRNA levels in the striatum was also observed after chronic treatment of rats with reserpine, indicating that reserpine increases enkephalin synthesis by eliminating the DAergic inhibition (Mocchetti et al. 1985).

In contrast to the dopaminergic system, application of serotonergic drugs and/or lesioning of the raphe nuclei with 5,7-dihydroxytryptamine failed to alter striatal PENK mRNA content (Mocchetti et al. 1984; Morris et al. 1988a).

Specific alterations in striatal levels of PENK mRNA have been found after increasing endogenous GABA levels. Chronic treatment of rats with the GABA transaminase inhibitor amino oxyacetic acid (AOAA) causes an increase in PENK mRNA levels in the striatum (Sivam and Hong 1986). On the other hand, our group observed a decrease in the level of PENK mRNA in rats chronically treated with the

GABA transaminase inhibitors AOAA and ethanolamine orthosulfate (EOS; S. Reimer and V. Höllt, unpubl.). In mice, the combined application of the GABA A agonist muscimol and diazepam for 3 days causes a decrease in striatal PENK mRNA levels. After injection for 1 week, however, a marked increase in PENK mRNA levels was observed (Llorens-Cortes et al. 1990). These results indicate that overstimulation of GABAergic transmission causes an initial decrease in striatal PENK gene expression, followed by an adaptive process which results in a progressive stimulation of PENK mRNA levels in the striatum.

Repeated administration of lithium to rats increases striatal PENK mRNA levels (Sivam et al. 1988). Similarly, electro-acupuncture of rats has been shown to markedly increase the levels of PENK mRNA in the striatum (Zheng et al. 1988).

Discrepant results have been reported with regard to the effect of opiates on striatal PENK mRNA levels. Chronic treatment with morphine was reported to cause a slight decrease in the levels of PENK mRNA in the striatum (Uhl et al. 1988), whereas we found no effect of chronic morphine treatment on striatal PENK mRNA levels (Höllt et al. 1989a). Inactivation of μ-opiate receptors by local injection of the irreversible μ-antagonist β-funaltrexamine into the striatum did not change the levels of the PENK mRNA in this structure. In contrast, local injection of the nonselective, irreversible antagonist β-chlornaltrexamine causes a marked increase in PENK mRNA levels at the site of injection (Morris et al. 1988c). This indicates that the activation of non-μ, possibly δ- and/or κ-opioid receptors, tonically suppress striatal PENK gene expression.

The prenatal development of PENK mRNA levels in the striatum has been studied in pigs. In this species, the PENK message increases to maximum levels around midgestation, followed by a decline (Pittius et al. 1987b). In rats, striatal PENK mRNA levels show a biphasic developmental profile with an initial peak at postnatal day 2, a decline to embryonic levels by day 7, and a second increase to adult levels over the second to fourth week after birth (Schwartz and Simantov 1988; Rosen and Polakiewicz 1989b).

Hypothalamus

High levels of PENK mRNA have also been found in the hypothalamus of various species (Jingami et al. 1984; Pittius et al. 1985; Romano et al. 1988). Cells expressing the PENK gene are concentrated in the ventromedial hypothalamus of the rat brain. In the bovine hypothalamus, PENK mRNA producing cells have been located in the lateral part of the paraventricular and the dorsal part of the supraoptic nucleus, subdivisions that were described to be predominantly oxytocinergic (Günther and Martin 1989).

Estrogen treatment of ovariectomized rats markedly increases PENK mRNA levels in the ventrolateral aspect of the ventromedial nucleus. This is due to both an increase in the number of PENK-expressing neurons and an increase in the amount of PENK mRNA per neuron (Romano et al. 1988). Time-course studies revealed that estrogen stimulates PENK mRNA levels in the ventromedial hypothalamus very rapidly (within 1 h). Following estrogen removal, PENK mRNA levels decline rapidly (Romano et al. 1989). This indicates that estrogen might stimulate PENK mRNA by affecting the rates of both mRNA appearance and degradation. In addition, progesterone treatment attenuated the decline of PENK mRNA levels after estrogen removal. This finding suggests that the ENKergic system in the hypothalamus may be regulated by both estrogen and progesterone during the estrous cycle. Treatment of rats with drugs that lower brain 5-hydroxytryptamine failed to alter PENK mRNA levels in the hypothalamus (Mocchetti et al. 1984).

Chronic morphine treatment has been reported to decrease PENK mRNA levels in the hypothalamus of rats (Uhl et al. 1988). Other groups failed to see any effect of morphine and/or NAL treatment (Lightman and Young 1988; Höllt et al. 1989b). However, NAL-precipitated withdrawal in morphine-tolerant rats resulted in a dramatic increase in PENK mRNA levels in cells of the paraventricular nucleus (Lightman and Young 1987a). These data suggest that the activation of hypothalamic PENK-producing neurons may be important in the neuroendocrine response to stress. In addition, stress due to administration of hypertonic saline stress to rats resulted in an increase of prooxytocin and PENK mRNAs in the magnocellular nuclei of the hypothalamus. This response was abolished in lactating rats, indicating that lactation is associated with selective inhibition of hypothalamic stress responses (Lightman and Young 1987b). Similarly, daily application of electroconvulsive shocks increased PENK mRNA in the hypothalamus of rats (Yoshikawa et al. 1985). It is tempting to speculate that an enhancement of PENK biosynthesis in areas of the limbic system might contribute to the antidepressive action of electroconvulsive shock treatment in humans.

Hippocampus and Cortex

PENK mRNA has also been found in the hippocampus of rats (Pittius et al. 1985; Naranjo et al. 1986a; White et al. 1987; Morris et al. 1988d; Moneta and Höllt 1990). Within the hippocampus, cells producing PENK-derived peptides have been localized in the granule cell layer of the dentate gyrus and in various interneurons within the hippocampal formation (Gall et al. 1981). Moreover, using in-situ hybridization techniques PENK mRNA has been located in the granule cells of the dentate gyrus after electrical stimulation (Morris et al. 1988d). Repetitive stimulation of the amygdala (kindling) of rats, which resulted in a progressive development of generalized seizures, causes a marked increase in hippocampal levels of PENK mRNA (Naranjo et al. 1986a). These data suggest that the PENK system might participate in the development of kindling phenomena. A similar increase in hippocampal PENK mRNA levels was found in rats with recurrent limbic seizures induced by contralateral lesions of the dentate gyrus hilus (Gall et al. 1981). Moreover, direct electrical stimulation of the dentate gyrus granule cells markedly increases levels of PENK mRNA at the site of stimulation (Morris et al. 1988d). In addition, repetitive electrical stimulation of the perforant pathway (tractus perforans) results in an increase in PENK mRNA levels in the ipsilateral dentate gyrus, indicating that PENK gene expression in granule cells can be induced transynaptically (Moneta et al. 1990).

Electrical stimulation of the amygdala also increases PENK mRNA levels in the entorhinal cortex, frontal cortex, nucleus accumbens, and the amygdala itself (Naranjo et al. 1986a). Similarly, repeated electroconvulsive shock increases levels of PENK mRNA in the hippocampus and the entorhinal cortex of rats (Xie et al.

1989). The degree of increase was highest in the entorhinal cortex which contains ENKergic neurons projecting into the dentate gyrus granule cells as part of the tractus perforans. It, thus, appears that several ENKergic neurons within the entorhinal-hippocampal formation are sensitive to seizure activity. The finding that PENK mRNA is also induced in other brain areas such as the frontal cortex implies that electrically induced recurrent seizures activate multisynaptic excitatory chains in several brain regions.

Spinal Cord and Lower Brainstem

PENK gene expression in the spinal cord was investigated by several groups (Iadarola et al. 1986, 1988a,b; Przewlocki et al. 1988; Weihe et al. 1989).

Inflammatory stimuli have been found to be associated with an increase in the level of PENK mRNA in the spinal cord of rats (Iadarola et al. 1986, 1988a), although the effects were much smaller than those seen for PDYN mRNA. Moreover, in arthritic rats suffering from chronic inflammation, no major change in the level of PENK mRNA has been found (Weihe et al. 1989). In addition, no significant increase was found in the levels of PENK mRNA (Northern blot analysis) in the spinal cord after spinal injury or after transection (Przewlocki et al. 1988). Measurements of mRNA using in-situ hybridization, however, indicated substantial, but localized alterations in the PENK gene expression in the spinal cord and lower brainstem after manipulation of primary afferent input. Thus, in-situ hybridization experiments revealed that the PENK gene is expressed in many neurons of lamina I and II of the nucleus caudalis of the trigeminal nuclear complex (Harlan et al. 1987; Nishimori et al. 1988). Levels of PENK mRNA were decreased in nucleus caudalis and spinal cord dorsal horn neurons following lesions of the primary afferents (Nishimori et al. 1988). Moreover, electrical stimulation of the trigeminal nerve elicited a rapid and dramatic induction of PENK mRNA in lamina I and II neurons (Nishimori et al. 1989). Taken together, the electrical stimulation, deafferentation, and inflammation experiments suggest a function-related plasticity in expression of PENK in these neurons following changes in primary afferent stimulation. Primary afferent stimulation can also enhance the expression of c-fos in subsets of lamina I and II neurons; this precedes the expression of PENK and PDYN (Hunt et al. 1987; Draisci and Iadarola 1989). The 5′-flanking region of the PENK gene contains a recognition sequence for the Fos/Jun heterodimeric complex (AP1) (Comb et al. 1986; Curran et al. 1988). This supports the hypothesis that the modulatory effects of primary afferent stimulation on PENK (and/or PDYN) gene expression in the spinal cord and nucleus caudalis is mediated via the transcription factor Fos.

Pituitary

In addition to nervous tissues, PENK mRNA has been reported in the pineal gland (Aloyo et al. 1990) and in the pituitary (Pittius et al. 1985; Schäfer et al. 1990). In-situ hybridization techniques have shown that within the rat pituitary PENK mRNA is localized in the anterior and neural lobes, but not in the intermediate lobe (Schäfer et al. 1990). The localization of PENK mRNA in the neural lobe suggests that pituicytes (a special class of astroglial cells) synthesize ENKs. In fact, PENK mRNA has recently been demonstrated in astrocyte cultures of neonatal rat brain (Vilijn et al. 1988; Schwartz and Simantov 1988).

Adrenal Medulla

PENK gene expression in the adrenal medulla of rats and in chromaffin cells of bovine adrenal medullary tissue has been extensively investigated. In-situ hybridization of sections derived from bovine adrenal medullae revealed that PENK mRNA was localized selectively in cells at the outer margin of the medulla, a region rich in epinephrine-containing cells (Bloch et al. 1985; Wan et al. 1989b). Incubation of primary cultures of chromaffin cells from bovine adrenal medulla with 8-Br-cAMP, forskolin, or cholera toxin resulted in an increase in PENK mRNA content (Eiden et al. 1984a; Quach et al. 1984; Kley et al. 1987a). In addition, substances which activate adenylate cyclase, such as vasointestinal peptide (VIP) have been shown to increase PENK mRNA levels in chromaffin cells (Wan and Livett 1989a). Membrane depolarization, induced by nicotine, high K^+, or veratridine, causes a marked increase in PENK mRNA levels in cultured bovine chromaffin cells (Eiden et al. 1984a; Kley et al. 1986; Naranjo et al. 1986b). The effect is inhibited by drugs which block voltage-dependent Ca^{2+} channels (e.g., verapamil, nifedipine, D_{600}) or by reduced Ca^{2+} ion concentration in the medium (Siegel et al. 1985; Kley et al. 1986; Naranjo et al. 1986b; Waschek et al. 1987; Waschek and Eiden 1988). These findings indicate that the influx of extracellular Ca^{2+} through voltage-dependent Ca^{2+} channels is a prerequisite for the induction of PENK gene expression. Ba^{2+} ions stimulate PENK mRNA synthesis and ENK release from chromaffin cells (Kley et al. 1986; Waschek et al. 1987). In low Ca^{2+} medium the effect of Ba^{2+} on PENK gene expression is blocked, but it still causes a release of ENKs, indicating that Ba^{2+} can substitute for extracellular Ca^{2+} in mediating peptide secretion. Ca^{2+} might act at two different targets to activate secretion versus biosynthesis.

Initially, glucocorticoids were reported to increase PENK mRNA levels in chromaffin cells. In addition, they were shown to exert a marked permissive action on the stimulation of PENK gene expression by depolarizing agents (Naranjo et al. 1986b). However, a recent study failed to find any glucocorticoid-induced changes in PENK mRNA levels in bovine chromaffin cells, although the steroids markedly induced the mRNA coding for phenyl ethanolamine N-methyl transferase (PNMT) in these cells (Wan and Livett 1989b).

Since the majority of drugs which affect PENK mRNA levels also release PENK peptides, it is possible that release per se might be the stimulus for increased PENK synthesis. This possibility, however, appears to be unlikely, since the nicotine-induced increase in PENK mRNA can be blocked by nicotine antagonists, such as hexamethonium or tubocurare, even if added 1 to 3 h after the addition of nicotine (C.J. Farin, N. Kley, and V. Höllt, submitted). Furthermore, although the microinjection of calmodulin antibodies into cultured chromaffin cells blocks catecholamine release in response to stimulation (Kenigsberg and Trifaro 1985), the addition of calmodulin antagonists, such as W_7, does not inhibit PENK mRNA levels in chromaffin cells (N. Kley, unpubl.). These findings indicate that the release, but not biosynthesis, of PENK peptides is calmodulin-dependent.

Histamine, bradykinin, or angiotensin-induced activation of phospholipase C, which gives rise to the second messengers diacylgycerol and inositol-3-phosphate, causes an increase in the levels of PENK mRNA in chromaffin cells (Bommer et al. 1987; Kley et al. 1987b; Wan et al. 1989a; Farin et al. 1990). Activation of protein kinase C appears to be an important trigger for the induction of PENK gene expression, since tumor-promoting phorbol esters increase PENK mRNA levels (Kley 1988), whereas inhibitors of protein kinase C, such as H_7 and staurosporin, decrease histamine-induced PENK mRNA levels (N. Kley, unpubl.). Although the activation of protein kinase C by phorbol esters can enhance PENK mRNA levels, phorbol esters inhibit the stimulatory effect of depolarizing agents on PENK biosynthesis (Kley 1988; Pruss and Stauderman 1988). Phorbol esters appear to inhibit PENK gene expression by inactivating voltage-dependent Ca^{2+} channels, indicating that a sustained elevation of intracellular Ca^{2+} is necessary to increase ENK synthesis in chromaffin cells. Although muscarine activates phospholipase C (Noble et al. 1986), it does not activate PENK gene expression. However, following coincubation of muscarine with the Ca^{2+} ionophore A 23187, at concentrations which do not increase PENK mRNA levels, a clear increase in the expression of the PENK was observed (Farin et al. 1990). Nuclear runoff experiments indicated that the increase of PENK mRNA by histamine is mediated by an increased rate of transcription. Measurements of the PENK mRNA half-life indicated that histamine does not alter the stability of the mRNA coding for PENK (Farin et al. 1990).

Depletion of catecholamine stores by reserpine cause a decrease in the levels of PENK mRNA (Eiden et al. 1984b). Since reserpine has been reported to block the release of ENKs from chromaffin cells, it has been proposed that the intracellular accumulation of these peptides may activate a negative feedback system, whereby PENK biosynthesis is turned off (Mocchetti et al. 1985; Naranjo et al. 1988).

In rat adrenal medulla, regulation of PENK mRNA appears to be different from that in primary cultures of bovine adrenal chromaffin cells. Thus, denervation, by sectioning of the splanchnic nerve, markedly increases the level of PENK mRNA in the adrenal gland of rats (Kilpatrick et al. 1984), indicating that splanchnic innervation exerts a tonic inhibitory influence on PENK gene expression. Such an innervation-dependent suppression is supported by experiments in which rat adrenal medullae were explanted. In such an organ culture system, levels of PENK mRNA rise 74-fold after 4 days (LaGamma et al. 1985; Inturrisi et al. 1988a). Depolarization with either elevated K^+ or veratridine inhibited this increase. Inhibition of Ca^{2+} influx by verapamil or D_{600} prevented the effect of the depolarizing agents (LaGamma et al. 1988). Calmodulin inhibitors had no effect, whereas trifluperazine, a drug which also inhibits protein kinase C, partially antagonized the effect of K^+ depolarization. These data suggest that the inhibitory effect of membrane depolarization on adrenal PENK gene expression occurs through Ca^{2+} and possibly through protein kinase C. An inhibitory effect of membrane depolarization on PENK gene transcription could be demonstrated using runoff assays (LaGamma et al. 1989).

cAMP derivatives also inhibit the increase of PENK mRNA during culture (LaGamma et al. 1988). In contrast, glucocorticoids have been shown to markedly increase PENK mRNA levels in adrenal medullary explants from control and hypophysectomized rats (LaGamma and Adler 1987; Inturrisi et al. 1988b).

Denervated or explanted adrenal medullae from neonatal rats do not show any increase in ENK content, indicating that there is an ontogenetic development of the factors regulating PENK gene expression induced by membrane depolarization (LaGamma et al. 1988). These in vitro results contrast to those obtained in rats in which increased splanchnic nerve activity, generated by insulin, caused a dramatic increase in the levels of adrenal medullary mRNA (Kanamatsu et al. 1986; Fischer-Colbrie et al. 1988). The effect of insulin was blocked by combined treatment of rats with chlorisondamine and atropine, or by bilateral transsection of the splanchnic nerves, indicating that insulin exerts its effect on PENK gene expression via splanchnic nerve activation.

Recently, we found that treatment of rats with morphine markedly increases the levels of PENK mRNA in the adrenal medulla (Höllt et al. 1989a). Moreover, morphine can also exert its effect after intracerebroventricular application, indicating that morphine activates PENK gene expression by increasing splanchnic nerve activity (Reimer and Höllt 1990). An explanation of the increase in PENK mRNA following denervation might be that the denervation causes an induction of PENK gene expression by stimulating the release of acetylcholine from splanchnic nerve endings, rather than by the unmasking of a tonic inhibition of the intact nerve.

In preliminary experiments with primary cultures of rat adrenal medullary cells, we found that depolarizing stimuli and 8-Br-cAMP increase PENK mRNA levels, indicating that rat chromaffin cells respond similarly to bovine cells. Moreover, depolarizing stimuli and 8-Br-cAMP have been shown to increase PENK mRNA in cultured fetal rat brain cells (R. Simantov and V. Höllt, submitted). Although glucocorticoids modulate PENK gene expression in bovine chromaffin cells (Naranjo et al. 1986b) and in rat adrenal medullary explants (Inturrisi et al. 1988b; Keshet et al. 1989), no specific alteration in PENK mRNA has been found in rats after hypophysectomy and/or glucocorticoid treatment (Fischer-Colbrie et al. 1988; Stachowiak et al. 1988).

Heart and Gastrointestinal Tract

High concentrations of PENK mRNA have been found in the rat heart ventricle. Although the levels of PENK mRNA are higher than in the brain, the content of PENK-derived peptides is only 3% of that in brain (Howells et al. 1986). PENK mRNA expression in the ventricular cardiac muscle is developmentally regulated. Low levels occur during the first postnatal week; these decrease during week 3, and then rise to reach maximum levels by adulthood (Springhorn et al. 1989). These findings suggest that ENKs may play a role in the differentiation of neonatal cardiac cells. In primary cardiac muscle cell cultures, PENK mRNA was induced by 8-Br-cAMP and 3-isobutyl-1-methylxanthine (IBMX), whereas the phorbol ester TPA only elicited a transient increase of PENK gene expression (Springhorn et al. 1989).

Large amounts of PENK mRNA were also found in stomach, duodenum, ileum, and colon (Zhang et al. 1989). Since the levels of PENK mRNA have been shown to change with age, the ENKs may have differential effects on gastrointestinal function during aging.

Gonads

Regulation of PENK gene expression has been studied in male and female gonads. A PENK mRNA species of 1450 nucleotides was found in the uterus, oviduct and ovary, and in testis, vas deferens, epididymidis, seminal vesicles and prostate of rats and hamsters (Kilpatrick et al. 1985; Kilpatrick and Millette 1986; Kilpatrick and Rosenthal 1986). Such data indicate that PENK-derived peptides are synthesized at multiple sites within the male and female reproductive tracts and may locally regulate reproductive function. In the rat testis, an additional mRNA species of 1700 to 1900 nucleotides was observed. This species contains a distinct 5′-nontranslated mRNA sequence derived by alternate splicing within intron A of the rat gene (Yoshikawa et al. 1989a). PENK mRNA is present in mouse spermatocytes and in spermatids, but not in extracts of mature sperm, suggesting that developing germ cells may be a major site of PENK synthesis in the testis and that PENK-derived peptides may function as germ cell-associated autocrine and/or paracrine hormones (Kilpatrick and Rosenthal 1986; Kilpatrick and Millette 1986). In rats, PENK mRNA levels are markedly changed during the estrous cycle in both the ovary and uterus. The highest concentrations occur at estrus in the rat ovary, and at metestrus and diestrus in the rat uterus (Jin et al. 1988). In the uterus of rhesus monkeys, PENK mRNA is expressed primarily in the endometrium of the uterus. PENK gene expression in the endometrium is induced by 17 β-estradiol, an action antagonized by progesterone (Low et al. 1989).

The regulation of PENK gene expression has been recently studied in cultured rat testicular peritubular cells. PENK mRNA was found to be increased by forskolin and by the phorbol ester TPA. Both drugs synergistically increase PENK mRNA in these cells. Moreover, glucocorticoids potentiated the effect of forskolin, but not that of TPA. These finding indicate that regulation of PENK gene expression in the testicular peritubular cells is similar to that in the chromaffin cells, and that PENK gene expression can be induced by activation of protein kinase C and by a cAMP-dependent pathway. Furthermore, they indicate that glucocorticoids exert a permissive effect (Yoshikawa et al. 1989b).

Immune System

Activation of mouse T-helper cells in vitro by concanavalin A results in a dramatic induction of PENK mRNA (Zurawski et al. 1986). In activated mouse T-helper cells as much as 0.4% of the total mRNA codes for PENK, allowing for the cloning of mouse PENK mRNA. Interestingly, the predicted amino acid sequence of mouse PENK contains a different sequence for the octapeptide (see above). PENK mRNA was also found in normal rat B-cells. The expression of PENK mRNA in these cells was markedly enhanced by LPS or *Salmonella typhimurium* in vitro (Rosen et al. 1989). In vivo, PENK is constitutively expressed at low levels in thymus and spleen of pathogen-free mice. Infection of the animals with murine hepatitis virus, a naturally occurring viral pathogen, caused two to five fold increase in the PENK mRNA levels in thymus and spleen (Linner et al. 1989). These findings indicate the existence of an autocrine/paracrine opioid system modulating normal lymphoid tissue.

Cell Lines

PENK mRNA has been detected in various cell lines. Low levels of pro-ENK mRNA have been found in PC12 cells, a cell line derived from a rat pheochromocytoma (Byrd et al. 1987). Although PENK mRNA levels in PC12 cells can be increased by treatment with sodium butyrate, they are still too low to make PC12 cells a useful tool for studying PENK gene expression. A higher abundance of PENK mRNA was found in cultured C6 rat glioma cells. In this cell line, activation of β-adrenergic receptors by norepinephrine increases cAMP levels and stimulates PENK mRNA. Glucocorticoids have no effect, but they potentiate the inducing effect of norepinephrine on PENK gene expression (Yoshikawa and Sabol 1986). In addition, PENK mRNA is also present in mouse neuroblastoma-rat glioma hybrid cells (NG108CC15). Treatment of the cells with opiates (etorphine, D-ala$_2$-D-met$_5$-ENK) increases PENK mRNA levels by mechanism not involving adenylate cyclase (Schwartz 1988). PENK mRNA has also been found in several human neuroblastoma cell lines (SK-N-MC; SH-SY5Y) (Folkesson et al. 1988). In addition, PENK mRNA has been found in human leukocytes from patients with chronic lymphoblastic leukemia (Monstein et al. 1986). In human neuroblastoma SK-N-MC cells, dibutyryl-cAMP and β-receptor agonists increased PENK mRNA (Folkesson et al. 1989). Glucocorticoids had no effect per se, but inhibited the effect of norepinephrine, in contrast to earlier results obtained with rat C6 glioma cells (Yoshikawa and Sabol 1986).

3.3 Regulatory Elements Within the PENK Gene

The regulatory structural elements of the PENK gene have been extensively studied in transfection experiments. The sequence requirements for transcription of the human PENK gene were analyzed after transfection of COS monkey cells with a fusion vector of 946 bp of the 5′-flanking region and 63 bp of the 5′-noncoding region with an SV 40 vector (Terao et al. 1983). In this system, the fusion product was efficiently transcribed. Deletion up to 172 bp upstream of the capping site had essentially no effect on the expression of the fusion gene. In contrast, deletions up to 145, 11, 81, and 67 bp upstream of the capping site resulted in a gradual decrease in the transcriptional efficiency, indicating that these sequences contain functional enhancer and/or promoter sites.

These sequences also have cAMP- and phorbol ester-inducible elements, as revealed by transfection experiments using a human PENK-CAT fusion gene (pENKAT-12). Transfection of this construct into monkey CV1 cells revealed that DNA sequences, required for regulation by both cAMP and phorbol ester, map to the same 37-bp region located at 107–71 bp 5′ to the mRNA capping site of the PENK gene and exert properties of transcriptional enhancers (Comb et al. 1986). Further studies provided evidence that two DNA elements (ENKCRE-1 and ENKCRE-2) are located within this enhancer region; these are responsible for the transcriptional response to cAMP and phorbol ester. The proximal promoter element, ENKCRE-2, is essential for both basal and regulated enhancer functions. The distal promoter element, ENKCRE-1, has no inherent capacity to activate transcription, but synergistically augments cAMP- and phorbol ester-inducible transcription in the presence of ENKCRE-2 (Comb et al. 1988; Hyman et al. 1988). Four different protein factors found in HeLa cell nuclear extract bind in vitro to the enhancer regions: a transcription factor, termed ENKFT-1, binds to the DNA region encompassing ENKCRE-1. AP-1 and AP-4 bind to overlapping sites spanning ENKCRE-2. A fourth transcription factor AP-2 binds to a site immediately downstream of ENKCRE-2 (Comb et al. 1988). This AP-2 DNA binding element acts synergistically with the enhancer elements to confer maximal response to cAMP and phorbol esters (Hyman et al. 1989). These finding demonstrate the synergistic interaction of three DNA elements within the human PENK gene. In addition, since none of these elements account for the response to a single messenger, it appears that the cAMP and phorbol ester responses converge on all three elements (Hyman et al. 1989). cAMP analogs also elevate expression of the pENKAT-12 fusion plasmid after transfection into mouse AtT-20 pituitary cells (Beaubien and Douglass 1989). Although TPA is consistently capable of elevating pENKAT-12 expression in CV1-tumor cells (Comb et al. 1986), this capacity is apparently cell-specific, since no induction of pENKAT-12 expression was seen in AtT-20 pituitary cells (Beaubien and Douglass 1989). This indicates the absence of protein kinase C-sensitive transcription factors in AtT-20 cells.

A 100% match between the ENKCRE-2 elements of the human and rat PENK genes exists (Rosen et al. 1984; LaGamma et al. 1989). On the other hand, the sequence of the ENKCRE-1 and the AP-2 are less conserved between the two genes. It has been proposed that such differences in DNA structure might be responsible for differences in the regulation of the human and rat PENK genes (LaGamma et al. 1989). In gel retardation studies using a 299-bp DNA derived from the 5′-flanking region of the rat PENK gene, various putative transcription factors have been found in adrenal medullary and striatal tissue. Some of them are specific for each tissue, indicating the involvement of such factors in tissue-specific regulation of PENK gene expression (LaGamma et al. 1989).

3.4 Conclusions

The PENK gene is expressed in the central nervous system and in several peripheral tissues, particularly in the adrenal medulla. A mRNA species of about 1400 nucleotides is translated into the PENK protein which, in turn, is processed to met-ENK, leu-ENK, and a variety of met-ENK-containing peptides. There is little evidence for size heterogeneity of the transcripts, although in the rat testis a larger transcript of about 1900 nucleotides was found.

The regulation of the PENK gene is regulated by factors which influence intracellular concentrations of Ca^{2+}, as well as by adenylate cyclase- or protein kinase C-dependent signal transduction mechanisms. These pathways also appear to be activated in vivo by application of stress, direct electrical stimulation, and kindling. Glucocorticoids have very little effect per se, but potentiate the effect of the above mentioned factors. cAMP- and protein kinase C (phorbol ester)-dependent elements on the PENK gene have been localized. In addition, a further element which synergistically augments cAMP and phorbol ester induction was identified. These elements are targets for a variety of transcription factors which might be regarded as "third messengers" in the coupling between stimulus and gene expression.

4 Prodynorphin (PDYN)

PDYN (PENK B) is a precursor of leu-ENK and of several opioid peptides which contain leu-ENK at the N-terminus, such as dynorphin A (DYN), leumorphin, dynorphin B (= rimorphin), α- and β-neoendorphin. The PDYN gene is predominantly expressed in the brain with a high concentration in the magnocellular nuclei of the hypothalamus and the granule cells of the dentrate gyrus of the hippocampus. In addition, PDYN-like mRNA has been found in various peripheral organs, such as heart, testis, gut, adrenal gland, and pituitary.

4.1 *Structure*

The sequence of PDYN has only been identified for pig (Kakidani et al. 1982), human (Horikawa et al. 1983), and rat (Civelli et al. 1985).

The size of the mRNA in pig hypothalamus is about 3200 nucleotides (Jingami et al. 1984). A slightly lower molecular size for PDYN mRNA in porcine brain (about 2800 bp) has been reported by others (Pittius et al. 1987a). The PDYN mRNA possesses a very long nontranslated 3′-terminal end of about 1500 nucleotides. PDYN mRNAs of the same size have also been found in the hypothalamus, hippocampus, anterior pituitary, and heart of pigs (Jingami et al. 1984; Pittuis et al. 1987a). Shorter mRNA species of about 1900 nucleotides have been localized in the duodenum, lungs, and heart ventricle. These species, however, appear to belong to another, closely related gene, as revealed by S1 mapping experiments (Pittius et al. 1987a). In rats, a slightly smaller PDYN mRNA species of about 2600 nucleotides was observed in the striatum (Civelli et al. 1985). PDYN mRNA has been shown to have a widespread distribution in the CNS and periphery. mRNA species which are about 100–200 nucleotides smaller have been found in the adrenal cortex (Day et al. 1990). These differences may be due to alternate splicing, resulting in the deletion of exon 2 of the PDYN gene (Day et al. 1990).

The complete structural organization of the human PDYN gene is known (Horikawa et al. 1983). The gene contains four exons separated by three introns: exon 1 (1.4 kb) intron A (1.2 kb), exon 2 (60 bp), intron B (9.9 kb), exon 3 (145 bp), intron C (1.7 kb), and exon 4 (2.2 kb); (see Fig. 1). Compared to the other opioid peptide genes, exons 1 and 4 (which contain nontranslated sequences) are very large. It is possible that the 5′-terminal region of this gene contains an additional intron (Horikawa et al. 1983). The 5′-flanking region of the human PDYN gene contains triple tandem-repeated sequences of 68 bp. It has been suggested that these sequences might be involved in the modulation of PDYN gene expression (Horikawa et al. 1983). As revealed by Southern and Northern blotting experiments, only a single gene appears to exist in man, pigs, and rats. The human gene was localized on chromosome 20 (Litt and Buder 1989).

4.2 Regulation

Regulation of PDYN gene expression has been studied in the hypothalamus, striatum, hippocampus, and spinal cord, as well as in the pituitary, adrenal gland, and gonadal tract.

Hypothalamus

In the rat hypothalamus, PDYN mRNA has been localized in the magnocellular divisions of the supraoptic and paraventricular nuclei (Morris et al. 1986; Sherman et al. 1986). With chronic osmotic challenge, PDYN mRNA levels in the supraoptic and paraventricular nuclei are increased, in parallel with those for provasopressin (Sherman et al. 1986). These findings indicate a coordinate regulation of mRNA expression of coexisting peptide systems. A proportional increase in the levels of PDYN and provasopressin mRNA in the hypothalamus has also been found in rats treated with nicotine (Höllt and Horn 1989; V. Höllt, unpubl.). The regulation of these mRNAs can also differ, however, since repeated electroconvulsive shocks concomitantly activate PDYN and provasopressin gene expression in the supraoptic, but not in the paraventricular, hypothalamic nuclei (Schäfer et al. 1989).

Striatum

The striatum contains many GABAergic neurons projecting into the substantia nigra and the external segment of the pallidum which express PDYN (Graybiel 1986). PDYN mRNA in the striatum, the nucleus accumbens, and the olfactory tubercle has been demonstrated by in-situ hybridization (Morris et al. 1986, 1989; Young et al. 1986).

Destruction of the substantia nigra with 6-hydroxy-DA resulted in a slight decrease (Young et al. 1986) or had no effect (Morris et al. 1989) on PDYN mRNA levels in the striatum, indicating that the mesostriatal DA system does not exert a major influence on PDYN synthesis in this brain region. Similarly, chronic treatment with the DA antagonist haloperidol did not change PDYN mRNA levels in the striatum (Morris et al. 1989). On the other hand, chronic treatment with the DA agonist apomorphine has been shown to increase the levels of PDYN mRNA in the striatum (Li et al. 1988). Moreover, a tendency of decreased PDYN mRNA levels in the striatum and nucleus accumbens was observed in rats after chronic application of the D_1-antagonist SCH 23 390 (Morris et al. 1988b). Recent results from our group showed that after lesion of the medial forebrain bundle, the PDYN mRNA levels in the nucleus accumbens, an area which contains a high proportion of D_1-receptors, are decreased (S. Reimer et al., unpubl.). These findings suggest that there is a slight tonic enhancement of PDYN synthesis in the basal ganglia via D_1-receptors.

The raphe-striatal serotonergic pathway tonically enhances PDYN gene expression in striatal target cells, since destruction of the dorsal raphe nucleus by microinjection of 5,7-dihydroxytryptamine caused significant reductions in PDYN mRNA levels in the medial nucleus accumbens and the caudomedial striatum, regions which contain a particularly dense serotonergic innervation (Morris et al. 1988a).

Enhancement of GABAergic transmission by chronic administration of GABA transaminase inhibitors (AOAA; EOS; γ-vinyl-GABA) causes a decrease in striatal levels of PDYN (S. Reimer et al., unpubl.). This might reflect a type of autoinhibition, since the dynorphinergic neurons in the striatum contain GABA as neurotransmitter.

In addition, irreversible inactivation of striatal opioid receptors by local application of β-chlornaltrexamine caused an increase in PDYN mRNA levels at the site of injection (Morris et al. 1988c). This indicates that endogenous opioids tonically inhibit striatal PDYN biosynthesis. This effect, however, appears to be mediated via δ- and κ-receptors rather than μ-opioid receptors, since the local injection of the irreversible antagonist β-funaltrexamine, as well as the administration of the μ-agonist morphine or the antagonist NAL have no effect on the striatal levels of PDYN (Morris et al. 1988c; V. Höllt, unpubl.).

PDYN biosynthesis in the striatum was also slightly increased after electrical kindling of the deep prepyriform cortex, indicating that the striatum participates in the neuronal circuits activated by cortical kindling (Lee et al. 1989).

The prenatal development of PDYN mRNA levels has been studied in pigs (Pittius et al. 1987b). PDYN mRNA levels increase until about midgestation and decline during the later prenatal development, in a manner similar to that of PENK mRNA (see above). The postnatal development of PDYN mRNA levels has been investigated in rat striatum (Rosen and Polakiewicz 1989b). Striatal PDYN mRNA levels peak at an early stage of postnatal development and then decline to adult levels with a time course similar to that displayed by PENK mRNA. These findings suggest a possible role of both opioid peptide systems during pre- and postnatal development.

Hippocampus

Within the hippocampus, DYN peptides are synthesized in the granule cells of the dentate gyrus and transported within the mossy fiber pathways which innervate the pyramidal cells (McGinty et al. 1983; Morris et al. 1986). Following brief trains of high-frequency electrical stimulation to the dentate gyrus of rats, the levels of PDYN mRNA were markedly decreased on the stimulated, but not on the unstimulated side (Morris et al. 1988d). Similarly, chronic electrical stimulation of the dentate gyrus, which resulted in stage 4 kindling seizures in rats, caused a marked decrease in PDYN mRNA levels in the granule cells of the dentate gyrus. This decrease was seen in the stimulated as well as in the unstimulated hemispheres (Morris et al. 1987). The contralateral decrease in PDYN mRNA levels might be due to the activation of commissural projections. A bilateral decrease in PDYN mRNA levels can also be induced by unilateral electrical stimulation of the perforant pathway which results in stage 3 kindling seizures in rats (Moneta and Höllt 1990), indicating that PDYN gene expression in the hippocampus can be altered transsynaptically. PDYN mRNA levels in the hippocampus are also decreased during the development of deep prepyriform cortex kindling (Lee et al. 1989) and after repeated electroconvulsive shocks (Xie et al. 1989). The altered PDYN mRNA levels in the hippocampus return to normal 1 to 6 weeks following cessation of stimulation (Lee et al. 1989; Moneta and Höllt 1990). However, a further single stimulus was still effective in producing kindling seizures. These findings indicate that opioid peptides may play a role in the development, but not maintenance, of kindling. There is electrophysiological evidence for an inhibitory action of PDYN peptides on hippocampal pyramidal cells (Henricksen et al. 1982). A decrease in PDYN biosynthesis might, therefore, contribute to the hyperexcitability found in the kindling state. Thus, in contrast to the PENK gene, the PDYN gene is negatively regulated by neuronal activity.

In aged rats an increase in hippocampal levels of PDYN mRNA was observed (Lacaze-Masmonteil et al. 1987). An age-related loss of perforant path afferents, which inhibit dynorphin biosynthesis in the granule cells, might be responsible for this effect. Whether or not PDYN-derived peptides are involved in behavioral impairments is not known as yet. In this context, however, it is notable that opiate antagonists have been found to improve learning in rats (Collier and Routtenberg 1984).

Spinal Cord

In the spinal cord, a prominent role for PDYN-derived peptides has been suggested by the observation that PDYN mRNA levels in the spinal cord are markedly enhanced by acute or chronic inflammatory processes (Iadarola et al. 1986, 1988a,b; Höllt et al. 1987; Weihe et al. 1989). In polyarthritic rats a pronounced elevation in PDYN mRNA was found in the lumbosacral spinal cord (Höllt et al. 1987; Weihe et al. 1989). In rats with unilateral inflammation, a pronounced increase in PDYN mRNA was observed only in those spinal cord segments that received sensory inputs from the affected limb (Iadarola et al. 1988a,b). These changes were rapid in onset, being significantly elevated as early as 4 h after the injection of the inflammatory agent into the paw (Draisci and Iadarola 1989). PDYN mRNA levels peaked after about 3 days, returning to normal after about 2 weeks (Iadarola et al. 1988b). In-situ hybridization revealed that the increase in the PDYN mRNA levels occurs in the superficial dorsal-horn laminae I and II, and in the deep dorsal-horn laminae V and VI (Ruda et al. 1988). There appears to be a population of neurons in laminae IV/V, but not in laminae I/II, which coexpress PDYN and PENK peptides (Weihe et al. 1988). These neurons appear to increase their biosynthetic activity in response to inflammation, a finding that might explain the small increase in PENK mRNA observed by some groups (Iadarola et al. 1988b). In addition to inflammation, traumatic injury and cord transection also increased PDYN mRNA levels in the spinal cord (Przewlocki et al. 1988). This marked increase in PDYN biosynthesis suggests a highly specific role for PDYN-derived peptides in the modulation of pain associated with inflammation and tissue injury.

Outside the central nervous system, PDYN mRNA has been found in the pituitary, heart, and gonadal tract.

Pituitary

PDYN mRNA has been localized in the anterior pituitary of rats (Civelli et al. 1985; Schäfer et al. 1990) and pigs (Pittius et al. 1987a). It is still unclear which cell type produces PDYN, although the parallel release of luteinizing hormone together with PDYN peptides in response to GnRH suggests that both peptides are produced by gonadotrophic cells. PDYN mRNA in the rat anterior pituitary is decreased by estrogens and increased by anti-estrogens (Spampinato et al. 1990). In the rat, in-situ hybridization experiments revealed that PDYN mRNA is also produced in melanotrophic cells of the intermediate pituitary (Schäfer et al. 1990). In the pig, however, no PDYN mRNA could be demonstrated in the neurointermediate lobe when Northern blot analysis was employed (Pittius et al. 1987a).

Peripheral Tissues

Northern blot and solution analysis revealed the presence of PDYN mRNA in the porcine heart ventricle, gut, and lung; these tissues contain low, but significant, levels of PDYN-derived peptides (Pittius et al. 1987a). However, the mRNA species found in the gut and lung are smaller than those found in the brain and heart ventricle and appear to derive from a different gene which possesses a high homology to the PDYN gene (Pittius et al. 1987a). The presence of PDYN-derived peptides in enterochromaffin cells indicates that the PDYN gene is expressed in the gastrointestinal tract (Cetin 1988).

In the rat adrenal gland, a unique PDYN mRNA species, which is about 350 bases shorter than the PDYN transripts in the hypothalamus (2100 vs 2400 nucleotides), has been found (Civelli et al. 1985). The observed size difference of the PDYN mRNA in the adrenal may be due to alternate splicing, resulting in the deletion of exon 2 of the PDYN gene (Day et al. 1990). In-situ hybridization studies localized PDYN mRNA in the zona fasciculata and zona reticularis of the adrenal cortex (Day et al. 1990). PDYN mRNA has been shown to be influenced by pituitary-dependent factors, since there is a dramatic loss of PDYN mRNA in the zona fasciculata following hypophysectomy in rats (Day et al. 1990).

In the gonads, PDYN mRNA has been found in the testes of rats, rabbits, and guinea pigs, and in the rat ovary and uterus (Civelli et al. 1985; Douglass et al. 1987). Within the rat testis, PDYN peptides have been found in Leydig cells. Moreover, PDYN mRNA is found in the R2C Leydig tumor cell line and (McMurray et al. 1989) and in cultured Sertoli cells. The testicular PDYN mRNA species is about 100 bases smaller than that found in the hypothalamus. This may be due to alternate splicing, leading to the loss of exon 2 of the PDYN gene. In R2C rat Leydig cells, the PDYN gene is positively regulated by cAMP analogs, whereas phorbol esters exert a slight negative regulation on PDYN mRNA levels. The presence of mRNA together with PDYN-derived peptides in the reproductive tract suggests that these peptides may exert paracrine and/or autocrine effects.

4.3 Regulatory Elements Within the PDYN Gene

Although the complete structure of the human gene has been known for several years (Horikawa et al. 1983), no regulatory elements of this gene have as yet been characterized. However, the triple tandem-repeated sequences at the 5′-flanking region were suggested to be involved in the regulation of PDYN gene expression (Horikawa et al. 1983).

An incomplete structure, comprising the main exon has been published for the rat PDYN gene (Civelli et al. 1985). Recently, however, transfection studies with chimeric fusion genes have been reported. In these studies the sequences of the 5′-flanking region of the rat PDYN gene were joined with a CAT-reporter gene and introduced into 2RC Leydig cells (McMurray et al. 1989). The fusion genes were positively regulated by cAMP analogs in these cells. The cAMP-responsive DNA sequence has been localized to a 210-bp fragment comprising 122 bp of the 5′-flanking sequences, and 88 bp of exon 1 of the rat PDYN gene. This sequence contains a DNA fragment which is 80% homologous to the cAMP consensus sequence located downstream to the capping-site. On the other hand, expression of the prodynrophin fusion plasmid was not induced by TPA, indicating that the 210-bp PDYN sequence may not contain any phorbol ester-inducible element.

5 Summary

In the present report, an attempt was made to summarize features which specifically relate to the structure and regulation of each of the three opioid peptide genes. In this final summary some common characteristics of the structure and regulation between these genes will be emphasized. The three genes (see Fig. 1) are strikingly similar in their general organization. They all contain a main exon which codes for the vast majority of the protein sequences. The signal peptide and the translational initiation site are coded by another exon. This exon has an almost identical site in all three genes. These similarities in structural organization suggest that the three opioid peptide genes may have evolved from a common ancestor. All three genes are positively regulated by cAMP, although only the PENK gene posseses an identified cAMP-responsive element within the 5′-flanking region. Moreover, all three genes are modulated in response to stress in intact animals, albeit in different tissues. This indicates that all three opioid peptide genes may play an important role under stressful conditions.

References

Adler S, Waterman ML, He X, Rosenfeld MG (1988) Steroid receptor-mediated inhibition of rat prolactin gene expression does not require the receptor DNA-binding domain. Cell 52:685–695

Affolter HU, Reisine T (1985) Corticotropin releasing factor increases proopiomelanocortin messenger RNA in mouse anterior pituitary tumor cells. J Biol Chem 260:15477–15481

Agid Y, Javoy-Agid F (1985) Peptides and Parkinson's disease. Trends Neurosci 9:30–35

Aloyo VJ, Lewis ME, Walker RF (1990) Opioid peptide mRNAs in the rat pineal gland. In: Quirion R, Jhamandas K, Giounalakis C (eds) The International Narcotics Research Conference (IRNC) '89. Liss, New York, pp 235–238

Angulo JA, Christoph GR, Manning RW, Burkhart BA, Davis LG (1987) Reduction of dopamine receptor activity differentially alters striatal neuropeptide mRNA levels. Adv Exp Med Biol 221:385–391

Antakly T, Eisen H (1984) Immunocytochemical localization of glucocorticoid receptor in target cells. Endocrinology 155:1984-1989
Antakly T, Sasaki A, Liotta AS, Palkovits M, Krieger DT (1985) Induced expression of the glucocorticoid receptor in the rat intermediate pituitary lobe. Science 229:277-279
Augood SJ, Emson PC, Mitchell IJ, Boyce S, Clarke CE, Crossman AR (1989) Cellular localisation of enkephalin gene expression in MPTP-treated cynomolgus monkeys. Brain Res Mol Brain Res 6:85-92
Autelitano DJ, Clements JA, Nikolaidis I, Canny BJ, Funder JW (1987) Concomitant dopaminergic and glucocorticoid control of pituitary proopiomelanocortin messenger ribonucleic acid and beta-endorphin levels. Endocrinology 121:1689-1696
Autelitano DJ, Lundblad JR, Blum M, Roberts JL (1989) Hormonal regulation of POMC gene expression. Annu Rev Physiol 51:715-726
Beaubien BC, Douglasss J (1989) Regulatory elements of proenkephalin and pro-opiomelanocortin genes: receptor-mediated responses. Adv Biosci 75:217-220
Beaulieu S, Gagne B, Barden N (1988) Glucocorticoid regulation of proopiomelanocortin messenger ribonucleic acid content of rat hypothalamus. Mol Endocrinol 2:727-731
Birnberg NC, Lissitzky JC, Hinman M, Herbert E (1983) Glucocorticoids regulate proopiomelanocortin gene expression in vivo at the levels of transcription and secretion. Proc Natl Acad Sci USA 80:6982-6986
Blanc D, Cupo A, Castanas E, Bourhim N, Giraud P, Bannon MJ, Eiden LE (1985) Influence of acute, subchronic and chronic treatment with neuroleptic (haloperidol) on enkephalins and their precursors in the striatum of rat brain. Neuropeptides 5:567-570
Bloch B, Le-Guellec D, de Keyzer Y (1985) Detection of the messenger RNAs coding for the opioid peptide precursors in pituitary and adrenal by "in situ" hybridization: study in several mammal species. Neurosci Lett 53:141-148
Blum M, Roberts JL, Wardlaw SL (1989) Androgen regulation of proopiomelanocortin gene expression and peptide content in the basal hypothalamus. Endocrinology 124:2283-2288
Boileau G, Barbeau C, Jeannotte L, Chretien M, Drouin J (1983) Complete structure of the porcine proopiomelanocortin mRNA derived from the nucleotide sequence of cloned cDNA. Nucleic Acids Res 11:8063-8071
Bommer M, Liebisch D, Kley N, Herz A, Noble E (1987) Histamine affects release and biosynthesis of opioid peptides primarily via H1-receptors in bovine chromaffin cells. J Neurochem 49:1688-1696
Bronstein D, Akil H (1990) Effects of electrical stimulation in the periaqueductal gray on POMC peptides and mRNA in the rat brain. In: Quirion R, Jhamandas K, Giounalakis C (eds) The International Narcotics Research Conference (IRNC) '89. Liss, New York, pp 219-222
Bruhn TO, Sutton RE, Rivier CL, Vale WW (1984) Corticotropin-releasing factor regulates proopiomelanocortin messenger ribonucleic acid levels in vivo. Neuroendocrinology 39:170-175
Buckingham JC (1982) Secretion of corticotrophin and its hypothalamic releasing factor in response to morphine and opioid peptides. Neuroendocrinology 35:111-116
Burnstein KL, Cidlowski JA (1989) Regulation of gene expression by glucocorticoids. Annu Rev Physiol 51:683-699
Buzzetti R, McLoughlin L, Lavender PM, Clark AJ, Rees LH (1989) Expression of proopiomelanocortin gene and quantification of adrenocorticotropic hormone-like immunoreactivity in human normal peripheral mononuclear cells and lymphoid and myeloid malignancies. J Clin Invest 83:733-737
Byrd JC, Naranjo JR, Lindberg I (1987) Proenkephalin gene expression in the PC12 pheochromocytoma cell line: stimulation by sodium butyrate. Endocrinology 121:1299-1305
Cetin Y (1988) Enterochromaffin (EC-) cells of the mammalian gastro-enteropancreatic (GEP) endocrine system: cellular source of pro-dynorphin-derived peptides. Cell Tissue Res 253:173-179
Charron J, Drouin J (1986) Glucocorticoid inhibition of transcription from episomal proopiomelanocortin gene promoter. Proc Natl Acad Sci USA 83:8903-8907
Chen CL, Madigan MB (1987) Regulation of testicular proopiomelanocortin gene expression. Endocrinology 121:590-596
Chen CL, Dionne FT, Roberts JL (1983) Regulation of the proopiomelanocortin mRNA levels in rat pituitary by dopaminergic compounds. Proc Natl Acad Sci USA 80:2211-2215
Chen CL, Mather JP, Morris PL, Bardin CW (1984) Expression of proopiomelanocortin-like gene in the testis and epididymis. Proc Natl Acad Sci USA 81:5672-5675
Chen CL, Chang CC, Krieger DT, Bardin CW (1986) Expression and regulation of proopiomelanocortin-like gene in the ovary and placenta: comparison with the testis. Endocrinology 118:2382-2389

Chowen-Breed JA, Fraser HM, Vician L, Damassa DA, Clifton DK, Steiner RA (1989a) Testosterone regulation of proopiomelanocortin messenger ribonucleic acid in the arcuate nucleus of the male rat. Endocrinology 124:1697–1702

Chowen-Breed JA, Clifton DK, Steiner RA (1989b) Regional specificity of testosterone regulation of proopiomelanocortin gene expression in the arcuate nucleus of the male rat brain. Endocrinology 124:2875–2881

Chronwall BM, Hook GR, Millington WR (1988) Dopaminergic regulation of the biosynthetic activity of individual melanotropes in the rat pituitary intermediate lobe: a morphometric analysis by light and electron microscopy and in situ hybridization. Endocrinology 123:1992–2002

Civelli O, Birnberg N, Herbert E (1982) Detection and quantitation of proopiomelanocortin mRNA in pituitary and brain tissues from different species. J Biol Chem 257:6783–6787

Civelli O, Douglass J, Goldstein A, Herbert E (1985) Sequence and expression of the rat prodynorphin gene. Proc Natl Acad Sci USA 82:4291–4295

Clark AJ, Lavender PM, Besser GM, Rees LH (1989) Proopiomelanocortin mRNA size heterogeneity in ACTH-dependent Cushing's syndrome. J Mol Endocrinol 2:3–9

Cochet M, Chang AC, Cohen SN (1982) Characterization of the structural gene and putative 5′-regulatory sequences for human proopiomelanocortin. Nature (Lond) 297:335–339

Collier TJ, Routtenberg A (1984) Selective impairment of declarative memory following stimulation of dentate gyrus granule cells: a naloxone-sensitive effect. Brain Res 310:384–387

Comb M, Herbert E, Crea R (1982) Partial characterization of the mRNA that codes for enkephalins in bovine adrenal medulla and human pheochromocytoma. Proc Natl Acad Sci USA 79:360–364

Comb M, Birnberg NC, Seasholtz A, Herbert E, Goodman HM (1986) A cyclic AMP- and phorbol ester-inducible DNA element. Nature (Lond) 323:353–356

Comb M, Mermod N, Hyman SE, Pearlberg J, Ross ME, Goodman HM (1988) Proteins bound at adjacent DNA elements act synergistically to regulate human proenkephalin cAMP inducible transcription. EMBO J 7:3793–3805

Cote TE, Felder R, Kebabian JW, Sekura RD, Reisine T, Affolter HU (1986) D-2 dopamine receptor-mediated inhibition of pro-opiomelanocortin synthesis in rat intermediate lobe. Abolition by pertussis toxin or activators of adenylate cyclase. J Biol Chem 261:4555–4561

Curran T, Rauscher FJ, Cohen DR, Franza BR Jr (1988) Beyond the second messenger: oncogenes and transcription factors. Cold Spring Harb Symp Quant Biol 53 Pt 2:769–777

Dave JR, Eiden LE, Karanian JW, Eskay RL (1986) Ethanol exposure decreases pituitary corticotropin-releasing factor binding, adenylate cyclase activity, proopiomelanocortin biosynthesis, and plasma beta-endorphin levels in the rat. Endocrinology 118:280–286

Dave JR, Eiden LE, Lozovsky D, Waschek JA, Eskay RL (1987) Calcium-independent and calcium-dependent mechanisms regulate corticotropin-releasing factor-stimulated proopiomelanocortin peptide secretion and messenger ribonucleic acid production. Endocrinology 120:305–310

Day R, Schäfer MK-H, Watson SJ, Akil H (1990) Effects of hypophysectomy on dynorphin mRNA and peptide content in the rat adrenal gland. In: Quirion R, Jhamandas K, Giounalakis C (eds) The International Narcotics Conference (IRNC) '89. Liss, New York, pp 207–210

DeBold CR, Menefee JK, Nicholson WE, Orth DN (1988a) Proopiomelanocortin gene is expressed in many normal human tissues and in tumors not associated with ectopic adrenocorticotropin syndrome. Mol Endocrinol 2:862–870

DeBold CR, Nicholson WE, Orth DN (1988b) Immunoreactive proopiomelanocortin (POMC) peptides and POMC-like messenger ribonucleic acid are present in many rat nonpituitary tissues. Endocrinology 122:2648–2657

DeBold CR, Mufson EE, Menefee JK, Orth DN (1988c) Proopiomelanocortin gene expression in a pheochromocytoma using upstream transcription initiation sites. Biochem Biophys Res Commun 155:895–900

de Keyzer Y, Bertagna X, Lenne F, Girard F, Luton JP, Kahn A (1985) Altered proopiomelanocortin gene expression in adrenocorticotropin-producing nonpituitary tumors. Comparative studies with corticotropic adenomas and normal pituitaries. J Clin Invest 76:1892–1898

de Keyzer Y, Rousseau-Merck MF, Luton JP, Girard F, Kahn A, Bertagna X (1989a) Proopiomelanocortin gene expression in human phaeochromocytomas. J Mol Endocrinol 2:175–181

de Keyzer Y, Bertagna X, Luton JP, Kahn A (1989b) Variable modes of proopiomelanocortin gene transcription in human tumors. Mol Endocrinol 3:215–223

Douglass J, Cox B, Quinn B, Civelli O, Herbert E (1987) Expression of the prodynorphin gene in male and female mammalian reproductive tissues. Endocrinology 120:707–713

Draisci G, Iadarola MJ (1989) Temporal analysis of increases in c-fos, preprodynorphin and preproenkephalin mRNAs in rat spinal cord. Brain Res Mol Brain Res 6:31–37
Drouin J, Chamberland M, Charron J, Jeannotte L, Nemer M (1985) Structure of the rat proopiomelanocortin (POMC) gene. FEBS Lett 193:54–58
Eberwine JH, Roberts JL (1984) Glucocorticoid regulation of proopiomelanocortin gene transcription in the rat pituitary. J Biol Chem 259:2166–2170
Eberwine JH, Jonassen JA, Evinger MJ, Roberts JL (1987) Complex transcriptional regulation by glucocorticoids and corticotropin-releasing hormone of proopiomelanocortin gene expression in rat pituitary cultures. DNA 6:483–492
Eiden LE, Giraud P, Dave JR, Hotchkiss AJ, Affolter HU (1984a) Nicotinic receptor stimulation activates enkephalin release and biosynthesis in adrenal chromaffin cells. Nature (Lond) 312:661–663
Eiden LE, Giraud P, Affolter HU, Herbert E, Hotchkiss AJ (1984b) Alternative modes of enkephalin biosynthesis regulation by reserpine and cyclic AMP in cultured chromaffin cells. Proc Natl Acad Sci USA 81:3949–3953
Elkabes S, Loh YP (1988) Effect of salt loading on proopiomelanocortin (POMC) messenger ribonucleic acid levels, POMC biosynthesis, and secretion of POMC products in the mouse pituitary gland. Endocrinology 123:1754–1760
Elkabes S, Loh YP, Nieburgs A, Wray S (1989) Prenatal ontogenesis of proopiomelanocortin in the mouse central nervous system and pituitary gland: an in situ hybridization and immunocytochemical study. Brain Res Dev Brain Res 46:85–95
Farin CJ, Höllt V, Kley N (1990) Proenkephalin gene expression in cultured chromaffin cells is regulated at the transcriptional level. In: Quirion R, Jhamandas K, Giounalakis C (eds) The International Narcotics Research Conference (IRNC) '89. Liss, New York, pp 239–242
Fischer-Colbrie R, Iacangelo A, Eiden LE (1988) Neural and humoral factors separately regulate neuropeptide Y, enkephalin, and chromogranin A and B mRNA levels in rat adrenal medulla. Proc Natl Acad Sci USA 85:3240–3244
Folkesson R, Monstein HJ, Geijer T, Pahlman S, Nilsson K, Terenius L (1988) Expression of the proenkephalin gene in human neuroblastoma cell lines. Brain Res 427:147–154
Folkesson R, Monstein HJ, Geijer T, Terenius L (1989) Modulation of proenkephalin A gene expression by cyclic AMP. Brain Res Mol Brain Res 5:211–217
Fremeau RT Jr, Lundblad JR, Pritchett DB, Wilcox JN, Roberts JL (1986) Regulation of proopiomelanocortin gene transcription in individual cell nuclei. Science 234:1265–1269
Fremeau RT Jr, Autelitano DJ, Blum M, Wilcox J, Roberts JL (1989) Intervening sequence-specific in situ hybridization: detection of the proopiomelanocortin gene primary transcript in individual neurons. Mol Brain Res 6:197–201
Friedman E, Krieger DT, Mezey E, Leranth C, Brownstein MJ, Palkovits M (1983) Serotonergic innervation of the rat pituitary intermediate lobe: decrease after stalk section. Endocrinology 112:1943–1947
Gagner JP, Drouin J (1985) Opposite regulation of proopiomelanocortin gene transcription by glucocorticoids and CRH. Mol Cell Endocrinol 40:25–32
Gall C, Brecha N, Karten HJ, Chang KJ (1981) Localization of enkephalin-like immunoreactivity to identified axonal and neuronal populations of the rat hippocampus. J Comp Neurol 198:335–350
Gee CE, Chen CL, Roberts JL, Thompson R, Watson SJ (1983) Identification of proopiomelanocortin neurons in rat hypothalamus by in situ cDNA-mRNA hybridization. Nature (Lond) 306:374–376
Gizang-Ginsberg E, Wolgemuth DJ (1985) Localization of mRNAs in mouse testes by in situ hybridization: distribution of alpha-tubulin and developmental stage specificity of proopiomelanocortin transcripts. Dev Biol 111:293-305
Gizang-Ginsberg E, Wolgemuth DJ (1987) Expression of the proopiomelanocortin gene is developmentally regulated and affected by germ cells in the male mouse reproductive system. Proc Natl Acad Sci USA 84:1600–1604
Graybiel AM (1986) Neuropeptides in the basal ganglia. Res Publ Assoc Res Nerv Ment Dis 64:135–161
Grino M, Young WS, Burgunder JM (1989) Ontogeny of expression of the corticotropin-releasing factor gene in the hypothalamic paraventricular nucleus and of the proopiomelanocortin gene in rat pituitary. Endocrinology 124:60–68
Gubler U, Seeburg P, Hoffman BJ, Gage LP, Udenfriend S (1982) Molecular cloning establishes proenkephalin as precursor of enkephalin-containing peptides. Nature (Lond) 295:206–208
Günther W, Martin R (1989) Localization of preproenkephalin mRNA-synthesizing neurons in the bovine hypothalamus by in situ hybridization. Adv Bioscience 75:297–300

Harlan RE, Shivers BD, Romano GJ, Howells RD, Pfaff DW (1987) Localization of preproenkephalin in the rat brain and spinal cord by in situ hybridization. J Comp Neurol 258:159–184
Hatfield JM, Daikh DI, Adelman JP, Douglass J, Bond CT, Allen RG (1989) In situ hybridization detection of marked differences in pre-proopiomelanocortin messenger ribonucleic acid content of individual corticotropes and melanotropes. Endocrinology 124:1359–1364
Henricksen SJ, Chouvet G, Bloom FE (1982) In vivo cellular responses to electrophoretically applied dynorphin in the rat hippocampus. Life Sci 31:1785–1788
Höllt V (1986) Opioid peptide processing and receptor selectivity. Annu Rev Pharmacol Toxicol 26:59–77
Höllt V, Haarmann I (1984) Corticotropin-releasing factor differentially regulates proopiomelanocortin messenger ribonucleic acid levels in anterior as compared to intermediate pituitary lobes of rats. Biochem Biophys Res Commun 124:407–415
Höllt V, Haarmann I (1985) Differential alterations by chronic treatment with morphine of proopiomelanocortin mRNA levels in anterior as compared to intermediate pituitary lobes of rats. Neuropeptides 5:481–484
Höllt V, Horn G (1989) Nicotine and opioid peptides. In: Nordberg A, Fuxe K, Holmstedt B, Sundwall A (eds) Nicotinic receptors in the CNS: their role in synaptic transmission. Progress in Brain Research vol 79. Elsevier, Amsterdam, pp 187–193
Höllt V, Sincini E (1988) Bombesin and structurally related peptides increase inositol-1-phosphate production in a corticotrophic cell line of the pituitary (AtT-20). Acta Endocrinol 117 (Suppl 287):206–206
Höllt V, Haarmann I, Seizinger BR, Herz A (1982) Chronic haloperidol treatment increases the level of in vitro translatable messenger ribonucleic acid coding for the beta-endorphin/adrenocorticotropin precursor proopiomelanocortin in the pars intermedia of the rat pituitary. Endocrinology 110:1885–1891
Höllt V, Przewlocki R, Haarmann I, Almeida OF, Kley N, Millan MJ, Herz A (1986) Stress-induced alterations in the levels of messenger RNA coding for proopiomelanocortin and prolactin in rat pituitary. Neuroendocrinology 43:277–282
Höllt V, Haarmann I, Millan MJ, Herz A (1987) Prodynorphin gene expression is enhanced in the spinal cord of chronic arthritic rats. Neurosci Lett 73:90–94
Höllt V, Haarmann I, Reimer S (1989a) Opioid gene expression in rats after chronic morphine treatment. Adv Biosci 75:711–714
Höllt V, Haarmann I, Reimer S (1986b) Opioid peptide gene expression in rats after chronic morphine treatment. Adv Bioscience 75:711–714
Horikawa S, Takai T, Toyosato M, Takahashi H, Noda M, Kakidani H, Kubo T, Hirose T, Inayama S, Hayashida H, Miyata T, Numa S (1983) Isolation and structural organization of the human preproenkephalin B gene. Nature (Lond) 306:611–614
Howells RD, Kilpatrick DL, Bhatt R, Monahan JJ, Poonian M, Udenfriend S (1984) Molecular cloning and sequence determination of rat preproenuephalin cDNA. Proc Natl Acad Sci USA 81:7651–7655
Howells RD, Kilpatrick DL, Bailey LC, Noe M, Udenfriend S (1986) Proenkephalin mRNA in rat heart. Proc Natl Acad Sci USA 83:1960–1963
Hunt SP, Pini A, Evan G (1987) Induction of c-fos-like protein in spinal cord neurons following sensory stimulation. Nature (Lond) 328:632–634
Hyman SE, Comb M, Lin YS, Pearlberg J, Green MR, Goodman HM (1988) A common trans-acting factor is involved in transcriptional regulation of neurotransmitter genes by cyclic AMP. Mol Cell Biol 8:4225–4233
Hyman SE, Comb M, Pearlberg J, Goodman HM (1989) An AP-2 element acts synergistically with the cyclic AMP- and phorbol ester-inducible enhancer of the human proenkephalin gene. Mol Cell Biol 9:321–324
Iadarola MJ, Douglass J, Civelli O, Naranjo JR (1986) Increased spinal cord dynorphin mRNA during peripheral inflammation. Natl Inst Drug Abuse Res Monogr Ser 75:406–409
Iadarola MJ, Douglass J, Civelli O, Naranjo JR (1988a) Differential activation of spinal cord dynorphin and enkephalin neurons during hyperalgesia: evidence using cDNA hybridization. Brain Res 455:205–212
Iadarola MJ, Brady LS, Draisci G, Dubner R (1988b) Enhancement of dynorphin gene expression in spinal cord following experimental inflammation: stimulus specificity, behavioral parameters and opioid receptor binding. Pain 35:313–326

Inturrisi CE, LaGamma EF, Franklin SO, Huang T, Nip TJ, Yoburn BC (1988a) Characterization of enkephalins in rat adrenal medullary explants. Brain Res 448:230–236
Inturrisi CE, Branch AD, Robertson HD, Howells RD, Franklin SO, Shapiro JR, Calvano SE, Yoburn BC (1988b) Glucocorticoid regulation of enkephalins in cultured rat adrenal medulla. Mol Endocrinol 2:633–640
Israel A, Cohen SN (1985) Hormonally mediated negative regulation of human proopiomelanocortin gene expression after transfection into mouse L cells. Mol Cell Biol 5:2443–2453
Jeannotte L, Trifiro MA, Plante RK, Chamberland M, Drouin J (1987) Tissue-specific activity of the proopiomelanocortin gene promoter. Mol Cell Biol 7:4058–4064
Jin DF, Muffly KE, Okulicz WC, Kilpatrick DL (1988) Estrous cycle- and pregnancy-related differences in expression of the proenkaphalin and proopiomelanocortin genes in the ovary and uterus. Endocrinology 122:1466–1471
Jingami H, Nakanishi S, Imura H, Numa S (1984) Tissue distribution of messenger RNAs coding for opioid peptide precursors and related RNA. Eur J Biochem 142:441–447
Johnson L, Nordeen SK, Roberts JL, Baxter JD (1980) Studies on the mechanism of glucocorticoid hormone action. In: Roy A, Clark J (eds) Gene regulation by steroid hormones. Springer, Berlin Heidelberg New York Tokyo, pp 153–187
Kakidani H, Furutani Y, Takahashi H, Noda M, Morimoto Y, Horise T, Asai M, Inayama S, Nakanishi S, Numa S (1982) Cloning and sequence analysis of cDNA for porcine beta-neo-endorphin/dynorphin precursor. Nature (Lond) 298:245–249
Kanamatsu T, Unsworth CD, Diliberto EJ Jr, Viveros OH, Hong JS (1986) Reflex splanchnic nerve stimulation increases levels of proenkephalin A mRNA and proenkephalin A-related peptides in the rat adrenal medulla. Proc Natl Acad Sci USA 83:9245–9249
Kenigsberg RL, Trifaro JM (1985) Microinjection of calmodulin antibodies into cultured chromaffin cells blocks catecholamine release in response to stimulation. Neuroscience 14:335–347
Keshet E, Polakiewicz RD, Itin A, Ornoy A, Rosen H (1989) Proenkephalin A is expressed in mesodermal lineages during organogenesis. EMBO J 8:2917–2923
Kessler U, Sincini E, Stalla GK, Höllt V (1989) Bombesin stimulates release of a-endorphin in corticotrophic pituitary cells in vitro. Acta Endocrinol 120 (Suppl 1):206–206
Kew D, Kilpatrick DL (1989) Expression and regulation of the proenkephalin gene in rat Sertoli cells. Mol Endocrinol 3:179–184
Khlebodarova TM, Karasik GI, Matveeva NM, Serov OL, Golovin SY, Bondar AA, Karginov VA, Morozov IS, Zelenin SM, Mertvetsov NP (1988) The mink proopiomelanocortin gene: characterization of cDNA and chromosomal localization. Genomics 2:185–188
Kilpatrick DL, Millette CF (1986) Expression of proenkephalin messenger RNA by mouse spermatogenic cells. Proc Natl Acad Sci USA 83:5015–5018
Kilpatrick DL, Rosenthal JL (1986) The proenkephalin gene is widely expressed within the male and female reproductive systems of the rat and hamster. Endocrinology 119:370–374
Kilpatrick DL, Howells RD, Fleminger G, Udenfriend S (1984) Denervation of rat adrenal glands markedly increases preproenkephalin mRNA. Proc Natl Acad Sci USA 81:7221–7223
Kilpatrick DL, Howells RD, Noe M, Bailey LC, Udenfriend S (1985) Expression of preproenkephalin-like mRNA and its peptide products in mammalian testis and ovary. Proc Natl Acad Sci USA 82:7467–7469
Kilpatrick DL, Borland K, Jin DF (1987) Differential expression of opioid peptide genes by testicular germ cells and somatic cells. Proc Natl Acad Sci USA 84:5695–5699
Kitahara N, Nishizawa T, Iida K, Okazaki H, Andoh T, Soma GI (1988) Absence of a gamma-melanocyte-stimulating hormone sequence in proopiomelanocortin mRNA of chum salmon *Oncorhynchus keta*. Comp Biochem Physiol B 91:365–370
Kley N (1988) Multiple regulation of proenkephalin gene expression by protein kinase C. J Biol Chem 263:2003–2008
Kley N, Loeffler JP, Pittius CW, Höllt V (1986) Proenkephalin A gene expression in bovine adrenal chromaffin cells is regulated by changes in electrical activity. EMBO J 5:967–970
Kley N, Loeffler JP, Pittius CW, Höllt V (1987a) Involvement of ion channels in the induction of proenkephalin A gene expression by nicotine and cAMP in bovine chromaffin cells. J Biol Chem 262:4083–4089
Kley N, Loeffler JP, Höllt V (1987b) Ca^{2+}-dependent histaminergic regulation of proenkephalin mRNA levels in cultured adrenal chromaffin cells. Neuroendocrinology 46:89–92

Knight RM. Farah JM. Bishop JF. O'Donohue TL (1987) CRF and cAMP regulation of POMC gene expression in corticotrophic tumor cells. Peptides 8:927–934
Kowalski C. Giraud P. Boudouresque F. Lissitzky JC. Cupo A. Renard M. Saura RM. Oliver C (1989) Enkephalins expression in striatal cell cultures. Adv Bioscience 75:225–228
Lacaze-Masmonteil T. de Keyzer Y. Luton JP. Kahn A. Bertagna X (1987) Characterization of proopiomelanocortin transcripts in human nonpituitary tissues. Proc Natl Acad Sci USA 84:7261–7265
LaGamma EF. Adler JE (1987) Glucocorticoids regulate adrenal opiate peptides. Mol Brain Res 2:125–130
LaGamma EF. White JD. Adler JE. Krause JE. McKelvy JF. Black IB (1985) Depolarization regulates adrenal preproenkephalin mRNA. Proc Natl Acad Sci USA 82:8252–8255
LaGamma EF. White JD. McKelvy JF. Black IB (1988) Second messenger mechanisms governing opiate peptide transmitter regulation in the rat adrenal medulla. Brain Res 441:292–298
La Gamma EF. Goldstein NK. Snyder SB Jr. Weisinger G (1989) Preproenkephalin DNA-binding proteins in the rat: 5'-flanking region. Mol Brain Res 5:131–140
Lee PHK. Zhao D. Xie CW. McGinty JF. Mitchell CL. Hong JS (1989) Changes of proenkephalin and prodynorphin mRNAs and related peptides in rat brain during the development of deep prepyriform cortex kindling. Mol Brain Res 6:263–273
Legon S. Glover DM. Hughes J. Lowry PJ. Rigby PW. Watson CJ (1982) The structure and expression of the preproenkephalin gene. Nucleic Acids Res 10:7905–7918
Levy A. Lightman SL (1988) Quantitative in situ hybridization histochemistry in the rat pituitary gland: effect of bromocriptine on prolactin and proopiomelanocortin gene expression. J Endocrinol 118:205–210
Li H. Risbridger GP. Funder JW. Clements JA (1989) Effect of ethane dimethane sulphonate on proopiomelanocortin (POMC) mRNA and POMC-derived peptides in the rat testis. Mol Cell Endocrinol 65:203–207
Li SJ. Sivam SP. McGinty SF. Jiang HK. Douglass J. Calavetta L. Hong JS (1988) Regulation of the metabolism of striatal dynorphin by the dopaminergic system. J Pharmacol Exp Ther 246:403–408
Lightman SL. Young WS (1987a) Changes in hypothalamic preproenkephalin A mRNA following stress and opiate withdrawal. Nature (Lond) 328:643–645
Lightman SL. Young WS (1987b) Vasopressin. oxytocin. dynorphin. enkephalin and corticotrophin-releasing factor mRNA stimulation in the rat. J Physiol Lond 394:23–39
Lightman SL. Young WS (1988) Corticotrophin-releasing factor. vasopressin and proopiomelanocortin mRNA responses to stress and opiates in the rat. J Physiol Lond 403:511–523
Linner K. Beyer HS. Shahabi N. Peterson P. Sharp B (1989) Viral regulation of preproenkephalin and proopiomelanocortin gene expression in murine thymus and spleen. IRNC. Ste-Adele. Quebec. Canada (Abstract)
Litt M. Buder A (1989) A frequent RFLP identified by a human proenkephalin genomic clone [HGM9 symbol PENK]. Nucleic Acids Res 17:465–465
Llorens-Cortes C. Giros B. Quach T. Schwartz J-C (1990) Adaptive changes in two indices of enkephalin neuron activity in mouse striatum following gabaergic stimulation. In: Quirion R. Jhamandas K. Gianoulakis C (eds) The International Narcotics Research Conference (INRC) '89. Liss. New York. pp 203–206
Loeffler JP. Kley N. Pittius CW. Höllt V (1985) Corticotropin-releasing factor and forskolin increase proopiomelanocortin messenger RNA levels in rat anterior and intermediate cells in vitro. Neurosci Let 62:383–387
Loeffler JP. Kley N. Pittius CW. Almeida OF. Höllt V (1986a) In vivo and in vitro studies of GABAergic inhibition of prolactin biosynthesis. Neuroendocrinology 43:504–510
Loeffler JP. Kley N. Pittius CW. Höllt V (1986b) Calcium ion and cyclic adenosine 3',5'-monophosphate regulate proopiomelanocortin messenger ribonucleic acid levels in rat intermediate and anterior pituitary lobes. Endocrinology 119:2840–2847
Loeffler JP. Kley N. Pittius CW. Höllt V (1986c) Regulation of proopiomelanocortin (POMC) mRNA levels in primary pituitary cultures. Natl Inst Drug Abuse Res Monogr Ser 75:397–400
Loeffler JP. Demeneix BA. Kley NA. Höllt V (1988) Dopamine inhibition of proopiomelanocortin gene expression in the intermediate lobe of the pituitary. Interactions with corticotropin-releasing factor and the beta-adrenergic receptors and the adenylate cyclase system. Neuroendocrinology 47:95–101
Loeffler JP. Kley N. Louis JC. Demeneix BA (1989) Ca^{2+} regulates hormone secretion and

proopiomelanocortin gene expression in melanotrope cells via the calmodulin and the protein kinase C pathways. J Neurochem 52:1279–1283

Lolait SJ, Clements JA, Markwick AJ, Cheng C, McNally M, Smith AI, Funder JW (1986) Proopiomelanocortin messenger ribonucleic acid and posttranslational processing of beta endorphin in spleen macrophages. J Clin Invest 77:1776–1779

Low KG, Nielsen CP, West NB, Douglass J, Brenner RM, Maslar IA, Melner MH (1989) Proenkephalin gene expression in the primate uterus: regulation by estradiol in the endometrium. Mol Endocrinol 3:852–857

Luini A, Lewis D, Guild S, Corda D, Axelrod J (1985) Hormone secretagogues increase cytosolic calcium by increasing cAMP in corticotropin-secreting cells. Proc Natl Acad Sci USA 82:8034–8038

Lundblad JR, Roberts JL (1988) Regulation of proopiomelanocortin gene expression in pituitary. Endocrinol Rev 9:135–158

Martens GJ (1986) Expression of two proopiomelanocortin genes in the pituitary gland of *Xenopus laevis*: complete structures of the two preprohormones. Nucleic Acids Res 14:3791–3798

Martens GJ, Herbert E (1984) Polymorphism and absence of Leu-enkephalin sequences in proenkephalin genes in Xenopus laevis. Nature (Lond) 310:251–254

Martens GJ, Weterings KA, van Zoest ID, Jenks BG (1987) Physiologically-induced changes in proopiomelanocortin mRNA levels in the pituitary gland of the amphibian *Xenopus laevis*. Biochem Biophys Res Commun 143:678–684

McGinty JF, Henriksen SJ, Goldstein A, Terenius L, Bloom FE (1983) Dynorphin is contained within hippocampal mossy fibers: immunochemical alterations after kainic acid administration and colchicine-induced neurotoxicity. Proc Natl Acad Sci USA 80:589–593

McMillen IC, Mercer JE, Thorburn GD (1988) Proopiomelanocortin mRNA levels fall in the fetal sheep pituitary before birth. J Mol Endocrinol 1:141–145

McMurray CT, Devi L, Calavetta L, Douglass JO (1989) Regulated expression of the prodynorphin gene in the R2C Leydig tumor cell line. Endocrinology 124:49–59

Melner MH, Young SL, Czerwiec FS, Lyn D, Puett D, Roberts JL, Koos RD (1986) The regulation of granulosa cell proopiomelanocortin messenger ribonucleic acid by androgens and gonadotropins. Endocrinology 119:2082–2088

Millington WR, Blum M, Knight R, Mueller GP, Roberts JL, O'Donohue TL (1986) A diurnal rhythm in proopiomelanocortin messenger ribonucleic acid that varies concomitantly with the content and secretion of beta-endorphin in the intermediate lobe of the rat pituitary. Endocrinology 118:829–834

Mishina M, Kurosaki T, Yamamoto T, Notake M, Masu M, Numa S (1982) DNA sequences required for transcription in vivo of the human corticotropin-beta-lipotropin precursor gene. EMBO J 1:1533–1538

Mocchetti A, Ritter A, Costa E (1989) Down-regulation of proopiomelanocortin synthesis and beta-endorphin utilization in hypothalamus of morphine-tolerant rats. J Mol Neurosci 1:33–38

Mocchetti I, Giorgi O, Schwartz JP, Costa E (1984) A reduction of the tone of 5-hydroxytryptamine neurons decreases utilization rates of striatal and hypothalamic enkephalins. Eur J Pharmacol 106:427–430

Mocchetti I, Guidotti A, Schwartz JP, Costa E (1985) Reserpine changes the dynamic state of enkephalin stores in rat striatum and adrenal medulla by different mechanisms. J Neurosci 5:3379–3385

Mocchetti I, Naranjo JR, Costa E (1987) Regulation of striatal enkephalin turnover in rats receiving antagonists of specific dopamine receptor subtypes. J Pharmacol Exp Ther 241:1120–1124

Moneta ME, Höllt V (1990) Perforant path kindling induces differential alterations in the mRNA levels coding for prodynorphin and proenkephalin in the rat hippocampus. Neurosci Lett 110:273–278

Monstein HJ, Geijer T (1988) A highly sensitive Northern blot assay detects multiple proenkephali A-like mRNAs in human caudate nucleus and pheochromocytoma. Biosci Rep 8:255–261

Monstein HJ, Folkesson R, Terenius L (1986) Proenkephalin A-like mRNA in human leukem leukocytes and CNS-tissues. Life Sci 39:2237–2241

Morris BJ, Haarmann I, Kempter B, Höllt V, Herz A (1986) Localization of prodynorphin messen RNA in rat brain by in situ hybridization using a synthetic oligonucleotide probe. Neurosci 69:104–108

Morris BJ, Moneta ME, ten Bruggencate G, Höllt V (1987) Levels of prodynorphin mRNA in rat de gyrus are decreased during hippocampal kindling. Neurosci Lett 80:298–302

Morris BJ, Reimer S, Höllt V, Herz A (1988a) Regulation of striatal prodynorphin mRNA levels raphe-striatal pathway. Brain Res 464:15–22

Morris BJ, Höllt V, Herz A (1988b) Dopaminergic regulation of striatal proenkephalin mRNA and prodynorphin mRNA. Neuroscience 25:525-532

Morris BJ, Höllt V, Herz A (1988c) Opioid gene expression in rat striatum is modulated via opioid receptors: evidence from localized receptor inactivation. Neurosci Lett 89:80-84

Morris BJ, Feasey KJ, ten Bruggencate G, Herz A, Höllt V (1988d) Electrical stimulation in vivo increases the expression of proenkephalin in mRNA and decreases the expression of prodynorphin mRNA in rat hippocampal granule cells. Proc Natl Acad Sci USA 85:3226-3230

Morris BJ, Herz A, Höllt V (1989) Location of striatal opioid gene expression, and its modulation by the mesostriatal dopamine pathway: an in situ hybridization study. J Mol Neurosci 1:9-18

Munemura M, Cote TE, Tsuruta K, Eskay RL, Kebabian JW (1980) The dopamine receptor in the intermediate lobe of the rat pituitary gland: pharmacological characterization. Endocrinology 107:1676-1683

Nakamura M, Nakanishi S, Sueoka S, Imura H, Numa S (1978) Effects of steroid hormones on the level of corticotropin messenger RNA activity in cultured mouse-pituitary-tumor cells. Eur J Biochem 86:61-66

Nakanishi S, Inoue A, Kita T, Nakamura M, Chang AC, Cohen SN, Numa S (1979) Nucleotide sequence of cloned cDNA for bovine corticotropin-beta-lipotropin precursor. Nature (Lond) 278:423-427

Nakanishi S, Teranishi Y, Watanabe Y, Notake M, Noda M, Kakidani H, Jingami H, Numa S (1981) Isolation and characterization of the bovine corticotropin/beta-lipotropin precursor gene. Eur J Biochem 115:429-438

Naranjo JR, Iadarola MJ, Costa E (1986a) Changes in the dynamic state of brain proenkephalin-derived peptides during amygdaloid kindling. J Neurosci Res 16:75-87

Naranjo JR, Mocchetti I, Schwartz JP, Costa E (1986b) Permissive effect of dexamethasone on the increase of proenkephalin mRNA induced by depolarization of chromaffin cells. Proc Natl Acad Sci USA 83:1513-1517

Naranjo JR, Wise BC, Mellstrom B, Costa E (1988) Negative feedback regulation of the content of proenkephalin mRNA in chromaffin cell cultures. Neuropharmacology 27:337-343

Nelson JF, Bender M, Schachter BS (1988) Age-related changes in proopiomelanocortin messenger ribonucleic acid levels in hypothalamus and pituitary of female C57BL/6J mice. Endocrinology 123:340-344

Nishimori T, Moskowitz MA, Uhl GR (1988) Opioid peptide gene expression in rat trigeminal nucleus caudalis neurons: normal distribution and effects of trigeminal deafferentation. J Comp Neurol 274:142-150

Nishimori T, Buzzi MG, Moskowitz MA, Uhl GR (1989) Proenkephalin mRNA expression in nucleus caudalis neurons is enhanced by trigeminal stimulation. Mol Brain Res 6:203-210

Noble EP, Bommer M, Sincini E, Costa T, Herz A (1986) H1-histaminergic activation stimulates inositol-1-phosphate accumulation in chromaffin cells. Biochem Biophys Res Commun 135:566-573

Noda M, Teranishi Y, Takahashi H, Toyosato M, Notake M, Nakanishi S, Numa S (1982a) Isolation and structural organization of the human preproenkephalin gene. Nature (Lond) 297:431-434

Noda M, Furutani Y, Takahashi H, Toyosato M, Hirose T, Inayama S, Nakanishi S, Numa S (1982b) Cloning and sequence analysis of cDNA for bovine adrenal preproenkephalin. Nature (Lond) 295:202-206

Normand E, Popovici T, Onteniente B, Fellmann D, Piatier-Tonneau D, Aufffray C, Bloch B (1988) Dopaminergic neurons of the substantia nigra modulate preproenkephalin A gene expression in rat striatal neurons. Brain Res 439:39-46

Notake M, Tobimatsu T, Watanabe Y, Takahashi H, Mishina M, Numa S (1983a) Isolation and characterization of the mouse corticotropin-beta-lipotropin precursor gene and a related pseudogene. FEBS Lett 156:67-71

Notake M, Kurosaki T, Yamamoto T, Handa H, Mishina M, Numa S (1983b) Sequence requirement for transcription in vitro of the human corticotropin/beta-lipotropin precursor gene. Eur J Biochem 133:599-605

Numa S (1984) Opioid peptide precursors and their genes. In: The peptides, vol 6. Academic Press, London, pp 1-23

Oates E, Herbert E (1984) 5′ Sequence of porcine and rat proopiomelanocortin mRNA. One porcine and two rat forms. J Biol Chem 259:7421-7425

Oertel WH, Mugnaini E, Tappaz ML, Weise VK, Dahl AL, Schmechel DE, Kopin IJ (1982) Central

GABAergic innervation of neurointermediate pituitary lobe: biochemical and immunocytochemical study in the rat. Proc Natl Acad Sci USA 79:675–679
Owerbach D, Rutter WJ, Roberts JL, Whitfeld P, Shine J, Seeburg PH, Shows TB (1981) The proopiocortin (adrenocorticotropin/beta-lipoprotein) gene is located on chromosome 2 in humans. Somatic Cell Genet 7:359–369
Pintar JE, Schachter BS, Herman AB, Durgerian S, Krieger DT (1984) Characterization and localization of proopiomelanocortin messenger RNA in the adult rat testis. Science 225:632–634
Pittius CW, Kley N, Loeffler JP, Höllt V (1985) Quantitation of proenkephalin A messenger RNA in bovine brain, pituitary and adrenal medulla: correlation between mRNA and peptide levels. EMBO J 4:1257–1260
Pittius CW, Kley N, Loeffler JP, Höllt V (1987a) Proenkephalin B messenger RNA in procine tissues: characterization, quantification, and correlation with opioid peptides. J Neurochem 48:586–592
Pittius CW, Ellendorff F, Höllt V, Parvizi N (1987b) Ontogenetic development of proenkephalin A and proenkephalin B messenger RNA in fetal pigs. Exp Brain Res 69:208–212
Pritchett DB, Roberts JL (1987) Dopamine regulates expression of the glandular-type kallikrein gene at the transcriptional level in the pituitary. Proc Natl Acad Sci USA 84:5545–5549
Pruss RM, Stauderman KA (1988) Voltage-regulated calcium channels involved in the regulation of enkephalin synthesis are blocked by phorbol ester treatment. J Biol Chem 263:13173–13178
Przewłocki R, Haarmann I, Nikolarakis K, Herz A, Höllt V (1988) Prodynorphin gene expression in spinal cord is enhanced after traumatic injury in the rat. Brain Res 464:37–41
Quach TT, Tang F, Kageyama H, Mocchetti I, Guidotti A, Meek JL, Costa E, Schwartz JP (1984) Enkephalin biosynthesis in adrenal medulla. Modulation of proenkephalin mRNA content of cultured chromaffin cells by 8-bromoadenosine 3′,5′-monophosphate. Mol Pharmacol 26:255–260
Reimer S, Höllt V (1990) Morphine increases proenkephalin gene expression in the adrenal medulla by a central mechanism. In: Quirion R, Jhamandas K, Giounalakis C (eds) The International Narcotics Research Conference (IRNC) '89. Liss, New York, pp 215–218
Reisine T, Rougon G, Barbet J, Affolter HU (1985) Corticotropin-releasing factor-induced adrenocorticotropin hormone release and synthesis is blocked by incorporation of the inhibitor of cyclic AMP-dependent protein kinase into anterior pituitary tumor cells by liposomes. Proc Natl Acad Sci USA 82:8261–8265
Roberts JL, Budarf ML, Baxter JD, Herbert E (1979) Selective reduction of proadrenocorticotropin/endorphin proteins and messenger ribonucleic acid activity in mouse pituitary tumor cells by glucocorticoids. Biochemistry 18:4907–4915
Roberts JL, Lundblad JR, Eberwine JH, Fremeau RT, Salton SR, Blum M (1987) Hormonal regulation of POMC gene expression in pituitary. Ann NY Acad Sci 512:275–285
Romano GJ, Shivers BD, Harlan RE, Howells RD, Pfaff DW (1987) Haloperidol increases proenkephalin mRNA levels in the caudate-putamen of the rat: a quantitative study at the cellular level using in situ hybridization. Brain Res 388:33–41
Romano GJ, Harlan RE, Shivers BD, Howells RD, Pfaff DW (1988) Estrogen increases proenkephalin messenger ribonucleic acid levels in the ventromedial hypothalamus of the rat. Mol Endocrinol 2:1320–1328
Romano GJ, Mobbs CV, Howells RD, Pfaff DW (1989) Estrogen regulation of proenkephalin gene expression in the ventromedial hypothalamus of the rat: temporal qualities and synergism with progesterone. Brain Res Mol Brain Res 5:51–58
Rosen H, Polakiewicz RD (1989a) Increase in hypothalamic proopiomelanocortin gene expression in response to prolonged low level lead exposure. Brain Res 493:380–384
Rosen H, Polakiewicz R (1989b) Postnatal expression of opioid genes in rat brain. Brain Res Dev Brain Res 46:123–129
Rosen H, Douglass J, Herbert E (1984) Isolation and characterization of the rat proenkephalin gene. J Biol Chem 259:14309–14313
Rosen J, Behar O, Abramsky O, Ovadia H (1989) Regulated expression of proenkephalin A in normal lymphocytes. J Immunol 143:3703–3707
Ruda MA, Iadarola MJ, Cohen LV, Young WS (1988) In situ hybridization histochemistry and immunocytochemistry reveal an increase in spinal dynorphin biosynthesis in a rat model of peripheral inflammation and hyperalgesia. Proc Natl Acad Sci USA 85:622–626
Samuels MH, Wierman ME, Wang C, Ridgway EC (1989) The effect of altered thyroid status on pituitary hormone messenger ribonucleic acid concentrations in the rat. Endocrinology 124:2277–2282

Schachter BS, Johnson LK, Baxter JD, Roberts JL (1982) Differential regulation by glucocorticoids of proopiomelanocortin mRNA levels in the anterior and intermediate lobes of the rat pituitary. Endocrinology 110:1442–1444
Schäfer MK-H, Day R, Herman JP, Kwasiborski V, Sladek CD, Akil H, Watson SJ (1989) Effects of electroconvulsive shock on dynorphin in the hypothalamic-neurohypophysial system of the rat. Adv Bioscience 75:599–602
Schäfer MK-H, Day R, Akil H, Watson SJ (1990) Identification of prodynorphin and proenkephalin cells in the neurointermediate lobe of the rat pituitary gland. In: Quirion R, Jhamadas K, Giounalakis C (eds) The International Narcotics Research Conference (IRNC) '89. Liss, New York, pp 231–234
Schwartz JP (1988) Chronic exposure to opiate agonists increases proenkephalin biosynthesis in NG108 cells. Brain Res 427:141–146
Schwartz JP, Simantov R (1988) Developmental expression of proenkephalin mRNA in rat striatum and in striatal cultures. Brain Res 468:311–314
Seger MA, van Eekelen JA, Kiss JZ, Burbach JP, de Kloet ER (1988) Stimulation of proopiomelanocortin gene expression by glucocorticoids in the denervated rat intermediate pituitary gland. Neuroendocrinology 47:350–357
Seizinger BR, Bovermann K, Höllt V, Herz A (1984a) Enhanced activity of the beta-endorphinergic system in the anterior and neurointermediate lobe of the rat pituitary after chronic treatment with ethanol liquid diet. J Pharmacol Exp Ther 230:455–461
Seizinger BR, Höllt V, Herz A (1984b) Effects of chronic ethanol treatment on the in vitro biosynthesis of proopiomelanocortin and its posttranslational processing to beta-endorphin in the intermediate lobe of the rat pituitary. J Neurochem 43:607–613
Sherman TG, Civelli O, Douglass J, Herbert E, Burke S, Watson SJ (1986) Hypothalamic dynorphin and vasopressin mRNA expression in normal and Brattleboro rats. Fed Proc 45:2323–2327
Shiomi H, Watson SJ, Kelsey JE, Akil H (1986) Pretranslational and posttranslational mechanisms for regulating beta-endorphin-adrenocorticotropin of the anterior pituitary lobe. Endocrinology 119:1793–1799
Shivers BD, Harlan RE, Romano GJ, Howells RD, Pfaff DW (1986) Cellular localization of proenkephalin mRNA in rat brain: gene expression in the caudate-putamen and cerebellar cortex. Proc Natl Acad Sci USA 83:6221–6225
Siegel RE, Eiden LE, Affolter HU (1985) Elevated potassium stimulates enkephalin biosynthesis in bovine chromaffin cells. Neuropeptides 6:543–552
Simard J, Labrie F, Gossard F (1986) Regulation of growth hormone mRNA and proopiomelanocortin mRNA levels by cyclic AMP in rat anterior pituitary cells in culture. DNA 5:263–270
Sivam SP, Hong JS (1986) GABAergic regulation of enkephalin in rat striatum: alterations of Met5-enkephalin level, precursor content and preproenkephalin messenger RNA abundance. J Pharmacol Exp Ther 237:326–331
Sivam SP, Strunk C, Smith DR, Hong JS (1986a) Proenkephalin-A gene regulation in the rat striatum: influence of lithium and haloperidol. Mol Pharmacol 30:186–191
Sivam SP, Breese GR, Napier TC, Mueller RA, Hong JS (1986b) Dopaminergic regulation of proenkephalin-A gene expression in the basal ganglia. Natl Inst Drug Abuse Res Monogr Ser 75:389–392
Sivam SP, Breese GR, Krause JE, Napier TC, Mueller RA, Hong JS (1987) Neonatal and adult 6-hydroxydopamine-induced lesions differentially alter tachykinin and enkephalin gene expression. J Neurochem 49:1623–1633
Sivam SP, Takeuchi K, Li S, Douglass J, Civelli O, Calvetta L, Herbert E, McGinty JF, Hong JS (1988) Lithium increases dynorphin A(1–8) and prodynorphin mRNA levels in the basal ganglia of rats. Brain Res 427:155–163
Smith EM, Morrill AC, Meyer WJ, Blalock JE (1986) Corticotropin releasing factor induction of leukocyte-derived immunocreative ACTH and endorphins. Nature (Lond) 321:881–882
Spampinato S, Bachetti T, Canossa M, Ferri S (1990) Prodynorphin messenger RNA expression in the rat anterior pituitary is regulated by estrogen. In: Quirion R, Jhamandas K, Giounalakis C (eds). The International Narcotics Research Conference (INRC) '89. Liss, New York, pp 211–214
Springhorn JP, Claycomb WC (1989) Preproenkephalin in mRNA expression in developing rat heart and in cultured ventricular cardiac muscle cells. Biochem J 258:73–78
Stachowiak MK, Lee PH, Rigual RJ, Viveros OH, Hong JS (1988) Roles of the pituitary-adrenocortical

axis in control of the native and cryptic enkephalin levels and proenkephalin mRNA in the sympathoadrenal system of the rat. Brain Res 427:263–273

Stalla GK, Stalla J, Huber M, Loeffler JP, Höllt V, von Werder K, Muller OA (1988) Ketoconazole inhibits corticotropic cell function in vitro. Endocrinology 122:618–623

Stalla GK, Stalla J, Mojto J, Oeckler R, Buchfelder M, Muller OA (1989a) Regulation of corticotrophic adenoma cells in vitro. Acta Endocrinol 120 (Suppl 1):209

Stalla GK, Stalla J, von Werder K, Muller OA, Gerzer R, Höllt V, Jakobs KH (1989b) Nitroimidazole derivatives inhibit anterior pituitary cell function apparently by a direct effect on the catalytic subunit of the adenylate cyclase holoenzyme. Endocrinology 125:699–706

Steenbergh PH, Hoppener JW, Zandberg J, Roos BA, Jansz HS, Lips CJ (1984) Expression of the proopiomelanocortin gene in human medullary thyroid carcinoma. J Clin Endocrinol Metab 58:904–908

Suda T, Tozawa F, Yamada M, Ushiyama T, Tomori N, Sumitomo T, Nakagami Y, Demura H, Shizume K (1988a) Effects of corticotropin-releasing hormone and dexamethasone on proopiomelanocortin messenger RNA level in human corticotroph adenoma cells in vitro. J Clin Invest 82:110–114

Suda T, Tozawa F, Yamada M, Ushiyama T, Tomori N, Sumitomo T, Nakagami Y, Shizume K (1988b) In vitro study on proopiomelanocortin messenger RNA levels in cultured rat anterior pituitary cells. Life Sci 42:1147–1152

Suda T, Tozawa F, Ushiyama T, Tomori N, Sumitomo T, Nakagami Y, Yamada M, Demura H, Shizume K (1989) Effects of protein kinase-C-related adrenocorticotropin secretagogues and interleukin-1 on proopiomelanocortin gene expression in rat anterior pituitary cells. Endocrinology 124:1444–1449

Takahashi H, Teranishi Y, Nakanishi S, Numa S (1981) Isolation and structural organization of the human corticotropin-beta-lipotropin precursor gene. FEBS Lett 135:97–102

Takahashi H, Hakamata Y, Watanabe Y, Kikuno R, Miyata T, Numa S (1983) Complete nucleotide sequence of the human corticotropin-beta-lipotropin precursor gene. Nucleic Acids Res 11:6847–6858

Tang F, Costa E, Schwartz JP (1983) Increase of proenkephalin mRNA and enkephalin content of rat striatum after daily injection of haloperidol for 2 to 3 weeks. Proc Natl Acad Sci USA 80:3841–3844

Terao M, Watanabe Y, Mishina M, Numa S (1983) Sequence requirement for transcription in vivo of the human preproenkephalin A gene. EMBO J 2:2223–2228

Tomiko SA, Taraskevich PS, Douglas WW (1983) GABA acts directly on cells of pituitary pars intermedia to alter hormone output. Nature (Lond) 301:706–707

Tozawa F, Suda T, Yamada M, Ushiyama T, Tomori N, Sumitomo T, Nakagami Y, Demura H, Shizume K (1988) Insulin-induced hypoglycemia increases proopiomelanocortin messenger ribonucleic acid levels in rat anterior pituitary gland. Endocrinology 122:1231–1235

Tremblay Y, Tretjakoff I, Peterson A, Antakly T, Zhang CX, Drouin J (1988) Pituitary-specific expression and glucocorticoid regulation of a proopiomelanocortin fusion gene in transgenic mice. Proc Natl Acad Sci USA 85:8890–8894

Uhl GR, Ryan JP, Schwartz JP (1988) Morphine alters preproenkephalin gene expression. Brain Res 459:391–397

Uhler M, Herbert E (1983) Complete amino acid sequence of mouse proopiomelanocortin derived from the nucleotide sequence of proopiomelanocortin cDNA. J Biol Chem 258:257–261

Uhler M, Herbert E, D'Eustachio P, Ruddle FD (1983) The mouse genome contains two nonallelic proopiomelanocortin genes. J Biol Chem 258:9444–9453

Vernier P, Julien JF, Rataboul P, Fourrier O, Feuerstein C, Mallet J (1988) Similar time course changes in striatal levels of glutamic acid decarboxylase and proenkephalin mRNA following dopaminergic deafferentation in the rat. J Neurochem 51:1375–1380

Vilijn MH, Vaysse PJ, Zukin RS, Kessler JA (1988) Expression of preproenkephalin mRNA by cultured astrocytes and neurons. Proc Natl Acad Sci USA 85:6551–6555

von Dreden G, Höllt V (1988) Vasopressin potentiates β-endorphin release but not the increase in the mRNA for proopiomelanocortin induced by corticotropin releasing factor in rat pituitary cells. Acta Endocrinol 117 (Suppl 287):124

von Dreden G, Loeffler JP, Grimm C, Höllt V (1988) Influence of calcium ions on proopiomelanocortin mRNA levels in clonal anterior pituitary cells. Neuroendocrinology 47:32–37

Wan DC, Livett BG (1989a) Vasoactive intestinal peptide stimulates proenkephalin A mRNA expression in bovine adrenal chromaffin cells. Neurosci Lett 101:218–222

Wan DC, Livett BG (1989b) Induction of phenylethanolamine N-methyltransferase mRNA expression by glucocorticoids in cultured bovine adrenal chromaffin cells. Eur J Pharmacol 172:107–115

Wan DC, Marley PD, Livett BG (1989a) Histamine activates proenkephalin A mRNA but not phenylethanolamine N-methyltransferase mRNA expression in cultured bovine adrenal chromaffin cells. Eur J Pharmacol 172:117–129

Wan DC, Scanlon D, Choi CL, Bunn SJ, Howe PR, Livett BG (1989b) Co-localization of RNAs coding for phenylethanolamine N-methyltransferase and proenkephalin A in bovine and ovine adrenals. J Auton Nerv Syst 26:231–240

Waschek JA, Eiden LE (1988) Calcium requirements for barium stimulation of enkephalin and vasoactive intestinal peptide biosynthesis in adrenomedullary chromaffin cells. Neuropeptides 11:39–45

Waschek JA, Dave JR, Eskay RL, Eiden LE (1987) Barium distinguishes separate calcium targets for synthesis and secretion of peptides in neuroendocrine cells. Biochem Biophys Res Commun 146:495–501

Weihe E, Millan MJ, Leibold A, Nohr D, Herz A (1988) Co-localization of proenkephalin- and prodynorphin-derived opioid peptides in laminae IV/V spinal neurons revealed in arthritic rats. Neurosci Lett 85:187–192

Weihe E, Millan MJ, Höllt V, Nohr D, Herz A (1989) Induction of the gene encoding pro-dynorphin by experimentally induced arthritis enhances staining for dynorphin in the spinal cord of rats. Neuroscience 31:77–95

White JD, Gall CM, McKelvy JF (1987) Enkephalin biosynthesis and enkephalin gene expression are increased in hippocampal mossy fibers following a unilateral lesion of the hilus. J Neurosci 7:753–759

Whitfeld PL, Seeburg PH, Shine J (1982) The human proopiomelanocortin gene: organization, sequence, and interspersion with repetitive DNA. DNA 1:133–143

Wiemann JN, Clifton DK, Steiner RA (1989) Pubertal changes in gonadotropin-releasing hormone and proopiomelanocortin gene expression in the brain of the male rat. Endocrinology 124:1760–1767

Wilcox JN, Roberts JL (1985) Estrogen decreases rat hypothalamic proopiomelanocortin messenger ribonucleic acid levels. Endocrinology 117:2392–2396

Xie CW, Lee PH, Takeuchi K, Owyang V, Li SJ, Douglass J, Hong JS (1989) Single or repeated electroconvulsive shocks alter the levels of prodynorphin and proenkephalin mRNAs in rat brain. Brain Res Mol Brain Res 6:11–19

Yoshikawa K, Sabol SL (1986) Expression of the enkephalin precursor gene in C6 rat glioma cells: regulation of beta-adrenergic agonists and glucocorticoids. Brain Res 387:75–83

Yoshikawa K, Williams C, Sabol SL (1984) Rat brain preproenkephalin mRNA. cDNA cloning, primary structure, and distribution in the central nervous system. J Biol Chem 259:14301–14308

Yoshikawa K, Hong JS, Sabol SL (1985) Electroconvulsive shock increases preproenkephalin messenger RNA abundance in rat hypothalamus. Proc Natl Acad Sci USA 82:589–593

Yoshikawa K, Maruyama K, Aizawa T, Yamamoto A (1989a) A new species of enkephalin precursor mRNA with a distinct 5′- untranslated region in haploid germ cells. FEBS Lett 246:193–196

Yoshikawa K, Aizawa T, Nozawa A (1989b) Phorbol ester regulates the abundance of enkephalin precursor mRNA but not of amyloid beta-protein precursor mRNA in rat testicular peritubular cells. Biochem Biophys Res Commun 161:568–575

Young WS, Bonner TI, Brann MR (1986) Mesencephalic dopamine neurons regulate the expression of neuropeptide mRNAs in the rat forebrain. Proc Natl Acad Sci USA 83:9827–9831

Zhang JS, Plevy JS, Albeck H, Culpepper-Morgan J, Friedman J, Kreek MJ (1989) Effects of age on distribution of preproenkephalin-like mRNA in the gastro-intestinal tract of the guinea pig. Adv Bioscience 75:349–350

Zheng M, Yang SG, Zou G (1988) Electro-acupuncture markedly increases proenkephalin mRNA in rat striatum and pituitary. Sci Sin B 31:81–86

Zurawski G, Benedik M, Kamb BJ, Abrams JS, Zurawski SM, Lee FD (1986) Activation of mouse T-helper cells induces abundant preproenkephalin mRNA synthesis. Science 232:772–775

CHAPTER 3

Distribution of Opioids in Brain and Peripheral Tissues

M.K.-H. Schafer, R. Day, S.J. Watson, and H. Akil

1 Introduction

Anatomical studies have played an important role in shaping the field of endogenous opioid peptides. Most probably, this is because very early on, multiple opioids were discovered almost simultaneously. The pentapeptides Met- and Leu-Enkephalin (ENK) were isolated together from brain tissue (Hughes et al. 1975) and β-endorphin (β-END) soon thereafter from pituitary extracts (Chretien et al. 1976; Li and Chung 1976). This immediately brought into question the issue of their biosynthetic relationships, their relative potencies, and their tissue-specific expression. Did these peptides represent two different classes of modulators, with β-END being a pituitary hormone, relatively long-acting, and ENKs short-acting classical neurotransmitters? Since the Met-ENK structure was contained in β-END, was the latter a precursor of the former? Or was Met-ENK an extraction artifact, with β-END being the "true" opioid? And what of Leu-ENK? Did it derive from a similarly large progenitor, a putative β-END with Leu in position five? Undoubtedly, such questions could have been eventually answered with biochemical, and later on, molecular biological tools. The answers, however, came surprisingly quickly, through immunohistochemistry. In retrospect, given the structural similarities of the endogenous opioids, the anatomical studies have been remarkably accurate.

Using antisera directed against the ENK and β-END, immunohistochemical studies showed many of the early assumptions to be incorrect. ENKs did not appear to be purification artifacts; antisera against them demonstrated punctate storage in specific neuronal pathways (Elde et al. 1976). β-END was not exclusively a pituitary hormone, as it was clearly expressed in the brain (Watson et al. 1977). Furthermore, β-END did not appear to be a precursor of Met-ENK, as the two were distributed differently in the CNS (Bloom et al. 1978; Watson et al. 1978). This was in sharp contrast to the inability of most immunohistochemical studies to separate Leu- from Met-ENK in the CNS (not surprising now, in view of the fact that both are coded for by the same proenkephalin precursor). Yet, some Leu-ENK antisera detected cell groups devoid of Met-ENK presaging the discovery of prodynorphin (PDYN) (Larsson et al. 1979).

These anatomical observations helped set the stage for subsequent biochemical and molecular biological studies. In turn, important biochemical discoveries prompted further useful anatomical studies. For example, the biosynthetic association between ACTH/MSH and β-LPH/β-END in the pituitary (Mains et al.

1977) was tested immunohistochemically and found to be true in pituitary and in brain (Pelletier et al. 1977; Watson et al. 1977; Weber et al. 1978; Pelletier 1979). The close association between biochemists and anatomists continued with the isolation of DYN from pituitary extracts (Goldstein et al. 1979). The observation that the immunoreactivity was in neural lobe fibers and that the cells of origin were the magnocellular hypothalamic neurons (Watson et al. 1981) pointed to the hypothalamus as the proper source of mRNA for subsequent molecular cloning of PDYN (Kakidani et al. 1982; Civelli et al. 1985). By 1982 with the help of recombinant cDNA technology, three members of the opioid family, the protein precursors proopiomelanocortin (POMC), proenkephalin (PENK), and PDYN had been identified, their primary amino acid sequence deduced from the mRNA structure (Nakanishi et al. 1979; Gubler et al. 1982; Kakidani et al. 1982; Noda et al. 1982), and many new potentially bioactive peptides could be predicted as a result of dibasic or monobasic cleavage sites and were eventually isolated.

Thus, in the case of opioid peptides, anatomy was concerned not only with expression of a peptide in particular cells, but with issues of biosynthesis, of differential processing, and of coexpression of multiple peptide products in the same neurons. We believe that this continues to be true. The interface between anatomical, biochemical, and cell biological approaches in the field of opioid and related species continues to reveal new systems and suggest new functions for opioids, and to bring into question old assumptions about the cell biology of peptide-expressing cells. The newer contributions rely on the combination of the more classical tools of immunohistochemistry with the newer tool of in situ hybridization. In situ hybridization detects not the peptide product, but rather the mRNA which codes for it. As such, it demonstrates cell bodies of origin. While immunohistochemical approaches had provided a static picture of opioid distribution in pituitary and brain, in situ hybridization has the great advantage of being relatively quantitative, as well as extremely sensitive (Valentino et al. 1987). This has allowed us to detect the expression of opioid peptide mRNAs in unexpected tissues and cell types. Given the structural complexity of the CNS, in situ hybridization has allowed us to begin studying gene regulation in the CNS at the single cell level, in an anatomical context.

In this chapter, we shall overview the now classical distribution of opioid peptides in brain and periphery. A detailed anatomical description is not in the scope of this article, since extensive reviews and research papers on this topic are available (see Watson et al. 1984; Khachaturian et al. 1985). Rather we shall focus on new findings which have emanated following the use of molecular biological techniques for the localization of gene expression, and confine ourselves to a few areas in the brain and some peripheral tissues to illustrate the functional implications for opioid peptides and their physiological role deduced from their tissue localization.

2 The Three Opioid Precursors in Brain and Pituitary

2.1 POMC

Cell Groups

POMC and its cleavage products are best characterized in the pituitary gland, the major site of POMC biosynthesis in the body. The POMC gene is expressed in corticotrophs in the anterior lobe and in all melanotrophs in the intermediate lobe. It gives rise to multiple peptides with different biological functions, including the highly potent opioid β-END, the stress hormone $ACTH_{1-31}$, and various MSH peptides (α-, β-, γ-MSH) (Nakanishi et al. 1979). While size and structure of the POMC precursor and its mRNA are identical across these two tissues, the post-translational processing pattern is quite different (Mains and Eipper 1981). In general, intermediate lobe cells, under strong dopaminergic inhibition, generate smaller peptides such as α-MSH and β-END, which also have undergone more posttranslational modifications. For example, β-END can be further processed by C-terminal shortening to β-END_{1-27} and β-END_{1-26}, resulting in a decreased receptor affinity (Akil et al. 1981). It is also N-terminally acetylated, virtually preventing any opioid activity. In fact, N-acetyl-β-END_{1-27}, a molecular species devoid of any opioid receptor activity, is the dominant product in the intermediate lobe. In contrast, most of the β-END-like immunoreactivity in the anterior lobe is, in fact, the biosynthetic intermediate β-LPH, whereas unmodified β-END represents the remainder of the stores.

The pattern of expression of ACTH and α-MSH in the pituitary gland was well established in the 1970s. However, the isolation of β-END as an endogenous ligand for opiate receptors in 1976 (Chretien et al. 1976; Li and Chung 1976) led to several immunocytochemical studies in the pituitary and the CNS. While the existence of ACTH and β-END in the pituitary was readily accepted, the presence of the same peptides in the CNS encountered more resistance. Several investigators initially believed that the peptides found in the brain were of pituitary origin. However, studies on hypophysectomized animals showed the continued presence of ACTH-like material in brain (Krieger et al. 1977). Lesion studies of the arcuate nucleus with mono-sodium glutamate caused marked reductions of β-END-like material in hypothalamic extracts (Krieger et al. 1979) finally confirming the immunocytochemical observations that the POMC gene was in fact expressed in the brain. Today, we know of only two distinct cell groups, in the adult animal, which express the POMC gene. The first cell group is located in the arcuate nucleus, in the posterior hypothalamus, where POMC-containing neurons are homogenously distributed over its entire rostrocaudal length, with some scattered cells in the medial-basal hypothalamus (Watson et al. 1977; Bloch et al. 1978; Bloom et al. 1978). A second smaller group of POMC neurons was described in the caudal part of the nucleus tractus solitarius (NTS). These small POMC-producing neurons are confined to the commissural portion of the NTS (Joseph et al. 1983; Schwartzberg and Nakane 1983). Their visualization of immunocytochemical means is not a trivial task, probably due to their low peptide content. High levels of colchicine injected directly into the ventricular system are a prerequisite for successful localization.

Therefore, the existence of this second POMC-containing cell group is still not commonly accepted. As mentioned above, the addition of in situ hybridization to the histochemical repertoire allows us to demonstrate the presence of any expressed gene by localization of its mRNA. The demonstration of POMC mRNA in arcuate cells by Gee et al. (1983) was, in fact, the pioneering study of mRNA localization in brain. While this POMC mRNA could be fairly easily detected in the arcuate nucleus, the detection of POMC neurons in the NTS both by Northern blot analysis (D.M. Bronstein, pers. comm.) and by in situ hybridization did not yield any positive results, casting doubt on the previously reported immunocytochemical observations. Recently, in our laboratory we have finally succeeded in demonstrating POMC mRNA in the caudal NTS neurons by in situ hybridization (Fig. 1), but the requirement for long exposure times of the autoradiograms suggests that the

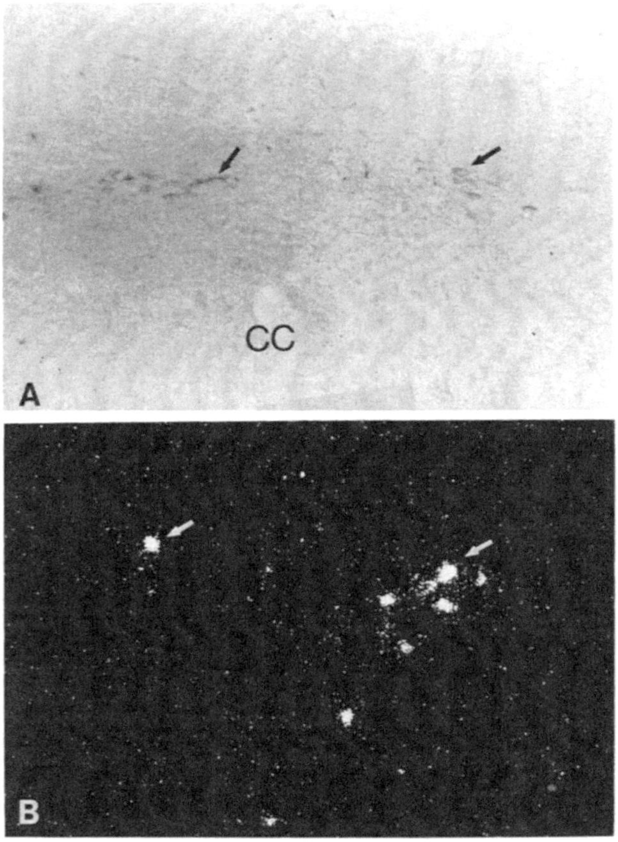

Fig. 1. Localization of POMC-containing neurons in the NTS. ACTH immunoreactive cell bodies (*arrows*) can be observed after injection of high doses of colchicine (1 μg/g body weight) in the lateral ventricle (**A**). Cells are localized in the caudal part of the NTS (commissural ncl.). In situ localization of POMC mRNA with a rat cRNA probe in the NTS (**B**). A few labeled cells (*arrows*) are detected only in the commissural ncl. of the NTS confirming the immunohistochemical findings (*CC* central canal; exposure time of autoradiogram: 5 weeks; mag. 200 ×)

POMC mRNA levels in NTS neurons are several fold lower than those observed in the arcuate nucleus. Whether this reflects a lower biosynthetic activity remains to be determined, particularly considering the differences in cell size between the two systems.

POMC Projections in the CNS

Both groups of POMC cells have extensive projections to other areas and a rich network of POMC fibers can be observed in brain as illustrated in Fig. 2. In general, POMC neurons in the arcuate nucleus extend their fibers in all directions and to many areas throughout the CNS, whereas NTS POMC neurons exhibit more local and caudal projections extending down to the spinal cord (Tsou et al. 1986; Palkovits et al. 1987; Joseph and Micheal 1988).

Fibers originating from the arcuate nucleus have been shown to innervate many areas throughout the brain. Rostral projections pass through periventricular and diencephalic regions and innervate hypothalamic nuclei and limbic structures, including the septum, bed nucleus of the stria terminalis, and the preoptic area. Laterally, POMC fibers extend through the medial basal hypothalamus ventrally

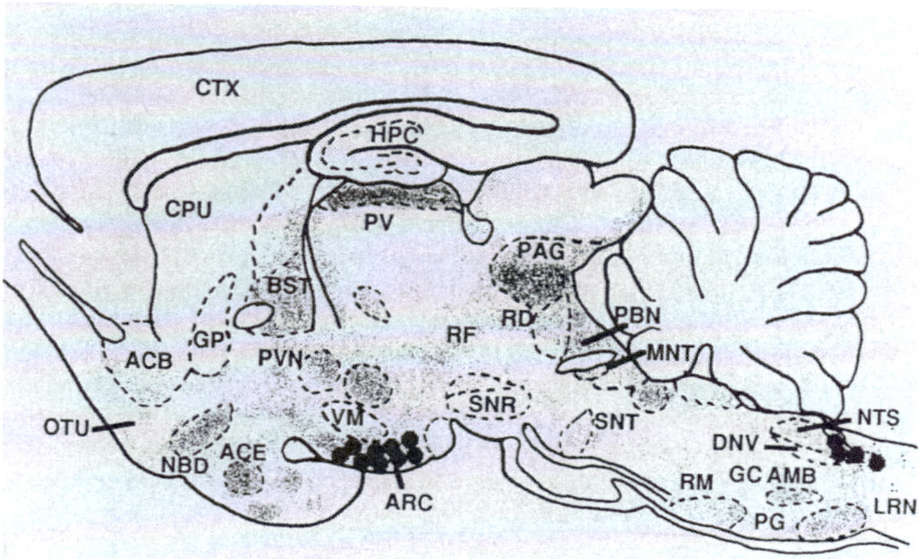

Fig. 2. Distribution of POMC in the rat CNS. Cell bodies (*black circles*) are located only in the arcuate nucleus (*ARC*) and in the caudal portion of the nucleus of the solitary tract (*NTS*). Terminal fields (*shaded areas*) of POMC projections are found in many brain nuclei. *ACB* ncl. accumbens septi; *ACE* central ncl. of amygdala; *AMB* ncl. ambiguus; *BST* bed ncl. of the stria terminalis; *CPU* caudate-putamen; *CTX* cortex; *DM* dorsomedial ncl. of hypothalamus; *DNV* dorsal motor nucleus of vagus; *GC* ncl. reticularis gigantocellularis; *GP* globus pallidus; *HPC* hippocampal formation; *LRN* lateral reticular nucleus; *MNT* mesencephalic ncl. of trigeminal; *NDB* ncl. of diagonal band; *OTU* olfactory tubercle; *PAG* periaqueductal gray; *PBN* parabrachial nucleus; *PG* ncl. reticularis paragigantocellularis; *PV* periventricular ncl. of the thalamus; *PVN* paraventricular ncl.; *RF* reticular formation; *RM* ncl. raphe magnus; *SNR* substantia nigra, pars reticulata; *SNT* sensory ncl. of trigeminal

and enter the amygdala. Caudal projections course through the dorsal diencephalon into the mesencephalon and brainstem, innervating the periventricular thalamus and the periaqueductal gray, areas associated with nociception and sensory integration. Projections to the brainstem regions include, ventrally, the reticular formation and, dorsally, the nuclei parabrachialis and ambiguus, the NTS, and the dorsal motor nucleus of the vagus, regions which are all involved in autonomic regulation of the respiratory and cardiovascular systems. Areas which show no or extremely few POMC-containing fibers are striatum, cortex, and hippocampus.

As mentioned, the small POMC cells in the NTS show a more caudal orientation forming a fiber bundle between the NTS and the ventrolateral medulla. The POMC fibers observed in the NTS seem to be of local origin, whereas the catecholamine-containing brainstem nuclei and the baroreceptor areas of the lower brainstem receive a double innervation from the arcuate nucleus and the NTS (Palkovits et al. 1987). POMC fibers observed in the spinal cord (Tsou et al. 1986) originate, at least in part, from the NTS, although the presence of local POMC circuits in the cord cannot be excluded.

While the physiological role of POMC-derived peptides in the pituitary is clearly characterized by its involvement in the stress-regulating circuitry, their functional role in the CNS is less clear. Given the potent analgesic properties of β-END, it is likely to play a role in pain modulation, including phenomena such as stimulation-produced analgesia (see Millan, this Vol.). Electrical stimulation of the periaqueductal gray, an area rich in POMC fibers, can produce profound behavioral analgesia in the rat (Akil et al. 1972, 1976), while stimulation of the medial thalamus, along the POMC bundle, can relieve pain in man (Richardson and Akil 1977). In addition, the best sites of stimulation-produced analgesia represent a pattern which coincides almost perfectly with the β-END pathway (Watson et al. 1978). More recently, stimulation of the NTS has also been shown to produce analgesia (Lewis et al. 1987). Finally, stimulation appears to lead to a depletion in β-END content and to be dependent on the presence of an intact arcuate nucleus (Millan et al. 1986).

While the involvement of β-END in pain contraol has both anatomical and functional validity, the expression of POMC in other cells suggests numerous other functions. For example, POMC projections to the paraventricular nucleus and the median eminence would suggest central involvement in anterior and posterior pituitary hormone regulation. Recently, regulation of POMC mRNA levels in the anterior subset of arcuate POMC cells by androgens was reported (Chowen-Breed et al. 1989) suggesting a central role of POMC in reproductive physiology.

2.2 PENK

PENK contains seven opiate-active peptides including the pentapeptides Met-ENK and Leu-ENK, their C-terminal extended forms Met-ENK Arg-Phe, Met-ENK-Arg-Gly-Leu, and longer opioids BAM 22, peptide E, and peptide F (Comb et al. 1982; Noda et al. 1982). Compared to POMC it is a far more complex opioid family, not only in terms of its precursor structure, but also in terms of its tissue distribution.

Neurons containing PENK-derived peptides can be found virtually at all levels of the neuraxis from the cerebral cortex down to the spinal cord. After the discovery of the first opiate-active peptides Met- and Leu-ENK, immunocytochemical studies by Elde (1976) and others (Bloom et al. 1978; Watson et al. 1978) showed a very similar distribution pattern for both peptides, which was quite different from that of POMC. These studies were followed by extensive mapping studies in various species (Hökfelt et al. 1977; Uhl et al. 1979; Finley et al. 1981; Haber and Elde 1982; Khachaturian et al. 1983; Williams and Dockray 1983; Merchenthaler et al. 1986) describing the extensive PENK-containing neuronal circuits in the CNS, some of them being confined to short local networks, others forming long tract projections. Most of these immunocytochemical studies are in good agreement regarding the widespread nature of these peptides. The use of the neurotoxin colchicine, which inhibits axonal transport of secretory granules, has aided investigators in the identification of PENK-expressing cell bodies. Perikarya were demonstrated in most regions of the telencephalon, including cerebral cortex, olfactory tubercle, piriform cortex (see Fig. 3), amygdala, striatum, septum, bed nucleus of stria terminalis, and the preoptic area. In the diencephalon, perikarya where seen in most hypothalamic nucleic, and in the periventricular and lateral geniculate nucleus of the thalamus. In the midbrain, ENK cells were found in the colliculi, periaqueductal gray, and interpeduncular nucleus. In the brainstem, perikarya were identified in the parabrachial, dorsotegmental, vestibular and raphe nuclei, nuclei gigantocellularis and paragigantocellularis, NTS, lateral reticular nucleus, and in the spinal cord dorsal gray. Figure 4 illustrates one of the best examples of a rich enkephalinergic fiber network, the woolly fibers in the globus pallidus. Here, the striatal fibers originating from neurons located in caudate-putamen, terminate in a dense network on dendrites of pallidal neurons.

Recently, the distribution PENK mRNA by in situ hybridization histochemistry was reported by Harlan et al. (1987) and compared to previous immunocytochemical data. The general distribution pattern of PENK peptide and mRNA-containing perikarya matched surprisingly well, particularly in the studies where high doses of colchicine were used. However, the in situ studies revealed even more PENK-containing neurons than observed in immunocytochemical studies. The most striking discrepancies existed in cortical areas, the olfactory tubercle, the piriform cortex, the nucleus accumbens, and in particular in the cerebellum, where a population of neurons in the granular cell layers, identified as Golgi II cells, were demonstrated. Given the widespread distribution of ENK neurons, it should be apparent that PENK may be involved in a variety of CNS functions including pain modulation, processing of sensory information, motor function, and endocrine modulation.

2.3 *PDYN*

In 1979, Goldstein and co-workers isolated a 13-amino acid peptide which contained the Leu-ENK sequence at its N-terminal end. Soon thereafter the full 17-amino acid sequence of this peptide was elucidated, and other Leu-ENK extended peptides

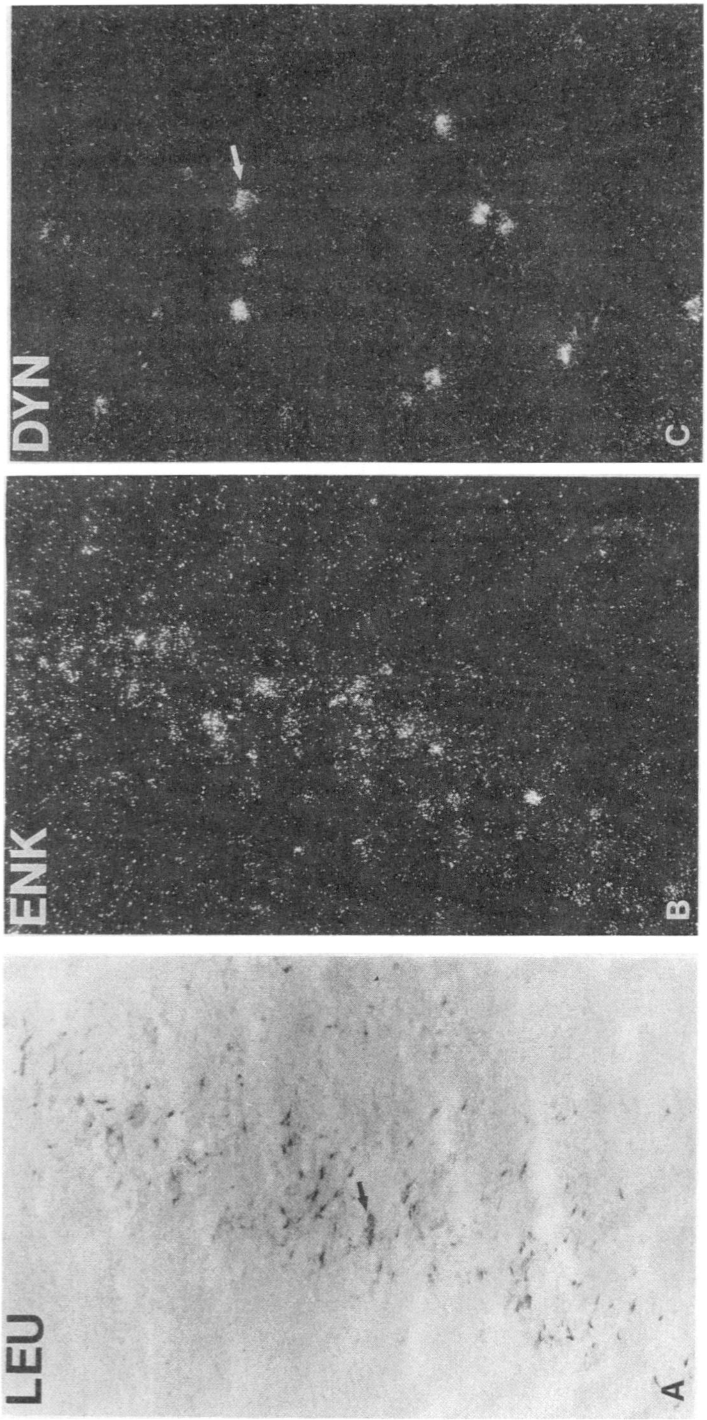

Fig. 3. Opioid gene expression in rat cortex. Leu-ENK immunoreactive cell bodies (*arrow*) and fibers can be demonstrated in cortical neurons of the piriform cortex only following colchicine pretreatment (**A**). In unhandled rats proenkephalin mRNA is localized with a rat cRNA probe in the same region in many cells of the piriform cortex (**B**; exposure time: 6 days). Prodynorphin mRNA-expressing cells (**C**; *arrow*) are located in parietal cortex throughout several cortical layers (exposure time: 10 days)

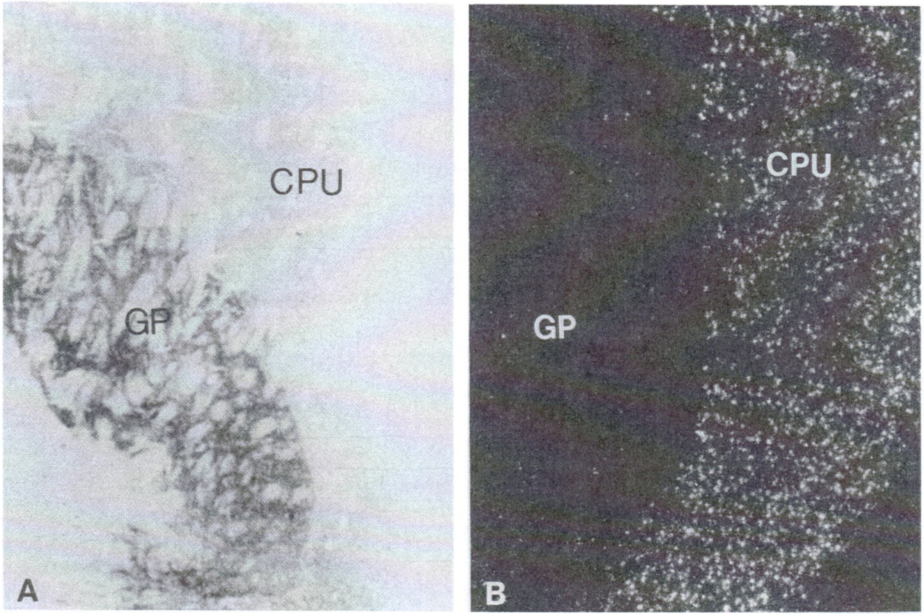

Fig. 4. PENK gene expression in the striato-pallidal pathway. Following immunostaining with an antibody to Leu-ENK, the globus pallidus (*GP*) shows one of the most intense ENKergic fiber networks in the rat brain (**A**). These "woolly fibers" surround the dendrites of pallidal neurons in a very fine, dense network. Cell bodies of origin are the striosomes located in caudate-putamen (*CPU*), not visible in **A**. Without requiring colchicine pretreatment, numerous cell bodies expressing PENK mRNA can be observed in the caudate-putamen by in situ hybridization (**B**), but virtually no positively labeled cells are seen in *GP*. (exposure time: 6 days; mag. 40 ×)

were found in hypothalamic extracts: a-neo-END, DYN B (for review, see Watson et al. 1984). Immunocytochemical studies showed these peptides to be contained in the same cells throughout various brain areas (Watson et al. 1982; Weber and Barchas 1983), and the characterization of the third opioid family, PDYN, began.

PDYN like PENK, contains several opiate-active peptides. These are DYN A, DYN B, α- and β-neo-END, as well as Leu-ENK which constitutes their N-terminal sequence (Kakidani et al. 1982). This precursor is also synthesized throughout the CNS and, compared to POMC and PENK, takes an intermediate position in terms of its tissue distribution (Khachaturian et al. 1982). Immunoreactive perikarya are distributed in several cortical areas, in the striatum, amygdala, the dentate gyrus of the hippocampal formation, hypothalamic nuclei including the supraoptic and paraventricular nuclei, midbrain periaqueductal gray, and many brainstem areas, including the parabrachial and spinal trigeminal nuclei, NTS, lateral reticular nucleus, and the dorsal horn of the spinal cord. Areas containing a high density of PDYN fibers are the globus pallidus, nucleus accumbens, the ventral pallidum, the substantia nigra pars reticulata, and the mossy fibers of the hippocampus. In general, PDYN could be found in many areas which also contain peptides derived from PENK. Their similar distribution pattern suggested the coexistence of both precursors in the same cells. However, by avoiding the use of antibodies raised against sequences shared by both precursors, such as Leu-ENK and peptide E, and by using midportion or longer fragment antibodies, it was demonstrated that the distribution of the two precursors was indeed different (Watson et al. 1982).

Two anatomical circuits containing high levels of PDYN-derived peptides have helped to shed some light on the functional significance of this class of opioids. These are the nigrostriato-nigral loop, which is described elsewhere in this volume in more detail, and the hypothalamic hypophysial neurosecretory system, where DYNs were first demonstrated by Watson et al. (1982). This system has been extensively characterized over the last decade. PDYN is coexpressed with the classical neuropeptide vasopressin in magnocellular neurons, stored in the same secretory granules (Whitnall et al. 1983), and transported to the terminals in the posterior lobe of the pituitary, where it is coreleased upon secretory stimuli such as dehydration and salt-loading (Höllt et al. 1981). Opioids derived from PDYN have recently been shown to inhibit the release of oxytocin from its terminals in the posterior lobe via kappa receptors (Bondy et al. 1988). In addition, they appear to inhibit vasopressin itself, leading to diuresis (Leander 1983).

Although the presence of PDYN-derived peptides in the anterior lobe of pituitary was known for some time (Seizinger et al. 1981), PDYN was only recently shown to be coexpressed within a subset of LH/FSH cells (Khachaturian et al. 1986) and the complete posttranslational processing pattern has been elucidated (Day and Akil 1989).

While PDYN mRNA localization studies have in general confirmed the immunocytochemical findings (Morris et al. 1986; Schafer et al. 1988), they have added valuable new information on the changes in PDYN gene expression. In fact, the hypothalamic magnocellular neurons, probably containing the highest levels of PDYN mRNA, have served as a model system to validate the quantitative character of in situ hybridization versus Northern blot or protection assays. Following salt-loading, dramatic increases were observed for both PDYN and vasopressin mRNA, strongly suggesting the coregulation of these coexpressed precursors at the transcriptional level (Sherman et al. 1986). The same dramatic increases of PDYN mRNA in these cells could be demonstrated in quantitative in situ hybridization studies (Watson et al. 1988), which showed that the in situ method was not only quite comparable to Northern analysis, but added much higher resolution to it. Other studies on PDYN gene expression followed in the striatonigral system and hippocampus (Morris et al. 1989), demonstrating a possible role of PDYN in seizure activity.

2.4 PENK Gene Expression in Glial Cells

Cells expressing opioids, as we have described above, have generally been considered to be of neuronal or neuroendocrine origin. However, several recent reports have suggested that PENK mRNA is expressed by glial cells. PENK mRNA, but not PDYN mRNA, could be isolated from tissue extracts of primary astrocyte cultures of neonatal rat brain (Vilijn et al. 1988). Another report demonstrated the presence of PENK mRNA in primary cultured astrocytes from monkey tissue (Saneto et al. 1989). We observed in our laboratory PENK mRNA in pituicytes, a special class of astroglial cells, in the posterior pituitary of adult rats by in situ hybridization (Schafer et al. 1990). To what extent PENK mRNA is expressed in situ by astrocytes in the adult CNS remains to be determined. However, preliminary results indicate that such expression is quite likely. This may not only lead to a reevaluation of PENK distribution in the CNS, but may also open an exciting new area of research on opioids in glia, the major cellular component of the CNS. Whether astrocytes translate PENK mRNA into protein or are capable of proper posttranslational processing and secretion of the cleavage products remains to be determined. The dramatic discrepancy between relatively high levels of mRNA and the extremely low levels of detectable stored peptides is fairly common for peripheral tissues. The lack of immunocytochemical evidence for PENK-derived opioids in astrocytes or pituicytes may reflect the subcellular structure of these cells, which lack the typical secretory granules observed in neurons and endocrine cells. Single glia lack the regulated secretory pathway. PENK, if expressed as a precursor at the protein level, may be rapidly and constitutively released. The functional implications are at present at best a matter of speculation, but the strategic localization of astrocytes and their topographical proximity to blood vessels via so-called end-feet processes, terminating on capillaries, may provide some clues.

3 Opioids in Peripheral Tissues

In the past several years, extensive studies have examined the distribution and regulation of opioid-related neuropeptides in the periphery. Peptide products of all three precursors have been detected by either immunocytochemistry or radioimmunoassay in a variety of peripheral systems. In general, the opioid peptide content is usually lower when compared to brain, pituitary, or spinal cord. Notable exceptions are the adrenal medulla, which in many species, contains high levels of ENKs, and the gastrointestinal tract and sympathetic ganglia which contain substantial amounts of ENK and DYN peptides. It is not within the scope of this chapter to extensively review these areas as they have been well described elsewhere (Goldstein 1984; Udenfriend and Kilpatrick 1984; Smith and Funder 1988). However, recent data, triggered by the cloning of the three opioid genes, has forced a reevaluation of the presence of opioid peptides in several peripheral organs where the low content of opioid peptides was sometimes presumed to be due to artifacts or false-positives. The analysis, detection, and quantification of PDYN, PENK, and POMC mRNAs in peripheral tissues by Northern blot analysis and in situ hybridization have led to some unexpected findings. High abundance of the three mRNAs has been detected in various tissues which contain very low peptide levels relative to brain. These observed discrepancies between mRNA and peptide cellular anatomy will be the focus of this section and will be discussed in terms of cell biological mechanisms. Several examples which have underlining common themes have been chosen, and include PENK in the heart, PDYN in the adrenal, and opioids derived from all three precursors in reproductive tissues.

3.1 PENK in the Heart

One of the earliest examples indicating a clear discrepancy between opioid peptide content and mRNA content was the detection of PENK mRNA in the rat heart (Howells et al. 1986). Northern blot analysis of mRNA extracted from both brain and heart revealed PENK mRNA of approximately 1500 bases. Surprisingly, the quantity of PENK mRNA found in heart exceeded that found in brain. However, it had previously been shown that only low levels of ENK-like peptide immunoreactivity could be detected in heart (Lang et al. 1983; Weihe et al. 1983). It has been estimated that the opioid peptide content of the heart is only 3% of that found in the brain, and unlike that in brain, the ENK immunoreactivity is found only in high molecular weight forms, indicating a lack of protein processing at the steady state in heart tissue (Howells et al. 1986). In brain tissue, PENK mRNA levels generally agree with PENK peptide levels.

PENK mRNA was also determined to originate from the rat ventricular myocardium, as opposed to the atrial myocardium (Howells et al. 1986). The source of PENK gene expression in the rat heart has been clarified by a recent study demonstrating that PENK mRNA is found in cultured ventricular cardiac muscle cells of the rat heart and not in neurons associated with this tissue. Furthermore, PENK mRNA expression is developmentally regulated and levels in cultured ventricular cardiac muscle cells can be modulated by cyclic AMP and 3-isobutyl-1-methylxanthine (Springhorn and Claycomb 1989). No correlation with peptide levels was attempted. In Fig. 5 we demonstrate the distribution of PENK mRNA in the rat heart by in situ hybridization. The results confirm a high abundance of that message in a large number of ventricular muscle cells, while little signal could be detected in atrial cells. Various explanations for the disparity between PENK mRNA and PENK peptide content in the heart have been put forth. It has been suggested that the message may not be efficiently translated, or if translated, may undergo a rapid turnover or degradation. While inefficient translation is a distinct possibility, the fact that the PENK mRNA is structurally identical to the brain mRNA argues against differential stability of the transcript. Polysomal loading studies should provide some clues to these questions. It is interesting to note that ventricular cardiac muscle cells do not contain classical secretory granules, such as those associated with typical neuroendocrine secretory cells (Jamieson and Palade 1964). This may indicate a low cellular storage capacity which would correlate with the low peptide content observed. The ultimate answer may be derived from studies examining secretion versus synthesis rates of PENK.

3.2 PDYN in the Adrenal

Opioid peptides were intially demonstrated in the adrenal gland by immunohistochemical detection of Leu-ENK and Met-ENK in chromaffin cells of the adrenal medulla (Schultzberg et al. 1978). Since then, several studies have examined the synthesis, regulation, and distribution of PENK mRNA and peptides at this level (Udenfriend and Kilpatrick 1984). While the rat adrenal medulla is not an abundant

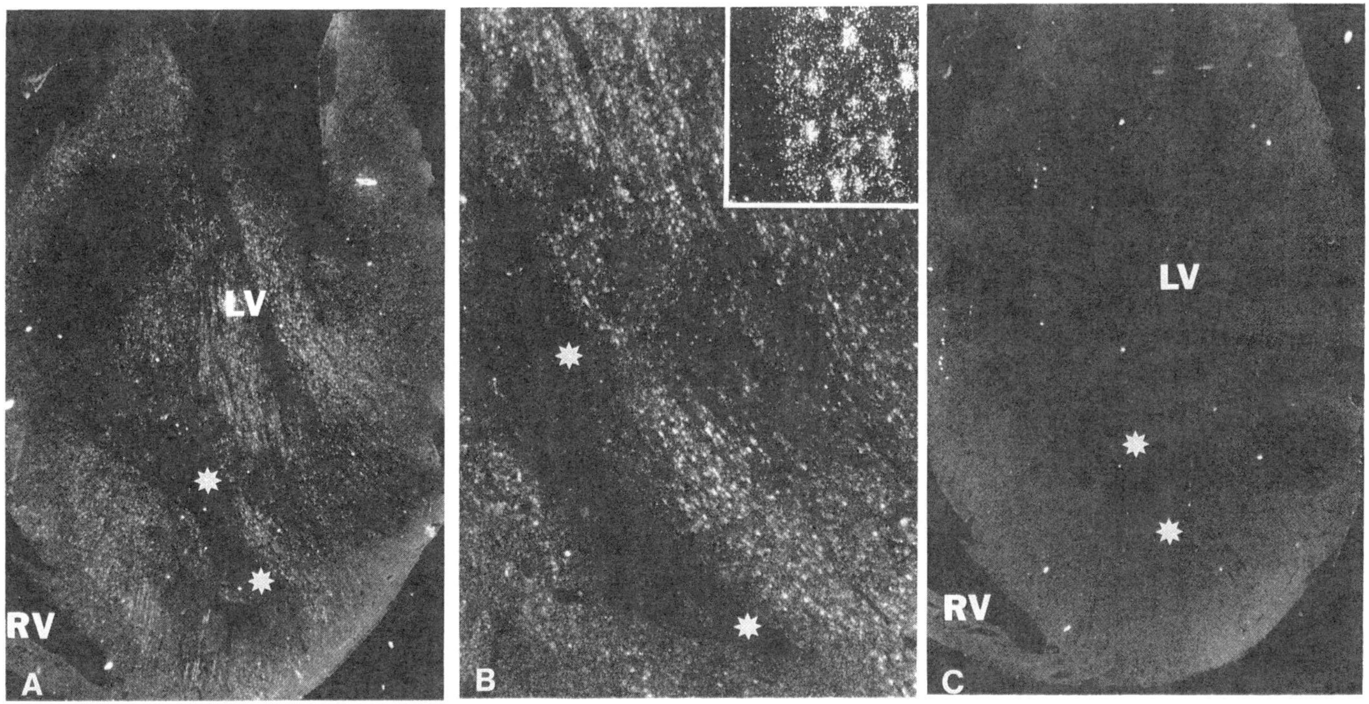

Fig. 5. Distribution of PENK mRNA in the rat heart. Hybridization of post-fixed, frozen, 20-μm-thick sections of rat heart with a PENK cRNA probe reveals several intensely labeled areas in the left ventricle (*LV*; **A**). RNAse pretreatment of an adjacent section prior to the hybridization abolishes all signals (**C**). Higher magnification (**B**; 60 ×) demonstrates the presence of PENK mRNA in heart muscle fibers. Single myocardial cells positively labeled with the proenkephalin cRNA probe are shown in **B**, *insert* (mag. 200 x). *RV* right ventricle; *asterisks* lumen of the *LV*)

source of PENK peptides when compared to other species, it can be rapidly up-regulated (Kanamatsu et al. 1986). However, PDYN peptide immunoreactivity, in the adrenal, has been somewhat of a mystery. Early reports (Spampinato and Goldstein 1983) detected little, if any, DYN immunoreactivity in whole rat adrenals. In recent studies we have determined that, although low in PDYN peptide immunoreactivity when compared to brain or pituitary, the rat contains approximately 0.5 pmol/adrenal pair and that this immunoreactivity corresponds to an authentic DYN-size peptide (Day et al. 1989).

Contrary to peptide levels, PDYN mRNA is highly abundant in the rat adrenal. This tissue is one of the most abundant sources of PDYN message in the rat, including all brain regions and pituitary. In this case, however, the PDYN mRNA is of an aberrant size, being approximately 100 bases shorter than the striatal transcript, as determined by Northern gel analysis (Civelli et al. 1985). This difference is thought to be due to a differential tissue splicing event, by analogy to the PDYN mRNA in the rat testis. Rat testicular PDYN mRNA lacks the exon 2 region of the gene (Collard et al. 1989). This difference in mRNA length reinforced to some extent the idea of inefficient translation, accounting for the low DYN peptide levels as opposed to the high DYN mRNA levels. It should be noted that this change in mRNA size does not affect the protein coding or signal peptide regions and does not result in an altered frameshift. Once again, polysomal loading studies in the adrenal gland may result in some interesting information in this regard.

In our laboratory, in situ hybridization studies localizing PDYN mRNA in the adrenal gland has yielded some surprising results as the highest concentration of PDYN mRNA was detected in the adrenal cortex (Day et al. 1989). This distribution did, however, correlate with PDYN peptide distribution in adrenal glands which were dissected into cortex and medulla, as higher concentrations of immunoreactivity were detected in the cortical extracts. As postulated with PENK in the heart, factors such as release, peptide stability, and/or translation efficiency may account for the discrepancy between PDYN peptide levels versus mRNA levels. It is interesting to note that adrenal cortical cells do not have storage granules. Their output is regulated at the biosynthetic level rather than at the secretory level. We are currently investigating the question of cellular regulation and the functional target of PDYN peptides in the adrenal cortex.

3.3 Opioids in Reproductive Tissues

It has recently been demonstrated that POMC (Bardin et al. 1987), PENK (Kilpatrick et al. 1985), and PDYN (Civelli et al. 1985) mRNAs are expressed in the testis. In all three cases low levels of peptide products are detected in this tissue relative to brain areas or pituitary, but mRNA levels are equally, or more abundant, in the testis. For example, when comparing POMC-derived peptide content in various tissues, it was observed that testicular levels are 50–100 times less abundant than brain levels, 500–1000 times less abundant than hypothalamic levels, and 1–3 million times less abundant than intermediate pituitary levels (Bardin et al. 1987). However, POMC mRNA levels in the testis are almost as high as those in the hypothalamus. Northern blot analysis reveals that in the testis, the sizes of the mRNA transcripts are not the same as those of the corresponding mRNA transcripts found in brain. Testicular POMC mRNA is about 400 bases shorter than hypothalamic POMC mRNA, PDYN mRNA is about 100 bases shorter than striatal PDYN mRNA, while PENK mRNA is about 350 bases longer than the brain PENK mRNA. Is there a relationship between the size of these aberrant molecules and the apparent low peptide content levels? Recent data concerning PDYN in the testis may provide some answers. As mentioned above, PDYN mRNA is shorter than brain forms of PDYN mRNA, but in the testis, this difference has been well characterized (Collard et al. 1989). The 60 bases which represent exon 2 are missing in the testis. This alternative splicing does not interfere with the protein coding region. Furthermore, polysome loading studies suggest that PDYN mRNA is efficiently translated in the testis. Interestingly, PDYN mRNA is associated with Sertoli cells. Although these cells do secrete a variety of proteins, they do not possess typical secretory granules associated with regulated release. Stimulation of cultured Sertoli cells with cAMP results in large increases in PDYN mRNA levels (six- to tenfold). This increase in the message pool correlates directly with increases in PDYN peptide immunoreactivity observed in the media of these cells, wherein a fourfold increase in peptide immunoreactivity is observed. However, the cellular PDYN peptide content was relatively unchanged, with only a 50–70% elevation (Collard et al. 1989). These studies suggest that, although aberrant in size, testicular PDYN mRNA is translated efficiently but that cellular content of PDYN peptide remains low, because the cells do not possess a storage mechanism and release their products, as they are made, in a constitutive manner.

If this is the case, then certain questions can be raised as to the importance of the observed differences in mRNA splicing. In the example of PDYN mRNA in testis versus brain, what is the importance of a missing exon 2? While polysomal loading studies show that both forms of PDYN mRNA are being translated, are the rates of translation equivalent? Is it possible that these deletions may regulate the stability of these messages? RNA regulatory elements have been described in other systems such as in the human ferritin mRNA (Casey et al. 1988) and, depending on their context in the mRNA, these regulating elements can control either translatability or message degradation. The mechanisms involved in alternate splicing are poorly understood, and it will be of great interest to determine whether differences observed in opioid mRNA transcripts are of functional significance.

4 Conclusions

The distribution of endogenous opioids throughout the body suggests a broad spectrum of actions for this family of peptides. The detection of peptide products, and more recently opioid mRNAs by in situ hybridization, has given us some clues as to the function of opioids in the CNS. These studies have pointed to the role of opioids in systems related to stress responsiveness, sexual function, water balance, autonomic control as well as pain responsiveness and addictive processes. However, the physiological role of opioids in the periphery is just beginning to be understood.

With the introduction of molecular biological techniques, our knowledge of peripheral opioid distribution has greatly profited. It is noted that several examples of apparent discrepancies between mRNA abundance and peptide cellular anatomy in peripheral tissues have been described. In general, when compared to brain regions, opioid mRNA levels in peripheral tissues are higher or equally abundant, whereas opioid peptide levels are relatively low. This does not occur in all peripheral tissues which contain opioids, as several examples where mRNA and peptide levels correlate can be shown. The discrepancies do, however, appear to occur in a variety of cell types which do not contain typical secretory granules associated with regulated release. Preliminary evidence suggests that opioid peptides may be released from these cells in a constitutive manner and that they are not stored intracellularly. This observation is not unique for peripheral tissues, but can also be made in the CNS, where the use of in situ hybridization techniques has permitted us to demonstrate the presence of opioid mRNA in nonneuronal cells such as astroglia. In addition, quantitative in situ hybridization has allowed the study of the regulation of mRNA levels at an anatomical level for both brain and periphery. Taken together, this body of new information on the expression and regulation of opioids may give us a new perspective on the cellular and physiological mechanisms of opioid action and may once again be exemplary for other peptide families.

Acknowledgments. This work was supported by NIDA DA 02265, NIMH MH422251, and the Teophile Raphael Fund. Robert Day is a fellow of the Medical Research Council of Canada. Martin Schafer is supported by the NIH DK07245 training grant.

References

Akil H, Mayer DJ, Liebeskind JC (1972) Comparison chez le rat entre l'analgésie iduite par stimulation de la substance grise péri-aqueducale et l'analgésie morphinique. CR Acad Sci Paris 274:3603–3605
Akil H, Mayer DJ, Liebeskind JC (1976) Antagonism of stimulation-produced analgesia by naloxone, a narcotic antagonist. Science 191:961–962
Akil H, Young E, Watson SJ, Coy DH (1981) Opiate binding properties of naturally occurring N- and C-terminus modified beta-endorphin. Peptides 2:289–292
Bardin CW, Chen CL, Morris PL, Gerendai I, Boitani C, Liotta AS, Margioris A, Krieger DT (1987) Proopiomelanocortin-derived peptides in testis, ovary, and tissues of reproduction. Recent Prog Hormone Res 43:1–28
Bloch B, Bugnon C, Fellman D, Lenys D (1978) Immunocytochemical evidence that the same neurons in the human infundibular nucleus and stained with anti-endorphins and antisera of other related peptides. Neurosci Lett 10:147–152
Bloom F, Battenberg E, Rossier J, Ling N, Guillemin R (1978) Neurons containing beta-endorphin in rat brain exist separately from those containing enkephalin: immunocytochemical studies. Proc Natl Acad Sci USA 75:1591–1595
Bondy CA, Gainer H, Russell JT (1988) Dynorphin A inhibits and naloxone increases the electrically stimulated release of oxytocin but not vasopressin from the terminals of the neural lobe. Endocrinology 122(4):1321–1327
Casey JL, Hentze MW, Koeller DM, Wright Caughman S, Rouault TA, Klausner RD, Harford JB (1988) Iron responsive elements: regulatory RNA sequences that control mRNA levels and translation. Science 240:924–928
Chowen-Breed J, Fraser HM, Vician L, Damassa DA, Clifton DK, Steiner RA (1989) Testosterone regulation of proopiomelanocortin messenger ribonucleic acid in the arcuate nucleus of the male rat. Endocrinology 124(4):1697–1702

Chretien M, Benjannet S, Dragon N, Seidah NG, Lis M (1976) Isolation of peptides with opiate activity from sheep and human pituitaries: relationship to β-LPH. Biochem Biophys Res Commun 72:472–478

Civelli O, Douglass J, Goldstein A, Herbert E (1985) Sequence and expression of the rat prodynorphin gene. Proc Natl Acad Sci USA 82:4291–4295

Collard MW, Garrett JE, Douglass J (1989) Prodynorphin mRNA synthesis and peptide secretion in cultured rat Sertoli cells. In: 71st Annu Meet Endocrine Soc, Seattle, USA 1212:325

Comb J, Seeburg PH, Adelman J, Eiden L, Herbert E (1982) Primary structure of the human Met- and Leu-enkephalin precursor and its mRNA. Nature (London) 295:663–667

Day R, Akil H (1989) The posttranslational processing of prodynorphin in the rat anterior pituitary. Endocrinology 124(5):2392–2405

Day R, Schafer MKH, Watson SJ, Akil H (1989) Distribution of opioid peptides and their mRNAs in the rat adrenal gland. In: Advances in gene technology: molecular neurobiology and neuropharmacology. Proc. of the 1989 Miami Bio/Technology Winter Symp, vol 9, pp 100–101

Elde R, Hökfelt T, Johansson O, Terenius L (1976) Immunohistochemical localization of enkephalin in the central nervous system of the rat. J Comp Neurol 198:541–565

Finley JCW, Maderdrut JL, Petrusz P (1981) The immunocytochemical localization of enkephalin in the central nervous system of the rat. J Comp Neurol 198:541–565

Gee C, Chen CLC, Roberts J, Thompson RC, Watson SJ (1983) Identification of proopiomelanocortin neurons in rat hypothalamus by in situ cDNA-mRNA hybridization. Nature (London) 306:374–376

Goldstein A (1984) Biology and chemistry of the dynorphin peptides. In: Udenfriend S, Meierhofer J (eds) The peptides, vol 6: The peptides, analysis, synthesis and biology. Academic Press, New York London, pp 95–145

Goldstein A, Tachibana S, Lownwey LI, Hunkapiller M, Hood L (1979) Dynorphin (1–13), an extraordinary potent opioid peptide. Proc Natl Acad Sci USA 76:6666–6670

Gubler U, Seeburg PH, Hoffman BJ, Gage LP, Udenfriend S (1982) Molecular cDNA cloning establishes proenkephalin as precursor of enkephalin containing peptides. Nature (London) 295:206–208

Haber SN, Elde R (1982) The distribution of enkephalin-immunoreactive fibers and terminals in the monkey central nervous system: an immunohistochemical study. Neuroscience 7:1049–1095

Harlan RE, Shivers BD, Romano GJ, Howells RD, Pfaff DW (1987) Localization of preproenkephalin mRNA in the rat brain and spinal cord by in situ hybridization. J Comp Neurol 258:159–184

Hökfelt T, Elde R, Johansson O, Terenius L, Stein L (1977) The distribution of enkephalin immunoreactive cell bodies in the rat central nervous system. Neurosci Lett 5:25–31

Höllt V, Haarmann I, Seizinger BR, Herz A (1981) Levels of dynorphin (1–13) immunoreactivity in rat neurointermediate pituitaries are concomitantly altered with those of leucine-enkephalin and vasopressin in response to various endocrine manipulations. Neuroendocrinology 33:333–339

Howells RD, Kilpatrick DL, Charles Bailey L, Noe M, Udenfriend S (1986) Proenkephalin in the heart. Proc Natl Acad Sci USA 83:1960–1963

Hughes J, Smith TW, Kosterlitz HW, Fothergill LA, Morgan MA, Morris HR (1975) Identification of two related pentapeptides from the brain with potent opiate agonist activity. Nature (London) 258:577–579

Jamieson JD, Palade GE (1964) Specific granules in atrial muscle cells. J Cell Biol 23:151–1172

Joseph SA, Micheal GJ (1988) Efferent ACTH-IR opiocortin projections from nucleus tractus solitarius: a hypothalamic deafferentation study. Peptides 9(1):193–201

Joseph SA, Pilcher WH, Bennet-Clarke C (1983) Immunocytochemical localization of ACTH perikarya in nucleus tractus solitarius: evidence for a second opiocortin neuronal system. Neurosci Lett 38:221–225

Kakidani H, Furutani Y, Takehashi H, Noda M, Morimoto Y, Hirone T, Asai M, Inayama S, Nakanishi S, Numa S (1982) Cloning and sequence analysis of cDNA for porcine beta-neo-endorphin, dynorphin precursor. Nature (London) 298:245–249

Kanamatsu T, Unsworth CD Diliberto EJ Jr, Viveros OH, Hong JS (1986) Reflex splanchnic nerve stimulation increases levels of proenkephalin A mRNA and proenkephalin A-related peptides in the rat adrenal medulla. Proc Natl Acad Sci USA 83:9245–9249

Khachaturian H, Watson SJ, Lewis ME, Coy D, Goldstein A, Akil H (1982) Dynorphin immunocytochemistry in the rat central nervous system. Peptides 3:941–954

Khachaturian H, Lewis ME, Watson SJ (1983) Enkephalin systems in diencephalon and brainstem of the rat. J Comp Neurol 219:310–320

Khachaturian H, Lewis ME, Schafer MKH, Watson SJ (1985) Anatomy of the CNS opioid systems. Trends Neurosci 8(3):111–119
Khachaturian H, Sherman TG, Lloyd RV, Civelli O, Douglass J, Herbert E, Akil H (1986) Prodynorphin is co-localized with LH and FSH in the gonadotrophs. Endocrinology 119:1409–1411
Kilpatrick DL, Howells RD, Noe M, Charles Bailey L, Udenfriend S (1985) Expression of pre-proenkephalin-like mRNA and its peptide products in mammalian testis and ovary. Proc Natl Acad Sci USA 82:7467–7469
Krieger DT, Liotta A, Brownstein MJ (1977) Presence of corticotropin in brain of normal and hypophysectomized rats. Proc Natl Acad Sci USA 74:648–652
Krieger DT, Liotta AS, Nicholson G, Kizer JS (1979) Brain ACTH and endorphin reduced in rats with monosodium glutamate-induced arcuate nuclear lesions. Nature (London) 278:562–564
Lang RE, Hermann K, Dietz R, Gaida W, Ganten D, Kraft K, Unger T (1983) Evidence for the presence of enkephalins in the heart. Life Sci 32(4):399–406
Larsson LI, Childers S, Snyder SH (1979) Met- and Leu-enkephalin immunoreactivity in separate neurons. Nature (London) 282:407–410
Leander JD (1983) A kappa opioid effect. Increased urination in the rat. J Pharm Exp Ther 224:89–94
Lewis JW, Baldrighi G, Akil H (1987) A possible interface between autonomic function and pain control: opioid analgesia and the nucleus tractus solitarius. Brain Res 424:65–70
Li CH, Chung D (1976) Isolation and structure of a triakontapeptide with opiate activity from camel pituitary glands. Proc Natl Acad Sci USA 73:1145–1148
Loh HH, Tseng LF, Wei E, Li CH (1976) β-endorphin as a potent analgesic agent. Proc Natl Acad Sci USA 73:2895–2898
Mains RE, Eipper BA (1981) Differences in the post-translational processing of β-endorphin in rat anterior and intermediate pituitary. J Biol Chem 256:5683–5688
Mains RE, Eipper BA, Ling N (1977) Common precursor to corticotropins and endorphins. Proc Natl Acad Sci USA 74:3014–3018
Mayer DJ, Wölfle TL, Akil H, Carder B, Liebeskind JD (1971) Analgesia from electrical stimulation in the brainstem of the rat. Science 174:1351–1354
Merchenthaler I, Maderdrut JL, Altschuler RA, Petrusz P (1986) Immunocytochemical localization of proenkephalin-derived peptides in the central nervous system of the rat. Neuroscience 17(2):325–348
Millan MH, Millan MJ, Herz A (1986) Depletion of central β-endorphin blocks midbrain stimulation produced analgesia in the freely moving rat. Neuroscience 18(3):641–649
Morris BJ, Haarmann I, Kempter B, Höllt V, Herz A (1986) Localization of prodynorphin messenger RNA in rat brain by in situ hybridization using a synthetic oligonucleotide probe. Neurosci Lett 69(1):104–108
Morris BJ, Herz A, Höllt V (1989) Localization of striatal opioid gene expression, and its modulation by the mesostriatal dopamine pathway: an in situ hybridization study. J Mol Neurosci 1:9–18
Nakanishi S, Inoue A, Kita T, Nukamura M, Chung ACY, Cohen SN, Numa S (1979) Nucleotide sequence of cloned cDNA for bovine corticotropin-beta lipotropin precursor. Nature (London) 278:423–427
Noda M, Furutani Y, Takahashi H, Toyosato M, Hirose T, Inayama S, Nakanishi S, Numa S (1982) Cloning and sequence analysis of cDNA for bovine adrenal prepro-enkephalin. Nature (London) 295:202–206
Palkovits M, Mezey E, Eskay RL (1987) Distribution and possible origin of beta-endorphin and ACTH in discrete brainstem nuclei of rats. Neuropeptides 9:123–137
Pelletier G (1979) Ultrastructural immunohistochemical localization of adreno-corticotropin and β-lipotropin in the rat. J Histochem Cytochem 27:1046–1048
Pelletier G, Leclerc R, LaBrie F, Cote J, Chretien M, Lis M (1977) Immunohistochemical localization of beta-LPH hormone in the pituitary gland. Endocrinology 100:770–776
Richardson DE, Akil H (1977) Pain reduction by electrical brain stimulation in man: part 1: acute administration in periaqueductal and periventricular sites. J Neurosurg 47:178–183
Saneto RP, Low KG, Nielsen CP, Melner MH (1989) Regulation of proenkephalin gene expression in type I astrocytes. In: Advances in gene technology: molecular neurobiology and neuropharmacology. Proc 1989 Miami Bio/Technology Winter Symposium, vol 9, p 50
Schafer MKH, Herman JP, Day R, Douglass J, Loats H, Akil H, Watson SJ (1988) The distribution of prodynorphin mRNA throughout the rat brain: a semi-quantitative mapping study. In: 18th Annu Meet Soc Neurosci, Toronto, vol 14, p 545

Schafer MKH, Day R, Ortega M, Akil H, Watson SJ (1990) Proenkephalin messenger RNA is expressed in both the anterior and posterior pituitary. Neuroendocrinology 51:444–448
Schultzberg M, Lundberg JM, Hökfelt T, Terenius L, Brandt J, Elde RP, Goldstein M (1978) Enkephalin-like immunoreactivity in gland cells and nerve terminals of the adrenal medulla. Neuroscience 3:1169–1186
Schwartzberg DG, Nakane PK (1983) ACTH-related peptide containing neurons within the medulla oblongata of the rat. Brain Res 276:351–356
Seizinger BR, Höllt V, Herz A (1981) Immunoreactive dynorphin in the rat adeno-hypophysis consists exclusively of 6000 dalton species. Biochem Biophys Res Commun 103:256–263
Sherman TG, Douglass J, Civelli O, Herbert E, Watson SJ (1986) Coordinate expression of hypothalamic prodynorphin and provasopressin mRNAs with osmotic stimulation. Neuroendocrinology 44:222–228
Smith AI, Funder JW (1988) Proopiomelanocortin processing in the pituitary, central nervous system, and peripheral tissues. Endocrine Rev 9:159–179
Spampinato S, Goldstein A (1983) Immunoreactive dynorphin in rat tissues and plasma. Neuropeptides 3:193
Springhorn JP, Claycomb WC (1989) Preproenkephalin mRNA expression in developing rat heart and in cultured ventricular cardiac muscle cells. Biochem J 258:73–78
Tsou K, Khachaturian H, Akil H, Watson SJ (1986) Immunocytochemical localization of proopiomelanocortin-derived peptides in the adult rat spinal cord. Brain Res 378:28–35
Udenfriend S, Kilpatrick DL (1984) Proenkephalin and the products of its processing: chemistry and biology. In: Udenfriend S, Meierhofer J (eds) The peptides, vol 6: The peptides, analysis, synthesis and biology. Academic Press, New York London, pp 25–68
Uhl GR, Goodman RR, Kuhar MJ, Childers SR, Snyder SH (1979) Immunocytochemical mapping of enkephalin containing cell bodies, fibers and nerve terminals in the brainstem of the rat. Brain Res 116:75–94
Valentino KL, Eberwine JH, Barchas JD (1987) In situ hybridization: applications to neurobiology. Oxford Univ Press, New York Oxford
Vilijn M, Vaysse PJ, Zukin RS, Kessler JA (1988) Expression of preproenkephalin mRNA by cultured astrocytes and neurons. Proc Natl Acad Sci USA 85:6551–6555
Watson SJ, Barchas JD, Li CHJ (1977) β-lipotropin: localization of cells and axons in rat brain by immunocytochemistry. Proc Natl Acad Sci USA 74(11):5155–5158
Watson SJ, Akil H, Richard CW, Barchas JD (1978) Evidence for two separate opiate peptide neuronal systems and the coexistence of beta-LPH, beta-endorphin and ACTH immunoreactivities in the same hypothalamic neurons. Nature (London) 275:226–228
Watson SJ, Akil H, Ghazarossian VE, Goldstein A (1981) Dynorphin immunocytochemical localization in brain and peripheral nervous system: preliminary studies. Proc Natl Acad Sci USA 78(2):1260–1263
Watson SJ, Khachaturian H, Akil H, Coy D, Goldstein A (1982) Comparison of the distribution of dynorphin systems and enkephalin systems in brain. Science 218:1134–1136
Watson SJ, Akil H, Khachaturian H, Young E, Lewis ME (1984) Opioid systems: anatomical, physiological and clinical perspectives. In: Hughes J, Collier HOJ, Rance MJ, Tyers MB (eds) Opioids, past, present, future. Taylor & Francis, London, pp 146–178
Watson SJ, Sherman TG, Schafer MKH, Patel P, Herman JP, Akil H (1988) Regulation of mRNA in peptidergic systems: quantitative and in situ studies. In: Chretien M, McKerns KW (eds) Molecular biology of brain and endocrine peptidergic systems. Plenum Press, New York, pp 225–241
Weber E, Barchas JD (1983) Immunohistochemical distribution of dynorphin B in rat brain: relation to dynorphin A and alpha-neo-endorphin systems. Proc Natl Acad Sci USA 80:1125–1129
Weber E, Voigt R, Martin R (1978) Concomitant storage of ACTH and endorphin-like immunoreactivity in the secretory granule of anterior pituitary corticotrophs. Brain Res 157:385–390
Weihe E, McKnight AT, Corbett AD, Hartschuh W, Reinecke M, Kosterlitz HW (1983) Characterization of opioid peptides in guinea-pig heart and skin. Life Sci 33 Suppl 1:711–714
Whitnall MH, Gainer H, Cox BM, Molineaux CJ (1983) Dynorphin-A(1–8) is contained within vasopressin neurosecretory vesicles in rat pituitary. Science 222:1137–1139
Williams RG, Dockray GJ (1983) Distribution of enkephalin-related peptides in rat brain: immunohistochemical studies using antisera to met-enkephalin and met-enkephalin Arg-Phe. Neuroscience 9:563–586

Section II

Opioid Receptor Biochemistry and Signal Transduction

CHAPTER 4

Multiple Opioid Receptors

E.J. Simon

1 Introduction

It is a distinct honor and a great pleasure to contribute to this book, a Festschrift in honor of Professor Albert Herz. As usual, this is a bitter-sweet event. On the one hand, it is high time to pay tribute to Albert Herz, an outstanding researcher whose laboratory has contributed enormously to the field of opiate/opioid pharmacology and who has been one of the leaders of the International Narcotic Research Conference (INRC) from its inception and is currently its Secretary. On the other hand, this marks Albert's retirement and therefore the closing down of what is clearly one of the most active laboratories in Psychopharmacology and especially in the area of the study of opiates and the endogenous opioid system. However, judging from Albert's vigor and enthusiasm and from analogy with other retirements, I expect great things from Albert after he has officially retired from the Max Planck Institute. I wish him the best of luck and continued good health for whatever activities he decides to pursue and wherever he decides to pursue them. I also extend best wishes for a long, healthy and happy life to his lovely wife, Marlise and to his family. My wife Irene and I are proud to have the Herzes as old and good friends. Two highlights of our long-standing friendship were our trip to India during which Albert and I spent many enjoyable hours together, and our planning together of the INRC meeting in Garmisch-Partenkirchen, a highly successful meeting and a most pleasant one. We hope that we shall have many equally enjoyable interactions in the future.

I have been asked to review the current state of our knowledge on the nature of multiple opioid receptors. In the pages allotted I shall have to select portions of this vast field and give only key references and review articles. Fortunately, many aspects of this research will be covered by other contributors. Let me state here that Albert Herz and his collaborators have contributed greatly to our knowledge of opioid receptor heterogeneity and I apologize if this fact is not as evident as it should be from the number of citations.

2 Historical Overview

When stereospecific binding of opiate drugs was first demonstrated in 1973 (Simon et al. 1973; Terenius 1973; Pert and Snyder 1973), it was generally thought that a single type of opiate receptor existed. The discovery of the endogenous opioid peptides (Hughes et al. 1975) and the finding that there are quite a number of such peptides (for a review, see Simon and Hiller 1989) led to the suggestion that there may be different receptors for the various peptides and, based on our knowledge of other neurotransmitters, probably more than one receptor per peptide. The notion that the endogenous opioid peptides are the true natural ligands of the receptors led to the now widely accepted appellation "opioid receptors", which we will adopt henceforth.

The first definitive evidence for multiple opioid receptors was obtained by Martin and co-workers (Martin et al. 1976; Gilbert and Martin 1976) who studied the pharmacology of morphine and its congeners in chronic spinal dogs. Based on their findings they postulated three receptors which they named μ (for morphine), κ (for ketocyclazocine) and σ (for SKF-10047, N-allylnormetazocine). The evidence that led them to this conclusion was the widely different pharmacological profiles of these three drugs as well as their inability to replace each other to alleviate withdrawal symptoms in dogs treated chronically with one of them. It is now well known that these receptor types have withstood the test of time remarkably well.

The discovery of the enkephalins (ENKs) by Hughes and Kosterlitz and their co-workers (Hughes et al. 1975) led to the postulate of another receptor type with preference for the ENKs. The evidence for this came from work in Kosterlitz's laboratory with isolated organ systems (Lord et al. 1977). ENKs were much less effective than morphine in inhibiting contraction of the isolated guinea pig ileum (GPI), whereas the reverse was true in the mouse vas deferens (MVD). The receptor that seemed to predominate in the latter tissue was named δ (for *d*eferens). Further support for this hypothesis came from competition binding studies and from the finding that the receptors in the MVD were significantly more resistant to the opiate antagonist naloxone (NAL) than those in the GPI, i.e., ten times as much NAL was required for the same degree of reversal of opioid action.

Since that time a number of other types of receptors have been postulated including an epsilon (ε) receptor, selective for β-endorphin (β-END), found in the rat vas deferens as well as lamda and iota receptors. There is also evidence for receptor subtypes such as μ_1, μ_2 and κ_1 and κ_2. This review will concentrate on the most widely studied and therefore best-established receptor types, μ, δ, and κ. The σ-receptor appears to be related or identical to the PCP receptor (Zukin and Zukin 1981) and may not be strictly speaking an opioid receptor, since its binding and function is not reversed by NAL.

3 Opioid Receptor Ligands

3.1 Putative Endogenous Ligands

The δ-receptor was, as stated, first defined as an ENK receptor and these pentapeptides are still thought to be the endogenous ligands for this receptor. The κ-receptor is felt to be the receptor for the peptides derived from prodynorphin, i.e., dynorphin (DYN) A and B and the neoendorphins. A number of candidates have been advanced as possible endogenous ligands for the μ-receptor. These include β-END, morphine, which has been reported recently to be present in animal brain (Goldstein et al. 1985; Donnerer et al. 1986), as well as the ENKs for which the μ-receptor could be an "isoreceptor" since their affinity for this receptor is only 10- to 20-fold lower than for the δ-receptor. In no case has the endogenous ligand-receptor relationship been proven. In fact, mapping studies have led to the finding that there is a mismatch between the distribution of opioid receptors and their putative endogenous ligands. The reason for the mismatch is not understood, but it is a further reason for caution regarding the hypotheses discussed. It should be pointed out that such mismatches have been seen with all neurotransmitter receptors that have been well localized by morphological techniques. For many of these, the identity of the endogenous ligand is well established (for review and discussion of possible explanations, see Herkenham 1987).

3.2 Synthetic Selective Ligands

The field has been enormously advanced by the availability of highly selective ligands that make possible the study of a particular receptor type with minimal interference from others.

For the μ-receptor an analog of ENK, D-ala^2-MePhe^4-gly-ol^5-ENK (DAGO), the first highly selective analog synthesized (Handa et al. 1981) and an analog of morphiceptin, PL017, are highly selective and widely used. For δ-receptors the earliest moderately selective ligand was Tyr-D-Ser-Gly-Phe-Leu-Thr (DSLET) (Gacel et al. 1980). D-penicilamine^2-D-or L-penicilamine^5-ENK (DPDPE and DPDLE) (Mosberg et al. 1983) and DSBuLET, an O-ter-Butyl-D-Ser analog of DSLET from Roques' laboratory, are, at present, the most selective δ-ligands available. A number of selective κ-agonists have been synthesized, U 50,488H and U 69,593 at Upjohn, Cambridge 20 by Hughes and co-workers at Parke-Davis, as well as two DYN analogs, Tyr-Gly-Gly-Phe-Leu-Arg-Arg-Ile-Arg-Pro-Arg-Leu-Arg-Gly-NH $(CH_2)_5$-NH_2 (DAKLI) by Goldstein's laboratory and D-Pro^{10}-DYN(1-11) (DPDYN) made by Gairin and co-workers in Toulouse. Selective antagonists are also finally becoming available. Cyclic analogs of somatostatin synthesized in Hruby's laboratory are selective μ-antagonists. For δ-receptors there is ICI 174,864 and a derivative of naltrexone, naltrindole, synthesized by Portoghese's laboratory. The latter group also synthesized nor-binaltorphimine (norBNI), a selective κ-antagonist.

A word should also be said about the selective irreversible ligands that have become very useful. These include β-funaltrexamine (β-FNA), the etonitazine derivative BIT and naloxonazine, all selective, irreversible μ-ligands and fentanylisothiocyanate (FIT), a selective, irreversible δ-ligand. The rather nonselective, irreversible opioid ligand, β-chloronaltrexamine (β-CNA) should also be mentioned, since it has been very useful and will be discussed further.

More and more powerful and selective ligands are becoming available and are catalyzing rapid progress in our understanding of the opioid receptor types.

4 Pharmacological Studies

I shall arbitrarily divide the discussion of multiple opioid receptors into pharmacological and biochemical studies.

Since the early work of Martin and Kosterlitz there have been many experiments in whole animals as well as in isolated peripheral organ systems and cell and tissue cultures that support the concept of multiple receptors. Herz's laboratory has clearly demonstrated in drug discrimination tests that animals can distinguish between receptor types. Thus, animals trained to recognize the discriminative stimulus induced by fentanyl, a selective ligand for the μ-receptor (Shearman and Herz 1982), generalize this effect to other μ-agonists as well as to δ-agonists but not to κ-type drugs. Animals trained to respond to ethylketocyclazocine (EKC), a κ-agonist, respond to other κ-agonists but not to μ- or δ-agonists. Studies from Holaday's laboratory (Holaday and Tortella 1984) have implicated δ-receptors in the cardiovascular effects of opioid peptides. These workers have also obtained evidence suggesting an allosteric relationship between μ- and δ-receptors. Such an interaction had been postulated earlier by Rothman and Westfall (1982) and has recently been independently supported by experiments of Schoffelmeer et al. (1988) with brain slices.

Very important evidence for multiple opioid receptors came again from Albert Herz's laboratory. They demonstrated that selective tolerance of a given receptor type could be induced in isolated organ systems. Thus, when the MVD was rendered tolerant to the selective μ-ligand, sufentanil, there was cross-tolerance to other μ-ligands but to neither δ- nor κ-ligands. A similar result was obtained for μ- and κ-receptors in the GPI but no δ-receptors could be detected. The δ-ligands were presumed to work via μ-receptors in this system (for a more detailed review of these studies, see Wüster et al. 1983).

Ontogenetic studies have also indicated the distinctiveness of the major receptor types since they can be shown to develop at different times. A number of groups have found that μ-receptors develop earlier than δ-receptors. There is some disagreement about the ontogeny of κ-receptors, although their developmental pattern is clearly different from those of the others, and most investigators find that κ too develops earlier than δ-sites (for a thorough review, see McDowell and Kitchen 1987).

Autoradiographic studies in several laboratories have resulted in the detailed mapping of the three major types of opioid receptors in the CNS of a number of species. It is evident that the distribution of μ-, δ-, and κ-sites, while overlapping, is quite distinct. For a full discussion of the mapping of opioid peptides and receptors, see chapters by Schafer et al. and Mansour et al. (this Vol.).

An anti-idiotypic monoclonal antibody against opioid binding sites has been prepared by the Herz laboratory (Gramsch et al. 1988). This was accomplished by using as antigen an antibody against β-END, which displayed opioid binding properties similar to those of opioid receptors. These antibodies react with both μ- and δ-receptors, suggesting considerable similarities between their binding sites, but they do not recognize κ-sites.

A very interesting study carried out by Schoffelmeer et al. (1988) shows a difference between receptor types with respect to their inhibition of neurotransmitter release from rat brain slices. Inhibition of the release of norepinephrine was observed in rat brain cortex slices with selective μ-ligands, in rat brain striatal slices the release of acetylcholine was observed with δ-agonists and that of dopamine with κ-agonists.

There has also been progress in the elucidation of the electrophysiological effects of opioid receptor activation. This work, done largely in the laboratories of Macdonald and North (see e.g. Gross and Macdonald 1987; North et al. 1987) is discussed by R.A. North (this Vol.), but I shall summarize the findings here very briefly. It has been found that the activation of either μ- or δ-receptors results in reduction of the duration of Ca^{2+}-dependent action potentials due to the opening of potassium channels. The activation of the κ-receptor also causes a reduction in a Ca^{2+} current but by a different mechanism, namely, by closing Ca^{2+} channels. The effect appears to be specifically on channels of the N-type. There is also some preliminary evidence suggesting that some of these effects may involve a direct interaction between G-proteins and ion channels.

The important question as to the major physiological roles and the possible functional specialization of the different types of opioid receptor has not yet been resolved. There is considerable evidence that suggests that analgesia can be mediated via all three of the major types of opioid receptors. There is also some evidence suggesting that dependence may be mainly due to μ-receptor activation. These findings have spurred activity in the synthesis of novel κ-type drugs for use as possible nonaddictive analgesics.

Some attempts have been made to distinguish between the types of analgesia mediated by different opioid receptors. Thus, there have been reports (Wood et al. 1981; Chaillet et al. 1984) that supraspinal analgesia may be mediated via μ- rather than δ- or κ-receptors. This should not, however, be regarded as definitive since some evidence for supraspinal δ (Porreca et al. 1987) and κ (Carr et al. 1982) analgesia also exists.

At the spinal level, analgesia against thermal pain stimuli seems to be mediated by separate μ- and δ-mechanisms, while κ-receptors seem to preferentially mediate analgesia against chemically induced visceral pain (Yaksh 1984).

The work of Han et al. (1984) suggests that κ-receptors may play a role in high frequency electroacupuncture, while δ-receptors may be important for lower frequency electroacupuncture.

Another area in which the role of different opioid receptor types has been examined is in feeding behavior. Studies employing the selective κ-agonists, U50, 488H and tifluadom, consistently support a role for κ-receptors (e.g., Morley and Levine 1983). Further support comes from the recent finding from our laboratory (Bak et al. 1988) that feeding induced by electrical brain stimulation is inhibited by the lateral ventricular injection of nor-BNI, the selective κ-antagonist.

5 Biochemical Studies

5.1 *Membrane-Bound Opioid Receptors*

Early biochemical evidence for receptor heterogeneity involved receptor binding studies in brain membranes. Thus, the Kosterlitz group (Lord et al. 1977) obtained support for the existence of separate μ- and δ-receptors from competition binding studies in guinea pig brain homogenates. It was observed that ENKs compete significantly better against a labeled ENK than against a labeled opiate alkaloid (tritiated NAL), while the opposite was true for competition of opiates such as morphine or NAL. When similar experiments were done to detect κ-binding, it proved to be more difficult. The reasons for this early problem have been clarified. Many early studies were done in rat brain which is now known to contain a very low concentration of κ-sites. Moreover, the κ-ligands used, EKC and bremazocine, turned out to be quite nonselective, binding almost as well to μ- and δ-receptors as they do to κ-receptors, although they act like κ-agonists in vivo. The latter fact can be explained by the observation that these drugs are antagonists at the other opioid binding sites. The first successful demonstration of κ-binding sites was in Kosterlitz's laboratory using guinea pig brain homogenate to bind EKC in the presence of saturating concentrations of selective μ- and δ-ligands. These results have been confirmed by competition experiments with selective κ-ligands such as U50, 488H, Cambridge 20 and DPDYN in many laboratories.

Such competition binding studies have been carried out in various brain regions of a number of species and have led, together with autoradiographic fine mapping, to the finding that the ratio of μ:δ:κ receptors varies widely from region to region and, for a given region, from species to species. A number of regions have been found to be highly enriched in a single receptor type. Thus, the guinea pig cerebellum is highly enriched in κ-receptors (80–90%), while rabbit cerebellum is enriched in μ-receptors, as is the thalamus of several species. No region highly enriched in δ-receptors is yet known. The cell line NG108-15, a neuroblastoma-glioma hybrid cell, is most frequently used as a source of δ-receptors.

Another, perhaps more convincing, approach to the differentiation of opioid receptor types is the ability to protect a given type selectively against inactivation by chemical reagents. This was first shown by Robson and Kosterlitz (1979) using phenoxybenzamine and by our laboratory (Smith and Simon 1980) with N-ethylmaleimide. Both reagents are irreversible inactivators of opioid receptors. The presence of μ-ligands such as morphine, NAL, or DAGO can selectively protect the μ-sites, while the presence of δ- or κ-ligands will protect the corresponding sites. James and Goldstein (1984) have further refined this approach. They utilized β-CNA, an opioid receptor-selective irreversible reagent, to produce, by selective protection, tissues that are highly enriched in a given receptor type.

Interesting reagents that can distinguish between receptor types are ethyl alcohol and related short chain alcohols (Hiller et al. 1984). These compounds are reversible inhibitors of opioid receptor binding and exert a much greater effect on δ-sites than on μ and κ. The inhibition is due to a change in the K_d with no effect on the B_{max}. The decrease in affinity is the result of an increase in the dissociation rate of the ligand-receptor complex. As expected, the effect is not competitive and is probably due to a change in receptor conformation resulting from a decrease in membrane fluidity, to which δ-receptors seem to be more sensitive than the others.

All the studies cited tend to support, but do not prove, the molecular distinctiveness of the three major opioid receptor types. To prove whether or not receptor types are distinct molecules requires their physical separation. The progress in this area will be summarized in the next section.

5.2 Separation and Isolation of Binding Sites

The work in this domain has been reviewed quite recently (Simon and Hiller 1988; Simonds 1988). Only the most recent references, not included in those reviews, will be given.

One way to establish whether receptor types represent distinct molecular species is to separate either active binding sites or binding subunits of the receptors. A separation of active binding sites has been achieved for κ-receptors. Dr. Itzhak in our laboratory was able to separate κ-sites from other opioid binding sites using sucrose density gradient centrifugation, while Chow and Zukin had success using a Sepharose column. Such a separation proved more difficult for μ- and δ-sites. However, a different approach, that of affinity cross-linking of labeled ligands, followed by separation of the labeled polypeptides by SDS-polyacrylamide electrophoresis (SDS-PAGE) has been successful.

For this purpose, Howard in our laboratory used radioiodinated human β-END (labeled in position 27) which has high affinity for both μ- and δ-receptors but virtually none for κ. This ligand permitted us to examine μ- and δ-receptors in the same tissue. Membranes were bound with ^{125}I-β-END and washed thoroughly. Cross-linking was achieved using one of the commercially available cross-linking reagents, such as bis[2-(succinimidooxycarbonyloxy)-ethyl]sulfone (BSCOES). This treatment resulted in the covalent linking of 40–50% of the ligand prebound to μ- and δ-receptors. The cross-linked binding sites were extracted from the membranes with SDS and separated by SDS-PAGE. Autoradiography of the gels resulted in the visualization of a number of labeled bands, the number and relative density varying from tissue to tissue. Results from a number of tissues are presented in Fig. 1. The tissues are arranged in the order of decreasing $\mu{:}\delta$ receptor ratio, ranging from very high levels of μ-receptors in the rat thalamus to pure δ-receptors in the NG108–15 cells. The most important result of these studies was that a band with an apparent molecular weight (Mr) of 65 kDa was present only in tissues known to contain μ-receptors and its density increased with the $\mu{:}\delta$ receptor ratio of the tissue. A band of apparent Mr of 53 kDa correlated well with the presence and proportion of δ-sites. These bands, as well as several others, were eliminated when the binding was carried out in the presence of excess unlabeled NAL or bremazocine, indicating that they are all subunits of opioid binding sites. The identity of bands other than the 65- and 53-kDa bands is not clear.

While these results suggested that the two proteins represent separate binding subunits of μ- and δ-receptors, further evidence was deemed necessary. Howard

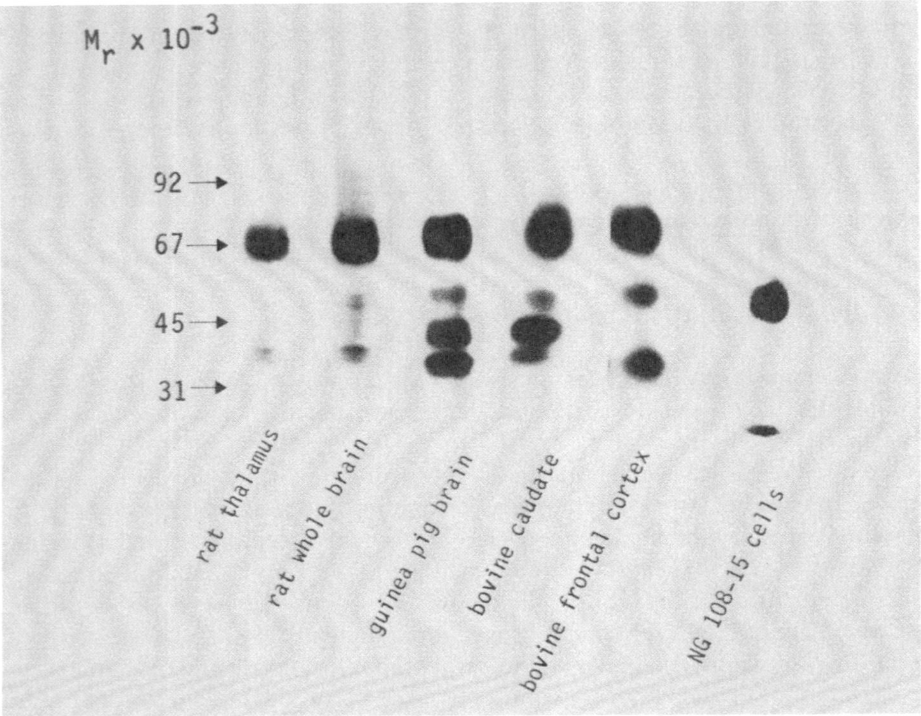

Fig. 1. Affinity cross-linking of ^{125}I-β-END$_h$ to membranes from brain tissues and from NG108-15 cells. Membranes prepared from the tissues indicated and from NG108-15 cells were incubated with ^{125}I-β-END$_h$ (2 nM), cross-linked, and solubilized in SDS-PAGE sample buffer. The extract was submitted to SDS-PAGE followed by autoradiography of the gel to visualize the radioactive protein bands. For details, see Howard et al. (1985)

obtained important evidence by performing binding prior to cross-linking in the presence of highly selective μ- and δ-ligands. It was found that DAGO selectively suppressed the labeling of the 65-kDa band at concentrations where it had little or no effect on the 53-kDa band, while DPDPE preferentially suppressed the labeling of the 53-kDa band. Figure 2 gives the results of a typical experiment carried out with membranes from bovine frontal cortex. These results support the hypothesis that the 65- and 53-kDa glycoprotein bands represent binding subunits of μ- and δ-receptors, respectively.

A similar cross-linking study has recently been performed with κ-receptors in our laboratory (Yao et al. 1989), in collaboration with the laboratory of Dr. J. Cros in Toulouse. DPDYN was obtained from Dr. Gairin and radioiodinated as described by him and his collaborators. A membrane fraction from guinea pig cerebellum (85–90% κ) was allowed to bind the labeled DPDYN in the absence or presence of an excess of unlabeled bremazocine (nonspecific binding), as well as in the presence of μ-, δ-, and κ-ligands. The membranes were washed and cross-linked, and the SDS extracts of the labeled membranes were submitted to SDS-PAGE followed by

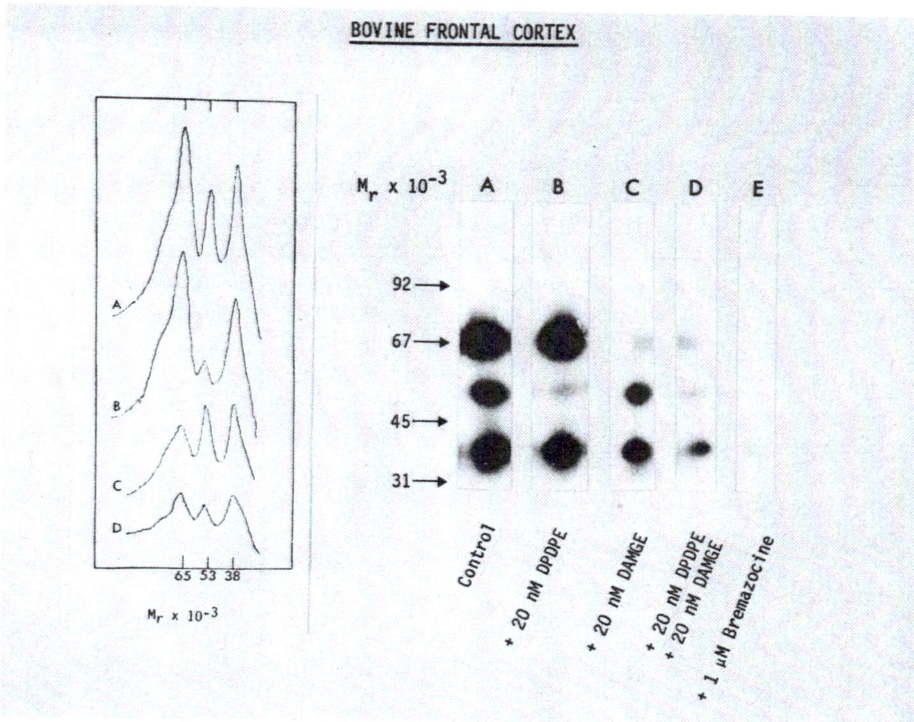

Fig. 2. The effect of site-selective ligands on the cross-linking of ^{125}-I-β-END to bovine frontal cortex membranes. Membranes were bound with ^{125}I-β-END in the presence and absence of competing ligands and cross-linked with BSCOES. After solubilization in SDS-PAGE sample buffer, the extracts were submitted to SDS-PAGE, followed by autoradiography of the gel. The *arrows* indicate the positions of the Mr standards. Also shown are densitometric scans of lanes *A-D*. *DAMGE* refers to the μ-selective ligand usually called DAGO (Howard et al. 1986)

autoradiography. Two labeled protein bands of apparent Mr of 55 and 35 kDa were obtained. Neither band was reduced significantly when binding was performed in the presence of μ-, or δ-ligands. However, κ-ligands such as U50,488H and DPDYN competed with ^{125}-I-DPDYN binding. The 55-kDa band disappeared completely, whereas the 35-kDa band was reduced in intensity but did not disappear, even at rather high concentrations of κ-competitors. We conclude that the 55-kDa band is a high affinity κ-binding protein. The 35-kDa band may be a low affinity κ-site or a mixture of a κ- and a nonopioid protein, the latter labeled nonspecifically.

The results so far discussed suggest that the three major types of opioid receptors are indeed separate molecular entities. We do not know whether they are separate gene products (polypeptides) or whether the differences reside in posttranslational modifications (sugar residues, phosphorylation, etc). This question will only be answered when the receptors have been cloned and sequenced.

Progress has been reported for the purification of all three major types of opioid receptors. The δ-receptor was purified to homogeneity by Klee and co-workers from NG108–15 cells in culture in a covalently linked (inactive) form. This was achieved by use of the δ-selective affinity labeling agent, ^{3}H-3-methylfentanyl-isothiocyanate (superFIT). A membrane fraction from NG108–15 cells was labeled with this reagent and solubilized with a detergent. Purification was achieved by chromatography on wheat germ agglutinin (WGA)-agarose, followed by chromatography on a column of immobilized antibodies to fentanylisothiocyanate (FIT). Preparative SDS-PAGE was the final purification step. A glycoprotein of Mr 58 kDa was purified 30 000-fold to a final specific binding activity of 21 000 pmol of superFIT bound per mg protein, close to the theoretically expected value for a single binding site per 58 000 daltons.

There has not yet been extensive purification of δ-sites from CNS tissue, nor has an active δ-binding site been purified from any source.

The first partial purification of μ-receptors was reported by Bidlack and co-workers, and the first purification to homogeneity of μ-binding sites was reported from our laboratory. The latter involved a two-step purification of digitonin-solubilized crude membranes from bovine striatum. The major step was ligand affinity chromatography on immobilized 6-desoxynaltrexylethylenediamine (NED) developed by Dr. Gioannini, which results in 3000–5000-fold purification in a single pass. This was followed by lectin affinity chromatography on immobilized WGA, earlier shown by us to retain applied opioid binding sites which can then be eluted with N-acetylglucosamine in a yield of 40–50% of applied receptor. The purification is ca. 65 000-fold with an overall yield of 3–6%. SDS-PAGE of the purified μ-binding site gave a single band of 65 kDa, as detected by autoradiography of a radioiodinated sample or by silver staining. Specific binding activity of the purified receptor is ca. 15 000 pmol/mg protein, near the theoretical value for a protein of this size with a single binding site. It should be noted that the purified receptor binds antagonists with high affinity but agonists with very low affinity (in the μ-molar range). This may be due to the absence of essential lipids or to uncoupling from the G-protein. Efforts to reconstitute high affinity agonist binding to the purified μ-receptor are in progress.

Purification to homogeneity of μ-receptors has since been reported by other laboratories, by Cho et al. in San Francisco and Demoliou-Mason and Barnard in England. An interesting difference between the result of Cho and ours is that the μ-receptor which they purified requires acidic lipids for high affinity binding, while ours does not (at least not for antagonist binding). The reason for this difference is, at present, not understood.

The κ-receptor has been purified fairly extensively from a nonmammalian source, the frog brain, by J. Simon and co-workers in Hungary. Very recently a report on the purification of κ-binding sites from human placenta has appeared (Ahmed et al. 1989). The authors do not claim homogeneity but specific binding activity is nearly 50% of the theoretical value. The purification procedure is remarkable in as much as no ligand affinity chromatography step was used. The technique consisted of covalent chromatography on an SH-containing Sepharose column followed by a WGA column. Neither column is specific for opioid receptors.

6 Status Report on Opioid Receptor Cloning and Sequencing

Attempts to obtain the amino acid sequence of the δ-receptor by expression cloning were reported by Law and collaborators (unpublished), but have not yet succeeded. Schofield et al. reported the cloning of a protein, derived from the purification of μ-receptor, by screening of brain cDNA libraries. Oligonucleotide probes were prepared from amino acid sequences obtained by microsequencing the purified protein. A clone of cDNA was sequenced but the resulting protein had properties that were not those expected of a G_1-protein-linked receptor (see Smith and Loh, this Vol.).

Our laboratory has been engaged in efforts to clone the μ-receptor. In collaboration with C. Strader at Merck we have fragmented the protein and sequenced two peptides, one 22 and one 15 amino acids in length. Neither of the peptides is present in any known protein in the computer data bases available. We are therefore confident that they are derived from the μ-receptor, though proof for this is still lacking. Screening of bovine brain cDNA libraries with oligonucleotides synthesized according to the codon assignments of one of these peptides by R. Dixon at Merck has to date resulted in a number of false-positive clones. This may be due to the high degeneracy of the amino acids in our peptides. Additional probes corresponding to less degenerate portions of the peptides have been synthesized and are being used for screening bovine brain cDNA libraries. Antibodies to a portion of one of the peptides have been prepared and are being tested for cross-reactivity with opioid receptors.

7 Concluding Remarks

This is far from being an exhaustive review, but the evidence presented is quite convincing that the three major types of opioid receptors are distinct molecules. Judging from work done on other receptors, likely to be in the same family with opioid receptors (e.g., noradrenergic and muscarinic receptors), the different opioid receptors may well be different gene products. Among four different muscarinic and several different noradrenergic receptors, all have been found to be distinct polypeptides, albeit with high levels of homology. They all have homology to rhodopsin and all have seven putative transmembrane sequences, i.e., they are presumed to traverse the membrane seven times. It is, therefore, expected that when the opioid receptors are cloned they will show these characteristics, since μ- and δ-receptors have been known to be G-protein-linked for some time and evidence for such coupling of the κ-receptor has recently been reported (Attali et al. 1989). Why success in the cloning of the opioid receptors has been so slow in coming to fruition is unclear, but it is hoped that success will be achieved in the near future, hopefully by the time this volume is published.

A word of apology may be in order for the neglect of other types and subtypes of opioid receptors. The author has little doubt that many of the other receptors will turn out to be important. Judging from the complexity of other receptor systems (e.g., the GABA receptor, for which different types seem to be discovered continuously), the discovery of additional types of opioid receptors is highly probable. The constraints of space have forced me to limit the discussion to the most studied receptors.

Finally, a word should be said about function. It is evident that the final answers are not yet available. It appears that there is considerable overlap between the functions of different receptor types, but some evidence for specialization is already coming in and more may be found in the future. Thus, different types of pain may be regulated by different types of receptors. The κ-receptor seems to be involved in water metabolism and the δ-receptor may be the major one playing a role in cardiovascular physiology. The μ-receptor seems to have an important role in the phenomena, which comprise what we call narcotic addiction, and there is some evidence for specialization with respect to other behaviors. This continues to be an exciting field in which the future looks hopeful for much important theoretical and practical information. We are confident that the Herz laboratory, though smaller than in the past, will continue to make important contributions.

References

Ahmed MS, Zhou D-H, Cavinato AG, Maulik D (1989) Opioid binding properties of the purified κ-receptor from human placenta. Life Sci 44:861–871

Attali B, Saya D, Vogel Z (1989) κ-Agonists inhibit adenylate cyclase and produce heterologous desensitization in rat spinal cord. J Neurochem 52:360–369

Bak T, Carr KD, Simon EJ, Portoghese PS (1988) Opposite effects of rostral and caudal ventricular infusions of nor-binaltorphimine on stimulation-induced feeding. Neurosci Abstr 14:1107

Carr KD, Bonnet KA, Simon EJ (1982) Mu and kappa opioid agonists elevate brain stimulation threshold for escape by inhibiting aversion. Brain Res 245:389–393

Chaillet P, Coulaud A, Zajac J-M, Fournie-Zaluskie MC, Constentin J, Roques BP (1984) The μ rather than the δ subtype of opioid receptors appears to be involved in enkephalin-induced analgesia. Eur J Pharmacol 101:83–90

Cherubini E, North RA (1985) Mu and kappa opioids inhibit transmitter release by different mechanisms. Proc Natl Acad Sci USA 82:1860–1863

Donnerer J, Oka K, Brossi A, Rice KC, Spector S (1986) Presence and formation of codeine and morphine in the rat. Proc Natl Acad Sci USA 83:4566–4567

Gilbert PE, Martin WR (1976) The effects of morphine- and nalorphine-like drugs in the nondependent, morphine-dependent and cyclazocine-dependent chronic spinal dog. J Pharmacol Exp Ther 198:66–82

Gacel G, Fournie-Zaluski M-C, Roques BP (1980) Tyr-D-Ser-Gly-Phe-Leu-Thr, a highly preferential ligand for delta opiate receptors. FEBS Lett 118:245–247

Goldstein A, Barrett RW, James IF, Lowney LI, Weitz CJ, Knipmeyer LL, Rapoport H (1985) Morphine and other opiates from beef brain and adrenal. Proc Natl Acad Sci USA 82:5203–5207

Gosnell BA (1988) Involvement of μ opioid receptors in the amygdala in the control of feeding. Neuropharmacology 27:319–326

Gramsch C, Schulz R, Kosin S, Herz (1988) Monoclonal antiidiotypic antibodies to opioid receptors. J Biol Chem 263:5853–5859

Gross RA, Macdonald RL (1987) Dynorphin A selectively reduces a large transient (N-type) calcium current of mouse dorsal root ganglion neurons in cell culture. Proc Natl Acad Sci USA 84:5469–5473

Han J-S, Xie GX, Zhou ZF (1984) Acupuncture mechanisms in rabbits studied with microinjection of antibodies against β-endorphin, enkephalin and substance P. Neuropharmacology 23:1–5

Handa BK, Lane AC, Lord JAH, Morgan BA, Rance MJ, Smith CFC (1981) Analogues of beta-LPH 61–64 possessing selective agonist activity at mu opiate receptors. Eur J Pharmacol 70:531–540

Herkenham M (1987) Mismatches between neurotransmitter and receptor localizations in brain: observations and implications. Neuroscience 23:1–38

Hiller JM, Angel LM, Simon EJ (1984) Characterization of the selective inhibition of the δ subclass of opioid binding sites by alcohols. Mol Pharmacol 25:249-255

Holaday JW, Tortella FC (1984) Multiple opioid receptors: possible physiological function of μ- and δ-binding sites in vivo. In: Muller EE, Genazzani AR (eds) Central and peripheral endorphins: basic and clinical aspects. Raven, New York, pp 237-245

Howard AD, La Baume S, Gioannini TL, Hiller JM, Simon EJ (1985) Covalent labeling of opioid receptors with radioiodinated human β-endorphin. J Biol Chem 260:10833-10839

Howard AD, Sarne Y, Gioannini TL, Hiller JM, Simon EJ (1986) Identification of distinct binding site subunit of μ and δ opioid receptors. Biochemistry 25:357-360

Hughes J, Smith TW, Kosterlitz HW, Fothergill LA, Morgan BA, Morris HR (1975) Identification of two related pentapeptides from the brain with potent opiate agonist activity. Nature (London) 258:577-579

James IF, Goldstein A (1984) Site directed alkylation of multiple opioid receptors. I. Binding selectivity. Mol Pharmacol 25:337-342

Jenck F, Quirion R, Wise RA (1987) Opioid receptor subtypes associated with ventral tegmental facilitation and periaqueductal gray inhibition of feeding. Brain Res 423:39-44

Lord JAH, Waterfield AA, Hughes J, Kosterlitz HW (1977) Endogenous opioid peptides: multiple agonists and receptors. Nature (London) 267:495-499

Martin WR, Eades CG, Thompson JA, Huppler RE, Gilbert PE (1976) The effects of morphine- and nalorphine-like drugs in the nondependent and morphine-dependent chronic spinal dog. J Pharmacol Exp Ther 197:517-532

McDowell J, Kitchen I (1987) Development of opioid systems: peptides, receptors and pharmacology. Brain Res Rev 12:397-421

Morley JE, Levine AS (1983) Involvement of dynorphin and the kappa receptor in feeding. Peptides 4:797-800

Mosberg HI, Hurst R, Hruby VJ, Gee K, Yamamura HI, Galligan JJ, Burks TF (1983) Bis-penicillamine enkephalins possess highly improved specificity toward δ opioid receptors. Proc Natl Acad Sci USA 80:5871-5874

North RA, Williams JT, Surprenant A, Christie MJ (1987) Mu and delta receptors belong to a family of receptors that are coupled to potassium channels. Proc Natl Acad Sci USA 84:5487-5491

Pert CB, Snyder SH (1973) Opiate receptor: demonstration in nervous tissue. Science 182:1359-1361

Porreca F, Heyman JS, Mosberg HI, Rice KC, Jacobsen AE, Wood JH (1987) Role of μ and δ receptors in the supraspinal and spinal analgesic effects of (D-Pen^2, D-Pen^5) enkephalin in the mouse. J Pharmacol Exp Ther 241:392-400

Robson LE, Kosterlitz HW (1979) Specific protection of the binding sites of D-Ala^2-D-Leu^5-enkephalin (δ receptors) and dihydromorphine (μ receptors). Proc R Soc London Ser B 205:425-432

Rothman RB, Westfall TC (1982) Allosteric coupling between morphine and enkephalin receptors in vitro. Mol Pharmacol 21:548-555

Schoffelmeer ANM, Rice KC, Jacobson AE, van Gelderen JG, Hogenboom F, Heijna MH, Mulder AH (1988) μ-, δ- and κ-receptor-mediated inhibition of neurotransmitter release and adenylate-cyclase activity in rat brain slices: studies with fentanyl isothiocyanate. Eur J Pharmacol 154:169-178

Schofield PR, McFarland KC, Hatlick JS, Wilcox JN, Cho TM, Roy S, Lee NM, Loh HH, Seeburg PH (1989) EMBO J 8:489-495

Schulz R, Wüster M, Krenss H, Herz A (1980) Lack of cross-tolerance on multiple opiate receptors in the mouse vas deferens. Mol Pharmacol 18:395-401

Schulz R, Wüster M, Rubini P, Herz A (1981) Functional opiate receptors in the guinea pig ileum: their differentiation by means of selective tolerance development. J Pharmacol Exp Ther 219:547-550

Shearman GT, Herz A (1982) Evidence that the discriminative stimulus properties of fentanyl and ethylketocyclazocine are mediated by an interaction with different opiate receptors. J Pharmacol Exp Ther 221:735-739

Simon EJ, Hiller JM (1988) Solubilization and purification of opioid binding sites. In: Pasternak GW (ed) The opiate receptors. Humana, Clifton, NJ, pp 165-194

Simon EJ, Hiller JM (1989) Opioid peptides and opioid receptors. In: Siegel GJ, Agranoff BW, Albers RW, Molinoff PB (eds) Basic neurochemistry: molecular, cellular, and medical aspects, 4th edn. Raven, New York, pp 271-285

Simon EJ, Hiller JM, Edelman I (1973) Stereospecific binding of the potent narcotic analgesic ^{3}H-etorphine to rat brain homogenate. Proc Natl Acad Sci USA 70:1947-1949

Simonds WF (1988) The molecular basis of opioid receptor function. Endocrine Rev 9:200–212
Smith JR, Simon EJ (1980) Selective protection of sterospecific enkephalin and opiate binding against inactivation by N-ethylmaleimide: evidence for two classes of opiate receptors. Proc Natl Acad Sci USA 77:281–284
Terenius L (1973) Stereospecific interaction between narcotic analgesics and a synaptic plasma membrane fraction of rat cerebral cortex. Acta Pharmacol Toxicol 32:317–329.pa
Wood PL, Rackham A, Richard J (1981) Spinal analgesia: comparison of the μ-agonist morphine and the κ-agonist ethylketazocine. Life Sci 28:2119–2125
Wüster M, Schulz R, Herz A (1983) A subclassification of multiple opiate receptors by means of selective tolerance development. J Receptor Res 3:199–214
Yaksh TL (1984) Multiple spinal opiate receptor systems in analgesia. In: Kruger L, Liebeskind J (eds) Advances in pain research and therapy, vol 6. Raven, New York, pp 197–215
Yao Y-H, Gairin J, Meunier J-C, Hiller JM, Gioannini TL, Cros J, Simon EJ (1989) Cross-linking of kappa receptors in the guinea pig cerebellum with D-Pro^{10}dynorphin (1–11). Adv Biosci 75:21–24
Zukin RS, Zukin SR (1981) Demonstration of [^{3}H]cyclazocine binding to multiple opiate receptor sites. Mol Pharmacol 20:246–254

CHAPTER 5

Molecular Approaches to Isolating Opioid Receptors

A.P. Smith and H.H. Loh

1 Introduction

Opioid receptors were first identified in mammalian brain in the early 1970s (Pert and Snyder 1973; Simon et al. 1973; Terenius 1973), and their pharmacological properties have since been extensively characterized (for reviews, see Iwamoto and Martin 1981; Smith and Loh 1981; Wood 1982). However, full structural and functional characterization of a receptor requires its purification and reconstitution in membrane preparations; for several reasons, progress in this area of opioid receptor research has been relatively slow.

First, opioid receptors have proved to be very difficult to solubilize in a form that retains binding activity. Ligand binding of opioid receptors is eliminated or greatly reduced by low concentrations of most nonionic detergents that have been used successfully to solubilize other cell surface receptors. At least part of this problem may be due to the more stringent criteria necessary to establish the existence of opioid receptors. The activity of most cell surface receptors is measured by specific, or displaceable, binding, which is defined as the difference between the amount of radioactive ligand bound in the presence and absence of a large amount of the same unlabeled ligand. A genuine opioid receptor, in contrast, must exhibit not only specific, but stereospecific binding – that which is selective for the *l*-forms of stereoisomeric agonists, such as levorphanol and levallorphan.

In fact, most laboratories that have reported solubilization of opioid receptors have demonstrated the existence of only specific binding in the preparations. Stereospecific binding has been either very low, or undetectable. While it is conceivable that opioid receptors lose stereospecificity upon solubilization, the inability to demonstrate it casts doubt on the pharmacological relevance of the preparation.

A second obstacle to purification of opioid receptors is that no simple biological process reflecting opioid receptor function has yet been identified. Opioid receptors have been classically defined as those promoting antinociception in mammals, or in some cases, by their effects in certain in vitro systems. Recent work has indicated an association between opioid receptors and several other functional molecules in cell membranes, including adenylate cyclase (Sharma et al. 1975; Law et al. 1981) and ion channels (J.T. Williams et al. 1982; Werz and MacDonald 1984, 1985; North and Williams 1985). However, none of these molecules has been demonstrated to play a role in the process of antinociception. Thus, there is still no definitive bioassay that can be used to confirm the functional relevance of a putative isolated opioid receptor.

Finally, opioid receptors are heterogeneous; at least three different classes, μ, δ and κ, which differ in their ligand selectivity and their pharmacological effects, are present in brain (Martin et al. 1976; Lord et al. 1977; Wood 1982). Moreover, other receptor types may exist, in brain and peripheral tissues (Schulz et al. 1981; Grevel and Sadee 1983), in addition to subtypes of μ (Nishimura et al. 1984; Loew et al. 1986) and κ (Iyengar et al. 1986) opioid receptors. Pharmacological characterization and isolation of a single receptor type are further complicated by the fact that most opioid ligands are not completely selective for a single receptor type.

Despite these problems, substantial progress has been made in the purification of opioid receptors in recent years. The introduction of new nonionic detergents, such as 3-(3-cholamidopropyl) dimethylammonio-1-propanesulfonate (CHAPS), has made possible solubilized preparations with higher yields of opioid binding activity. Purification of specific opioid receptor types has been facilitated by the use of tissue sources which are pure or enriched in one type of opioid receptor, such as NG108-15 neuroblastoma-glioma hybrid cells (δ-opioid receptors) and human placenta (κ-opioid receptors), as well as by the development of synthetic ligands with relatively high specificity for a particular opioid receptor type. Synthesis of several new covalent opioid ligands that are useful in identifying and purifying opioid binding proteins in the presence of denaturing detergents has circumvented the problem of maintaining ligand binding activity during purification. Also, idiotypic and anti-idiotypic antibodies have been used, in addition to the more conventional affinity ligands, to purify opioid receptors which retain binding activity.

As purified or partially purified preparations of opioid receptors have become available, investigators have also begun to characterize them, deploying the full armory of tools made available by modern molecular biology. Several groups have now prepared antibodies to opioid receptors, which can be used to map the distribution of receptors in the brain, to determine the regions in the receptor molecule involved in ligand binding and other functions, and to purify further the receptor. Several groups are also attempting to clone the receptor, using standard techniques for isolating cDNA and transfecting it into a homogeneous cell line.

In this chapter, we briefly discuss and evaluate these developments. We first consider studies reporting purification of opioid receptors and then look at attempts to characterize these receptors.

2 Purification of Opioid Receptors

Two basic approaches are used to purify cell surface receptors. One is to solubilize the receptor in a nondenaturing detergent that preserves ligand binding activity; the receptor is then isolated from other membrane components by various fractionation steps, such as affinity chromatography, with binding activity being monitored at each step. The alternative approach is to label the receptor with a radioactive, covalently bound ligand; the receptor can then be solubilized with a denaturing detergent, and purification followed by means of the radioactive label.

The first approach is generally preferred, as it is necessary to retain binding activity if the purified receptor is to be reconstituted and tested for function. An additional advantage of this approach, as applied to opioid receptors, is that by using several affinity columns, which differ in their selectivity for different opioid receptor types, it may be possible to purify more than one type from a single solubilized preparation. When opioid receptors are covalently labeled, the selectivity of purification is largely dependent upon the selectivity of the covalent ligand, through some additional selectivity may be obtained by protecting particular receptor types with reversibly bound ligand during the labeling process. In any case, once the covalent labeling process has occurred, no further selectivity is possible.

On the other hand, if a suitable covalent label can be found, the second approach to purification is usually faster and easier, and the molecular weight of the putative receptor can be determined by SDS gel electrophoresis (even without purification). In addition to synthesizing and testing compounds capable of reacting covalently with opioid receptors, some researchers have taken advantage of the fact that many opioid receptor ligands are peptides, which can be covalently attached to the receptor by means of a variety of bifunctional cross-linking reagents. Although covalently labeled opioid receptors lack ligand binding activity, they can in principle be used to purify active receptors indirectly, by means of cDNA cloning techniques. This will be discussed further below.

2.1 Purification of Opioid Receptors Retaining Ligand Binding

Bidlack et al. (1981) solubilized rat brain membranes with Triton X-100, then isolated opioid binding material from the solubilized preparation by means of affinity chromatography using 14-β-bromoacetamido-morphine. The resulting preparation bound dihydromorphine and several other opioids with affinities in the nanomolar range, but the specific binding activity (cpm/mg protein) was at least an order of magnitude lower than that theoretically expected for a pure opioid receptor. SDS gel analysis revealed three polypeptides of molecular weights between 25–50 kDa.

Simon and his colleagues solubilized opioid receptors from toad and also from rat and other mammalian brain preparations with digitonin (Howells et al. 1982). Using opioid ligand affinity chromatography, coupled with lectin affinity chromatography, they were able to purify an opioid binding protein from bovine striatum to theoretical homogeneity (Gioannini et al. 1985). However, protein estimation was difficult, due to the small amounts isolated, and the presence of detergent. The opioid binding material had a molecular weight of 300–350 kDa under nondenaturing conditions, and 65 kDa on SDS gels. The μ-binding properties of this material were not demonstrated directly, but were suggested by its ability to bind the μ-selective ligand Tyr-D-Ala-Gly-NMe-Phe-Gly-ol (DAGO) as well as β-endorphin (B-END), and also by the fact that the material was isolated from a source low in δ-opioid receptors.

Maneckjee et al. (1985) reported partial purification (500-fold) of a μ-specific opioid receptor from rat brain, using CHAPS as solubilizer; SDS gel analysis indicated the presence of three peptides, of molecular weights 94, 42 and 35 kDa. Subsequently, they prepared a more highly purified receptor preparation from bovine striatum (Maneckjee et al. 1987), using a combination of affinity chromatography and hydroxyapatite chromatography. Studies with polyclonal antibodies against this protein (discussed further below) indicated that the 94 kDa component was involved in opioid binding.

Using a combination of affinity chromatography, lectin chromatography, and gel filtration, Cho et al. (1986) reported purification from bovine brain of a protein selective for opioid alkaloid ligands. A novel feature of this protein was that it required acidic lipids which possess unsaturated fatty acids in order to manifest binding activity; neither the protein nor the lipids alone showed significant opioid binding (Hasegawa et al. 1987). The binding affinities of ligands to the reconstituted material were lower than the corresponding values for binding to brain membranes, but the rank order of binding affinities to the two preparations was highly correlated (Cho et al. 1986; Hasegawa et al. 1987).

Ueda et al. (1987, 1988), using a similar method, also obtained a protein consisting primarily of a 58 kDa band. When this preparation was reconstituted with the purified G-proteins G_i or G_o, a large increase in displacement of ^{3}H-naloxone binding by the μ-agonist DAGO was observed, and this increase was sensitive to GTP. Mu, but not δ- or κ-agonists also stimulated GTP binding and GTPase activity in this preparation. This work thus suggests that μ-opioid receptors may exert their effects in vivo through a G-protein.

As is apparent from the above discussion, many of the solubilized opioid receptor preparations initially reported did not bind κ-ligands. However, Chow and Zukin (1983) reported that CHAPS-solubilized rat brain membranes contained κ- as well as μ-binding species, as assayed by ^{3}H-bremazocine binding. The material eluted as two peaks of 50 and about 250 kDa. Itzhak and colleagues (Itzhak et al. 1984a,b; Itzhak and Pasternak 1986) solubilized guinea pig brain membranes using digitonin in the presence of high NaCl concentrations, and found that δ-binding was associated with material of 750–875 kDa, while κ-binding was associated with material of about 400 kDa.

DeMoliou-Mason and Barnard (1984), using a digitonin extraction in the presence of 10 mM $MgCl_2$, were able to solubilize material that bound dynorphin$_{1-9}$ (DYN$_{1-9}$), a putative κ-ligand, as well as μ- and δ-ligands. Subsequently, they characterized this binding material by gel filtration (DeMoliou-Mason and Barnard 1986). Ligands of the δ-type were found to bind material over 500 kDa, while both μ-alkaloids and DYN$_{1-9}$ bound both this high molecular weight material and lower molecular weight material.

Simon et al. (1987) purified κ-opioid receptors from digitonin-solubilized frog brain membranes. An affinity column consisting of D-ala^2-D-leu^5-enkephalin (DADLE) coupled to Sepharose-6B was used to isolate μ-, δ- and κ-receptors from the solubilized preparation, and κ-receptors were separated from the other two receptor types by gel filtration. The extent of purification was over 4000-fold (based on pmol of ^{3}H-ethylketocyclazocine bound/mg protein). SDS gel analysis revealed two bands, of 65 and 58 kDa.

Cavinato et al. (1988) purified κ-opioid receptors from human placental tissue, which is thought to be enriched in this receptor type. The protein was partially purified by preparative SDS gel electrophoresis, and had a molecular weight of 63 kDa.

2.2 Purification of Covalently Labeled Opioid Receptors

Some of the earliest attempts to purify opioid receptors were carried out using the covalent labeling approach. However, in most cases, it could not be demonstrated that the label specifically bound to opioid receptors (Bidlack et al. 1982). More recently, a number of covalent opioid ligands, which have proven useful for selectively eliminating specific opioid receptor types, have been synthesized. These include β-funaltrexamine (β-FNA), β-chlornaltrexamine (β-CNA), 2-(4-ethoxybenzyl)-1-diethylaminoethyl-5-isothiocyantobenzimidoazole (BIT), N-phenyl-N-[1-(4-isothiocyanato) phenylethyl-4-piperidinyl]-propanamide (FIT), 6-desoxy-oxymorphone (FOXY), and 6-desoxy-6β-fluoronaltrexone (cycloFOXY) (Caruso et al. 1980; Rice et al. 1983; Rothman et al. 1988a,b).

Using a similar technique, Bero et al. (1988) found that ^{125}I-β-END labeled three peptides with molecular weights of 108, 73 and 49 kDa. However, the correspondence of these three species to μ-, δ- and κ-receptors was not clear. Preincubation with unlabeled β-END was capable of blocking much or most of the cross-linking to each of these species, but other ligands were much less effective. Thus, etorphine (which binds to μ-, δ- and κ-opioid receptors) blocked 60% of the cross-linking to the 73 kDa band, but $<40\%$ to either of the other two bands. The μ-selective ligand DAGO and the κ-selective ligand trans-3,4 dichloro-N-methyl-N[7-(1-pyrrolidinyl)-cyclohexyl] benzeneacetamide (U-50,488H) blocked 30–50% of the binding to the 108 and 73 kDa bands, while the δ-selective ligand DPDPE was ineffective in blocking cross-linking to any of the three species.

The most interesting result of this study was the finding that an antibody prepared against a purified opioid binding protein (see below) was capable of blocking labeling to all three bands. This band also inhibited binding of ^{3}H-DYN to brain membranes.

Newman and Barnard (1984) also performed cross-linking studies on brain membranes, using D-Ala^{2}-Leu-enkephalin-chloromethyl ketone (DALECK), an enkephalin derivative selective for μ-opioid receptors. They observed a single band of 58 kDa and SDS gels.

In summary, there is considerable variety in the molecular size of purified or partially purified opioid binding proteins, as reported by different investigators. In those instances when the molecular weight was determined under nondenaturing conditions, association of the binding proteins with other membrane proteins, as well as with detergent micelles, may have contributed to the discrepancies. Differences in molecular weights as determined on SDS gels, on the other hand, are more difficult to reconcile, as they must reflect distinctly different species.

It is possible that different opioid receptor types have different molecular weights: some support for this is provided by cross-linking experiments in which μ- or δ-selective ligands have been used to protect sites from cross-linking (Howard et al. 1986; Bero et al. 1988). However, discrepancies in the reported molecular weights of putative opioid receptors of the same type remain. Furthermore, a molecular weight of approximately 60 kDa has been found by some groups for all three major opioid receptor types (Newman and Barnard 1984; Gioannini et al. 1985; Simonds et al. 1985; Cho et al. 1986; Howard et al. 1986; Simon et al. 1987; Ueda et al. 1987; Cavinato et al. 1988; S. Roy et al. 1988a). This suggests that all three of these opioid receptors types may be similar molecules – perhaps constituting identical gene products that are modified differentially after translation.

3 Characterization of Opioid Receptors

Purification of a cell surface receptor is only the first step towards its molecular characterization. Once the purified receptor is available, its physicochemical structure, including amino acid sequence and conformation, can be determined. Antibodies to the receptor can be prepared, and used to map the functional domains of the receptor, as well as to detemine its cellular and brain regional localization, for example. Most importantly, the receptor can also be reconstituted back into a membrane, and its functional activity evaluated. Because opioid receptors have only recently been purified, work in these areas is just beginning. However, some significant progress has already been reported.

3.1 Preparation of Antibodies to Purified Opioid Receptors

Several groups have now prepared antibodies to opioid receptor preparations that were previously purified in their respective laboratories. As discussed earlier, Bidlack et al. (1981) reported purification of an opioid receptor from Triton-solubilized rat brain membranes, using affinity chromatography. Subsequently, this group prepared a monoclonal antibody (mAb) to this material, which proved to be directed against the 35 kDa band observed on SDS gels. This mAb was capable of partially inhibiting opioid binding to the solubilized preparation (Bidlack and Denton 1985), though long incubation periods were required. Moreover, Fab fragments prepared from the antibody rapidly and completely inhibited binding of μ- and δ-opioid ligands to brain membranes (Bidlack and O'Malley 1986).

Maneckjee et al. (1987) prepared a polyclonal antibody to the highly purified opioid binding protein they had previously isolated from bovine striatum. This antibody selectively inhibited the binding of μ-opioids to rat brain membranes, and also selectively precipitated the 94 kDa band.

S. Roy et al. (1988a,b) prepared both monoclonal and polyclonal antibodies to the 58 kDa opioid binding protein purified from bovine brain by Cho et al. (1986). The polyclonal antibodies selectively bound to a 58 kDa protein in rat brain membranes, but to a 45 kDa protein in membranes of NG 108-15 cells. The monoclonal antibody inhibited opioid binding to the purified opioid binding protein (bovine-derived), while Fab fragments prepared from the mAb inhibited opioid binding to rat brain membranes. Interestingly, binding of μ-, δ- and κ-ligands was inhibited, suggesting that the antibody was not directed against the binding site, but toward some other component of the receptor that is shared by all three receptor types. This conclusion was supported by the finding that the inhibition was noncompetitive in each case. Most significant, this group recently demonstrated that the polyclonal antibodies could inhibit opioid effects in vivo (P. Green et al., pers. commun.). When the antibodies were injected intracerebroventricularly, they reduced morphine-induced analgesia. In contrast, the antibodies had no effect on morphine-induced analgesia when injected intrathecally.

A somewhat different immunological approach was introduced by B.F. Roy et al. (1988). These investigators detected and isolated anti-idiotypic antibodies to β-END (i.e., antibodies to anti-β-END antibodies) from serum of patients suffering from a major depressive disorder. Thus, these antibodies presumably contain a site similar in conformation to β-END itself. Binding of the antibodies to brain membranes was inhibited by opioid ligands. In addition, the anti-idiotype bound to a 60 kDa peptide in Western immunoblots.

3.2 Cloning of Opioid Binding Proteins

As discussed earlier, Cho et al. (1986) purified an opioid binding protein from bovine brain that was selective for alkaloids. The cDNA coding for this opioid binding protein was recently cloned, using standard procedures (Schofield et al. 1989). The amino acid sequence of a portion of the opioid binding protein was determined, oligonucleotide probes corresponding to these sequences synthesized, and these probes were used to screen a cDNA library from bovine brain. A single DNA sequence was isolated that contained all the sequences of these probes. This sequence was determined, and translated to give the amino acid sequence of the opioid binding protein.

The protein consists of 345 amino acids, with a calculated molecular weight of 37.9 kDa. The discrepancy between this value and the molecular weight of the originally purified opioid binding protein (58 kDa) suggests that the protein might be glycosylated, a conclusion supported by two other observations: (1) the opioid binding protein binds to lectin affinity columns (Cho et al. 1986); and (2) there are six potential glycosylation sites in the protein's amino acid sequence, consisting of an asparagine residue in close proximity to serine or threonine residues.

While it has not yet been possible to show that transfection of the cDNA into cells lacking opioid receptors confers binding activity, other, indirect evidence indicates the pharmacological relevance of this opioid binding protein. An antibody to a portion of the predicted amino acid sequence was used to construct an affinity column. This column was capable of specifically binding protein from solubilized brain membranes that bound opioids in in vitro assays (Schofield et al. 1989). Moreover, a monoclonal antibody raised to the purified protein (S. Roy et al. 1988b) inhibited opioid binding to this affinity-purified material. These results indicate that the cloned sequence indeed codes for the purified opioid binding protein.

A search of the NBRF-PIR data base revealed that this opioid binding protein had significant homologies to several proteins, all of which are members of the immunoglobulin (Ig) superfamily (Schofield et al. 1989; A.P. Smith, unpubl. data). This is a group of proteins characterized by repeating domains flanked by cysteine residues (A.F. Williams 1987). The highest degree of homology is to several cell adhesion molecules, including neural cell adhesion molecule, myelin-associated glycoprotein, and fasciclin II; the amino acid sequence of each of these proteins shows about a 20% homology with that of the opioid binding protein. Somewhat lower significance values, also suggestive of an evolutionary relationship, were found for several peptide receptors, including those for platelet-derived growth factor (PDGF) and interleukin-6.

Sequence analysis is therefore consistent with the notion that this protein functions as a neuropeptide receptor, while suggesting that it could also play a role in cell adhesion. Accordingly, it has been named OBCAM, or opioid binding cell adhesion molecule. A possible role of this molecule in cell adhesion is also consistent with recent evidence that opioids can modulate cell-cell interactions of immune cells (Stefano et al. 1989).

Nevertheless, the homology of this opioid binding protein with members of the Ig superfamily is somewhat surprising, because some opioid receptors, namely, the δ-type in NG108-15 neuroblastoma × glioma hybrid cells and in the mammalian striatum, are coupled via a G-protein to adenylate cyclase (Sharma et al. 1975; Koski et al. 1982; see below). This has suggested that these receptors are structurally and functionally related to a group of cell surface receptors that are associated with G-proteins. These include the β_1-, β_2-, and α_2-adrenergic, m_1 and m_2 muscarinic, serotonin 1a and 1c, and substance K receptors (Dixon et al. 1986; Kubo et al. 1986; Kobilka et al. 1987; Peralta et al. 1987; Julius et al. 1988). All of these receptors have seven hydrophobic regions in their interior, which are thought to span the membrane.

While the second messenger(s) mediating the actions of OBCAM has not yet been determined, its sequence suggests that either it does not interact with G-proteins or, if it does, this interaction is quite distinct from that of those receptors with multiple membrane-spanning regions. In fact, on the basis of its sequence homologies, it seems more likely that it would function as a type III tyrosine kinase growth factor receptor, such as the PDGF receptor. Activation of this latter receptor results in phosphorylation of several intracellular substrates, and of the receptor itself, and induces a wide variety of other events within the cell (L.T. Williams 1989). However, OBCAM does not possess the consensus sequence that is thought to contain the tyrosine kinase activity.

An attempt has also been made to clone the δ-receptor of NG108-15 hybrid cells, without beginning with purified receptor (P.Y. Law, pers. commun.). Total mRNA was isolated from NG108-15 cells, and used to synthesize a "library" of cDNA, that is, DNA containing all of the genomic sequences found in NG108-15 cells. This cDNA library was fractionated by size, and then inserted (by means of a plasmid vector) into a glioma cell line that expresses no detectable opioid binding. A key to the success of these studies was the construction of the subtraction probes, which were used to enrich or "prescreen" the initial colonies for opioid receptor cDNA. To make these probes, we took advantage of the fact that several treatments can selectively alter the number of opioid receptors in clonal cells. Prolonged treatment of NG108-15 hybrid cells with an opioid agonist (e.g., 100 nM DADLE for 24 h) results in up to an 80% decrease, or downregulation, in receptor number (Law et al. 1983). In addition, treatment of the PC-12 cell subtype PC12h (which ordinarily contains undetectable levels of opioid receptors) with nerve growth factor for 14 days results in the appearance of high levels of δ-opioid receptors. Accordingly mRNA was prepared from these cells under both high receptor (normal NG108-15 cells, or

NGF-treated PC12h cells) and low receptor (downregulated NG108-15 hybrid cells or normal PC12h cells) number conditions. The mRNA from the high receptor cells was used to make cDNA's, which were then hybridized with mRNA from low receptor cells. This hybridization in effect "subtracts" all the common mRNA from the high and low receptor conditions, and in theory, leaves cDNA highly enriched in opioid receptor DNA. After such screening, the colonies that tested positive for both subtraction probes were screened for opioid binding. Several candidate cDNAs that bestowed both opioid binding and opioid-mediated adenylate cyclase inhibition on these cells are currently under investigation.

4 Conclusions

Though opioid receptors have proved to be more difficult to purify and characterize than other cell surface receptors, significant progress has been made in the last few years. At least a dozen groups have now reported purification of opioid binding proteins, either in a form that retains ligand binding properties, or in a covalently bound form. While there are some discrepancies in the molecular weights of these proteins, it is significant that many investigators have reported a molecular weight of about 60 kDa for the receptor, regardless of whether it is of the μ-, δ- or κ-type. This suggests that these three major opioid receptor types may be highly similar.

With purified opioid receptors available, characterization may begin. Several groups have prepared polyclonal or monoclonal antibodies to their material. These should be useful in mapping of opioid receptors in the brain, in determining the regions in the peptide required for ligand binding and association with second messengers, and in further purification schemes. One group has also cloned the cDNA for a purified opioid binding protein, which will open the door to detailed structural studies.

References

Bero LA, Roy S, Lee NM (1988) Identification of endogenous opioid receptor components in rat brain using a monoclonal antibody. Mol Pharmacol 34:614-620

Bidlack JM, Denton RR (1985) A monoclonal antibody capable of inhibiting opioid binding to rat neural membranes. J Biol Chem 260:15655-15661

Bidlack JM, O'Malley WE (1986) Inhibition of μ and δ but not κ opioid binding to membranes by Fab fragments from a monoclonal antibody directed against the opioid receptor. J Biol Chem 261:15844-15849

Bidlack JM, Abood LG, Osei-Gyimah P, Archer S (1981) Purification of the opiate receptor from rat brain. Proc Natl Acad Sci USA 78:636-639

Bidlack JM, Abood LG, Munemitsu SM, Archer S, Gala D, Kreilick RW (1982) Affinity labeling and purification of the opiate receptor from rat brain. In: Costa E, Trabucchi M (eds) Regulatory peptides: from molecular biology to function. Raven, New York, pp 301-309

Bochet P, Icard-Liepkalns C, Pasquini F, Garbay-Jaureguiberry C, Beaudet A, Roques B, Rossier J (1988) Photoaffinity labeling of opioid δ receptors with an iodinated azido-ligand: [^{125}I] [D-Thr^2, pN3Phe^4, Leu^5]enkephalyl-Thr^6. Mol Pharmacol 34:436-443

Caruso TP, Larson DL, Portoghese PS, Takemori AE (1980) Isolation of selective ^{3}H-chlornaltrexamine-bound complexes, possible opioid receptor components in brains of mice. Life Sci 27:2063-2069

Cavinato AG, MacLeod RM, Ahmed MS (1988) A non-denaturing gel electrophoresis system for the purification of membrane-bound proteins. Prep Biochem 18:205–216
Cho TM, Hasegawa J, Ge BL, Loh HH (1986) Purification to apparent homogeneity of a μ-specific opioid receptor from rat brain. Proc Natl Acad Sci USA 83:4138–4142
Chow T, Zukin RS (1983) Solubilization and preliminary characterization of mu and kappa opiate receptor subtypes from rat brain. Mol Pharmacol 24:203–212
DeMoliou-Mason Cd, Barnard EA (1984) Solubilization in high yield of opioid receptors retaining high affinity delta, mu and kappa binding sites. FEBS Lett 170:378–382
DeMoliou-Mason CD, Barnard EA (1986) Characterization of opioid receptor subtypes in solution. J Neurochem 46:1129–1136
Dixon RAF, Kobilka BK, Strader DJ, Benovic JL, Dohlman HG, Frielle T, Bolanowski MA, Bennett CD, Rands E, Diehl RE, Mumford RA, Slater EE, Sigal IS, Caron MG, Lefkowitz RJ, Strader CD (1986) Cloning of the gene and cDNA for mammalian β-adrenergic receptor and homology with rhodopsin. Nature (London) 321:75–79
Gioannini TL, Howard AD, Hiller JM, Simon EJ (1985) Purification of an active opioid-binding protein from bovine striatum. J Biol Chem 260:15117–15121
Grevel JT, Sadee W (1983) An opiate binding site in the rat brain is highly selective for 4,5-epoxymorphinans. Science 221:1198–1201
Hasegawa J, Loh HH, Lee NM (1987) Lipid requirement for μ opioid receptor binding. J Neurochem 49:1007–1012
Howard AD, de Le Baume S, Gioannini TL, Hiller JM, Simon EJ (1985) Covalent labeling of opioid receptors with radioiodinated human β-endorphin: identification of binding site subunits. J Biol Chem 260:10833–10839
Howard AD, Sarne Y, Gioannini TL, Hiller JM, Simon EJ (1986) Identification of distinct binding site subunits of μ and δ opioid receptors. Biochemistry 25:357–360
Howells RD, Gioannini TL, Hiller JM, Simon EJ (1982) Solubilization and characterization of active opiate binding sites from mammalian brain. J Pharmacol Exp Ther 222:629–634
Itzhak Y, Pasternak GW (1986) κ-Opiate binding to rat brain and guinea-pig cerebellum: sensitivity towards ions and nucleotides. Neurosci Lett 64:81–84
Itzhak Y, Hiller JM, Simon EJ (1984a) Solubilization and characterization of κ opioid binding sites from guinea-pig cerebellum. Neuropeptides 5:201–204
Itzhak Y, Hiller JM, Simon EJ (1984b) Solubilization and characterization of μ, δ and κ opioid binding sites from guinea-pig brain: physical separation of κ receptors. Proc Natl Acad Sci USA 81:4217–4222
Iwamoto ET, Martin WR (1981) Multiple opioid receptors. Med Res Rev 1:411–440
Iyengar S, Kim HS, Wood PL (1986) Effects of kappa opiate agonists on neurochemical and neuroendocrine indices: evidence for kappa receptor subtypes. J Pharmacol Exp Ther 238:429–436
Julius D, MacDermott AB, Axel R, Jessell TM (1988) Molecular characterization of a functional cDNA encoding to serotonin 1c receptor. Science 241:558–564
Kobilka BK, Matsui H, Kobilka TS, Yang-Feng TL, Francke U, Caron MG, Lefkowitz RJ, Regan JW (1987) Cloning, sequencing and expression of the gene coding for the human platelet α-adrenergic receptor. Science 238:650–656
Koski G, Streaty RA, Klee WA (1982) Modulation of sodium-sensitive GTPase by partial opiate agonist: an explanation for the dual requirement for Na+ and GTP in inhibitory regulation of adenylate cyclase. J Biol Chem 257:14035–14040
Kubo T, Fukuda K, Mikami A, Maeda A, Takahashi H, Mishina M, Haga T, Haga K, Ichiyama A, Kangawa K, Kojima M, Matsuo H, Hirose T, Numa S (1986) Cloning, sequencing and expression of complementary DNA encoding the muscarinic acetylcholine receptor. Nature (London) 323:411–416
Law PY, Wu J, Koehler JE, Loh HH (1981) Demonstration and characterization of opiate inhibition of the striatal adenylate cyclase. J Neurochem 36:1834–1846
Law PY, Hom DS, Loh HH (1983) Opiate receptor down-regulation and desensitization in neuroblastoma × glioma NG108-15 hybrid cells are two separate cellular adaptation processes. Mol Pharmacol 25:413–424
Loew G, Keys C, Luke B, Polgar W, Toll L (1986) Structure-activity relationships of morphiceptin analogs: receptor binding and molecular determinants of μ-affinity and selectivity. Mol Pharmacol 29:546–553

Lord JAH, Waterfield AA, Hughes J, Kosterlitz HW (1977) Endogenous opioid peptides: multiple agonists and receptors. Nature (London) 267:495–499

Maneckjee R, Zukin RS, Archer S, Michael J, Osei-Gyimah P (1985) Purification and characterization of the μ opioid receptor from rat brain using affinity chromatography. Proc Natl Acad Sci USA 82:594–598

Maneckjee R, Archer S, Zukin RS (1987) Characterization of a polyclonal antibody to the μ opioid receptor. J Neuroimmunol 17:199–208

Martin WR, Eades CG, Thompson JA, Huppler RE, Gilbert PE (1976) The effects of morphine and nalorphine-like drugs in the nondependent and morphine-dependent chronic spinal dog. J Pharmacol Exp Ther 197:517–532

Newman PL, Barnard EA (1984) Identification of an opioid receptor subunit carrying the μ binding site. Biochemistry 23:5385–5389

Nishimura SC, Recht LD, Pasternak GW (1984) Biochemical characterization of high-affinity ^{3}H-opioid binding. Further evidence for *mu*$_1$ sites. Mol Pharmacol 25:29–37

North RA, Williams JT (1985) On the potassium conductance increased by opioids in rat locus coeruleus neurons. J Physiol 364:265–280

Peralta EG, Winslow JW, Peterson GL, Smith DH, Ashkenazi A, Ramachandran J, Schimerlik MI, Capon DJ (1987) Primary structure and biochemical properties of an M_2 muscarinic receptor. Science 236:600–605

Pert CB, Snyder SH (1973) Opiate receptor: its demonstration in nervous tissue. Science 179:1011–1014

Rice KC, Jacobson AE, Burke TR Jr, Bajwa BS, Streaty RW, Klee WA (1983) Irreversible ligands with high selectivity toward μ or δ opiate receptors. Science 220:314–316

Rothman RB, Bykov V, Reid A, DeCosta BR, Newman AH, Jacobson AE, Rice KC (1988a) A brief study of the selectivity of norbinaltorphimine, (−)-cycloFOXY, and (+)-cycloFOXY among opioid receptor subtypes in vitro. Neuropeptides 12:181–187

Rothman RB, Bykov V, Rice KC, Jacobson AE, Kooper JN, Bowen WD (1988b) Tritiated 6-β-fluoro-6-desoxy-oxymorphone (3H-FOXY): a new ligand and affinity probe for the μ opioid receptors. Neuropeptides 11:1–6

Roy BF, Bowen WD, Frazier JS, Rose JW, McFarland HF, McFarlin DE, Murphy DL, Morihasa JM (1988) Human antiidiotypic antibody against opiate receptors. Ann Neurol 24:57–63

Roy S, Zhu YX, Lee NM, Loh HH (1988a) Different molecular weight forms of opioid receptors revealed by polyclonal antibodies. Biochem Biophys Res Commun 150:237–244

Roy S, Zhu YX, Loh HH, Lee NM (1988b) A monoclonal antibody that inhibits opioid binding to rat brain membranes. Biochem Biophys Res Commun 154:688–693

Schofield PR, McFarland KC, Hayflick JS, Wilcox JN, Cho TM, Roy S, Lee NM, Loh HH, Seeburg PH (1989) Molecular characterization of a new immunoglobulin superfamily protein with potential roles in opioid binding and cell contact. EMBO J 8:489–495

Schulz R, Wüster M, Herz A (1981) Pharmacological characterization of the ε-opiate receptor. J Pharmacol Exp Ther 216:604–606

Sharma S, Nirenberg M, Klee W (1975) Morphine receptors are regulators of adenylate cyclase activity. Proc Natl Acad Sci USA 72:590–594

Simon EJ, Hiller JM, Edelman J (1973) Stereospecific binding of the potent narcotic analgesic [^{3}H]Etorphine to rat brain homogenate. Proc Natl Acad Sci USA 70:1947–1949

Simon J, Benyhe S, Hepp J, Khan A, Borsodi A, Szucs M, Medzihradsky K, Wolleman M (1987) Purification of a κ-opioid receptor subtype from frog brain. Neuropeptides 10:19–28

Simonds WF, Burke TR, Rice KC, Jacobson AE, Klee WA (1985) Purification of the opioid receptor of NG108-15 neuroblastoma × glioma hybrid cells. Proc Natl Acad Sci USA 82: 4774–4778

Smith AP, Loh HH (1979) Multiple molecular forms of stereospecific opiate binding. Mol Pharmacol 16:757–766

Smith AP, Loh HH (1981) The opiate receptor. In: Li CH (ed) Hormonal proteins and peptides. Academic Press, New York, pp 89–170

Stefano GB, Leung MK, Zhao X, Scharrer B (1989) Evidence for the involvement of opioid neuropeptides in the adherence and migration of immunocompetent invertebrate hemocytes. Proc Natl Acad Sci USA 86:626–630

Terenius L (1973) Stereospecific interaction between narcotic analgesics and a synaptic plasma membrane fraction of rat cerebral cortex. Acta Pharmacol 32:317–320

Ueda H, Harada H, Misawa H, Nozaki M, Takagi H (1987) Purified opioid μ-receptor is of a different molecular size than δ- and κ-receptors. Neurosci Lett 75:339–344

Ueda H, Harada H, Nozaki M, Katada T, Ui M, Satoh M, Takagi H (1988) Reconstitution of rat brain μ-opioid receptors with purified guanine nucleotide-binding regulatory proteins, G_i and G_o. Proc Natl Acad Sci USA 85:7013–7017

Werz MA, MacDonald RL (1984) Dynorphin reduces voltage-dependent calcium conductance of mouse dorsal root ganglion neurons. Neuropeptides 5:253–256

Werz MA, MacDonald RL (1985) Dynorphin and neoendorphin peptides decrease dorsal root ganglion neuron calcium-dependent action potential duration. J Pharmacol Exp Ther 234:49–56

Williams AF (1987) A year in the life of the immunoglobulin superfamily. Immunol Today 8:298–303

Williams LT (1989) Signal transduction by the platelet-derived growth factor receptor. Science 243:1560–1564

Williams JT, Egan TM, North RA (1982) Enkephalin opens potassium channel on mammalian central neurons. Nature (London) 229:74–77

Wood PL (1982) Multiple opiate receptors: support for unique mu, delta and kappa sites. Neuropharmacology 21:487–497

Yeung CW (1987) Photoaffinity labelling of opioid receptor of rat brain membranes with ^{125}I(D-Ala^2, p-N3-Phe^4-Met^5) enkephalin. Arch Biochem Biophys 254:81–91

Zukin RS, Kream R (1979) Chemical crosslinking of a solubilized enkephalin macromolecular complex. Proc Natl Acad Sci USA 76:1593–1597

CHAPTER 6

Interactions Between Opioid Receptors and Guanine Nucleotide Regulatory Proteins in Intact Membranes: Studies in a Neurotumor Cell Line

T. Costa, S. Ott, L. Vachon, and F.-J. Klinz

1 The Family of G-Protein-Coupled Receptors

Opioid receptors belong to a large family of receptors (Rodbell 1980; Lefkowitz and Caron 1988) which share the ability to modify the function of membrane enzymes or ion channels by interacting with one or more guanine nucleotide regulatory proteins (G-proteins). G-proteins are ubiquitous GTPases of the cell plasma membrane and act as transducers of the signal originated upon binding of an agonist to one of this group of receptors (for recent reviews: Gilman 1987; Birnbaumer et al. 1988; Neer and Clapham 1988; Pfeuffer and Helmreich 1988). An activated G-protein exhibits an increased rate constant of dissociation for GDP and, thus, an enhanced turnover number for GTPase activity (Gilman 1987). In their active state G-proteins can interact with enzymes and ion channels in either a stimulatory or inhibitory fashion, and thus transmit the signal originating from the receptor on to several effector systems. Although the general features of signal transduction pathways involving G-proteins were first elucidated for the regulation of adenyl cyclase and cyclic GMP phosphodiesterase, there is now compelling evidence that ion channels such as certain types of K^+ channels (Pfaffinger et al. 1985; Yatani et al. 1987, 1988; Dunnet et al. 1989; Mattera et al. 1989a), and voltage-dependent Ca^{2+} channels (Hescheler et al. 1987, 1988; Brown and Birnbaumer 1989; Yatani and Brown 1989) are also effector systems under control of G-proteins. Another system in which receptor-operated signaling may be mediated by intervening G-proteins is the regulation of phospholipid catabolism by enzymes such as the phosphatidylinositol-specific phospholipase C (Ui 1986) and phospholipase A_2 (Bokoch and Gilman 1984; Okajima and Ui 1984; Burch et al. 1986).

Because a single type of receptor (R) can interact with more than one type of G-protein (G) (Murayama and Ui 1985; Haga et al. 1989; Mattera et al. 1989a) and the same type of G-protein can interact with more than one effector (E) system (Yatani et al. 1988; Mattera et al. 1989b), a complex algebra of negative and positive interactions between different types of R, G, and E may exist in any given cell.

However, regardless of the type of effector system involved and the type of signal, negative or positive, that is actually transduced, the activation of a G-protein is the first measurable biochemical event that follows receptor recognition by an agonist. This first step may be regarded as the biochemical equivalent of the pharmacological concept of stimulus generated on the receptor by the agonist, on which much of the theoretical framework of molecular pharmacology is based. Thus, the elucidation of the mechanism underlying receptor G-protein interaction is of importance for the description, in molecular terms, of fundamental concepts such as the nature of efficacy and partial agonism, and goes beyond the mathematical and pragmatic approaches used to date.

2 Mechanism of Receptor-G-protein Interaction

Since receptor and G-protein, upon binding to each other, can mutually modify the physico-chemical properties of the binding sites for their respective ligands (i.e., agonist and guanine nucleotide), the direct interaction between G-protein and receptor can be studied either as receptor-mediated acceleration of nucleotide turnover on the guanine nucleotide binding site of the G-protein, or as G-protein-mediated enhancement of agonist dissociation from the recognition site of the receptor. For several reasons, both these events are best studied in a reconstituted system, in which receptor and G-protein, highly purified, are coinserted into liposomes.

Using this strategy, the interaction between β-adrenergic receptors and G_s (the G-protein mediating stimulation of adenyl cyclase) was extensively studied (Pedersen and Ross 1982; Brandt et al. 1983; Cerione et al. 1984; Asano et al. 1984; Hekman et al. 1984; May et al. 1985; Lefkowitz et al. 1985; Gilman 1987). Similar studies have been also extended to receptors that interact with G_i and G_o, such as muscarinic receptors (Haga et al. 1985, 1986, 1989; Florio and Sternweis 1985), D_2-dopamine receptors (Senogles et al. 1987), α_2-adrenergic receptors (Cerione et al. 1986), and opioid receptors (Ueda et al. 1988; Fujioka et al. 1988).

There is disagreement on the details of the mechanism of G-protein subunit dissociation and exchange, and on the relative contribution of G-protein subunits in effector signaling (Birnbaumer 1987; Gilman 1987; Levitzki 1987a). Instead, there is agreement on the fact that the two reactants, receptor and G-protein, exist in the membrane as dissociated units, the interation between which is favored or hindered by their ligands, agonist, and GTP, respectively. The kinetics of this interaction follows a "collision-coupling" mechanism (Hekman et al. 1984; Levitzki 1987a,b) that allows R to activate catalytically several molecules of G (Fig. 1). According to this model, G is not only a transducer but also a preamplifier of the weak signal generated by agonist binding to R (Gilman 1987), and the apparent affinity of the agonist for activation of G is higher than the true affinity of the agonist for the receptor binding site (Asano et al. 1984; Cerione et al. 1984; Hekman et al. 1984).

Such a model of R-G interaction has several implications for the theoretical pharmacologist. One is that the phenomenon of "spare receptors" or "receptor reserve" (i.e., the finding that maximal biological responses are observed with small degrees of receptor occupancy) does not result from a stoichiometric excess of R over G (a concept that, in fact, conflicts with the reality of the membrane, where G-proteins are in large excess over receptors); instead, it results from a dynamic relationship. A second implication is that the first amplification step of the hormonal signal operates at the very beginning of the transduction pathway (Gilman 1987).

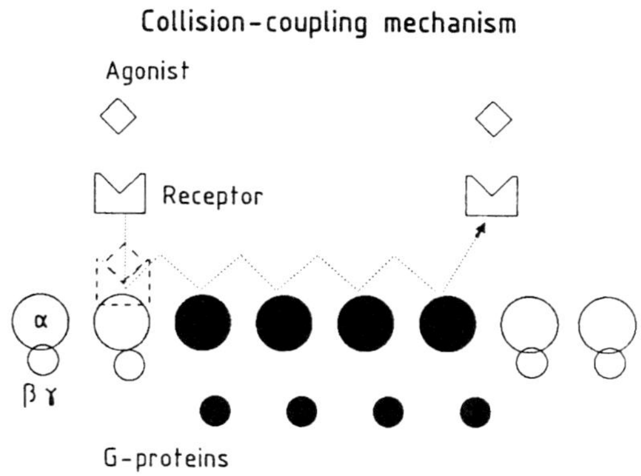

Fig. 1. Schematic view of the catalytic activation of G-proteins by a receptor. According to the collision-coupling concept (Tolkovsky and Levitzki 1978), the complex HRG is transient under physiological concentrations of GTP. Thus, a single HR complex can activate several molecules of G. Activation of G is portrayed as dissociation into α- and $\beta\gamma$-subunits

Such a mechanism of activation is well established for rhodopsin-mediated activation of transducin (Stryer 1986). For hormone receptors, kinetic evidence for a "collision-coupling" mechanism of R-mediated activation of G has been mainly derived from studies with reconstituted systems (Hekman et al. 1984) or intact membranes of erythrocytes (Tolkovsky and Levitzki 1978), and mainly for the β-adrenergic receptor, a G_s-coupled receptor. It is still an open question as to whether the basic features of this mechanism can be extrapolated to all the families of G-protein-interacting receptors including those interacting with the G_i/G_o group of G-proteins, and as to whether data obtained from a reconstituted system in lipidic vesicles are representative of the intact cell membrane.

In this chapter, we will briefly summarize our attempts to address these questions using δ-opioid receptors present in the intact membranes of the neurotumor cell line NG108-15 (Hamprecht 1977). In this cell line opioid receptors regulate at least two different effector systems, adenyl cyclase (Sharma et al. 1979) and voltage-dependent Ca^{2+} channels (Hescheler et al. 1987), both of which are coupled to receptor activation in an inhibitory fashion via pertussis toxin-sensitive G-proteins (Costa et al. 1983; Kurose et al. 1983; Hescheler et al. 1987). Evidence for a large receptor reserve in opioid receptor-mediated cyclic AMP response in NG108-15 cells has been documented, either by the use of an irreversible antagonist (Fantozzi et al. 1981), or by the analysis of occupancy-activity relationships in intact cells (Law et al. 1983; Costa et al. 1985).

However, since in all those studies receptor occupancy was compared with agonist-mediated inhibition of cAMP accumulation in intact cells, rather than with receptor-dependent responses measured in the membranes, there is little information on which step of signal transmission is responsible for the amplification mechanism in opioid receptor-mediated signaling.

3 Receptor-G-Protein Interactions in Intact Membranes

As mentioned above, several serious methodological problems afflict those wishing to study R- and G-interaction in an intact membrane. Receptor-mediated stimulation of high-affinity GTPase activity can be detected in membranes, as pioneered by Cassel and Selinger (1976) in erythrocyte membranes, but the signal-to-noise ratio of this response does not allow for kinetic analysis. Even if NG108-15 cells provide a membrane preparation whose GTPase response to opioid is unusually large, as discovered by Klee and co-workers (Koski and Klee 1981), the limited and subsaturating range at which GTP concentrations can be varied, the requirement for a complex incubation mixture, the lack of knowledge of the preexisting stoichiometry, and the need to subtract contaminating nonspecific nucleotidase activity, are all factors that nullify any effort to apply standard kinetic analysis to this agonist-dependent reaction. Likewise, several problems limit the interpretation of the study of GTP-mediated regulation of receptor affinity. One of them is the instability of guanine nucleotide analogs, even those that are commonly referred to as "stable". We reported recently, for example, that both GTPγS and GppNHp, two guanine nucleotides that cannot be hydrolyzed by the intrinsic GTPase and G-proteins, are quickly modified in membrane preparations by both hydrolytic and transphosphorylating reactions (Ott and Costa 1989). A second problem is the complexity of the model required to correctly analyze G-protein-mediated allosteric effects on the receptor. What is generally referred to as "GTP regulation of receptor affinity" indicates the fact that guanine nucleotide occupation of the α-subunit of the G-protein reduces the stability constant of the high-affinity ternary complex HRG (where H is the hormone agonist) and induces the formation of low-affinity HR complexes (DeLean et al. 1980; Cerione et al. 1984). Mathematical modeling according to this "ternary complex model" of agonist binding data has been successfully applied to the β-adrenergic receptor in erythrocytes (DeLean et al. 1980), and has the potential to yield such important parameters as the fraction of coupled receptors and the stability constant for R-G interaction in native membranes. However, whereas for β-adrenoceptors in red cells, the addition of guanine nucleotides leads to a complete interconversion of receptors to the low affinity form (DeLean et al. 1980; Cerione et al. 1984), for opioid receptors in NG108-15 cells, and most likely for many other receptors interacting with pertussis toxin-sensitive G-proteins, the effect of guanine nucleotides is incomplete and mostly additive to that of Na^+ ions (Blume 1978). By studying the effect of Na^+ and nucleotides on the kinetics of [^{3}H]diprenorphine binding in NG108-15 cell membranes, we found that at least three forms of receptor can be distinguished on the basis of their rate constants of association and dissociation (Ott et al. 1986). Three affinity states of receptors for opioids (Law et al. 1985a) and muscarinics (Birdsall et al. 1980) have been also reported on the basis of equilibrium binding studies. These findings are not compatible with a ternary complex model in its simple form. The limits and the need for extension of the ternary complex model when applied to a G_i/G_o-coupled receptor have been emphasized by Wregget and DeLean (1984) in their study of the D_2-dopamine receptor.

3.1 An Indirect Approach

An alternative approach to probe R-G interactions in native membranes is indirect. First, we alter the normal stoichiometry of receptor and G-proteins in intact cells.

Next, membranes are prepared and the effect of this alteration on coupling is studied either as receptor-mediated activation of GTPase, or as G-protein-mediated interconversion of receptors into low-affinity forms.

The pattern of loss of these responses is diagnostic of the type of coupling between R and G in the intact membranes. For instance, if the receptor acts as a catalyst, by activating a large pool of G-proteins, as the collision coupling mechanism of activation would predict, then gradual reductions in receptor density would diminish the efficiency of the system before any reduction in the total degree of activation is detectable. That is, a reduction in receptor number will first lead to reduction of the apparent affinity of the agonist for activation of GTPase, and only later will it diminish the maximal responsiveness, a situation which, in pharmacological terms, is described as receptor reserve (Fig. 2).

3.2 Electron Inactivation Studies

A first method to alter the normal stoichiometry of R and G in an intact membrane is radiation inactivation. The method is based on the loss of biochemical activity in molecules exposed to high energy electrons. The theory and the methodological constraints of the technique have been reviewed (Kepner and Macey 1968; Kempner and Schlegel 1979; Kempner 1988). Table 1 summarizes briefly some theoretical principles. How target size analysis can be applied to the study of R-G interactions is illustrated schematically in Fig. 3. High-affinity, GTP-sensitive

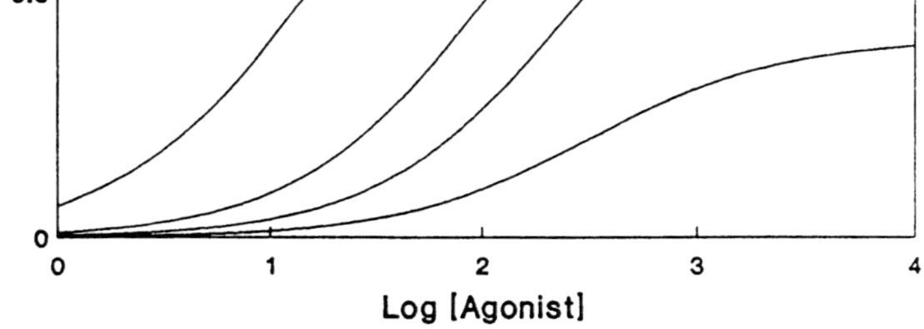

Fig. 2. Receptor reserve in receptor-mediated activation of G-protein. Theoretical concentration-response curves following a progressive decrease of receptor density (from *right to left*) for a system exhibiting receptor reserve. It is assumed that there is nonlinear coupling between binding and activation. The more efficient the receptor is in activating a correspondingly larger population of G, the wider is the discrepancy between binding and activation, and the larger is the shift of IC_{50} for the agonist produced by reductions in receptor density

Table 1. Principles of target size analysis[a]

1. A primary ionization occurring anywhere within the molecule of an enzyme results in its complete inactivation.
2. The probability for an enzyme to survive inactivation depends on the dose of irradiation and on the molecular mass of the enzyme.
3. The fraction of molecules of an enzyme that survives inactivation as the dose of irradiation increases is an exponential function, the slope of which is related to the molecular mass of the enzyme ("*target size*").
4. For multisubunit enzymes, there is not sufficient energy transfer to cause spreading of the radiation damage beyond the subunit within which the primary ionization occurs. Therefore, the target size of an enzyme defines its "*functional unit*", i.e., the minimal assembly of subunit that is required for its function.

[a] References: Kepner and Macey (1968); Kempner and Schlegel (1979); Kempner (1988).

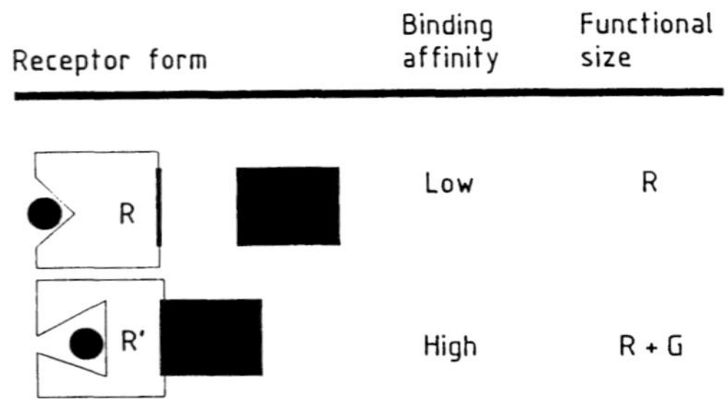

Fig. 3. Application of radiation inactivation analysis to the study of receptor affinity forms in membranes. The low-affinity binding site represents the HR complex, the high-affinity form is the HRG complex. Thus, the minimal assembly of subunits required for high-affinity binding (receptor + α-subunit of G-protein) is larger than that necessary for low-affinity binding

receptor binding requires the formation of the R-G complex, whereas low-affinity binding is exhibited by receptor units dissociated from the G-protein. Therefore, the minimal assembly of subunits necessary for high-affinity binding is larger and should yield a target size greater than that measured for low-affinity binding (Fig. 3). However, three sets of our results were in sharp contrast with this prediction (Ott et al. 1988). First, addition of guanine nucleotide to the binding reaction did not alter the target size of opioid receptors. Second, the irreversible inactivation of G, either prior to irradiation, usig pertussis toxin, or following irradiation using N-ethylmaleimide, left the apparent target size of opioid receptors unchanged. Third, when we measured directly, by dissociation kinetics, the size of the low- and high-affinity states of the receptor induced by GTP we found an identical target size of 98 kDa for both functions, which is considerably larger than the molecular weight of 58–65 kDa determined by affinity labeling (Klee et al. 1982; Newman and Barnard 1984; Howard et al. 1985). Although one possible explanation for this discrepancy is that the target size measured as inactivation of binding activity represents a functional dimer of the receptor (Ott et al. 1988), an alternative interpretation is that opioid receptors and G-protein do not exist as freely dissociable subunits in the membrane, but rather as tight complexes. The binding of agonist to R or guanine nucleotide to G may regulate, allosterically and reciprocally, the affinity of the two components without inducing their dissociation. In this case, primary ionizations of either R or G will invariably inactivate high- or low-affinity binding with identical, apparent target sizes (Ott et al. 1988).

3.3 Agonist-Induced Reduction in Density of Functional Receptors

When intact NG108-15 cells are exposed to an opioid agonist, the responsiveness of G-proteins to opioid receptor stimulation is gradually lost. This loss of responsiveness is a biexponential function of time (Vachon et al. 1987a). A fraction of agonist-stimulated GTPase activity disappears quickly with an apparent half-life of 5-10 min which is in agreement with the time course for the loss of GTP-sensitive high-affinity binding. Another and equivalent proportion of agonist-stimulated activity disappears more slowly with a time constant (50-80 min) similar to that for the agonist-induced disappearance of opioid binding sites (downregulation). A minor portion of surviving activity does not disappear at all. The fact that two mechanisms apparently affect two distinguishable components of receptor-operated (GTPase activity suggest that they could operate on two discrete populations of receptors. However, we found that the two components were experimentally induced. In fact, when desensitization was studied without depriving the cells of the agonist (that is, when cells were harvested, membranes prepared, and GTPase assayed under a constant concentration of agonist identical to that offered to the intact cells), the fast component of desensitization was predominant, and no residual activity could be detected (Vachon et al. 1987a). Thus, we concluded that the multiphasic pattern of desensitization was generated by rapid reversal occurring anywhere during washing of the cells, preparation of the membranes, and the time course of the enzymatic assay. In fact, when we directly studied the rate of reversal of desensitization, we found that part of the agonist-stimulated GTPase reappeared quickly upon withdrawal of the agonist, with an apparent time constant of 5-10 min, while the remaining desensitized activity was restored with a very slow time constant, consistent with that expected for recovery from downregulation. It is thus likely that the rapid decline of responsiveness involves a rapidly reversible mechanism, as for example receptor phosphorylation, which is known to occur for rhodopsin (Kühn et al. 1973), β-adrenergic (Benovic et al. 1987), and muscarinic receptors (Kwatra et al. 1989); instead, the slow process involves receptor downregulation (Chang et al. 1982). In view of the temporal relationship between these two events, it can be extrapolated that, in the intact cell exposed to a constant concentration of agonist, the primary event that quenches responsiveness is the rapidly reversible one, followed by the more durable reduction of receptor units. Does this reduction of opioid responsiveness involve an alteration of G-proteins? Most likely not. It was shown by Law et al. (1985a) that opioid receptors exposed to an agonist internalize without G-proteins. Using site-specific antibodies raised against subunits of G-protein (Mumby et al. 1986; Lang and Costa 1987), we also found that in cells exposed to opioid agonist there was no change in the measurable levels of G-proteins (Lang and Costa 1989). Thus, agonist-induced reduction of opioid receptor responsiveness in NG108-15 cells primarily affects the receptor and represents a situation in which functional receptor units in the membranes are decreased with respect to the population of G-proteins, When we examined the pattern of this loss of responsiveness to opioid receptors for GTPase, we found a diminished receptor-induced maximal stimulation with no change in the apparent affinity for the agonist (Vachon et al. 1987a), regardless of the duration of exposure to the agonist

(Vachon et al. 1987b). Thus, once again, there was no evidence for an amplification mechanism between receptor and G-protein in these intact membranes. In contast, receptor-mediated inhibition of adenyl cyclase disappeared first as a loss of apparent affinity, and then as a reduction of maximal inhibition (Vachon et al. 1987a), indicating that amplification may rather occur at the step of G-protein-mediated inhibition of this enzyme.

3.4 Reduction in Density of Functional Receptors Induced by an Alkylating Antagonist

Another means to induce a reduction in available R without affecting the population of G is offered by an irreversible receptor ligand. Three requirements are necessary for such a ligand. First, it should be able to bind to the receptor with reasonably high affinity. Second, it should be devoid of intrinsic activity, i.e., upon binding it should not produce any significant activation of G. Third, it should bear a reactive group able to establish a covalent interaction with nucleophilic groups present in, or very close to, the binding site. β-Chlornaltrexamine (CNA), an analog of naltrexone (Portoghese et al. 1979), appears to meet all these requirements. The advantage of such an approach is that receptor sites can be inactivated in intact cells simply by exposing them to the ligand in amine-free buffers at low temperature, and then the consequence of this manipulation can be examined in membranes. Using these conditions, the disappearance of binding activity following CNA treatment was solely accounted for by a reduction in the apparent number of binding sites, which is indicative of irreversible antagonism.

We found a very close relationship between the diminution of the number of sites and the decrease of maximal GTPase stimulation following exposure of cells to increasing concentrations of CNA; in contrast, the decrease in the maximal inhibitory effect on adenyl cyclase required much higher concentrations of CNA (Fig. 4). Examination of concentration-response curve indicated that the change in responsiveness to the agonist for GTPase was only due to a decrease in the maximal stimulation, while for adenyl cyclase, there was an initial loss of sensitivity (i.e., increase of apparent IC_{50}) which was followed by a reduction in the maximal inhibitory effect (Costa et al. 1988). Thus, once again, the consequence of a decrease in receptor density is not consistent with the existence of an amplification step between receptor and G-protein, and suggests that these two molecules are tightly associated in intact membranes. Rather, the amplification mechanism may operate at the step of inhibition of adenyl cyclase. This would be consistent with the idea that the α-subunits of pertussis toxin-sensitive G-proteins do not directly inhibit the enzyme, but $\beta\gamma$-subunits dissociated upon activation may exert indirect inhibition of adenyl cyclase by interacting with the α-subunit of G_s (Gilman 1987).

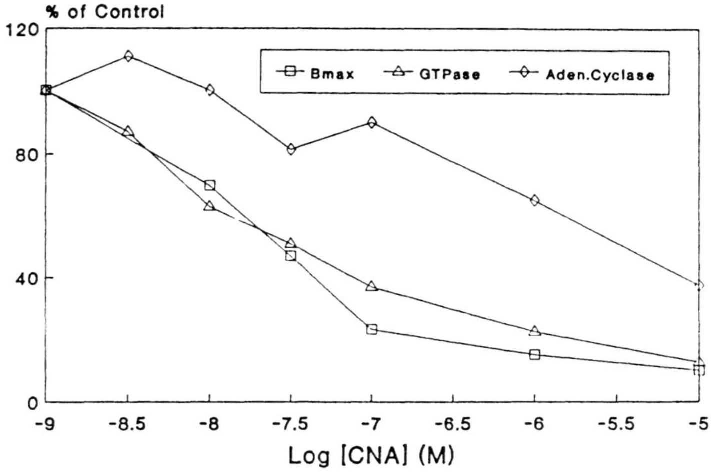

Fig. 4. Relation between reduction of maximal receptor-mediated effects (on GTPase and adenyl cyclase) and diminution of opioid binding sites as a function of the concentration of alkylating antagonist. Maximal stimulation of GTPase (*triangles*) and inhibition of adenyl cyclase (*diamonds*) were measured at a saturating concentration of agonist (10 μM). Then density of binding sites (B_{max}, *squares*) was determined by saturation analysis. Data are expressed as percent of the control value measured in cells treated in the absence of CNA. Experimental details are given in Costa et al. (1988)

3.5 Spontaneous Differences of the Ratio Receptor/G-Proteins Between Subcellular Membranes

It has been known for some years that opioid receptors in NG108-15 cells are present in both 'heavy' and 'light' membrane fractions (Sweat and Klee 1985). If membranes are separated according to a simple subcellular fractionation protocol into heavy (sedimenting at 25 000 × g) and light (> 100 000 × g) particles, the total number of opioid binding sites is approximately equivalent in the two fractions (Ott et al. 1989). However, these two populations of receptors differ in several ways. One difference is in their apparent distribution into affinity states: heavy fraction receptors exist as a virtually single class of high-affinity sites, whereas in light membranes, only 40% of receptors are in high-affinity form. Treatment with pertussis toxin converts most of the receptors into low-affinity form in both fractions (Ott et al. 1989). A second difference between the two membrane fractions is the magnitude of receptor-induced activation of G-proteins. In light membranes, receptor-stimulated high-affinity GTPase is only 20–30% of that detected in heavy membranes (Ott et al. 1989). Unlike the total number of opioid receptor molecules, which is identical in the two fractions, the amount of G-proteins is clearly different: Western blot analysis indicated that 50–70% less α (both α_{i2} and α_o) and β-subunits of pertussis toxin-sensitive G-proteins were present in light membrane fractions (Ott et al. 1989). Thus, the smaller amount of G-proteins present in light membranes seems to be responsible for the apparent reduced coupling of the receptor in this membrane fraction. A comparison of the amount of pertussis toxin substrate, evaluated by quantitative Western blotting (20–30 pmol/mg of protein in heavy membrane fraction) with the total number of receptors (1.2–1.4 pmol/mg of protein) as determined by agonist saturation analysis (Ott et al. 1989), indicates that there is a 15–20-fold excess of G over R in heavy membranes, but only a 4–7-fold excess in light membranes. Thus, even if in both cases G is in stoichiometric excess over R, it appears that a more than ten fold molar excess is required to stabilize the bulk of the receptor population into the high-affinity form. This is in agreement with reconstitution studies using muscarinic receptors (Haga et al. 1986), where it was shown that a 20-fold molar excess of G over R was essential for the maximal induction of the high-affinity GTP-sensitive form of the receptor. Despite the difference in R/G ratios, the concentration-response curves for agonist-induced stimulation of GTPase in light and heavy membrane fractions only differed in maximal effects and displayed identical apparent affinities for the agonist (Fig. 5). Thus, differences in the receptor/G-protein ratio, sufficient to alter the proportion of the high-affinity form of the receptor, only affects the maximal GTPase activity that can be stimulated by the receptor, but not the apparent sensitivity for stimulation. These results are once again in contrast with a collision-coupling mechanism of G-protein activation.

3.6 Antagonists with Negative Intrinsic Activity

The data reviewed so far indicate that receptor and G-protein establish a very tight interaction in the intact membrane, and suggest that they do not behave as freely floating units which interact reversibly according to the law of mass action. The data also suggest that a large proportion of opioid receptors are precoupled to G-proteins in the membranes, although it is not clear whether this precoupled high-affinity

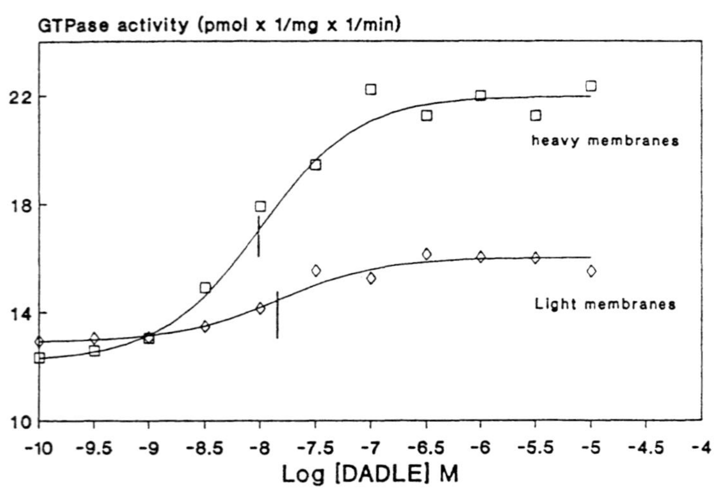

Fig. 5. Concentration-response curves for agonist-mediated stimulation of GTPase activity in light and heavy membranes. Experimental details are in Ott et al. (1989). Note that the IC_{50} for the agonist, [D-Ala^2,D-Leu^5]enkephalin, does not differ between the two membrane fractions

complex has increased GTPase activity, nor whether it represents a truly spontaneously coupled "empty" receptor, or rather an occupied receptor, bound to the agonist in very slow dissociating form.

Additional information comes from studies with competitive antagonists. If part of the basal GTPase activity present in the membrane reflects spontaneous coupling between R and G, it is conceivable that some antagonists may inhibit this basal activity by inducing a conformation of R less able to activate G. We recently found that two types of opioid antagonists can be distinguished on the basis of their effects on basal GTPase activity: those with null intrinsic activity, and those with negative intrinsic activity. The first type of antagonists include compounds like MR 2266 and naltrexone. Upon binding to the receptor, they do not alter the basal level of GTPase, but inhibit in a competitive fashion the stimulation elicited by the agonist. Their binding isotherms are consistent with a homogeneous population of binding sites either in the presence or absence of Na^+ and GTP, and in membranes obtained from cells pretreated with pertussis toxin.

The second type includes peptidergic antagonists like ICI 174864 or ICI 154129. These compounds inhibit basal GTPase in a concentration-dependent fashion. Their effect on GTPase can be antagonized in a stereospecific fashion by a neutral antagonist, indicating that this inhibition is receptor-mediated and not the consequence of a nonspecific, direct interaction with the G-protein.

Does the inhibitory effect of these antagonists depend on their interaction with an empty opioid receptor which spontaneously activates the G-protein or rather, does it result from their ability to "displace" an endogenous agonist bound to the receptor with very high affinity? Three lines of evidence are in favor of the first explanation. First, we found that in membranes prepared from cells that had been maintained for 24 h in the presence of a saturating concentration (100 μM) of the neutral antagonist MR 2266, thereby preventing the binding of any endogenous agonist to the receptor, the inhibitory effect of the negative antagonist was identical to that in control membranes. Second, the degree of negative intrinsic activity of opioid antagonists, determined as the maximal inhibitory effect on basal GTPase, was not correlated with their receptor affinities: such a correlation should instead be expected if the inhibition resulted from competition with an endogenous agonist which stimulated the receptor. Third, the apparent affinities (pA_2 values) of the neutral antagonist MR 2266 for blocking the stimulation of the agonist and the inhibition of the negative antagonist were very similar, which would be impossible if a slowly dissociating form of agonist-bound receptor was involved in the spontaneous activation of GTPase.

Unlike neutral antagonists, negative antagonists display complex binding isotherms consistent with two affinity forms of the receptor; treatment of membranes with pertussis toxin converts these sites into a single high-affinity form, a situation which is opposite to that observed for agonists.

A further indication that negative antagonists interfere with the ability of the receptor to induce an active state of the G-protein comes from studies on receptor-stimulated ADP-ribosylation of G_i/G_o catalyzed by cholera toxin. Although this toxin preferentially modifies G_s, substrates of pertussis toxin can also be modified by cholera toxin under special experimental conditions (Abood et al. 1982; Graves et al. 1983; Owens et al. 1985). As shown more recently (Gierschik and Jakobs 1987;

Milligan and McKenzie 1988), cholera toxin-catalyzed ADP-ribosylation of G_i is potentiated by receptor occupancy, which suggests that the preferential substrate of the toxin is the G-protein complexed to the receptor. If so, the fraction of G_i which is labeled by cholera toxin in the absence of ligand may be the subpopulation of G-proteins spontaneously coupled to receptors, and should be reduced by a negative antagonist. Indeed, we found that cholera toxin-catalyzed basal labeling of the 40-kDa band, corresponding to G_i/G_o, in NG108-15 membranes was markedly reduced by ICI 174864. Naloxone, did not affect the labeling but suppressed the inhibitory effect of the peptide, while its inactive enantiomer, (+)naloxone, did not antagonize the inhibition (Klinz and Costa 1989).

Collectively, these data indicate that unoccupied receptors are not inactive, but can elicit activation of G-protein, and that at least three distinct forms of receptor can exist in the membrane: an activated form, which is that stabilized by a full agonist; a neutral form, which is that of unoccupied receptors; and an inactivated form, which is that stabilized by a negative antagonist.

3.7 The Role of Na^+ Ions

The fact that Na^+ alters the binding properties of opioid receptors was an early discovery in the history of opioid receptor research (Pert and Snyder 1973; Simon and Groth 1975). The finding that Na^+ differentially affected agonist and antagonist binding (Pert and Snyder 1974) gave to this phenomenon the dignity of functional relevance, despite a lack of knowledge on the underlying molecular mechanism(s). Later, the findings that many G-protein-coupled receptors share agonist-specific Na^+ effects (Michel et al. 1980; Rosenberger et al. 1980; Limbird et al. 1982; Minuth and Jakobs 1986), and that at these receptors Na^+ regulates both agonist-induced activation of GTPase (Koski et al. 1982) and agonist-mediated inhibition of adenyl cyclase (Aktories et al. 1979; Blume 1978; Jakobs et al. 1984), lent generality to this phenomenon and suggested that the R-G complex is the molecular target of Na^+ ions.

The role of Na^+ in the regulation of opioid receptor-mediated stimulation of GTPase has been explained as an ability of this ion to selectively suppress receptor-independent basal GTPase activity, leaving receptor-dependent activity unchanged (Koski et al. 1982). Taking advantage of the negative antagonist ICI 174864, which allows one to measure empty receptor-mediated activity (i.e., activity in the absence of ligand which can be inhibited by the negative antagonist), we have recently reexamined the effect of Na^+. Na^+ regulation of GTPase was: (1) absent in G-proteins purified from bovine brain, a clear indication that the true basal activity of G is not affected by the ion; (2) abolished in membranes by pertussis toxin at concentrations identical to those necessary to suppress agonist-mediated stimulation and antagonist-mediated inhibition; (3) dependent on concentrations of free Mg^{2+}, similar to those required for receptor-dependent regulation of the cyclase; (4) reconstituted in membranes pretreated with pertussis toxin by purified G-proteins at concentrations identical to those necessary to reconstitute receptor-dependent activity (Costa et al. 1989). Collectively, these data indicate that Na^+ ions selectively inhibit the GTPase activity stimulated by empty receptors and do not affect the activity elicited by ligand-bound receptor.

Where and how does Na^+ act? The target of Na^+ is probably not a domain of the G-protein α-subunit. In fact, inactivation of G-protein-mediated regulation of receptor affinity by either N-ethylmaleimide (Limbird and Speck 1983) or by pertussis toxin (Wüster et al. 1984) does not suppress Na^+ regulation. The fact that pertussis toxin abolishes Na^+-mediated regulation of basal GTPase but not Na^+ effects on receptor binding is a strong indication that Na^+ can alter the conformation of receptor molecules bypassing the interface R-G, and that this alteration results in the loss of the ability of receptors to establish the agonist-independent, spontaneous interaction with G-protein. Na^+ might interact with the receptor itself, or with another, still unknown, receptor-associated component. In favor of the first possibility is the observation that affinity purified α_2-adrenergic receptors maintain sensitivity to Na^+ (Repaske et al. 1987). Others have shown, however, that even after a high degree of purification, α_2-adrenergic receptors are still associated to G-proteins (Regan et al. 1986), suggesting that a Na^+-sensitive molecule could also copurify with receptors during affinity chromatography. In favor of the presence of a "Na^+ unit" distinct from the receptor are studies on target-size analysis of opioid receptor binding showing that an allosteric inhibitor that mediates the effect of Na^+ on agonist binding exists in the membrane and has an apparent molecular size larger than the receptor itself (Ott et al. 1988). Indeed, electron inactivation appears to be the only available means to date to abolish Na^+-mediated regulation of receptor affinity in the membrane without inactivating receptor binding itself.

When we examined the effect of Na^+ on the GTPase measured in the presence of several agonists and antagonists (Costa et al. 1990), we found that the more the intrinsic activity of the ligand differs from zero, either in a positive or negative direction, the less the GTPase elicited in the presence of that ligand is sensitive to Na^+ (Fig. 6). If the effect of Na^+ resulted from the ability of the cation to inhibit receptor-independent activity more than receptor-dependent activity (Koski et al. 1982), then GTPase should be minimally sensitive to Na^+ in the presence of the agonist, and maximally sensitive in the presence of the negative antagonist. Instead, we found that Na^+ sensitivity is maximal in either the absence of ligand or the presence of a partial agonist or antagonist, but minimal in the presence of either an agonist or negative antagonist (Fig. 6) (Costa et al. 1990). Thus, these data may be a further indication that the mechanism of receptor-mediated activation of G-proteins in membranes involves performed R-G complexes that are activated by agonists and inactivated by antagonists rather than "freely floating" dissociated components whose ability to interact with each other is enhanced by agonists and disrupted by negative antagonists. Where and how Na^+ acts, however, are still open questions.

4 Epilogue

In this chapter we have reviewed our efforts to gain information on the interaction between opioid receptors and G-proteins as it occurs in intact membranes of

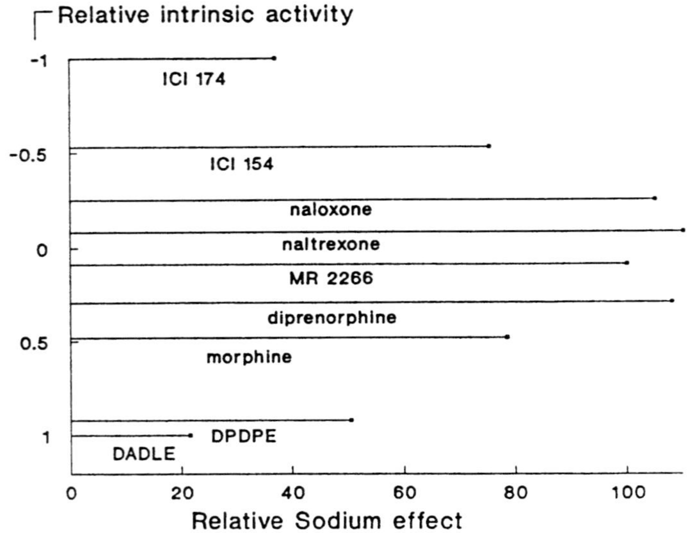

Fig. 6. Relation between relative intrinsic activity and sodium effect on GTPase. Relative intrinsic activity is the ratio between maximal effect on GTPase of each ligand and the maximal effect induced by [D-Ala^2,D-Leu^5]enkephalin (DADLE), for stimulation (positive values), and by ICI 174864 (ICI 174), for inhibition (negative values), respectively. Relative sodium effect is the difference in GTPase activity in the presence of each ligand upon replacement of KCl by NaCl (both 150 mM) in the reaction. It is expressed as percent of the sodium effect observed in the absence of any ligand (basal activity)

NG108-15 cells. As often happens, these studies have raised new questions rather than provided definite answers.

First, we find that the mechanism of opioid receptor-mediated stimulation of the high-affinity GTPase of G-protein in intact membranes is not consistent with the idea that a single receptor molecule can activate several molecules of G-proteins within the life span of one activation cycle. This suggests that the active complex between occupied receptor and G-protein might be much less transient than expected from studies on β-adrenergic receptor and G_s in reconstituted systems. It is not clear whether this discrepancy reflects a difference in the experimental system (intact membranes vs liposomes), or a difference in the mechanism of activation between different types of G-proteins. In the first case, it is possible that other constituents of the membrane constrain and regulate R-G interactions. In the second case, it is conceivable that the stability constant for the HRG complex, and thus its lifetime, might represent a fundamental difference between stimulatory (i.e., G_s or transducin-coupled) and inhibitory (i.e., G_i- or G_o-coupled) receptors, and, in turn, might determine the magnitude of amplification necessary in each of these transduction systems. The data summarized here suggest that opioid receptor and G-protein can exist as tightly performed complexes in the membrane and that the step of interaction between G-protein and effector system, rather than between receptor and G-protein, might be that responsible for the receptor reserve in opioid-mediated responses. One pharmacological implication of this tight interaction is that the high-affinity binding site measured in a radioligand binding assay represents a receptor-G-protein complex. High-affinity binding is used to characterize and identify multiple types of opioid receptors and the degree of selectivity by which different opioid agonists can discriminate among them. It is likely, however, that these studies primarily characterize differences due to multiple receptor-G-protein complexes, rather than only to binding sites of different receptor subtypes.

A second question stems from the finding that a large proportion of opioid receptors in these membranes appear to be active in the absence of agonist. This observation explains why pertussis toxin-catalyzed ADP-ribosylation does not alter the basal rate of GTP hydrolysis of purified G-proteins, but instead results in a decrease in basal GTPase activity of G-proteins in membranes (Costa et al. 1988). The physiological meaning of this spontaneous receptor activity, however, remains unclear. In principle, the most important regulatory feature of an efficient information transfer system is suppression of background noise. For receptor-operated transmembrane signaling systems, noise is receptor activity in the absence of agonist. Birnbaumer (1987) has proposed that the main physiological role of the $\beta\gamma$-subunit of G-proteins is that of a noise attenuator, since it suppresses the interaction of unoccupied receptors with G-proteins, without affecting that of agonist-occupied receptors. Thus, the stoichiometric relation between $\beta\gamma$-subunits and the number of receptors in the membrane might be a critical factor in determining the degree of G-protein activation by empty receptors in a given membrane. One approach to directly test this hypothesis would be to determine the magnitude of antagonist effect in membranes preequilibrated with increasing concentrations of $\beta\gamma$-subunits.

A third question, strictly related to the point just discussed above, is the effect of Na^+, and its physiological meaning. We found that Na^+ ions selectively suppress spontaneous agonist-independent receptor activity, but not the activity induced by occupied receptors. Does this mean that Na^+, either acting on the receptor domain or on another membrane component, is the physiological suppressor of receptor noise? Intracellular Na^+ concentrations are very low under resting conditions; therefore, to be sound, this hypothesis would require an extracellular site for the Na^+ effect. However, there is compelling evidence from intact platelets (Conolly and Limbird 1983; Motulsky and Insel 1983) indicating that Na^+ acts intracellularly. We have shown (Costa et al. 1990) that the suppression of agonist-independent receptor activity occurs at $[Na^+]/[K^+]$ ratios that are well within the physiological intracellular range. It might be speculated that, in vivo under resting conditions, inhibitory receptors (i.e., opioid and similar receptors, the activation of which results in the reduction of cellular excitability) are meant to exert a certain degree of spontaneous activity which in turn results in tonic inhibition. Depolarization and the consequent increase in intracellular Na^+ concentration would suppress this spontaneous activity and optimize the dynamic range of receptor responsiveness to inhibitory extracellular signals. Alternatively, the Na^+ effect could be only an artifact of biochemical studies in vitro, however, still an important one, because it directs our attention to the existence of additional domains involved in the regulation of receptor-G-protein interactions, and unveils the fact that the receptor undergoes a set of conformational changes during activation and inactivation. Clearly, reconstitution studies with purified receptors and G-proteins will be necessary to further investigate the site and the mechanism of action of Na^+.

Acknowledgements. We are most grateful to Prof. Albert Herz for encouragement, support, and constructive criticism over many years of research spent together. Christine Gless contributed impeccable technical assistance for much of our experimental work reviewed here. Ursula Bäuerle provided precious help with cell culture. The work was funded by the Deutsche Forschungsgemeinschaft, Bonn.

References

Abood ME, Hurley JB, Pappone MC, Bourne HR, Stryer L (1982) Functional homology between signal-coupling proteins. Cholera toxin inactivates the GTPase activity of transducin. J Biol Chem 257:10540–10543

Aktories K, Schultz G, Jakobs KH (1979) Inhibition of hamster fat cell adenylate cyclase by prostaglandin E_1 and epinephrine: requirement for GTP and soium ions. FEBS Lett 107:100–104

Asano T, Pedersen SE, Scott CW, Ross EM (1984) Reconstitution of catecholamine-stimulated binding of guanosine 5′-O-(3-thiotriphosphate) to the stimulatory GTP-binding protein of adenylate cyclase. Biochemistry 23:5460–5467

Benovic JL, Strasser RH, Caron MG, Lefkowitz RJ (1987) β-Adrenergic receptor kinase. Identification of a novel protein kinase that phosphorylates the agonist-occupied form of the receptor. Proc Natl Acad Sci USA 83:2797

Birdsall NJM, Hulme EC, Burgen ASV (1980) The character of muscarinic receptors in different regions of the rat brain. Proc R Soc London Ser B 207:1–12

Birnbaumer L (1987) Which G protein subunits are the active mediators in signal transduction? Trends Pharmacol Sci 8:209–211

Birnbaumer L, Codina J, Mattera R, Yatani A, Brown AM (1988) G proteins and transmembrane signalling. In: Cooke BA, King RJB, van der Molen HJ (eds) Hormones and their actions, pt 2. Elsevier, Amsterdam, pp 1–46

Blume AJ (1978) Interactions of ligands with opiate receptors of brain membranes: regulation by ions and nucleotides. Proc Natl Acad Sci USA 75:1713–1717

Bokoch GM, Gilman AG (1984) Inhibition of receptor-mediated release of arachidonic acid by pertussis toxin. Cell 39:301–308

Brandt DR, Asano T, Pedersen SE, Ross EM (1983) Reconstitution of catecholamine-stimulated guanosinetriphosphatase activity. Biochemistry 22:4357–4362

Brown AM, Birnbaumer L (1989) Direct G protein gating of ion channels. Am J Physiol 254:H401–410

Burch RM, Luini A, Axelrod J (1986) Phospholipase A_2 and phospholipase C are activated by distinct GTP-binding proteins in response to α_1-adrenergic stimulation in FRTL-5 cells. Proc Natl Acad Sci USA 83:7201–7205

Cassel D, Selinger Z (1976) Catecholamine-stimulated GTPase activity in turkey erythrocyte membranes. Biochim Biophys Acta 452:538–551

Cerione RA, Codina J, Benovuc JL, Lefkowitz RJ, Birnbaumer L, Caron MG (1984) The mammalian β-adrenergic receptor: reconstitution of functional interactions between pure receptor and pure stimulatory nucleotide binding protein of the adenylate cyclase system. Biochemistry 23:4519–4525

Cerione RA, Regan JW, Nakata H, Codina J, Benovic JL, Gierschik P, Somers RL, Spiegel AM, Birnbaumer L, Lefkowitz RJ, Caron MG (1986) Functional reconstitution of the α_2-adrenergic receptor with guanine nucleotide regulatory proteins in phospholipid vesicles. J Biol Chem 261:3901–3909

Chang K-J, Eckel RW, Blanchard SG (1982) Opioid peptides induce reduction of enkephalin receptors in cultured neuroblastoma cells. Nature (London) 296:446–448

Conolly TM, Limbird LE (1983) The influence of sodium on the α_2-adrenergic receptor system of human platelets. A method for removal of extraplatelet Na^+. Effect of sodium removal on aggregation: secretion and cAMP accumulation. J Biol Chem 258:3907–3912

Costa T, Herz A (1989) Antagonists with negative intrinsic activity at δ opioid receptors coupled to GTP-binding proteins. Proc Natl Acad Sci USA 86:7321–7325

Costa T, Aktories K, Schultz G, Wüster M (1983) Pertussis toxin decreases opioid receptor binding and adenylate cyclase inhibition in a neuroblastoma × glioma hybrid cell line. Life Sci (Suppl 1) 33:132–136

Costa T, Wüster M, Gramsch C, Herz A (1985) Multiple states of opioid receptors may modulate adenylate cyclase in intact neuroblastoma hybrid cells. Mol Pharmacol 28:146–154

Costa T, Klinz FJ, Vachon L, Herz A (1988) Opioid receptors are coupled tightly to G proteins but loosely to adenylate cyclase in NG108-15 cell membranes. Mol Pharmacol 34:744-754
Costa T, Lang J, Gless Ch, Herz A (1990) Spontaneous association between opioid receptors and G proteins in native membranes. Specific regulation by antagonists and sodium ions. Mol Pharmacol (in press)
DeLean A, Stadel JM, Lefkowitz JR (1980) A ternary complex model explains the agonist-specific binding proteins of the adenylate cyclase coupled β-adrenergic receptor. J Biol Chem 255:7108-7117
Dunnett MJ, Bullet MJ, Li GD, Wollheim CB, Petersen OH (1989) Galanin activates nucleotide-dependent K^+ channels in insulin-secreting cells via a pertussis toxin-sensitive G protein. EMBO J 8:413-420
Fantozzi R, Mullkin-Kirkpatrick D, Blume AJ (1981) Irreversible inactivation of opiate receptors in the neuroblastoma × glioma hybrid NG108-15 by chlornaltrexamine. Mol Pharmacol 20:8-15
Florio VA, Sternweis PC (1985) Reconstitution of resolved muscarinic receptors with purified GTP-binding proteins. J Biol Chem 260:3477-3483
Fujioka T, Inoue F, Sumita S, Kuriyama M (1988) Reconstitution of partially purified opioid receptors with a GTP-binding protein. Biochem Biophys Res Commun 156:54-60
Gierschik P, Jakobs KH (1987) Receptor-mediated ADP-ribosylation of a phospholipase C-stimulating G protein. FEBS Lett 244:219-223
Gilman AG (1987) G proteins: transducers of receptor-generated signals. Annu Rev Biochem 56:615-649
Graves CB, Klaven NB, McDonáld JM (1983) Effects of guanine nucleotides on cholera toxin-catalyzed ADP-ribosylation in rat adipocyte plasma membranes. Biochemistry 22:6291-6296
Haga K, Haga T, Ichiyama A, Katada T, Kurose H, Ui M (1985) Functional reconstitution of purified muscarinic receptors and inhibitory guanine nucleotide regulatory proteins. Nature (London) 316:731-733
Haga K, Haga T, Ichiyama A (1986) Reconstitution of the muscarinic acetylcholine receptor. J Biol Chem 261:10133-10140
Haga K, Uchiyama H, Haga T, Ichiyama A, Kangawa K, Matsuo H (1989) Cerebral muscarinic acetylcholine receptors interact with three kinds of GTP-binding proteins in a reconstituted system of purified components. Mol Pharmacol 35:286-294
Hamprecht B (1977) Structural, electrophysiological, biochemical and pharmacological properties of neuroblastoma × glioma cell hybrids in cell culture. Int Rev Cytol 49:99-170
Hekman M, Feder D, Keenan AK, Gal A, Klein HW, Pfeüffer T, Levitzki A, Helmreich EJ (1984) Reconstitution of β-adrenergic receptor with components of adenylate cyclase. EMBO J 3:3339-3345
Hescheler J, Rosenthal W, Trautwein W, Schultz G (1987) The GTP-binding protein G_o regulates neuronal calcium channels. Nature (London) 325:445-447
Hescheler J, Rosenthal W, Hinsch K-D, Wulfern M, Trautwein W, Schultz G (1988) Angiotensin II-induced stimulation of voltage-dependent Ca^{2+} currents in an adrenal cortical cell line. EMBO J 3:619-624
Howard AD, de La Baume S, Gioannini TL, Hiller JM, Simon EJ (1985) Covalent labeling of opioid receptors with radioiodinated human β-endorphin. J Biol Chem 260:1083-1089
Jakobs KH, Minuth M, Aktories K (1984) Sodium regulation of adenylate cyclase. J Receptor Res 4:443-458
Kempner ES (1988) Molecular size determination of enzymes by radiation inactivation. Adv Enzymol 61:107-147
Kempner ES, Schlegel W (1979) Size determination of enzymes by radiation inactivation. Anal Biochem 92:2-10
Kepner GR, Macey RI (1968) Membrane enzyme systems. Molecular size determination by radiation inactivation. Biochim Biophys Acta 163:188-203
Klee WA, Simonds WF, Sweat FW, Burke TR, Jacobson AE, Rice KC (1982) Identification of a Mr 58000 glycoprotein subunit of the opiate receptor. FEBS Lett 150:125-128
Klinz FJ, Costa T (1989) Cholera toxin ADP-ribosylates the receptor-coupled form of pertussis toxin-sensitive G proteins. Biochem Biophys Res Commun 165:554-560
Koski G, Klee WA (1981) Opiates inhibit adenylate cyclase by stimulating GTP hydrolysis. Proc Natl Acad Sci USA 78:4185-4189
Koski G, Streaty RA, Klee WA (1982) Modulation of sodium-sensitive GTPase by partial opiate agonists. An explanation for dual requirements of Na^+ and GTP in inhibitory regulation of adenylate cyclase. J Biol Chem 257:14035-14040

Kühn H, Cook JH, Dreyer WJ (1973) Phosphorylation of rhodopsin in bovine photoreceptor membranes. A dark reaction after illumination. Biochemistry 12:2495-2502
Kurose H, Katada T, Amano T, Ui M (1983) Specific uncouping by islet activating protein, pertussis toxin, of signal transduction via α_2-adrenergic, cholinergic, and opiate receptors in neuroblastoma × glioma hybrid cells. J Biol Chem 258:4870-4875
Kwatra MM, Ptasienski J, Hosey MM (1989) The porcine heart M2 muscarinic receptor. Agonist-induced phosphorylation and comparison of properties with the chick heart receptor. Mol Pharmacol 35:553-558
Lang J, Costa T (1987) Antisera against the 3-17 sequence of rat $G_{i\alpha}$ recognize only a 40 kDa G protein in brain. Biochem Biophys Res Commun 148:838-848
Lang J, Costa T (1989) Chronic naloxone treatment of NG108-15 cells alters the function but not the amount of pertussis toxin substrate. Adv Biosci 75:703-709
Law PY, Hom DS, Loh HH (1983) Opiate regulation of adenosine 3',5'-cyclic monophosphate levels in neuroblastoma × glioma NG108-15 hybrid cells: relationship between receptor occupancy and effect. Mol Pharmacol 23:26-35
Law PY, Hom DS, Loh HH (1985a) Multiple affinity states of opioid receptors in neuroblastoma × glioma cells. J Biol Chem 260:3561-3569
Law PY, Louie AK, Loh HH (1985b) Effect of pertussis toxin treatment on the down-regulation of opioid receptors of neuroblastoma × glioma NG108-15 hybrid cells. J Biol Chem 260:14818-14823
Lefkowitz RJ, Caron MG (1988) Adrenergic receptors. Models for the study of receptors coupled to guanine nucleotide regulatory proteins. J Biol Chem 263:4993-4996
Lefkowitz RJ, Cerione RA, Codina J, Birnbaumer L, Caron MG (1985) Reconstitution of the β-adrenergic receptor. J Membrane Biol 87:1-12
Levitzki A (1987a) Regulation of hormone-sensitive adenylate cyclase. Trends Pharmacol Sci 8:299-303
Levitzki A (1987b) Regulation of adenylate cyclase by hormones and G proteins. FEBS Lett 211:113-118
Limbird LE, Speck JL (1983) N-ethylmaleimide, elevated temperature, and digitonin solubilization eliminate guanine nucleotide but not sodium effects on human platelets α_2-adrenergic receptor-agonist interaction. J Cyclic Nucleotide Protein Phosphorylat Res 9:191-201
Limbird LE, Speck JL, Smith SK (1982) Sodium ion modulates agonist and antagonist interactions with the human platelets α_2-adrenergic receptor in membrane and solubilized preparations. Mol Pharmacol 21:609-617
Mattera R, Yatani A, Kirsch GE, Graf R, Okabe K, Olate J, Codina J, Brown AM, Birnbaumer L (1989a) Recombinant α_{i-3} subunit of G protein activates G_k-gated K^+ channels. J Biol Chem 264:465-471
Mattera R, Graziano MP, Yatani A, Zhou Z, Graf R, Codina J, Birnbaumer L, Gilman AG, Brown AM (1989b) Splice variants of the a-subunit of the G protein G_s activate both adenylyl cyclase and calcium channels. Science 243:804-807
May DC, Ross EM, Gilman AG, Smigel MD (1985) Reconstitution of catecholamine-stimulated adenylate cyclase activity using three purified proteins. J Biol Chem 260:15829-15833
Michel T, Hoffman BB, Lefkowitz RJ (1980) Differential regulation of the α_2-adrenergic receptor by sodium and guanine nucleotides Nature (London) 288:709-711
Milligan G, McKenzie FR (1988) Opioid peptides promote cholera toxin-catalysed ADP-ribosylation of the inhibitory guanine nucleotide binding protein (G_i) in membranes of neuroblastoma × glioma hybrid cells. Biochem J 252:369-373
Minuth M, Jakobs KH (1986) Sodium regulation of agonist and antagonist binding to β-adrenoceptors in intact and N_s-deficient membranes. Naunyn-Schmiedeberg's Arch Pharmacol 333:124-129
Motulsky HJ, Insel PA (1983) Influence of sodium on the α_2-adrenergic receptor system of human platelets. A role for intraplatelet sodium in receptor binding. J Biol Chem 258:3913-3919
Mumby SM, Kahn RA, Manning DR, Gilman AG (1986) Antisera of designed specificity for subunits of guanine nucleotide-binding proteins. Proc Natl Acad Sci USA 83:265-269
Murayama T, Ui M (1985) Receptor-mediated inhibition of adenylate cyclase and stimulation of arachidonic acid release in 3T3 fibroblasts. Selective susceptibility to islet-activating protein, pertussis toxin. J Biol Chem 260:7226-7233
Neer EJ, Clapham DE (1988) Roles of G protein subunits in transmembrane signalling. Nature (London) 333:129-134
Newman EL, Barnard EA (1984) Identification of an opioid receptor subunit carrying the μ-binding site. Biochemistry 23:5385-5389
Okajima F, Ui M (1984) ADP-ribosylation of a specific membrane protein by pertussis toxin associated

with inhibition of a chemotactic peptide-induced archidonate release in neutrophils. A possible role of the toxin substrate in Ca^{2+} biosignalling. J Biol Chem 259:13863–13871
Ott S, Costa T (1989) Enzymatic degradation of GTP and its "stable" analogues produce apparent isomerization of opioid receptors. J Rec Res 9:43–64
Ott S, Costa T, Herz A (1986) Effects of sodium and GTP on the binding kinetics of [^{3}H]diprenorphine in NG108-15 cell membranes. Naunyn-Schmiedeberg's Arch Pharmacol 334:444–451
Ott S, Costa T, Herz A (1988) Sodium modulates opioid receptors through a membrane component different from G proteins. Demonstration by target size analysis. J Biol Chem 263:10524–10533
Ott S, Costa T, Herz A (1989) Opioid receptors of neuroblastoma cells are in two domains of the plasma membrane that differ in content of G proteins. J Neurochem 52:619–626
Owens JR, Frame LT, Ui M, Cooper DM (1985) Cholera toxin ADP-ribosylates the islet-activating protein substrate in adipocyte membranes and alters its function. J Biol Chem 260:15946–15952
Pedersen SE, Ross EM (1982) Functional reconstitution of β-adrenergic receptors and the stimulatory GTP-binding protein of adenylate cyclase. Proc Natl Acad Sci USA 79:7228–7232
Pert CB, Snyder SH (1973) Opiate receptor: demonstration in nervous tissue. Science 179:1011–1014
Pert CB, Snyder SH (1974) Opiate receptor binding of agonist and antagonists affected differentially by sodium. Mol Pharmacol 10:868–879
Pfaffinger PJ, Martin JM, Hunter DD, Nathanson NM, Hille B (1985) GTP-binding proteins couple cardiac muscarinic receptors to K-channels. Nature (London) 317:536–538
Pfeuffer T, Helmreich EJM (1988) Structural and functional relationships of guanosine triphosphate binding proteins. Curr Top Cell Regul 29:129–215
Portoghese PS, Larson DL, Jiang JB, Caruso TP, Takemori AE (1979) Synthesis and pharmacological characterization of an alkylating analogue (chlornaltrexamine) of naltrexone with ultra-lasting narcotic antagonist properties. J Med Chem 22:168–173
Regan JW, Nakata H, DeMarinis RM, Caron MG, Lefkowitz RJ (1986) Purification and characterization of the human platelet α_2-adrenergic receptor. J Biol Chem 261:3890–3894
Repaske MG, Nunnari JM, Limbird LE (1987) Purification of the α_2-adrenergic receptor from porcine brain using a yohimbine agarose affinity matrix. J Biol Chem 262:12381–12386
Rodbell M (1980) The role of hormone receptors and GTP-regulatory proteins in membrane transduction. Nature (London) 284:17–22
Rosenberger LB, Yamamura HI, Roeske WR (1980) Cardiac muscarinic receptor binding is regulated by Na^{+} and guanyl nucleotides. J Biol Chem 255:820–823
Senogles SE, Benovic JL, Amlaiky N, Unson C, Milligan G, Vinitsky R, Spiegel AM, Caron MG (1987) The D_2-dopamine receptor is functionally associated with a pertussis toxin-sensitive G protein. J Biol Chem 262:4860–4867
Sharma S, Nirenberg M, Klee W (1979) Morphine receptors as regulators of adenylate cyclase activity. Proc Natl Acad Sci USA 76:5626–5630
Simon EJ, Groth J (1975) Kinetics of opiate receptor inactivation by sulfhydryl reagents: evidence for conformational change in presence of sodium ions. Proc Natl Acad Sci USA 72:2404–2407
Stryer L (1986) Cyclic GMP cascade of vision. Annu Rev Neurosci 9:87–119
Sweat FW, Klee WA (1985) Adenylate cyclase in two populations of membranes purified from neuroblastoma × glioma hybrid (NG108-15) cells. J Cyclic Nucleotide Protein Phosphor Res 10:565–578
Tolkovsky A, Levtzki A (1978) Mode of coupling between the β-adrenergic receptor and adenylate cyclase in turkey erythrocytes. Biochemistry 17:3795–3810
Ueda H, Harada H, Nozaki M, Katada T, Ui M (1988) Reconstitution of rat brain μ opioid receptors with purified G proteins, G_i and G_0. Proc Natl Acad Sci USA 85:7013–7017
Ui M (1986) Pertussis toxin as a probe of receptor coupling to inositol lipid metabolism. Receptor Biochem Methodol 7:163–195
Vachon L, Costa T, Herz A (1987a) GTPase and adenylate cyclase desensitize at different rates in NG108-15 cells. Mol Pharmacol 31:159–168
Vachon L, Costa T, Herz A (1987b) Opioid receptor desensitization in NG108-15 cells: differential effects of a full and a partial agonist on the opioid-dependent GTPase. Biochem Pharmacol 36:2889–2897
Wregget KA, DeLean A (1984) The ternary complex model. Its properties and application to ligand interactions with the D_2-dopamine receptor of the anterior pituitary gland. Mol Pharmacol 26:214–227

Wüster M, Costa T, Aktories K, Jakobs KH (1984) Sodium regulation of opioid agonist binding is potentiated by pertussis toxin. Biochem Biophys Res Commun 123:1107–1115
Yatani A, Brown AM (1989) Rapid β-adrenergic modulation of cardiac calcium channel currents by a fast G protein pathway. Science 245:71–74
Yatani A, Codina J, Brown AM, Birnbaumer L (1987) Direct activation of mammalian atrial muscarinic potassium channels by GTP regulatory protein G_k. Science 235:207–211
Yatani A, Mattera R, Codina J, Graf R, Okabe K, Padrell E, Iyengar R, Brown AM, Birnbaumer L (1988) The G protein-gated atrial K^+ channel is stimulated by three distinct G_i α-subunits. Nature (London) 336:680–682

CHAPTER 7

Guanine Nucleotide-Binding Proteins and Their Coupling to Opioid Receptors

J. Lang

1 Introduction

Several mechanisms have been reported for the transmission of signals from membrane receptors to effector systems. Knowledge about the transducing and modulatory role of guanine nucleotide-binding proteins (G-proteins)[1] in several signal pathways has increased greatly since the early demonstration that guanine nucleotides influence the binding of hormones to some receptors (Rodbell et al. 1971), and that their action is mediated by proteins which are distinct from the receptor molecule (Pfeuffer 1977; Hildebrandt et al. 1983). Although many of the components involved in this phenomenon have been well characterized, questions about the exact molecular mechanism, the specificity of the system, and its adaptive regulation are still unresolved.

This review discusses the molecular characteristics of G-proteins, with particular emphasis on the pertussis toxin-sensitive proteins; it will also focus on G-protein coupling to opioid receptors. Other reviews, which include more detailed discussions of stimulatory receptors, the rhodopsin-transducin system (Stryer and Bourne 1986; Gilman 1987; Chabre and Deterre 1989), G-protein reconstitution (Neer and Clapham 1988), and on the identification of G-proteins (Milligan 1988) have been published.

2 Guanine Nucleotide-Binding Proteins

2.1 General Characteristics

The G-proteins which transduce and modulate signal transduction encompass a group of heterotrimers each consisting of an α-, β-, and γ-subunit[2]. A general characteristic of these heterotrimers is the presence of an α-subunit which is unique to each subtype and which contains the site for NAD-dependent ADP-ribosylation

[1] Abbreviations used: DADLE = [D-Ala[2], D-Leu[5]]-enkephalin; G-proteins = guanine nucleotide-binding proteins, where α, β, and γ denote the subunits of the heterotrimer; G_i / G_o comprises the pertussis toxin (PTX)-sensitive G-proteins, whereas G_s denotes the G-protein initially believed to stimulate only the adenylate cyclase; retinal G-proteins are called transducin (Td); GTP = guanosine triphosphate; NAD = nicotine amide dinucleotide.

[2] Some G-proteins such as the ras oncogene product are formed by an α-subunit only (see Sect. 2.4); however, coupling of these G-proteins to opioid receptors has not been demonstrated.

by certain bacterial toxins and for GTPase activity (Sternweis et al. 1981; Northup et al. 1982; Ui 1984). The initial classification of extraretinal G-proteins and the attribution of a given function to a certain G-protein was based upon the differential abilities of two bacterial toxins to catalyze the covalent attachment of an ADP-ribose moiety to Gα, and to alter G-protein-mediated signal transduction from receptors to the effector enzyme, adenylate cyclase. Incubation with pertussis toxin (PTX) leads to the uncoupling of receptors from G_i, which mediates the inhibition of the adenylate cyclase (Ui 1984). After treatment with cholera toxin (CTX) constant activation of the stimulatory protein G_s, which then remains independent from receptor control, is observed (Cassel and Pfeuffer 1978). A related, but distinct, PTX substrate had been termed G_o (o stands for "other") since, initially, no specific function could be attributed to this subtype. It should be kept in mind that the classification as "stimulatory" or "inhibitory" has been coined for this effector enzyme and might differ in other systems.

2.2 Functional States of G-Proteins

A generally accepted model for the activation/deactivation cycle of the G-protein heterotrimer, as derived from studies with purified components (for review, see Gilman 1987), is shown in Fig. 1. Binding of an agonist (H) to receptor (R) leads to an interaction with a G-protein ($\alpha/\beta\gamma$). The inactive, GDP-binding state of G-proteins consists of the heterotrimer in which α and $\beta\gamma$ are firmly bound. The stability of this complex is well demonstrated by the fact that most of the purified G-protein consists of the GDP-bound form, the presence of this heterogeneous population of guanine nucleotides, however, considerably complicates the interpretation of kinetic data obtained from reconstitution studies if not removed prior to analysis (Ferguson et al. 1986).

Release of GDP (step 1, Fig. 1) leads to a tightly coupled complex between receptors and G-proteins (not depicted), which exhibits high affinity for the agonist. Binding of GTP to Gα induces dissociation of the heterotrimer into α- and $\beta\gamma$-subunits (step 2, Fig. 1), which regulate different effector systems such as ion channels, phospholipase A_2, and adenylate cyclase. These steps are critically

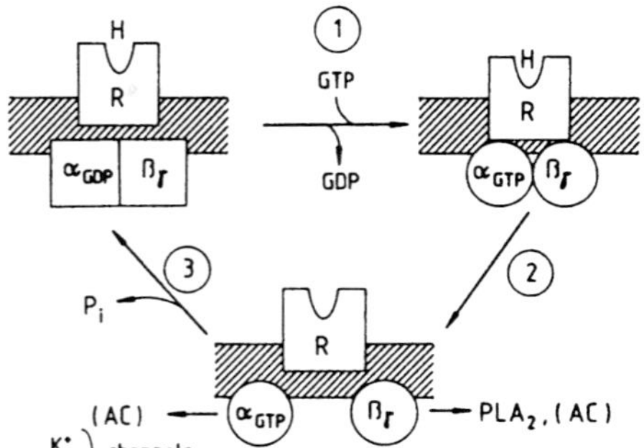

Fig. 1. Activation/deactivation cycle of G-proteins. See text (Sect. 2.4) for details. *H* Hormone; *R* receptor; α_{GDP} the GDP-bound state of Gα; α_{GTP} the GDP-bound state; possible interactions with some effector systems such as phospholipase A_2 (PLA_2), adenylate cyclase (*AC*), and ion channels (K^+, Ca^{2+}) are indicated

dependent upon the presence of different concentrations of Mg^{2+} (Higashijima et al. 1987). The activation cycle is terminated by hydrolysis of GTP to GDP due to the GTPase activity of $G\alpha$ (step 3, Fig. 1), resulting in a reassociation of $\alpha/\beta\gamma$. This step is probably favored by the presence of $G\beta\gamma$. Because hydrolysis proceeds rather slowly, as measured in purified components and membranes, the relatively long survival of activated $G\alpha$ allows for significant signal amplification, on the one hand, but does not match the rapid inhibition caused by G_i. The presence of a stimulatory factor might therefore be necessary to reach physiological rate constants of hydrolysis as has been found for $p21^{ras}$ (Adari et al. 1988).

The depicted model explains the transition from high to low receptor affinity upon incubation with GTP. It also offers several ways to differentiate signal transduction since the α- and $\beta\gamma$-subunits may exert an effect of their own and, additionally, the dissociation of $G\beta$ from one G-protein subtype might force another $G\alpha$ into the inactive heterotrimer state ($G\alpha\beta\gamma$). As shown in Fig. 2 these different mechanisms might, for example, be responsible for the inhibition of adenylate cyclase (Katada et al. 1986), which is under stimulatory control by dissociated $G\alpha_s$ (Smigel 1986): thus, (1) free $\beta\gamma$ can drive stimulatory $G\alpha_s$ into the inactive heterotrimer state, especially since $G\alpha_s$ is endowed with a higher affinity for $\beta\gamma$ than $G\alpha_i$; (2) activated $G\alpha_i$ may compete with $G\alpha_s$ for the binding site on the enzyme; and (3) $\beta\gamma$ may directly inhibit the enzyme. This last interpretation is, however, controversial since it might reflect the action of $\beta\gamma$ on residual $G\alpha_s$ contaminating the purified adenylate cyclase (Smigel 1986). Another interesting implication of this activation/deactivation cycle is that an alteration of G_i might change the characteristics of G_s-coupled processes and vice versa, providing that β-subunits can interchange freely between G_s and G_i and exert their effect by mass action. Indeed, interruption of G_i activation in membranes by PTX does alter the affinity of the stimulatory β-adrenoceptor (Marbach et al. 1988).

Although this model of subunit dissociation seems to be valid in membranes, at least for G_s (Iyengar et al. 1988), and demonstrates a crucial role of $\beta\gamma$ in the inhibition of adenylate cyclase in vivo (Bokoch 1987), it is probably subject to further modifications. Receptor-regulated phosphotransfer, providing GTP for the activation of G_i (Jakobs and Wieland 1989) and specific inactivation of $G\alpha_i$ by protein kinase C (Katada et al. 1985), is operational, at least in vitro. The existence of an antagonistic protein, which, like arrestin in the retina, might favor deactivation of $G\alpha$ (Wilden et al. 1986), has not yet been demonstrated for G_i/G_o.

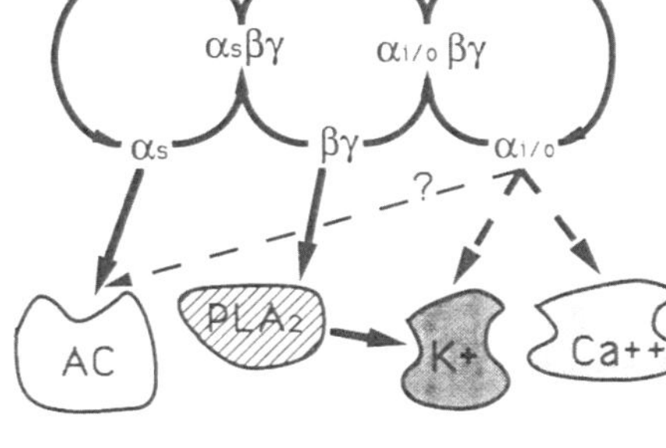

Fig. 2. Regulation of some second messengers by G-protein subunits. See text (Sects. 2.4 and 2.6) for details. *AC* Adenylate cyclase; *PLA_2* Phospholipase A_2; *K^+* and *Ca^{2+}* indicate respective ion channels. Regulation of ion channels by $G\alpha_s$ has not been included in this scheme

2.3 Structural Organization of the α-Subunit

Tentative definitions of different functions to epitopes in Gα (Fig. 3) became possible through the use of immunological techniques, tryptic peptides, point mutations of G-protein subunits, and comparison with data obtained from other types of G-proteins, like ribosomal elongation factor or p21^{ras}, the protein product of ras oncogenes.

Evidence from enzymatic digestion studies indicates that the amino terminus of Gα serves as a contact site for the β-subunit (Neer et al. 1988). As this part of Gα is also subject to myroistylation (Buss et al. 1987), a transformation which alters membrane binding in the related G-protein p21^{ras} (Buss et al. 1989), it may additionally be implicated in an analogous function. Epitopes for interaction with guanine nucleotides and the probable binding site for the phosphoryl moiety have been assigned by sequence comparison with bacterial and eukaryotic transcription factors (McCormick et al. 1985): the amino-terminal phosphoryl-binding loop (P, in Fig. 3) and the residues probably involved in GDP-binding (G, in Fig. 3). As would be expected for the active site of an enzyme, these regions are characterized by a high degree of hydrophilicity and are thus probably located on the surface of the peptide.

The designation of regions thought to interact with effectors (E, in Fig. 3) is based on prediction models showing a high degree of heterogeneity among sequences of different Gα-subtypes, and clustering in discrete loop-forming parts in analogy to p21^{ras} (Holbrook and Kim 1989). Alternatively, the long insert found in $G\alpha_s$, which is absent from the PTX-sensitive Gα-subunits, might be involved in coupling to effectors; however, direct experimental evidence proving the involvement of the latter is still lacking. The site of covalent modification of $G\alpha_s$ by CTX catalyzed ADP-ribosylation has been identified as arginine 191 in the $G\alpha_s$ peptide (Manning et al. 1984), whereas PTX catalyzes the ribosylation of a carboxy-terminal cystein (C, in Fig. 3) residue in $G\alpha_i/G\alpha_o$, which results in uncoupling of the receptor/G-protein complex (Ui et al. 1984). The blockade of the C-terminus of transducin by antibodies also hinders the coupling of the corresponding receptor (Hamm et al. 1988). Further evidence for the importance of the C-terminal amino acids of Gα in signal transmission stems from the observation that a C-terminal mutation of $G\alpha_s$ abolishes coupling to stimulatory receptors (Sullivan et al. 1987), whereas a $G\alpha_i$ chimera, containing the carboxy-terminus of $G\alpha_s$, binds to stimulatory β-adrenoceptors (Masters et al. 1988); these data suggest that the C-terminus determines receptor-specific coupling.

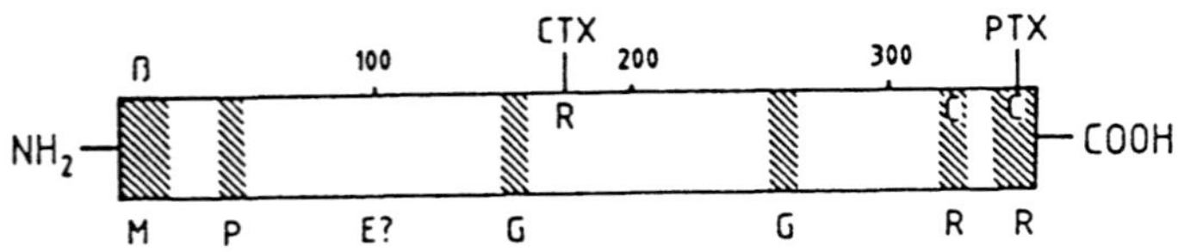

Fig. 3. Functional organization of Gα-subunits. Possible functional sites of Gα, derived from studies cited in the text, are shown. The sites of ADP-ribosylation by CTX (*CTX*) and pertussis toxin (*PTX*), the possible sites of interaction with the β-subunit (*β*), the receptor (*R*), the plasma membrane (*M*), and effectors (*E*) are indicated. The binding sites for the phosphoryl and guanosine moiety of GTP are indicated as *P* and *G*, respectively; *numbers* designate length in amino acids

2.4 Heterogeneity of Subunits

GTP-binding proteins in the broader sense encompass a ramified family of peptides including elongation factors, the so-called small molecular weight G-proteins (including the protein products of certain oncogenes), and the heterotrimeric substrates of bacterial toxins. Protein purification soon led to the appraisal that PTX ribosylates more than one Gα-peptide, and a number of distinct Gα-subtypes which exhibit a high degree of homology, with little, if any, species differences have now been identified (Table 1).

Three different Gα_i peptides have been purified (Katada et al. 1987; Yatani et al. 1987) and sequenced (Jones and Reed 1987): Gα_{i1}, Gα_{i3} (both 41 kDa) and Gα_{i2} (40 kDa). Although only one cDNA has been cloned for Gα_o, purification studies indicate the presence of two subforms (Goldsmith et al. 1988); these probably correspond to different peptide sequences with discrete distributions (Lang 1989). Information on the function and sequence of a 43 kDa PTX-sensitive α-subunit, present in low amounts in several tissues, is still lacking (Iyengar et al. 1987).

The CTX substrate Gα_s represents at least four subtypes with M_r of 45 and 52 kDa, which are generated from a single mRNA by alternative splicing, resulting in total or partial omission of a stretch of 15 amino acids (Bray et al. 1986). A cDNA sequence encoding a 41 kDa Gα designated Gα_z has been isolated; it does not form a substrate for CTX or PTX (Fong et al. 1988; Premont et al. 1989) and is generally regarded as a candidate for PTX-insensitive signal transmission by G-proteins. The

Table 1. G-protein subunits and subtypes[a]

Subtype	Approx. MW[b] (kDa) (SDS-PAGE)	Toxins[c]
α_{s1-4}	45–52	CTX
Tdα_{1-2}	39.5–40	CTX, PTX
α_{43}	43	PTX (CTX ?)
α_{i1}	41	PTX (CTX)
α_{i2}	40	PTX (CTX)
α_{i3}	41	PTX (CTX)
α_o[d]	38.5–39	PTX (CTX)
α_z	41	?
β_1[d]	36	–
β_2[d]	35	–
γ[d]	5–10	–

[a] With the exception of Gα_z and Gα_{43}, only purified and cloned CTX- or PTX-sensitive G-proteins are listed; the so-called small molecular weight Gα-subunits are not.

[b] Apparent molecular weights, as found in commonly used SDS-PAGE systems.

[c] Toxins in parentheses indicate ability to ADP-ribosylate in vitro.

[d] Multiple subtypes probably exist.

so-called small molecular weight G-proteins are comprised of α-subunits within the 20 kDa range, which mostly do not form heterotrimers and some of them are susceptible to certain bacterial toxins, but their exact classification and physiological function remain to be defined.

Two subtypes of the $G\beta$-subunit have been purified and cloned (Amatruda et al. 1988): whereas $G\beta_1$ (36 kDa) is present in retinal and extraretinal G-proteins, $G\beta_2$ (35 kDa) was only found in the latter. A marked heterogeneity among γ-subunits suggested by early peptide mapping (Hildebrandt et al. 1985) has been substantiated by cDNA cloning and immunological techniques (Robishaw et al. 1989), but its functional implication has still to be elucidated.

The continuous discovery of new subtypes implies that the number of species listed here might still be growing. As the differences among subtypes are often minor, even numerous chromatographic steps might not completely resolve them and those minor contaminants may always complicate the interpretation of reconstitution studies. Recombinant proteins were recently used to circumvent this problem; however, their rather low intrinsic activity presents a major drawback (Birnbaumer et al. 1988).

2.5 Identification and Quantification of G-Proteins

PTX or CTX catalyzed ADP-ribosylation using radioactive or fluorescent labeled NAD (α^{32}-NAD or ε-NAD) as ADP-ribose carrier, followed by gel chromatography, allows direct visualization of the α-subunits (Ui 1984; Hingorani and Ho 1988). Although this method has proved remarkably accurate in distinguishing between several subtypes (see Fig. 4) in many cases, it is generally recognized that toxins alone do not constitute a sufficient criterion for exact classification, e.g., the retinal G-proteins $Td\alpha_1$ and $Td\alpha_2$ are not the only proteins to be ribosylated by both CTX

Fig. 4. Identification of G-proteins. **A** Autoradiogram of bovine brain PTX substrate which was separated on SDS-PAGE after ADP- ribosylation with [^{32}P]-NAD and PTX. Molecular weights calculated from standards are given on the *left*. **B** Protein stain and immunoblot of purified G-proteins using antisera directed against synthetic peptides according to a consensus sequence of PTX-sensitive $G\alpha$-subunits (anti-αcommon), or specific for $G\beta$ (anti-β), $G\alpha_o$ (anti-α_o), and $G\alpha_{i2}$ (anti-α_{i2})

and PTX; under certain conditions $G\alpha_i$, and possibly $G\alpha_o$, can also be ribosylated by these toxins (Owens et al. 1985). Furthermore, the different $G\alpha$-peptides are not always separable on the basis of molecular weight, as is evident by their similar M_r-values (Table 1). Some authors have advocated alkylation prior to SDS-PAGE to increase separation. In our own experience, however, general conditions of gel electrophoresis, e.g., sample load, gel gradient, and length, are more important factors to consider, but none ensures complete resolution. The use of gel electrophoresis with urea gradients also provides high resolution separations (Scherer et al. 1987), but in this system the correlation between migration and subtypes remains to be established.

Immunological identification by antisera, in contrast to ADP-ribosylation, offers the advantage of directly identifying all subunits (α, β, and γ). Initial utilization of antisera generated against even highly purified proteins was often complicated by the high degree of sequence homology among $G\alpha$ or $G\beta$ subtypes, respectively, resulting in a considerable degree of cross-reactivity. Antisera raised against synthetic peptides have been proven to be powerful tools in many circumstances (for review, see van Regenmortel 1989). Mumby et al. (1986) were the first to demonstrate their suitability in distinguishing among subtypes for G-proteins and they probably constitute the most reliable tool available at the moment (see also Fig. 4). This approach allowed the identification of the 40 kDa $G\alpha$ as $G\alpha_{12}$ (Blacksmith et al. 1987; Lang and Costa 1987) and, as a consequence, the different cDNAs for $G\alpha_i$ could be related to distinct proteins. Additionally, the generation of an antiserum directed against a peptide sequence common to all CTX- and PTX-sensitive $G\alpha$-proteins has shown its utility in identifying new G-protein α-subunits. Nevertheless, one must be aware that these antibodies may also bind to peptides which do not share the sequence of the original antigen but which are recognized as they might mimic spatial arrangements and that specificities might be different in solid phase assays like ELISA or Western blot, as compared to interactions in solution (van Regenmortel 1989).

Quantification of G-proteins can be achieved by either measuring ADP-ribosylation or immunoreactivity. The use of toxins is unfortunately hampered by several factors. The amount of toxin-catalyzed incorporation of ADP-ribose depends critically on the presence of $G\beta\gamma$ coupled to $G\alpha$; the affinity of the $\beta\gamma$-complex is higher for $G\alpha_i$ than for $G\alpha_o$, giving erroneous low levels of the latter (Huff and Neer 1986). Additionally, some preparations contain a high concentration of endogenous NADase activity, resulting in erroneous low levels of incorporation measurable (Moss et al. 1986), and ADP-ribosylation is rather incomplete in membrane preparations as compared to the toxin-catalyzed reaction in vivo for some still unknown reasons (Lang and Costa 1989a). Lastly, tissue levels of GTP might directly alter the activation of PTX in analogy to its role in purified components (Mattera et al. 1986). However, in light of the $G\alpha/G\beta\gamma$ precoupling required for ribosylation, this approach may be used to determine functional states of G-protein/receptor coupling. The generation of highly reactive antisera allowed the development of radioimmunoassays for $G\alpha_o$ (Asano et al. 1987) and $G\alpha_s$ (Ransnäs and Insel 1988); these are able to detect femtomolar levels. Quantitative immunoblotting techniques developed for several subunits and subtypes circumvent the necessity of extraction and allow direct control of possible cross-reactivity with comparable detection limits (Lang and Schulz 1989).

2.6 Distribution of G-Proteins

G-proteins have been found in every organ or cell examined so far, including invertebrates (Tsuda et al. 1986) and yeast (Nakafuku et al. 1987). The distribution of the different PTX-sensitive subtypes varies considerably according to the source of plasma membranes. Contrary to numerous claims, several subtypes can always be found, although the amount present might be rather small and therefore can only be detected in at least partially purified preparations. Subcellular fractionation revealed that at least $G\alpha$ is not only confined to plasma membranes but can also be detected in organelles (Audigier et al. 1988) and cytoplasm (Murray et al. 1988; Gabrion et al. 1989).

As a general rule, $G\alpha_o$ seemes to be preponderant in neuronal tissue (Asano et al. 1988) but is absent in blood cells (Mumby et al. 1988; Lang and Costa 1989b), and its slower migrating form seems to represent the only $G\alpha_o$ form in adipocytes (Rouot et al. 1989), transverse tubules from skeletal muscle (Toutant et al. 1988), and the neuroblastoma × glioma cell line NG108-15 (Lang 1989). Contrary to $G\alpha_i$, no cell line has been described containing only $G\alpha_o$ as PTX substrate. In erythrocytes $G\alpha_{i3}$ represents the most abundant subtype (Iyengar et al. 1987; Codina et al. 1988) and the considerable amounts of $G\alpha_{i2}$ reported are probably due to contamination by platelets. $G\alpha_{i2}$ forms the major substrate in neutrophils (Gierschik et al. 1987), thrombocytes, and white blood cells, but not in the leukemic cell line S49 cyc^- (Lang and Costa 1989b). Considerable amounts of $G\alpha_{i1}$ have up to now only been described in brain tissue (Mumby et al. 1988), but it is probably also present in other tissues (Jones and Reed 1987). Subcellular fractionation of cardiac ventricular muscle demonstrated the presence of $G\alpha_o$ and rather low amounts of $G\alpha_{i2}$ without any 41 kDa $G\alpha$ in sarcolemma; however, only 41 kDa $G\alpha_i$ was detectable in sarcoplasmic reticulum (Lang and Costa 1989b). In contrast to the $G\alpha$-subtypes, no differential distribution was reported for the two β-peptides with the exception of human placenta (Hinsch et al. 1989).

Several reports suggest that the relative amounts of G-proteins and those of the corresponding mRNA do not always correlate. Indeed, comparison of results from several authors with the data of Brann et al. (1987) indicates that an inverse relationship exists between the relative abundance of $G\alpha$-protein subtypes in brain membranes ($G\alpha_o > G\alpha_i > G\alpha_s$) and the relative abundance of their mRNA's ($G\alpha_s > G\alpha_o > G\alpha_i$). This hints at posttranscriptional tissue-specific regulation of G-proteins; this probably also applies to the β-subunits (Hinsch et al. 1989).

2.7 Specificity of Signal Transduction

G-proteins transduce signals to a variety of effector molecules like adenylate cyclase, phosphodiesterase, ion channels, phospholipase A_2, phospholipase C, and NADPH-oxidase, although, in the latter two cases, the exact identity of the G-protein involved has yet to be elucidated (Boyer et al. 1989; Seifert and Schultz 1989). A major problem in signal transduction research is the elucidation of mechanisms that may ensure the specificity of signal transfer from receptor to effector. Reconstitution studies in phospholipid vesicles were generally inconclusive in demonstrating any preference among the G_i/G_o- or G_s-subtypes in coupling to a given receptor and provided the initially troublesome demonstration that the β-adrenergic receptor, which stimulates adenylate cyclase, can reconstitute in phospholipid vesicles with G_i (Asano et al. 1984). These data favored the view of an unexpected type of promiscuity among receptors and G-proteins; however, comparison of purified preparations for intrinsic activities (Katada et al. 1986) as well as affinity for $G\beta$ (Huff and Neer 1986) demonstrated differences among some $G\alpha$-peptides tested, which could allow differential activation. Further results obtained by reconstitution in membranes, patch-clamp or whole cell preparations in which coupling to second messengers other than adenylate cyclase was also investigated, offer several possible mechanisms of specificity concerning subtypes and subunits. However, current knowledge is still far from giving a coherent picture and certain precautions are necessary in the interpretation of these data: (1) with few exceptions (Birnbaumer et al. 1988), the G-protein preparations used are probably contaminated to varying degrees by different subtypes as demonstrated by many purification studies; (2) since some effectors like ion channels or phospholipases have not yet been used as purified components in reconstitution assays, these data do not formally rule out the participation of further intermediates, despite the pharmacological blockage of potential candidates.

Different possibilities of coupling between receptor and effector via G-proteins are demonstrated in Fig. 5. Interactions with second messenger systems seem to be

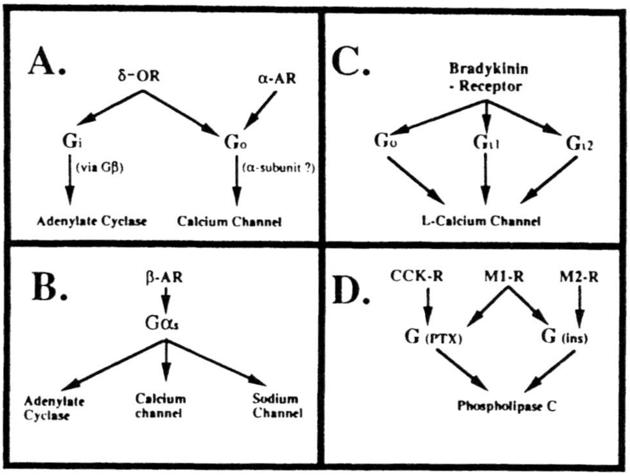

Fig. 5A-D. Receptor/effector coupling by G-proteins. Indicated are some known examples of transducing pathways; for details, see text (Sect. 2.7). *α-AR* and *β-AR* = α- and β-adrenoceptor, respectively; *δ-OR* = δ-opioid receptor; *CCK-R* = cholecystokinin receptor; *M1-R* and *M2-R* designate two of the muscarinic acetylcholine receptor subtypes; $G_{(PTX)}$ and $G_{(ins)}$ are the PTX-sensitive and -insentive G-proteins

specific to a certain degree for a given PTX substrate (Figs. 2 and 5A): whereas G_i couples directly or indirectly to adenylate cyclase which is stimulated by $G\alpha_s$ (see Sect. 2.2), $G\alpha_o$ does not seem to be involved (Katada et al. 1986; Liang and Galper 1988). In contrast, G_o restores (in the neuroblastoma × glioma hybrid cell line NG108-15) the opiate-mediated inhibition of Ca^{2+} channels (Hescheler et al. 1987); this interaction however, has not been further characterized. The role of $G\alpha_o$ in the inhibitory control of ion channels by receptors has also been established in other circumstances; e.g., the α-adrenergic-mediated control of Ca^{2+} currents in the above mentioned cell line (McFadzean et al. 1989) and, as shown in Fig. 6, muscarinic inhibition of K^+ channels in hippocampus (Toselli et al. in press). In contrast, $G\alpha_i$ activates K^+ channels in atrial myocytes (Yatani et al. 1987) and probably Ca^{2+} channels also in adrenal cortical or pituitary cell lines (Rosenthal et al. 1988).

The evidence mentioned suggests that a given G-protein subtype exerts a specific action on a given second messenger. However, $G\alpha_s$ has been shown (see Fig. 5B) to mediate not only β-adrenergic activation of adenylate cyclase and certain Ca^{2+} channels, but also inhibition of Na^+ channels (Schubert et al. 1989). Therefore, one subunit may interact differentially with several effectors. Additionally, reports that certain cells are devoid of some of the known G-protein subtypes (see Sect. 2.5) without necessarily lacking the functions which are attributed to the corresponding subtype have to be taken into account (Brass et al. 1988).

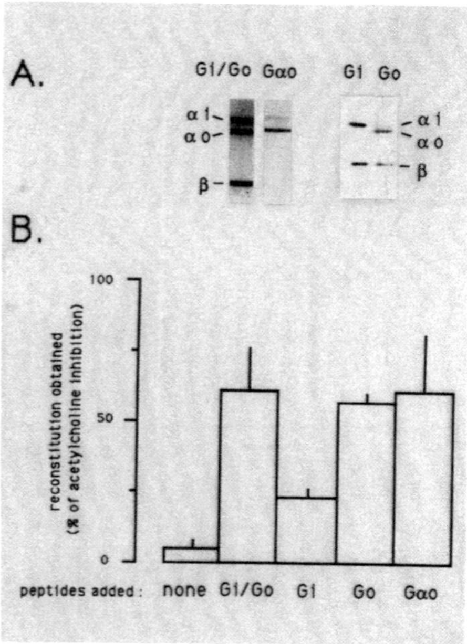

Fig. 6. Reconstitution of acetylcholine inhibition of high voltage activated Ca^{2+} channels in hippocampal neurons pretreated with PTX. **A** Protein stain of the peptides used for reconstitution. **B** Effects of G-protein peptides (1 nM) intracellularly infused on acetylcholine-mediated inhibition of high voltage activated Ca^{2+} (expressed as percentage of acetylcholine inhibition in cells which were not treated with PTX)

As far as the receptor/G-protein interaction is concerned, there is no evidence that one receptor interacts with only one specific G-protein. Indirect results suggest this kind of coupling only for transduction from receptors to a given second messenger, as in the case of certain Ca^{2+} channels (Rosenthal et al. 1988). Surprisingly enough, the contrary can also be observed (Fig. 5C): the bradykinin receptor may use several different G-protein subtypes with comparable efficiency to inhibit the L-type Ca^{2+} channel in dorsal root ganglia (Ewald et al. 1989), although indirect transmission has not been ruled out. A comparable situation was reported for K^+ channels in atrial myocytes, where all three known subtypes of $G\alpha_i$ are equally effective at least when direct, receptor-independent activation is recorded (Yatani et al. 1988). A more rigorous specificity cannot, of course, be excluded in the light of newly emerging receptor subtypes. It has indeed been shown (Ashkenazi et al. 1989) that muscarinergic receptor subtypes can couple via different G-proteins (PTX-sensitive and -insensitive) to the same effector in the same cell (Fig. 5D): whereas the M2-type regulates phospholipase C through a PTX-insensitive G-protein and the cholecystokinin receptor transduces via a PTX-substrate, the M1-subtype can use both pathways. It will be interesting to determine whether the M1- and M2-acetylcholine receptors use the same or different $G\alpha$-subtypes within the PTX-sensitive pathway.

Another possible way in which specificity might be conferred was initially proposed by Rodbell (1985): $G\alpha$ may be liberated from plasma membranes and be directed to specific targets after secondary transformations. This idea has been supported by the observations in vivo of (1) a translocation from plasma membranes into the cytoskeleton after the activation of adrenergic (Crouch et al. 1989) or FMLP receptors (P. Naccache and J. Lang, unpubl. observ.); (2) redistribution among subcellular particles upon specific stimulation (Rotrosen et al. 1988); and (3) the in vitro release of $G\alpha_{i2}$ from membranes upon activation of opioid receptors (Milligan et al. 1988). However, the time course observed for the release of $G\alpha_{i2}$ in vitro is rather slow and the functional importance of these translocations remains to be elucidated.

The $G\beta\gamma$-subunit performs at least two roles (see Fig. 2): it activates atrial muscarinic K^+ channels via phospholipase A_2 (Kim et al. 1989), and inhibits the stimulatory pathway by favoring the inactive G_s holotrimer configuration (see Sect. 2.2). Since β_1- and β_2-subtypes interact equally with α_i/α_o (at least in supporting PTX-catalyzed ADP-ribosylation), whereas only β_2 seems to interact with G_s (Hekman et al. 1987), specific effects may be achieved via one or the other subtype. Additionally, differences in affinity of the $\beta\gamma$-peptide to $G\alpha$ have to be considered ($G\alpha_s > G\alpha_i > G\alpha_o$). The rather hydrophobic nature of $G\beta\gamma$, and its requirement to insert $G\alpha$ into phospholipid vesicles (Sternweis 1986), support the view that $G\beta\gamma$, and in particular the γ-subunit, is involved in membrane anchoring of the heterotrimer (see Sect. 2.1 for $G\alpha$ and membrane binding). Interestingly enough, recent data additionally suggest a direct contact between $G\beta\gamma$ and the receptor (Kelleher and Johnson 1988).

The specific transmission of signals from one of the ever-growing number of receptors and their subtypes may depend not only on qualitative parameters like a specific G-protein subtype or subunit, but also on the degree of activation and their interaction (Hatta et al. 1986). This may, in turn, be influenced by the specific ionic environment or secondary transformations, and G-protein subtypes may exhibit different affinities for both (see Sect. 2.3). Perhaps still undefined components play a crucial role. The importance of cytoskeletal elements in the transfer of guanine nucleotides to G-proteins has recently been pointed out (Rasenick and Wang 1988). It should also be noted here that effector molecules might be more heterogeneous than previously thought. In the case of adenylate cyclase, at least two forms of the enzyme have been immunologically distinguished (Mollner and Pfeuffer 1988). Early signal transmission therefore probably proceeds via an amplification pathway with multiple possibilities of cooperation at the level of G-proteins and second messengers.

3 G-Proteins and Opioid Receptors

3.1 Mechanism and Specificity of Coupling

It is generally accepted that opioid receptors, upon stimulation, activate GTPase activity representing PTX-sensitive G-proteins (Kurose et al. 1983). The coupling between inhibitory receptors (including the opioid receptors) and G-proteins seems to be rather tight. Analysis of molecular size (Mollereau et al. 1988), receptor inactivation (Costa et al. 1988) and reconstitution with purified G-proteins (Costa et al. in press) indeed demonstrate precoupling in the absence of receptor ligands; this precoupling is under control of a Na^+-sensitive site different from the G-protein. Preliminary evidence also suggests that the basal GTPase activity of the β-adrenergic receptor-linked G_s reflects the presence of receptor/G-protein complexes in the absence of agonists (Murray and Keenan 1989), although its physiological importance remains a matter of debate (see Chabre 1977). This precoupling may also explain the lack of complete reconstitution of G_i/G_o-mediated events after pretreatment with PTX in vivo observed by many groups, as some ribosylated G-proteins might still be physically attached to the receptor and full replacement by externally added peptides cannot occur. In contrast to the receptor, coupling between G-proteins and the effector enzyme adenylate cyclase seems to be less tight (Vachon et al. 1987; Costa et al. 1988). This mechanism may ensure a rapid inhibitory control of different effector systems.

As outlined above (see Sect. 2.6), it is still unclear which element ensures specificity of transmission. Reconstitution of μ-receptor binding, using purified components (Ueda et al. 1988), demonstrated an equal capacity of G_i and G_o to couple with the receptor (precautions, see Sect. 2.6). It has been reported that an antiserum specific for the carboxy-terminus of $G\alpha_i$ blocks opioid receptor-mediated GTPase activity (McKenzie et al. 1988) and that the CTX-catalyzed ADP-ribosylation of a 40 kDa peptide is enhanced by opioid peptides in NG108-15 cell membranes (Milligan and McKenzie 1988). This suggested an involvement of only $G\alpha_{i2}$ in these opioid receptor-mediated effects. However, this cell line is endowed with the still unsequenced, slowly migrating form of $G\alpha_o$ which is difficult to separate from $G\alpha_{i2}$ (Lang and Costa 1989a; Lang 1989), and $G\alpha_o$ peptides can also form a substrate for CTX (Owens et al. 1985). In contrast, clear differences were found in coupling to effector systems. G_o is endowed with a higher efficiency than a mixture of $G\alpha_i$-subtypes in transducing the inhibitory signal from δ-opioid receptors to Ca^{2+} channels in neuroblastoma × glioma cells (Hescheler et al. 1987).

3.2 G-Proteins and Chronic Activation of Opioid Receptors

Investigation of the role of G-proteins in long-term adaptive processes has yielded a rather diffuse picture. Chronic activation of δ-receptors in the neuroblastoma × glioma cell line NG 108-15 with the agonist [D-Ala^2, D-Leu^5]-enkephalin (DADLE) leads to rapid down-regulation (Chang et al. 1982) which occurs independently from G-protein function (Law et al. 1985). Receptor phosphorylation by cAMP-dependent protein kinase might constitute the initial step in the uncoupling of receptors from G-proteins (Harada et al. 1989), in analogy to the β-adrenoceptor; however, this transformation is not sufficient to explain the interruption of signal transduction in the adrenergic system (Keenan et al. 1987; Cheung et al. 1989; also see Wilden et al. 1986). The exposure to the partial agonist morphine does not lead to down-regulation of the receptor (Vachon et al. 1987) but induces homologous desensitization, i.e., an increase in the specific activity of the effector enzyme adenylate cyclase (Sharma et al. 1979; for review, see Thomas and Hoffmann 1987), which is under the inhibitory control of the opioid receptor mediated by G_i (Kurose et al. 1983). Down-regulation and desensitization are distinct adaptive processes (Law et al. 1983) and, in contrast to the former, intact receptor/G-protein coupling is a prerequisite for initiation of desensitization (Louie et al. 1986). Studies with PTX pointed towards a loss in the tonic inhibitory regulation of adenylate cyclase by G_i in desensitization (Griffin et al. 1985) but, surprisingly, no change in protein levels of G_i and G_o α/β-subunits, ADP-ribosylation, or hormone-dependent GTPase activity was found (Lang and Costa 1989a). This suggests that the molecular alteration responsible for the desensitization of adenylate cyclase occurs between G-proteins and the effector enzyme, although one cannot completely exclude changes in a minor subset of G_i- or G_o-proteins. Since the concentration of these proteins in this cell line exceeds the number of opioid receptors by a factor of 50 (Lang and Costa 1989a), alterations in an opioid receptor-linked G-protein subset may easily escape detection. Further studies should clarify whether (1) only secondary alterations of G_i/G_o occur; (2) G_s, but not G_i, is involved, although direct stimulation of G_s is not altered as compared to native cells (Griffin et al. 1985); or (3) the changes are related only to the effector enzyme.

The results obtained from studies with whole animals differ considerably from those cited above: this may highlight differences between cell cultures and functional neuronal networks, which may show additional differences depending on the tissue examined. Interpretation is further complicated by the cellular and subcellular distribution of G-proteins (see Sect. 2.5); values obtained from crude extracts may obscure the real situation. If one assumes an indirect inhibitory role of the β-subunit on adenylate cyclase by complexing with $G\alpha_s$ to yield the heterotrimer (see Fig. 2), as has also been demonstrated in vivo (Watkins et al. 1987), an enhancement of specific enzyme activity should be accompanied by a decrease in $\beta\gamma$-subunits or an increase in PTX-sensitive Gα-peptides to scavenge G$\beta\gamma$. Nestler and co-workers (1989) indeed reported a specific increase in levels of Gα_o without alterations in the amount of Gβ after long-term exposure to opiates. This change was confined to the rat striatum and no effect was observed in the neocortex. Different results were obtained when investigating a preparation of guinea pig myenteric plexus. Similarly to rat striatum, this tissue contains only low levels of Gα_{i2} and mostly Gα_o as PTX-substrate. The latter, however, is not solely confined to plasma membranes, but is also located in the cytosol (Schulz et al. 1989). As demonstrated in Fig. 7, only a small increase in Gα_o (and probably Gα_i) was found in guinea pig myenteric plexus preparations. In contrast, a two fold greater up-regulation of Gβ levels was observed (Lang and Schulz 1989). These effects were localized to the nerve somata and absent from nerve terminals. This considerable increase in the β-subunit is stoichiometrically difficult to reconcile with the observed compensatory increase in the specific activity of adenylate cyclase (see Fig. 2). If the Gβ-peptide exerts its effect on adenylate cyclase only, a pronounced inhibition instead of supersensitivity should result. It must also be noted that an increase in Gα_o will not necessarily produce more complexed G_s ($\alpha\beta\gamma$-heterotrimer), and hence reduce stimulation of adenylate cyclase, since Gα_o is endowed with the least affinity of all Gα-subtypes for the β-peptide. Therefore, other effectors may play an important role and, indeed, several lines of evidence indicate a role of K^+ channels in the expression of tolerance/dependence (Christie et al. 1987). They might be under physiological regulation by the Gβ-peptide, in analogy to the cardiac muscarinic K^+ channels (Kim et al. 1989). Also, Gα_o couples directly to certain ion channels, but not to

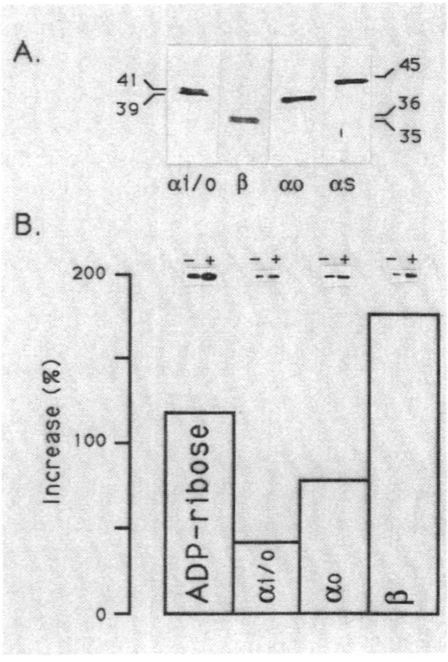

Fig. 7. Effect of chronic fentanyl treatment in vivo on G-protein subunits in a nerve somata preparation of guinea pig myenteric plexus. **A** Immunoblot of myenteric plexus proteins demonstrating the presence of Gα_i, Gβ, Gα_o, and Gα_s. **B** Increase in the amount of ADP-ribose incorporated in a 39/41 kDa protein and tissue levels of indicated G-protein peptides after treatment with the agonist fentanyl in vivo for 6 days. Details of representative immunoblots from controls (−) and treated animals (+) are shown as *inset*. For details, see Lang and Schulz (1989)

adenylate cyclase (see Sect. 2.6) and the alteration of such a signal pathway might be important in tolerance/dependence. A third, still different observation, has been reported from spinal cord-dorsal root ganglion cocultures, where the κ-opioid receptor inhibits Ca^{2+} influx via a PTX-sensitive G-protein by cAMP-dependent and -independent pathways; constant activation of the κ-receptor leads to heterogeneous desensitization of this ion flux and a decrease in G_i (Attali et al. 1989). The possibility of secondary transformations of G-proteins has not yet been investigated in chronic opioid receptor activation; however, the intact ADP-ribosylation observed in rat striatum and guinea pig ileum does not favor them.

These observations confirm that the levels of G-proteins are under control of receptor stimulation, at least in vivo, as has been demonstrated for other hormones. Additionally, together with functional studies, they suggest that adenylate cyclase may not constitute the only mediator of tolerance/dependence and that the observed supersensitivity of the enzyme may be independent of alterations in G-proteins. Clearly, more informations is needed to clarify these points and according to what is currently known from other systems, it would not be surprising if different mechanisms are operational, depending on the cell type, effectors, G-protein, and receptor subtypes involved.

Conclusion

Considerable progress has been made in acquiring knowledge about signal transduction by G-protein-linked membrane receptors. As is obvious from this review, the molecular mechanisms responsible for the action of opioids are just beginning to be unveiled and little is known about this system as compared to others (e.g., adrenergic receptors). Since G-proteins, opioid receptors, and certain effectors are now available as pure components, a considerable array of methods has been developed and sequences will probably be soon at hand; it should therefore be possible to elucidate the precise mechanisms of coupling and the regulatory mechanisms operating during acute and long-term activation of opioid receptor signal transmission. However, this knowledge should not obscure the fact that the "receptor/G-protein" may be only one part of the signal pathway, and may represent an over-simplification of the mechanisms ensuring the specific initiation of a signal cascade.

References

Adari H, Lowy DR, Willumsen BM, Der CJ, McCormick F (1988) Guanosine triphosphate activating protein (GAP) interacts with the p21ras effector binding domain. Science 240:518–521

Amatruda III TT, Gautam N, Fong HKW, Northup JK, Simon MI (1988) The 35- and 36-kDa β-subunits of GTP-binding regulatory proteins are products of separate genes. J Biol Chem 263:5008–5011

Asano T, Katada T, Gilman AG, Ross EM (1984) Activation of the inhibitory GTP-binding protein of adenylate cyclase, G_i, by the β-adrenergic receptors in reconstituted phospholipid vesicles. J Biol Chem 259:9351–9354

Asano T, Semba R, Ogasawara N, Kato K (1987) Highly sensitive immunoassay for the α-subunit of the GTP-binding protein G_o and its regional distribution in bovine brain. J Neurochem 48:1617–1623

Ashkenazi A, Peralta EG, Winslow JW, Ramachandran J, Cooper DJ (1989) Functionally distinct G proteins selectively couple different receptors to PI hydrolysis in the same cell. Cell 56:487–493

Attali B, Saya D, Nah SY, Vogel Z (1989) κ-Opiate agonists inhibit Ca^{2+} influx in rat spinal cord-dorsal root ganglion cocultures. Involvement of a GTP-binding protein. J Biol Chem 264:347–353

Audigier Y, Nigam SK, Blobel G (1988) Identification of a G protein in rough endoplasmic reticulum of canine pancreas. J Biol Chem 263:16352–16357

Birnbaumer L, Codina J, Mattera R, Yatani A, Graf R, Olate J, Sanford J, Brown AM (1988) Receptor-effector coupling by G proteins: purification of human erythrocyte G_{i2} and G_{i3} and analysis of effector regulation using recombinant α subunits synthesized in *Escherichia coli*. Cold Spring Harbor Symp Quant Biol 53:229–239

Bokoch G (1987) The presence of free G-protein β/γ subunits in human neutrophils results in suppression of adenylate cyclase activity. J Biol Chem 262:589–594

Boyer JL, Hepler JR, Harden TK (1989) Hormone and growth factor receptor-mediated regulation of phospholipase C activity. Trends Pharmacol Sci 10:360–364

Brann MR, Collins RM, Spiegel A (1987) Localization of mRNAs encoding the α-subunit of signal transducing G-proteins within rat brain and among peripheral tissues. FEBS Lett 22:191–198

Brass LF, Woolkalis MJ, Manning DR (1988) Interactions in platelets between G proteins and the agonists that stimulate phospholipase C and inhibit adenylyl cyclase. J Biol Chem 263:5338–5355

Bray PA, Carter C, Simons V, Guo C, Puckett J, Kamholz A, Spiegel A, Nirenberg M (1986) Human cDNA clones for four species of $G\alpha_s$ signal transduction protein. Proc Natl Acad Sci USA 83:8893–8897

Buss JE, Mumby SM, Casey PC, Gilman A, Sefton BM (1987) Myroistylated α-subunits of guanine nucleotide binding regulatory proteins. Proc Natl Acad Sci USA 84:7493–7497

Buss JE, Solski PA, Schaeffer JP, MacDonald MJ, Der CJ (1989) Activation of the cellular proto-oncogen product p21Ras by addition of a myristylation signal. Science 243:1600–1603

Cassel D, Pfeuffer T (1978) Mechanism of cholera toxin action: covalent modification of the guanyl nucleotide-binding protein of the adenylate cyclase system. Proc Natl Acad Sci USA 75:2669–2673

Chabre M (1987) Receptor-G protein precoupling: neither proven nor needed. Trends Neurosci 10:355–356

Chabre M, Deterre P (1989) Molecular mechanism of visual transduction. Eur J Biochem 179:255–266

Chang KJ, Eckel RW, Blanchard SG (1982) Opioid peptides induce reduction of enkephalin receptors in cultured neuroblastoma cell. Nature (London) 296:446–448

Cheung AH, Sigal IS, Dixon RAF, Strader CD (1989) Agonist-promoted sequestration of the β_2-adrenergic receptor requires regions involved in functional coupling with G_s. Mol Pharmacol 34:132–138

Christie MJ, Williams JT, North RA (1987) Cellular mechanisms of opioid tolerance: studies in single brain neurons. Mol Pharmacol 32:633–638

Codina J, Olate J, Abramowitz J, Mattera R, Cook RG, Birnbaumer L (1988) α_i-3 cDNA encodes the α-subunit of G_k, the stimulatory G-protein of receptor regulated K^+ channels. J Biol Chem 263:6746–6750

Costa T, Klinz FJ, Vachon L, Herz A (1988) Opioid receptors are coupled tightly to G-proteins but loosely to adenylate cyclase in NG108-15 cell membranes. Mol Pharmacol 34:744–754

Costa T, Lang J, Gless C, Herz A (1989) Spontaneous association between opioid receptors and G-proteins in native membranes. Specific regulation by sodium ions. Mol Pharmacol 37:383–394

Crouch MF, Winegar DA, Lapetina EG (1989) Epinephrine induces changes in the subcellular distribution of the inhibitory GTP-binding protein $G\alpha_{i2}$ and a 38-kDa phosphorylated protein in human platelet. Proc Natl Acad Sci USA 86:1776–1780

Ewald DA, Pang IH, Sternweis PC, Miller RJ (1989) Differential G protein-mediated coupling of neurotransmitter receptors to Ca^{2+} channels in rat dorsal root ganglion neurons in vitro. Neuron 2:1185–1193

Ferguson MF, Higashijima T, Smigel MD, Gilman AG (1986) The influence of bound GDP on the kinetics of guanine nucleotide binding to G-proteins. J Biol Chem 261:7393–7399

Fong HK, Yoshimoto KK, Eversole Cire P, Simon MI (1988) Identification on a GTP-binding protein alpha subunit that lacks an apparent ADP-ribosylation site for pertussis toxin. Proc Natl Acad Sci USA 85:3066–3070

Gabrion J, Brabet Ph, Dao BNT, Homburger V, Dumnis A, Jebben M, Rouot B, Bockaert J (1989) Ultrastructural localization of the GTP-binding protein G_o in neurons. Cell Signal 1:107–123

Gierschik P, Sidiropoulos D, Spiegel A, Jakobs KH (1987) Purification and immunochemical characterization of the major pertussis-toxin-sensitive guanine-nucleotide-binding protein of bovine neutrophil membranes. Eur J Biochem 165:185–194

Gilman A (1987) G-proteins: transducers of receptor generated signals. Annu Rev Biochem 56:615–649

Goldsmith P, Backlund PS, Rossiter K, Carter A, Milligan G, Unson CG, Spiegel A (1988) Purification of heterotrimeric GTP-binding proteins from brain: identification of a novel form of $G_o\alpha$. Biochemistry 27:7085–7090

Griffin MT, Law PY, Loh HH (1985) Involvement of both inhibitory and stimulatory guanine nucleotide binding proteins in the expression of chronic opiate regulation of adenylate cyclase activity in NG108-15 cells. J Neurochem 45:1585–1589

Hamm EH, Deretic D, Arendt A, Hargrave PA, Koenig B, Hofmann KP (1988) Site of G-protein binding to rhodopsin mapped with synthetic peptides from the α-subunit. Science 241:832–835

Harada H, Ueda H, Wada Y, Katada T, Ui M, Satoh M (1989) Phosphorylation of μ-opioid receptors – a putative mechanism of selective uncoupling of receptor-G_i interaction, measured with low-K_m GTPase and nucleotide-sensitive agonist binding. Neurosci Lett 100:221–226

Hatta S, Marcus MM, Rasenick MM (1986) Exchange of guanine nucleotide between GTP-binding proteins that regulate neuronal adenylate cyclase. Proc Natl Acad Sci USA 83:3776–3780

Hekman M, Holzhöfer A, Gierschik P, Im MJ, Jakobs KH, Pfeuffer T, Helmreich EJM (1987) Regulation of signal transfer from β_1-adrenoceptor to adenylate cyclase by $\beta\gamma$-subunits in a reconstituted system. Eur J Biochem 169:431–439

Hescheler J, Rosenthal W, Trautwein W, Schultz G (1987) The GTP-binding protein, G_o, regulates neuronal calcium channels. Nature (London) 325:445–447

Higashijima T, Ferguson KM, Sternweis PC, Smigel MD, Gilman AG (1987) Effects of Mg^{2+} and the $\beta\gamma$-subunit complex on the interactions of guanine nucleotides with G-proteins. J Biol Chem 262:762–766

Hildebrandt JD, Sekura RD, Codina J, Iyengar R, Manclark CR, Birnbaumer L (1983) Stimulation and inhibition of adenylyl cyclase mediated by distinct regulatory proteins. Nature (London) 302:706–709

Hildebrandt JD, Codina J, Rosenthal W, Birnbaumer L, Neer E, Yamazaki A, Bitensky MW (1985) Characterization by two-dimensional peptide mapping of the γ-subunits of N_s and N_i, the regulatory proteins of adenylyl cyclase, and of transducing, the guanine nucleotide-binding protein of rod outer segment of the eye. J Biol Chem 260:14867–14872

Hingorani VN, Ho YK (1988) Fluorescent labeling of signal transducing G-proteins. Pertussis toxin-catalyzed etheno-ADP ribosylation of transducin. J Biol Chem 263:19804–19808

Hinsch KD, Tychowiecka I, Gausepohl H, Frank R, Rosenthal W, Schultz G (1989) Tissue distribution of β_1- and β_2-subunits of regulatory guanine nucleotide-binding proteins. Biochim Biophys Acta 1013:60–67

Holbrook SR, Kim SH (1989) Molecular model of the G protein α subunit based on the crystal structure of the HRAS protein. Proc Natl Acad Sci USA 86:1751–1755

Huff RM, Neer EJ (1986) Subunit interactions of native and ADP-ribosylated α_{39} and α_{41}, two guanine nucleotide-binding proteins from bovine cerebral cortex. J Biol Chem 261:1105–1110

Iyengar R, Rich KA, Herberg JT, Grenet D, Mumby S, Codina J (1987) Identification of a new GTP-binding protein. A M_r = 43,000 substrate for pertussis toxin. J B Chem 262:9239–9245

Iyengar R, Rich KA, Herberg JT, Premont RT, Codina J (1988) Glucagon receptor-mediated activation of G_s is accompanied by subunit dissociation. J Biol Chem 263:15348–15353

Jakobs KH, Wieland T (1989) Evidence for receptor-regulated phosphotransfer reactions involved in activation in the adenylate cyclase inhibitory G protein in human platelet membranes. Eur J Biochem 183:115–121

Jones DT, Reed RR (1987) Molecular cloning of five GTP binding protein cDNA species from rat olfactory neuroepithelium. J Biol Chem 262:14241–14249

Katada T, Gilman AG, Watanabe Y, Bauer S, Jakobs KH (1985) Protein kinase C phosphorylates the inhibitory guanine-nucleotide-binding regulatory component and apparently suppresses its function in hormonal inhibition of adenylate cyclase. Eur J Biochem 151:431–437

Katada T, Oinuma M, Ui M (1986) Mechanisms for the inhibition of the catalytic activity of adenylate cyclase by the guanine nucleotide-binding proteins serving as the substrate of islet activating protein, pertussis toxin. J Biol Chem 261:5215–5221

Katada T, Oinuma M, Kusakabe K, Ui M (1987) A new GTP-binding protein in brain tissues serving as the specific substrate of islet-activating protein, pertussis toxin. FEBS Lett 213:353–358

Keenan AK, Cooney D, Holzhöfer A, Dees C, Hekman M (1987) Unimpaired coupling of phosphorylated, desensitized β-adrenoceptor to G_s in a reconstitution system. FEBS Lett 217:287–291
Kelleher DJ, Johnson GL (1988) Transducin inhibition of light-dependent rhodopsin phosphorylation: evidence for $\beta\gamma$ subunit interaction with rhodopsin. Mol Pharmacol 34:452–460
Kim D, Lewis DL, Grazidei L, Neer EJ, Bar-Sagi D, Clapham DE (1989) G-protein $\beta\gamma$-subunits activate the cardiac muscarinic K^+-channels via phospholipase A_2. Nature (London) 337:557–560
Kurose H, Katada T, Amano T, Ui M (1983) Specific uncoupling by islet-activating protein, pertussis toxin, of negative signal transduction via α-adrenergic, cholinergic, and opiate receptors in neuroblastoma × glioma hybrid cells. J Biol Chem 258:4870–4875
Lang J (1989) Purification and characterization of subforms of the guanine nucleotide-binding proteins $G\alpha_i$ and $G\alpha_o$. Eur J Biochem 183:687–692
Lang J, Costa T (1987) Antisera against the 3-17 sequence of rat $G\alpha_i$ recognize only a 40 kDa G-protein in brain. Biochem Biophys Res Commun 148:838–848
Lang J, Costa T (1989a) Chronic exposure of NG108-15 cells to opiate agonists does not alter the amount of the guanine-nucleotide binding proteins G_i and G_o. J Neurochem 53:1500–1506
Lang J, Costa T (1989b) Distribution of the α-subunit of the guanine nucleotide-binding protein G_{i2} and its comparison to $G\alpha_o$. J Recept Res 9:313–331
Lang J, Schulz R (1989) Chronic opiate receptor activation in vivo alters the level of G-protein subunits in guinea-pig myenteric plexus. Neuroscience 32:503–510
Law PY, Hom DS, Loh HH (1983) Opiate receptor down-regulation and desensitization in neuroblastoma × glioma NG108-15 hybrid cells are two separate cellular adaptation processes. Mol Pharmacol 24:413–424
Law PY, Louie AK, Loh HH (1985) Effects of pertussis toxin treatment on the down-regulation of the opiate receptors in neuroblastoma × glioma NG108-15 hybrid cells. J Biol Chem 260:14818–14823
Liang BT, Galper JB (1988) Differential sensitivity of α_o and α_i to ADP-ribosylation by pertussis toxin in the intact cultured embryonic chick ventricular myocyte. Biochem Pharmacol 37:4549–4555
Louie AK, Law PY, Loh HH (1986) Cell-free desensitization of opioid inhibition of adenylate cyclase in neuroblastoma × glioma NG108-15 hybrid cell membranes. J Neurochem 47:733–737
Manning DR, Fraser BA, Kahn RA, Gilman AG (1984) ADP-ribosylation of transducin by islet activating protein. Identification of asparagine as the site of ADP-ribosylation. J Biol Chem 259:749–756
Marbach I, Shiloach J, Levitzki A (1988) G_i affects the agonist-binding properties of β-adrenoceptors in the presence of G_s. Eur J Biochem 172:239–246
Masters SB, Sullivan KA, Miller RT, Beidermann B, Lopez NG, Ramachandran J, Bourne HR (1988) Carboxy terminal domain of $G\alpha_s$ specifies coupling of receptors to stimulation of adenylyl cyclase. Science 241:448–451
Mattera R, Codina J, Sekura R, Birnbaumer L (1986) The interaction of nucleotides with pertussis toxin. Direct evidence for a nucleotide binding site on the toxin regulating the rate of ADP-ribosylation of N_i, the inhibitory regulatory component of adenylyl cyclase. J Biol Chem 261:11173–11179
McCormick F, Clark BF, La Cour TF, Kjeldgaard M, Norskov-Lauritsen L, Nybord J (1985) A model of the tertiary structure of p21, the product of the ras oncogene. Science 230:78–82
McFadzean I, Mullaney I, Brown DA, Milligan G (1989) Antibodies to the GTP binding protein, G_o, antagonize noradrenaline-induced calcium current inhibition in NG108-15 hybrid cells. Neuron 3:177–182
McKenzie FR, Kelly ECH, Unson CG, Spiegel AM, Milligan G (1988) Antibodies which recognise the C-terminus of the inhibitory guanine nucleotide-binding protein (G_i) demonstrate that opioid peptides and fetal calf serum stimulate the high affinity GTPase activity of two separate pertussis toxin substrates. Biochem J 249:653–659
Milligan G (1988) Techniques used in the identification and analysis of function of pertussis toxin-sensitive guanine nucleotide binding proteins. Biochem J 255:1–13
Milligan G, McKenzie FR (1988) Opioid peptides promote cholera-toxin-catalyzed ADP-ribosylation of the inhibitory guanine-nucleotide-binding protein (G_i) in membranes of neuroblastoma × glioma hybrid cells. Biochem J 252:369–373
Milligan G, Mullaney I, Unson CG, Marshall L, Spiegel AM, McArdle H (1988) GTP analogues promote release of the α subunit of the guanine nucleotide binding protein, G_{i2}, from membranes of rat glioma C6 BU1 cells. Biochem J 254:391–396
Mollereau C, Pascaud A, Baillat G, Mazarguil H, Puget A, Meunier JC (1988) 5′-Guanylyl-

imidophosphate decreases affinity for agonists and apparent molecular size of frog brain opioid receptor in digitonin solution. J Biol Chem 263:18003–18008
Mollner S, Pfeuffer T (1988) Two different adenylyl cyclases in brain distinguished by monoclonal antibodies. Eur J Biochem 171:265–271
Moss J, Oppenheimer NJ, West RE, Stanley SJ (1986) Amino acid specific ADP-ribosylation: substrate specificity of an ADP-ribosylarginine hydrolase from turkey erythrocytes. Biochemistry 25:5408–5414
Mumby SM, Kahn RA, Manning DR, Gilman AG (1986) Antisera of designed specificity for subunits of guanine nucleotide-binding regulatory proteins. Proc Natl Acad Sci USA 83:265–269
Mumby S, Pasng IH, Gilman AG, Sternweis PC (1988) Chromatographic resolution and immunologic identification of the α_{40} and α_{41} subunits of guanine-nucleotide binding proteins from bovine brain. J Biol Chem 263:2020–2026
Murray R, Keenan AK (1989) The β-adrenoceptor is precoupled to G_s in chicken erythrocyte membranes. Cell Signal 1:173–180
Nakafuku M, Itoh H, Nakamura S, Kaziro Y (1987) Occurrence in *Saccharomyces cervisiae* of a gene homologous to the cDNA coding for the α subunit of mammalian G proteins. Proc Natl Acad Sci USA 84:2140–2144
Neer EJ, Clapham DE (1988) Roles of G protein subunits in transmembrane signalling. Nature (London) 333:129–134
Neer EJ, Pulsifer L, Wolf LG (1988) The amino terminus of G protein α subunits is required for interaction with $\beta\gamma$. J Biol Chem 263:8996–9000
Nestler EJ, Erdos JJ, Terwilliger R, Duman RS, Tallman JF (1989) Regulation of G proteins by chronic morphine in the rat locus coeruleus. Brain Res 476:203–209
Northup JK, Smigel MD, Gilman AG (1982) The guanine nucleotide activating site of the regulatory component of adenylate cyclase. Identification by ligand binding. J Biol Chem 257:11416–11423
Owens JR, Frame LT, Ui M, Cooper DMF (1985) Cholera toxin ADP-ribosylates the islet-activating protein substrate in adipocyte membranes and alters its function. J Biol Chem 260:15946–15952
Pfeuffer T (1977) GTP-binding proteins in membranes and the control of adenylate cyclase activity. J Biol Chem 252:7224–7234
Premont RT, Buku A, Iyengar R (1989) The $G\alpha_z$ gene product in human erythrocytes. Identification as a 41 kDa protein. J Biol Chem 264:14960–14964
Ransnäs LA, Insel PA (1988) Quantitation of the guanine nucleotide binding regulatory protein G_s in S49 cell membranes using antipeptide antibodies to α_s. J Biol Chem 263:9482–9485
Rasenick MM, Wang N (1988) Exchange of guanine nucleotides between tubulin and GTP-binding proteins that regulate adenylate cyclase: cytoskeletal modulation of neuronal signal transduction. J Neurochem 51:300–311
Robishaw JD, Kalman VK, Moomaw CR, Slaughter CA (1989) Existence of two γ subunits of the G proteins in brain. J Biol Chem 264:15758–15761
Rodbell M (1985) Programmable messengers: a new theory of hormone action. Trends Biochem Sci 10:461–464
Rodbell M, Krans HMJ, Pohl SL, Birnbaumer L (1971) The glucagon sensitive adenyl cyclase system in plasma membranes of rat liver. J Biol Chem 2246:14872–14876
Rosenthal W, Hescheler J, Trautwein W, Schultz G (1988) Receptor- and G-protein-mediated modulations of voltage-dependent calcium channels. Cold Spring Harbor Symp Quant Biol 53:247–254
Rotrosen D, Gallin JI, Spiegel AM, Malech HL (1988) Subcellular localization of $G_i\alpha$ in human neutrophils. J Biol Chem 263:10958–10964
Rouot B, Carrette J, Lafontan M, Lan Tran P, Fehrentz JA, Bockaert J, Toutant M (1989) The adipocyte $G_o\alpha$-immunoreactive polypeptide is different from the α subunit of the brain G_o protein. Biochem J 260:307–310
Scherer NM, Toro MJ, Entman M, Birnbaumer L (1987) G-protein distribution in canine cardiac sarcoplasmic reticulum and sarcolemma: comparison to rabbit skeletal muscle membranes and to brain and erythrocyte G-protein. Arch Biochem Biophys 259:431–440
Schubert B, VanDongen AMJ, Kirsch GE, Brown AM (1989) β-Adrenergic inhibition of cardiac sodium channels by dual G-protein pathways. Science 245:516–519
Schulz R, Lang J, Sinowatz F (1989) Chronic treatment of guinea pigs increases the amount of G-proteins

in the myenteric plexus. In: Cros J, Meunier J-C, Hamon M (eds) Progress in opioid research. Adv Biosci 75:707–710
Seifert R, Schultz G (1989) Involvement of pyrimidinoceptors in the regulation of cell function by uridine and by uracil nucleotides. Trends Pharmacol Sci 10:365–369
Sharma SK, Nirenberg M, Klee W (1979) Morphine receptors as regulators of adenylate cyclase activity. Proc Natl Acad Sci USA 76:5256–5630
Smigel MD (1986) Purification of the catalyst of adenylyl cyclase. J Biol Chem 261:1976–1982
Sternweis PC (1986) The purified α-subunits of G_o and G_i from bovine brain require βγ for association with phospholipid vesicles. J Biol Chem 261:631–637
Sternweis PC, Northup JK, Smigel MD, Gilman AG (1981) The regulatory component of adenylate cyclase. Purification and properties. J Biol Chem 256:11517–11526
Stryer L, Bourne HR (1986) G-proteins: a family of signal transducers. Annu Rev Cell Biol 2:391–419
Sullivan KA, Miller RT, Masters SB, Beiderman B, Heidemann W, Bourne HR (1987) Identification of receptor contact site involved in receptor G-protein coupling. Nature (London) 330:758–759
Thomas JM, Hoffman BB (1987) Adenylate cyclase supersensitivity; a general means of cellular adaptation to inhibitory agonists? Trends Pharmacol Sci 8:308–311
Toselli M, Lang J, Costa T, Lux HD (1989) Direct modulation of voltage-dependent calcium channels by muscarinic activation of a pertussis-toxin sensitive G-protein in hippocampal neurons. Pflueugers Arch Physiol 415:255–261
Toutant M, Barhanin J, Bockaert J, Rouot B (1988) G-proteins in skeletal muscle: evidence for a 40 kDa pertussis-toxin substrate in purified transverse tubules. Biochem J 254:405–409
Tsuda M, Tsuda T, Teramaya Y, Fukada Y, Akino T, Yamanaka G, Stryer L, Katada T, Ui M, Ebrey T (1986) Kinship of cephalopod photoreceptor G-protein with vertebrate transducin. FEBS Lett 198:5–10
Ueda H, Hirada H, Nozaki M, Katada T, Ui M, Satoh M, Takagi H (1988) Reconstitution of brain μ-receptors with purified guanine nucleotide-binding regulatory proteins, G_i and G_o. Proc Natl Acad Sci USA 85:7013–7017
Ui M (1984) Islet-activating protein, pertussis toxin: a probe for the functions of the inhibitory guanine nucleotide regulatory component of adenylate cyclase. Trends Pharmacol Sci 277–279
Vachon L, Costa T, Herz A (1987) GTPase and adenylate cyclase desensitize at different rates in NG 108–15 cells. Mol Pharmacol 31:159–168
VanDongen AMJ, Codina J, Olate J, Mattera R, Joho R, Birnhaumer L, Brown AM (1988) Newly identified brain potassium channels gated by the guanine nucleotide binding protein G_o. Science 242:1433–1437
Van Regenmortel MHV (1989) Stuctural and functional approaches to the study of protein antigenicity. Immunol Today 10:253–282
Watkins DC, Northup JK, Malbon CC (1987) Pertussis toxin treatment in vivo is associated with a decline in G-protein β-subunits. J Biol Chem 264:4184–4194
Wilden U, Hall SW, Kuhn H (1986) Phosphodiesterase activation by photoexcited rhodopsin is quenched when rhodopsin is phosphorylated and binds the intrinsic 48-kDa protein of rod outer segments. Proc Natl Acad Sci USA 83:1174–1178
Yatani A, Codina J, Brown AM, Birnbaumer L (1987) Direct activation of mammalian atrial muscarinic potassium channels by GTP regulatory protein G_k. Science 235:207–211
Yatani A, Mattera R, Codina J, Graf R, Okabe K, Padrell E, Iyengar R, Brown AM, Birnbaumer L (1988) The G-protein-gated atrial K^+ channels is stimulated by three distinct $G_i\alpha$-subunits. Nature (London) 336:680–682

Chapter 8

Opioid Receptors and Ion Channels

R.A. North

1 Introduction

At least three basic kinds of membrane receptor for transmitters and hormones can now be distinguished. The first type incorporates an ion channel (e.g., nicotinic acetylcholine receptor, γ-aminobutyric acid$_A$ receptor, 5-hydroxytryptamine$_3$ receptor); agonist binding greatly increases the probability of channel opening (Fig. 1A). The second kind has an internal domain which binds guanosine 5′-triphosphate (GTP) binding protein (G-protein); agonist binding results in a much increased rate of synthesis of activated G-protein, which then interacts with different cellular effectors (Fig. 1B). The third class itself has enzyme activity on the internal domain, which is altered by agonist binding to an external domain (e.g., insulin and atrial naturietic factor receptors).

Despite the slow progress in determining the molecular structure of the opioid receptors, considerable evidence exists to place them in the second class. Agonist binding shows a typical shift in affinity with added GTP; agonists stimulate GTPase activity; μ- and δ-receptors copurify with G-proteins; and μ-receptors can be reconstituted with both G_i and G_o (Ueda et al. 1988). The activated G-proteins have been shown to couple to at least three cellular effectors: adenylate cyclase is inhibited, K^+ channels are opened, and Ca^{2+} channels are closed. The consequence of opening K^+ channels is inhibition of cell firing; such an action on neurons in the pain transmission pathway probably underlies the analgesic actions of opioids. The consequences of reducing Ca^{2+} currents are less well understood but may include presynaptic inhibition as well as less direct effects on intracellular metabolism such as alterations in the phosphorylation states of proteins.

2 K^+ Conductance Increase

2.1 *Hyperpolarization*

Opioids inhibit the firing of neurons in many regions of the mammalian nervous system (Duggan and North 1984); where excitation has been observed (e.g., hippocampus, ventral tegmental area), it occurs through disinhibition, because a population of tonically active inhibitory interneurons are themselves inhibited by the opioids (Zieglgänsberger et al. 1979; Madison and Nicoll 1988). The direct inhibition of action potentials results from membrane hyperpolarization; the

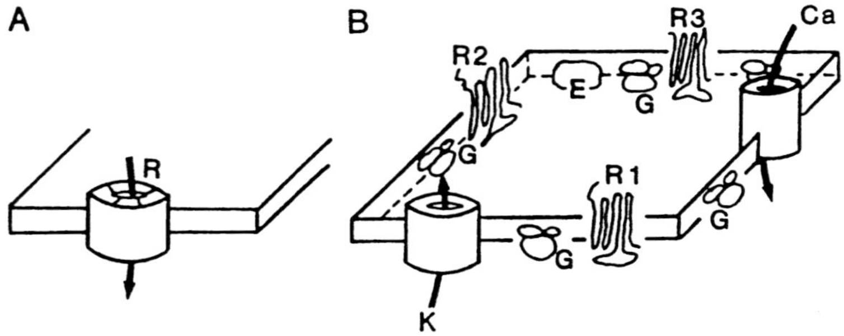

Fig. 1a,b. Two ways in which cell membrane receptors (*R*) couple to membrane ion channels (*arrows* through membrane). **A** The agonist binds directly to a subunit of the channel protein. Channel opening occurs in less than 1 ms. Examples in mammalian nerves include nicotinic, glutamate, 5-HT_3, and a type of ATP receptor (channels are permeable to cations) as well as $GABA_A$ and glycine receptors (channel is permeable to anions). **B** Agonist binding to the receptor catalyzes the formation of activated G-protein, which then interacts with one or more nearby channels. Different receptors (*R1*, *R2*, *R3*) can thereby couple to common effectors (e.g., K^+ or Ca^{2+} channels). The activated G-proteins can also combine with enzymatic effectors such as adenylate cyclase (*E*); the resultant changes in the level of product can open or close ion channels at more remote parts of the cell

membrane hyperpolarizes because a set of K^+ channels opens allowing ions to leave the cell (North 1986). Tissues in which this action has been shown include the locus coeruleus (Pepper and Henderson 1980), myenteric plexus (Morita and North 1982), submucous plexus (Mihara and North 1986), and hypothalamus (Charpak et al. 1988) of the guinea pig; the locus coeruleus (Williams et al. 1982; North and Williams 1985; North et al. 1987), substantia gelatinosa (Yoshimura and North 1983), parabrachialis (Christie and North 1988a), hippocampal interneurons (Madison and Nicoll 1988), and substantia nigra secondary neurons (Lacey et al. 1989) of the rat; and dorsal root ganglion cells of the mouse (Werz and Macdonald 1983).

2.2 *Properties of the K^+ Conductance*

The conductances activated by both μ-receptors (in rat locus coeruleus) and δ-receptors (in guinea pig submucous plexus) have similar macroscopic properties. They show inward rectification, and are very sensitive to external Ba^{2+}. These properties are similar to those of "the" inward rectifier K^+ conductance that has been widely characterized in egg cells, skeletal muscle, and neurons (see Hille 1984), except that the conductance is activated at less negative potentials. "The" inward rectifier is not significantly open until the membrane potential approaches the K^+ equilibrium potential (E_K), and thus plays a significant role in cells which normally have a resting potential fairly close to E_K. The neurons in which opioids have been shown to cause large conductance increases normally have resting potentials around –55 to –65 mV, potentials at which "the" inward rectifier is not significantly activated. However, this potential is about the midpoint for activation of the conductance that opens in the presence of opioids, ensuring that these agonists can cause significant hyperpolarization (Williams et al. 1988).

2.3 Coupling Through G-Proteins

In both the locus coeruleus and the submucous plexus, a pertussis toxin-sensitive G-protein has been shown to be involved in the coupling between receptor and K^+ channel (North and Williams 1985; Aghajanian and Wang 1986; North et al. 1987). The precise identity of the G-protein(s) used by the opioids to open K^+ channels is not known. Both G_i and G_o will reconstitute with purified μ-receptors, with about two times more G_i than G_o coupling on a molar basis (Ueda et al. 1988). In submucous plexus neurons pretreated with pertussis toxin, the response to opioid agonists (in this case at δ-receptors) is lost; however, the coupling between receptor and channel can be restored if purified G_i or G_o is introduced into the cell interior, and there is little difference in their relative effectiveness (Tatsumi et al. 1990).

2.4 Single K^+ Channels

The individual K^+ channels opened by opioids have been studied in the locus coeruleus; they have a unit conductance of about 45 pS when the K^+ concentration on both sides of the patch is about equal (Miyake et al. 1989). In the presence of agonist, the channels show periods of activity lasting for several seconds or even minutes, separated by similarly long closed periods. Within the periods of activity the distributions of open and closed times suggest that there may be two closed states (mean durations about 1 and 30 ms) and one open state. The duration of the open state is longer when higher agonist concentrations are used (e.g., 1 μM instead of 100 nM DAGO). The finding suggests that an activated G-protein might act cooperatively to open the K^+ channel, perhaps by binding to several channel subunits.

The electrophysiological work on single channels indicates that μ-opioid receptors are spread fairly uniformly over the surface of the cell; based on the dimensions of the cell and the dimensions of the tip of the patch clamp electrode, one can calculate that the density would be of the order of 1000 per cell. The density of receptors cannot be estimated because nothing is known about the stoichiometry of receptor to G-protein to channel; however, because the receptors and channels are fairly intimately located with respect to each other, it follows that the μ-receptors are distributed over the entire cell soma.

2.5 The Same Channels Are Opened by Agonists at Other Receptors

It is important to realize that the same K^+ channels that are opened by opioids are also opened by agonists at many other receptors; it is likely, but not proven, that the other receptors share the same pool of G-protein. Examples of cells expressing other receptors which share channels with opioid receptors are guinea pig submucous plexus (δ, adrenergic α_2, somatostatin), rat locus coeruleus (μ, α_2), rat parabrachialis (μ, $GABA_B$, muscarinic M_2), and rat substantia nigra secondary cells (μ, $GABA_B$) (for review see North 1989). The properties of the single channels opened by acetylcholine acting at M_2 receptors in the heart (Soejima and Noma 1984) are similar to those opened by opioids in rat locus coeruleus. In both cases, the agonist is effective only when applied to the membrane within a few square micrometers of the channel, implying that there is not a freely diffusible second messenger involved. K^+ channels having similar properties can also be opened by applying activated G-protein α-subunit directly to the inner surface of the cell membrane in inside-out patches (van Dongen et al. 1988).

Not all receptors on a neuron couple equally well to the K^+ channels. Both μ-opioids and α_2-receptor agonists increase the K^+ conductance of rat locus coeruleus neurons, but muscarine does not. However, muscarine does have other actions on the neurons which indicate that these cells do express muscarinic M_2 receptors (Egan and North 1986; Christie and North 1988b). In the closely adjacent neurons of the dorsal parabrachial nucleus, both μ-opioids (acting at μ-receptors) *and* muscarinic agonists (acting at M_2 receptors) will cause an increase in K^+ conductance. In neurons of the guinea pig submucous plexus, agonists at δ-opioid, α_2-adrenergic, and somatostatin receptors all cause an increase in K^+ conductance; however, during several hours of recording from the same cell, the responses to opioid agonists often disappear, whereas those to somatostatin remain reproducible.

2.6 *Lack of Synaptic Potentials Is Enigmatic*

Some of the receptors that share K^+ channels with opioid receptors have been shown to be involved in synaptic transmission. Thus, synaptic potentials mediated by noradrenaline (Hirst and McKirdy 1975; Egan et al. 1983; Surprenant and North 1988), by 5-hydroxytryptamine (Yoshimura and Higashi 1985; Pan et al. 1989) and by acetylcholine (Hartzell et al. 1977; Dodd and Horn 1983) have been recorded in various vertebrate neurons. On the other hand, similar synaptic potentials mediated by opioids have not been reported – even in tissues which are dense with terminals containing enkephalin, and loaded with cells expressing opioid receptors that respond well to exogenous agonists (for example, guinea pig submucous plexus neurons, North et al. 1987; rat pelvic ganglion neurons, Jiang Z-G and North RA, unpubl. observ.). The reason for this is not altogether clear, but may be related to our ignorance regarding the factors necessary to release the opioid peptides.

3 Ca^{2+} Conductance Decrease

A depression of Ca^{2+} current by opioid peptides has been extensively studied in both chick and mouse dorsal root ganglion cells. The first experiments measured the shortening of the duration of the action potential (Mudge et al. 1979; Werz and Macdonald 1983; Cherubini and North 1985), but this left open the possibility that a K^+ conductance increase, rather than a Ca^{2+} conductance decrease, was responsible. More recently, voltage-clamp recordings have shown a direct reduction in Ca^{2+} current in neuroblastoma × glioma hybrid cells (presumably δ-receptors: Hescheler et al. 1987), mouse dorsal root ganglion cells (κ-receptors: Macdonald and Werz 1986), and guinea pig submucous plexus neurons (δ-receptors: North et al. 1988). In the case of the dorsal root ganglion cells, the κ-receptor is involved in the inhibition of the Ca^{2+} current; in the neuroblastoma × glioma hybrid and submucous neurons the δ-receptor is responsible. Inhibition of Ca^{2+} currents by activating μ-receptors has also recently been described for a human neuroblastoma cell line (E. Seward and G. Henderson pers. commun.). In most of the cells in which a decrease in Ca^{2+} currents has been observed, there is no effect on K^+ conductance. However, guinea pig submucous neurons show both the increase in K^+ conductance and the decrease in Ca^{2+} conductance in the same cells; both appear to result from δ-receptor activation. On the other hand, rat locus coeruleus neurons show a large increase in K^+ conductance when μ-opioids are applied but there is no effect on Ca^{2+} currents – even though another agonist (muscarine) will inhibit the Ca^{2+} currents in these cells (M.J. Christie, J.T. Williams and R.A. North unpubl. observ.).

3.1 Transient and Sustained Ca^{2+} Currents

The particular properties of the Ca^{2+} conductance affected by opioids have not been worked out at the single channel level (Tsien et al. 1988). Whole cell recordings from dorsal root ganglion cells indicate that there are two current components that can be differentiated primarily by the degree to which they inactivate during a depolarizing pulse (typically from −60 to 0 mV) continuing for several 100 ms. Both the transient and sustained currents are reduced by dynorphin (DYN), the maximum inhibition of either component is about 40 or 50% (Macdonald and Werz 1986; Gross and Macdonald 1986). A very similar result was obtained for guinea pig submucous plexus neurons, in which δ-receptors were involved (R.A. North, K.-J. Shen, M. Shimerlik, H. Tatsumi, and A. Surprenant, unpubl. observ.). Larger and more selective inhibitions of the transient Ca^{2+} current were observed in neuroblastoma × glioma hybrid cells (Hescheler et al. 1987).

3.2 G-Proteins Couple Receptor to Channel

The inhibition of Ca^{2+} currents by opioids is lost after treatment of the cells with pertussis toxin (shown for the effects of δ-receptor agonists on neuroblastoma × glioma cells and guinea pig submucous plexus neurons). In both cell types, the effectiveness of the opioids is restored if G-proteins are allowed to enter the cell from the patch-clamp recording pipette. In the case of the neuroblastoma × glioma hybrid cells, G_o was about ten times more effective than G_i in restoring the response to opioids.

3.3 The Same Currents Are Depressed by Agonists at Other Receptors

Agonists at α_2-adrenoceptors, somatostatin receptors, and receptors depress the same Ca^{2+} current in guinea pig submucous plexus neurons (North et al. 1988). Both adrenaline and D-Ala-D-Leu-enkephalin depress the Ca^{2+} currents in neuroblastoma × glioma hybrid cells (NG108-15; Hescheler et al. 1987). However, selectivity in coupling receptor to effector pertains here as for the K^+. Carbachol, a muscarinic agonist, inhibits adenylate cyclase and increases K^+ conductance in these cells but does not depress Ca^{2+} currents (Hescheler et al. 1987).

3.4 Functional Role of Ca^{2+} Current Inhibition

The functional significance of the reduction in Ca^{2+} current is not known. It is often suggested that one consequence would be inhibition of transmitter release but there are several difficulties with this suggestion. First, it has not been shown that the kind of Ca^{2+} channels affected by opioids play a role in transmitter release (but see Miller 1987). Second, the currents studied in voltage-clamp experiments typically reach their peak in 10–20 ms, whereas the action potential in nerve terminals presumably reaches its peak much more rapidly. Third, an increase in K^+ conductance in nerve terminals would be expected to reduce action potential duration [as has been shown in cell bodies for μ- and δ-receptors (North and Williams 1983; Werz and Macdonald 1983)], and hence transmitter release. Fourth, a pertussis toxin-sensitive G-protein is known to be activated by opioid agonists; this may inhibit release by an action unrelated to any effects on ion channels, such as inhibition of Ca^{2+} release from internal stores (Gill et al. 1986).

It was mentioned above that opioids have three actions in several tissues: inhibition of adenylate cyclase, increase in K^+ conductance, and decrease in Ca^{2+} conductance. In these respects also, opioid receptors belong to a family that includes one receptor type for each of the major neurotransmitters (North et al. 1987). Experiments on submucous neurons of the guinea pig indicate that both the increase in K^+ conductance and the decrease in Ca^{2+} conductance can occur in the same cell (North et al. 1988). It is difficult to show adenylate cyclase inhibition on single cells, but the neuroblastoma × glioma cell line (NG108-15), in which the cyclase inhibition was shown (Sharma et al. 1975), is the same as that in which Ca^{2+} currents are inhibited. Several studies support the view that the inhibition of adenylate cyclase is independent of the effects on either K^+ or Ca^{2+} channels (North and Williams 1985; Hescheler et al. 1987; North et al. 1987), and presumably leads to other, perhaps longer-term, consequences of opioid action. The simplest explanation is that agonist binding to opioid receptors results in the activation of one or more G-proteins, which can then change the activity of one or more effectors. Such divergence of transmitter action within the cell may serve to bring about a concerted response to agonist action: K^+ conductance increase will inhibit cell firing, whereas reduced Ca^{2+} entry and inhibited adenylate cyclase might bring about longer-term adaptations to the inhibition of firing, such as a reduction in transcription of certain genes.

4 Tolerance and Dependence

Much of Prof. Herz's work has been directed toward an understanding of the long-term changes that accompany opiate administration. The coupling to K^+, Ca^{2+} channels and adenylate cyclase is shared by many other receptors, and this raises the question "What is unique about the opioids which results in the long-term changes leading to tolerance?" Tolerance can be shown with respect to inhibition of adenylate cyclase (Sharma et al. 1975) and K^+ conductance increase at the level of single cells (Christie et al. 1987). Experiments with irreversible antagonists indicate that tolerance is functionally equivalent to a loss of receptors, or to a relative uncoupling of receptor to G-proteins (these two possibilities have not been distinguished) (Chavkin and Goldstein 1984; Christie et al. 1987).

To shed light on this problem, rats were treated over several days with morphine, and intracellular recordings subsequently made from locus coerulues neurons in a brain slice (Christie et al. 1988). There were three important findings. First, the tolerance that was observed to the actions of the applied morphine was accompanied by a reduction in the maximum K^+ current evoked by morphine. However, there was no reduction in the maximum current elicited by [Met^5]enkephalin or Tyr-D-Ala-Gly-MePhe-Gly-ol. In this respect, the effect of chronic exposure to morphine was not experimentally distinguishable from the effects of 30-min exposure of normal cells to β-funaltrexamine (β-FNA). If β-funaltrexamine is assumed to act by irreversibly inactivating a fraction of the receptors on the cell surface, this result implies that morphine maximally opens the K^+ channels on a given cell only by occupying about 100% of the receptors, whereas the opioid peptides can cause the same K^+ conductance increase at a significantly lower fractional occupancy. The two possible explanations for the changes brought about by long-term exposure to morphine are (1) a reduction in the number of opioid receptors on the cell surface and (2) impaired coupling between activated μ-receptors and K^+channels. These electrophysiological studies could not differentiate between the two possibilities. There was no change in the dissociation equilibrium constant for naloxone on the neurons, suggesting that whatever change that had occurred in the receptor did not affect its affinity for naloxone.

The second finding was that there was no change in the sensitivity of the cells to agonists acting at α_2 receptors (either the full agonist UK14304 or the partial agonist clonidine). Since these agonists open the same population of K^+ channels as those opened by the opioids, this result puts the primary change occurring in tolerance either at the μ-receptor or at a G-protein uniquely accessed by μ-receptors.

The third finding of interest was that naloxone (NAL) had little or no effect on the neurons taken from morphine-treated animals, except to reverse the actions of any morphine which was present in the perfusing solution. In other words, there was no evidence for "dependence" at the level of the single cell. The excitation by NAL of single neurons taken from morphine-treated animals results mostly from the increased synaptic inputs to the cell either in vivo (Aghajanian 1978) or in vitro (Cherubini et al. 1988).

5 Conclusions

Presently available results suggest that opioid receptors couple through G-proteins to at least three effector molecules (Fig. 1B). The first is a K^+ channel; when it opens, K^+ ions leave the cell and hyperpolarization results. The second is a Ca^{2+} channel; inward Ca^{2+} currents are reduced as a result of opioid action. The third is adenylate cyclase, which is inhibited. Opioid receptors share these effectors with an extended family of receptors for other transmitters (North 1989). Future studies of opioid actions on ion conductances must proceed in several directions. One will be the extension of present studies into new brain regions, with direct demonstration of the receptor types involved and the ion conductances affected on identified neurons. A second will be an increased understanding of the neuronal connections responsible for given behaviors, so that the actions of opioids on sets of neurons can be interpreted. This will be particularly important for an understanding of the manifestations of withdrawal, which require sets of interconnected neurons. A third area of critical importance is the molecular cloning of the receptors, so that the regulation of their synthesis and expression can be investigated. Several types of G-protein mRNA are available, and a start has been made with the cloning of K^+ channels. In the not too distant future it may be possible to express receptor, G-protein, and channel in the amphibian oocyte or a mammalian cell. Such experiments promise to take our understanding of opioid actions of ion conductances to the molecular resolution which may be necessary to appreciate the changes responsible for tolerance.

References

Aghajanian GK (1978) Tolerance of locus coeruleus neurones to morphine and suppression of withdrawal response by clonidine. Nature (London) 276:186–188
Aghajanian GK, Wang YY (1986) Pertussis toxin blocks the outward currents evoked by opiate and α_2 agonists in locus coeruleus neurones. Brain Res 371:390–394
Charpak S, Dubois-dauphin M, Raggenbass M, Dreifuss JJ (1988) Direct inhibition by opioid peptides of neurones located in the ventromedial nucleus of the guinea pig hypothalamus. Brain Res 450:124–130
Chavkin C, Goldstein A (1984) Opioid receptor reserve in normal and morphine-tolerant guinea-pig ileum myenteric plexus. Proc Natl Acad Sci USA 81:7253–7257
Cherubini E, North RA (1985) μ and κ opioids inhibit transmitter release by different mechanisms. Proc Natl Acad Sci USA 82:1860–1863
Cherubini E, North RA, Tokimasa T (1988) Action of naloxone on myenteric neurons removed from morphine-treated guinea pigs. J Pharmacol Exp Ther 247:830–838
Christie MJ, North RA (1988a) Agonists at μ opioid, M_2 muscarinic and $GABA_B$ receptors increase the same potassium conductance in rat lateral parabrachial neurones. Br J Pharmacol 95:896–902
Christie MJ, North RA (1988b) Cardiac type M_2 receptors mediate both the muscarinic excitation of locus coeruleus and hyperpolarization of dorsal parabrachial neurones. Trends Phamacol Sci Suppl 89
Christie MJ, Williams JT, North RA (1987) Cellular mechanisms of opioid tolerance: studies in single brain neurones. Mol Pharmacol 32:633–638
Dodd J, Horn JP (1983) Muscarinic inhibition of sympathetic C neurones in the bullfrog. J Physiol 334:271–291
Duggan AW, North RA (1984) Electrophysiology of opioids. Pharmacol Rev 35:219–281
Egan TM, North RA (1986) Acetylcholine acts on muscarinic M_2 receptors to excite rat locus coeruleus neurones. Br J Pharmacol 85:733–735

Egan TM, Henderson G, North RA, Williams JT (1983) Noradrenaline-mediated synaptic inhibition in locus coeruleus neurones. J Physiol 345:477–488
Gill DL, Ueda T, Chueh S-H, Noel MW (1986) Ca^{2+} release from endoplasmic reticulum is mediated by a guanine nucleotide regulatory mechanism. Nature (London) 320:461–463
Gross RA, Macdonald RL (1987) Dynorphin A selectively reduces a large transient (N-type) calcium current of mouse dorsal root ganglion neurones in cell culture. Proc Natl Acad Sci USA 84:5469–5473
Hartzell HC, Kuffler SW, Stickgold R, Yoshikami D (1977) Synaptic excitation and inhibition resulting from direct action of acetylcholine on two types of chemoreceptors on individual amphibian parasympathetic neurones. J Physiol 271:817–846
Hescheler J, Rosenthal W, Trautwein W, Schultz G (1987) The GTP-binding protein, G_o, regulates neuronal calcium channels. Nature (London) 325:445–447
Hille B (1984) Ionic channels in excitable membranes. Sinauer, Sunderland, Massachusetts
Hirst GDS, McKirdy HC (1975) Synaptic potentials recorded from some neurones of the submucous plexus of the guinea-pig small intestine. J Physiol 249:369–385
Lacey MG, Mercuri NB, North RA (1989) Two cells types in rat substantia nigra zona compacta distinguished by membrane properties. J Neurosci 9:1233–1241
Macdonald RL, Werz MA (1986) Dynorphin A decreases voltage-dependent calcium conductance in mouse dorsal root ganglion neurones. J Physiol 377:237–250
Madison DV, Nicoll RA (1988) Enkephalin hyperpolarizes interneurones in the rat hippocampus. J Physiol 398:123–130
Mihara S, North RA (1986) Opioids increase potassium conductance in guinea-pig submucous plexus neurones by activating δ receptors. Br J Pharmacol 88:315–322
Miller RJ (1987) Multiple calcium channels and neuronal function. Science 235:46–52
Miyake M, Christie MJ, North RA (1989) Single potassium channels opened by opioids in rat locus coeruleus neurones. Proc Natl Acad Sci USA 86:3419–3422
Morita K, North RA (1982) Opiate activation of potassium conductance of myenteric neurones: inhibition by calcium ions. Brain Res 242:145–150
Mudge AW, Leeman SE, Fischbach GD (1979) Enkephalin inhibits release of substance P from sensory neurones in culture and decreases action potential duration. Proc Natl Acad Sci USA 76:526–530
North RA (1986) Opioid receptor types and membrane ion channels. Trends Neurosci 9:114–117
North RA (1989) Drug receptors and the inhibition of nerve cells. Br J Pharmacol 98:13–28
North RA, Williams JT (1983) How do opiates inhibit neurotransmitter release. Trends Neurosci 6:337–339
North RA, Williams JT (1985) On the potassium conductance increased by opioids in rat locus coeruleus neurones. J Physiol 364:265–280
North RA, Williams JT, Surprenant A, Christie MJ (1987) μ and δ opioid receptors both belong to a family of receptors which couple to a potassium conductance. Proc Natl Acad Sci USA 84:5487–5491
North RA, Surprenant A, Tatsumi H (1988) Potassium conductance increase and calcium conductance decrease both evoked by α_2 adrenaline and δ opioid receptor agonists in the same guinea pig submucous plexus neurones. J Physiol 406:179P
Pan ZZ, Colmers WF, Williams JT (1989) 5-HT mediated synaptic potentials in the dorsal raphe nucleus: interactions with excitatory amino acid and GABA neurotransmission. J Neurophysiol 62:481–486
Pepper CM, Henderson G (1980) Opiates and opioid peptides hyperpolarize locus coeruleus neurones in vitro. Science 209:394–396
Sharma SK, Klee WA, Nirenberg M (1975) Dual regulation of adenylate cyclase accounts for narcotic dependence and tolerance. Proc Natl Acad Sci USA 72:3092–3096
Soejima M, Noma A (1984) Mode of regulation of the ACh-sensitive K^+-channel by the muscarinic receptor in rabbit atrial cells. Pflügers Arch 266:324–334
Surprenant A, North RA (1988) Mechanism of synaptic inhibition by noradrenaline acting at α_2 adrenoceptors. Proc R Soc London Ser B 234:85–114
Tatsumi H, Costa M, Schimerlik, North RA (1990) Potassium conductance increased by noradrenaline, opioids, somatostatin and G-proteins: whole-cell recording from guinea-pig submucous plexus neurons. J Neurosci (in press)
Tsien RW, Lipscombe D, Madison DV, Bley KR, Fox AP (1988) Multiple types of neuronal calcium channels and their selective modulation. Trends Neurosci 11:431–437
Ueda H, Harada H, Nozaki M, Katada T, Ui M, Satoh M, Takagi H (1988) Reconstitution of rat brain μ opioid receptors with purified guanine nucleotide binding regulatory proteins, G_i and G_o. Proc Natl Acad Sci USA 85:7013–7017

van Dongen AMJ, Codina J, Olate J, Mattera R, Joho R, Birnbaumer L, Brown AM (1988) Newly identified brain potassium channels gated by the guanine nucleotide binding protein G_o. Science 242:1433-1437
Werz MA, Macdonald RL (1983) Opioid peptides with differential affinity for mu and delta receptors decrease sensory neurone calcium-dependent action potentials. J Pharmacol Exp Ther 227:394-402
Williams JT, Egan TM, North RA (1982) Enkephalin opens potassium channels in mammalian central neurones. Nature (London) 299:74-76
Williams JT, North RA, Tokimasa T (1988) Inward rectification of resting and receptor-linked potassium currents in rat locus coeruleus neurones. J Neurosci 8:4299-4306
Yoshimura M, Higashi H (1985) 5-Hydroxytryptamine mediates inhibitory postsynaptic potentials in rat dorsal raphe neurones. Neurosci Lett 53:69-74
Yoshimura M, North RA (1983) Substantia gelatinosa neurones in vitro hyperpolarized by enkephalin. Nature (London) 305:529-530
Zieglgänsberger W, French ED, Siggins GR, Bloom FE (1979) Opioid peptides may excite hippocampal pyramidal neurons by inhibiting adjacent inhibitory interneurons. Science 205:415-417

CHAPTER 9

Cellular Signaling Mechanisms Regulating Opioid Peptide Gene Expression

N.A. Kley, C.-J. Farin, and J.P. Loeffler

1 Introduction

Since the discovery of the first endogenous opioid peptides, methionine (Met-) and leucine-enkephalin (Leu-ENK) (Hughes et al. 1975), a large number of biologically active peptides, all possessing the N-terminal sequence Tyr-Gly-Gly-Phe-Met/Leu, have been identified. Early biochemical and immunohistochemical studies suggested that the diversity of opioid peptides results from tissue-specific processing of distinct high molecular weight precursors: their characterization was subsequently achieved by the molecular cloning of three opioid peptide precursors; proopiomelanocortin (POMC) (Nakanishi et al. 1979), proenkephalin (PENK) (Noda et al. 1982), and prodynorphin (PDYN) (Kakidani et al. 1982). POMC and PENK are the best studied of the three opioid peptide precursor genes, both at the structural and functional level. They are expressed in several distinct brain regions and a multitude of peripheral tissues. Their major sites of expression are the pituitary gland and the adrenal gland, respectively.

There is currently much interest in understanding the molecular basis of tissue-specific gene expression and the mechanisms enabling neuronal cells to adapt to specific environmental inputs by regulating the synthesis and release of neuroactive substances. The fact that the POMC and PENK genes are expressed in a number of different cell types make them a good model to study the basis of signaling mechanisms and regulatory factors involved in the transduction of signals triggering adaptive changes in genomic events. However, the complexity of neuronal networks is a limiting factor in the identification of the molecular mechanisms controlling gene expression in the CNS. Thus, as a first approach, simple in vitro model systems derived from neuroendocrine and endocrine tissues have been used: melanotrophs and corticotrophs of the intermediate and anterior lobe of the pituitary, respectively, to study POMC expression; chromaffin cells of the adrenal medulla to analyze PENK gene expression.

A multitude of second messenger generating systems are known to be involved in the intracellular transduction of signals first received by specific receptors in the plasma membrane. Thus, receptor-induced alterations in the activity of adenylate cyclase, phospholipase C, ion pumps, and gating of specific ion channels lead to changes in intracellular messengers such as cAMP, diacylglycerol (DAG), inositolphosphates, and Ca^{2+} (Fig. 1). Second messenger activated protein kinases, in turn, modify intracellular target proteins, thereby triggering the initiation of complex cellular responses. cAMP, Ca^{2+}, and DAG are known to activate cAMP-

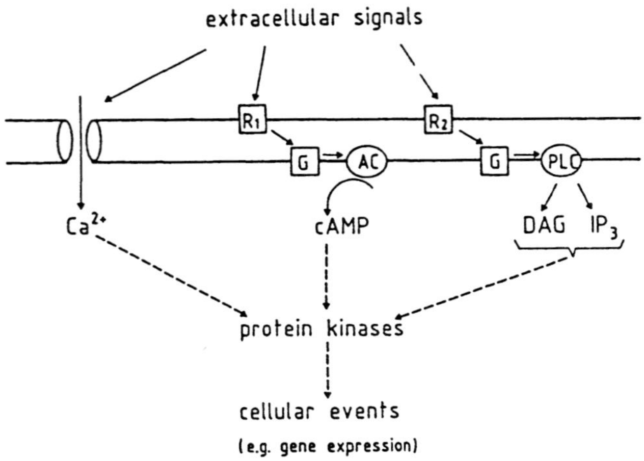

Fig. 1. Signal transduction pathways involved in the regulation of cellular events including changes in gene expression. *R* Receptor; *G* G-protein; *AC* adenylate cyclase; *PLC* phospholipase C; *DAG* diacylglycerol; IP_3 inositol triphosphate

dependent protein kinases, Ca^{2+}/calmodulin-dependent protein kinases, and/or Ca^{2+}phospholipid-dependent protein kinase C isozymes. These protein kinases regulate a variety of cellular responses ranging from electrical events to specific gene expression. However, in most instances the steps beyond protein kinase activation, and the final effectors which mediate changes in transcriptional activity are, as yet, poorly understood.

In this review, an attempt is made to report on recent progress in the study of signal transduction mechanisms involved in the regulation of POMC and PENK gene expression; in particular in melanotrophs and corticotrophs of the pituitary and the chromaffin cells of the adrenal medulla. The development of new approaches which may increase our understanding of the steps beyond protein kinase activation will also be discussed.

2 POMC Gene Expression in Melanotroph Cells

POMC is a polypeptide that is processed in a tissue-specific manner to yield a number of biologically active peptides. In melanotrophs, POMC gives rise primarily to melanocyte-stimulating hormone (MSH) and acetylated β-endorphin (β-END), whereas in corticotrophs ACTH or the opioid peptide β-END are predominant (for review, see Eipper and Mains 1980). Tissue-dependent differential regulation of POMC expression is regulated both at the posttranslational and at the genomic level. This tissue-specific regulation of POMC gene expression is mediated in part by the cell specific microenvironment, receptor complements, and signal transduction pathways. In addition, neurotransmitter and hormone effects on POMC gene expression in melanotrophs and corticotrophs indicate regulation of this gene by tissue-specific factors regulating transcriptional events. Indeed, a major effort is currently being made to identify the promotor sequences and transcription factors that may mediate these effects.

2.1 Regulation by Neurotransmitters

Unlike the anterior lobe (AL), the neuro-intermediate lobe (NIL) of the pituitary is poorly vascularized. The melanotrophs are innervated by the tubero-infundibular system, and receive direct inputs of neurotransmitter substances such as dopamine (DA) and γ-amino-butyric acid (GABA) (Oertel et al. 1982). Regulation of intermediate lobe cells by DA has been extensively studied, both in vivo and in vitro. Höllt and colleagues (1981) first demonstrated that chronic in vivo blockade of NIL-DA receptors by haloperidol results in an increase in translatable POMC mRNA in the NIL. Gee et al. (1983) used in situ hybridization techniques to demonstrate that DA depresses POMC mRNA levels under physiological conditions. Using intermediate lobe explants and isolated melanotrophs it was shown that the inhibition was mediated by D_2-receptors located on the melanotrophs (Cote et al. 1986; Loeffler et al. 1988). DA-ergic agonists and antagonists had no effect on POMC expression in AL cultures. It is likely that the ability of DA to inhibit POMC mRNA results, at least in part, from a decrease in gene transcription, since chronic in vivo administration of haloperidol has been reported to increase POMC transcription three- to four fold in the rat NIL (Pritchett and Roberts 1987).

GABA, like DA, inhibits POMC mRNA expression in the NIL. In vitro studies showed that inhibition of the GABA catabolic enzyme GABA-transaminase (which results in an increase in hypothalamic and pituitary levels of GABA) was correlated with a decrease in POMC mRNA levels in the NIL (Loeffler et al. 1986a). Subsequent in vitro studies showed this effect to be mediated by a direct action of GABA on melanotrophs (Loeffler et al. 1986a).

These results clearly implicate DA and GABA as inhibitory transmitters in the hypothalamic control of POMC gene expression in the NIL. A stimulatory role for hypothalamic-derived neurotransmitters appears to be played by corticotropin-releasing factor (CRF) and catecholamines. Both CRF and tyrosine hydroxylase have been identified in axons innervating the NIL (Vuillez et al. 1987). In vitro studies showed that both CRF and β-adrenergic receptor activation leads to an increase in NIL POMC mRNA levels (Loeffler et al. 1985, 1988; Dave et al. 1987). It is, however, unclear whether these effects occur at the genomic level. Eberwine and colleagues (1987) demonstrated that CRF increases POMC gene transcription in NIL cell cultures, whereas Gagner and Drouin (1987) failed to show any effect. However, long-term administration of CRF in vivo was shown to decrease NIL levels of POMC mRNA (Höllt and Haarmann 1984), whereas a large increase in POMC mRNA expression was observed in the AL. The basis for the opposite effects of CRF on NIL POMC gene expression in vivo and in vitro is unclear, but it may be that CRF has both direct and indirect actions on NIL cells.

2.2 Signaling Mechanisms

Both DA and GABA have been shown to exert an inhibitory influence on the secretion and de novo synthesis of POMC-derived peptides in melanotrophs (Douglas and Taraskevich 1978; Tomiko et al 1983; Cote et al. 1986). Biochemical and electrophysiological studies showed that these effects were correlated with a decrease in intracellular cAMP levels and/or electrical activity (Douglas and Taraskevich 1978; Cote et al. 1982), suggesting that cAMP and intracellular Ca^{2+} (Ca_i^{2+}) are the messengers coupling receptor activation to changes in secretory events. This is supported by the findings that the increase in both cAMP and Ca_i^{2+} levels is correlated with the increase in hormone α-MSH secretion and synthesis (Munemura et al. 1980; Tsuruta et al. 1982; Miyazaki et al. 1984a,b; Tomiko et al. 1984). Recent reports suggest that, at least in the initial steps, similar mechanisms are involved in the long-term regulation of POMC gene expression. An involvement of cAMP in the control of POMC gene expression was first shown by treating NIL explants or dispersed melanotrophs with forskolin (an activator of the catalytic subunit of the adenylate cyclase), the cAMP analog 8BrcAMP, cholera toxin (CT, which irreversibly activates the cyclase stimulatory G-protein: G_s), and phosphodiesterase inhibitors (Cote et al. 1986; Loeffler et al. 1986b). All these agents promoted a time-dependent increase in POMC mRNA levels. Similarly, it was demonstrated that intracellular Ca^{2+} (Ca_i^{2+}) regulates POMC mRNA expression: a reduction in external Ca^{2+} (Taleb et al. 1986) or inhibition of Ca^{2+} channels by the dihydrophyridine nifedipine resulted in a decrease in POMC mRNA levels (Loeffler et al. 1986b). Increasing Ca^{2+}-entry by the dihydropyridine BayK8644 promoted POMC mRNA expression (Loeffler et al. 1986b; Eberwine et al. 1987).

Recent studies suggest that inhibition of adenylate cyclase and a decrease in intracellular cAMP levels may not be the sole mechanism whereby DA inhibits POMC gene expression. In these studies it was shown that bypassing receptor-mediated regulation of cAMP levels by direct activation of the catalytic subunit of the cyclase with forskolin, or incubation with the cAMP analog 8brcAMP, did not completely prevent the DAergic inhibition of POMC mRNA (Loeffler et al. 1988). However, treatment with pertussis toxin (PTX, a bacterial toxin which ADP-ribosylates and inactivates G_i/G_o-like G-proteins) led to a complete loss of DA inhibition of POMC mRNA levels. These findings suggest the involvement of a separate inhibitory pathway, the effects of which are independent of changes in intracellular cAMP levels. Recent reports have shown that D_2-receptors may be coupled to voltage-sensitive ion channels via PTX-sensitive G-proteins (for review, see Vallar and Meldolesi 1989). Since Ca_i^{2+} plays a central role in the regulation of POMC gene expression (see above) in melanotrophs, and in view of the fact that DA reduces the electrical activity of these cells, such a D_2-receptor-coupled mechanism may mediate the inhibitory actions of DA. Further studies are needed to demonstrate this.

The GABAergic trans-synaptic inhibition of POMC gene expression is also likely to be mediated by changes in intracellular Ca^{2+}, since, as a result of opening of GABA-A receptor-associated Cl^- channels, a decrease in Ca^{2+} entry occurs due to electrical shunting. It is as yet unclear what role, if any, the GABA-B receptor plays in the GABAergic modulation of POMC gene expression.

In contrast to the D_2-receptor, the CRF and β-adrenergic receptors are positively coupled to the adenylate cyclase and an increase in cAMP levels appears to mediate their effects on POMC gene expression (Loeffler et al. 1985, 1988; Dave et al. 1987; Eberwine et al. 1988).

Receptors coupled to the phospholipase C appear to represent another class of receptors through which POMC gene expression may be regulated. Recent studies showed that activation of thyrotropin-releasing hormone (TRH) receptors, which is associated with an increase in phosphoinositide breakdown, leads to an induction of POMC mRNA (Loeffler et al., unpubl. observ.). It is likely that an IP_3-mediated increase in intracellular Ca^{2+} levels contributes to the effect on POMC mRNA. A possible role for protein kinase C activation will be discussed below.

2.3 Second Messenger Targets

In the previous section, evidence demonstrating that both cAMP and Ca^{2+} may serve as intracellular second messengers mediating effects of neurotransmitters on POMC gene expression was discussed. We now consider the intracellular targets for these messengers. Reisine et al. (1985) recently demonstrated that a protein kinase A (PKA) inhibitor peptide blocks cAMP-mediated stimulation of POMC gene expression in a cell line derived from the anterior pituitary (AtT20). Although such evidence for a role of PKA has not yet been reported for melanotrophs, cAMP-dependent protein kinase has been identified in these cells, suggesting that PKA may also mediate the effects of cAMP on POMC gene expression in melanotrophs.

Ca^{2+}/calmodulin-dependent protein kinases and Ca^{2+}/phospholipid-dependent protein kinase C are known to mediate a number of Ca^{2+} regulated processes, including specific gene expression (Nishizuka 1984; Morgan and Curran 1986; Jost et al. 1986). The role of Ca^{2+}/calmodulin-dependent protein kinases and protein kinase C (PKC) as intracellular targets for Ca^{2+} mediating effects on POMC gene expression has been investigated by treatment of melanotrophs with inhibitors and activators of the respective kinases. Loeffler et al. (1989) showed that treatment of cultured melanotrophs with the calmodulin inhibitor W7 decreases POMC mRNA, suggesting a role for Ca^{2+}/calmodulin-dependent protein kinases in maintaining basal POMC gene expression. However, it should be kept in mind that W7, especially at higher concentrations, may cross-react with, and also inhibit, protein kinase C. Thus, a clear demonstration of a specific role for Ca^{2+}/calmodulin-dependent protein kinases awaits further investigation. Indeed, PKC appears to play a role in regulating POMC mRNA levels in melanotrophs. Incubation of intermediate lobe cultures with the phorbol ester PMA, an activator of PKC, was shown to produce a decrease in POMC mRNA levels (Loeffler et al. 1989). This treatment also resulted in a rapid (within 30 min) down-regulation of PKC, suggesting that the subsequent decrease in POMC mRNA expression may be due to the down-regulation of the enzyme. At the present time, however, it is unclear whether the decrease in POMC mRNA levels is simply due to the down-regulation of the enzyme. There are reports indicating that PKC may modulate adenylate cyclase and ion-channel activity in a number of cell types (Kazmarek 1987). Thus, one cannot exclude the possibility that PMA treatment activates PKC to produce effects on these systems in melanotrophs, thereby modulating intracellular cAMP- and/or Ca^{2+} levels which are known to regulate POMC gene expression. Further studies are required to clarify these aspects.

3 PENK Gene Expression in Chromaffin Cells

The opioid peptide precursor PENK is tissue specifically processed to a number of biologically active peptides, including met- and leu-ENK (for review, see Höllt 1983), as well as larger opioid peptides with the characteristic opioid N-terminus. The adrenal medulla has attracted much attention because of its central role in the physiological response to stress and because these neural crest-derived cells serve as a useful model for studying neurosecretory mechanisms and signals mediating the transdifferentiation of these cells from a neuroendocrine to a neuronal phenotype. As such, they may also serve as a useful model system to study PENK gene expression in various states of cellular differentiation and to study, in parallel, the differential expression of various neuronal characteristics that may underly long-term changes in neuronal plasticity. This section will focus on the signaling mechanisms involved in the regulation of PENK gene expression in bovine chromaffin cells of the adrenal medulla, in which the PENK gene is highly expressed (Pittius et al. 1985).

3.1 Activity-Dependent Regulation

Early studies have shown that nicotinic receptor stimulation and membrane depolarization promote the secretion of catecholamines and opioid from chromaffin cells. Ca^{2+} has been proposed to serve as the crucial link in this stimulus-secretion coupling process (Douglas et al. 1967; Viveros et al. 1979). In 1984, Eiden et al. demonstrated that long-term nicotinic receptor activation leads to an increase in PENK gene expression and opioid peptide synthesis. An increased influx of Ca^{2+} through voltage-sensitive Ca^{2+} channels was shown to be required in this process. Thus, activation of nicotinic acetylcholine (Ach) receptors and a rise in intracellular Ca^{2+} levels promotes both acute effects on secretion and adaptive changes at the genomic level. It was subsequently demonstrated that changes in electrical activity, a phenomenon occurring as a consequence of nicotinic receptor activation, result in an increase in PENK gene expression by a similar mechanism: membrane depolarization induced by KCl or veratridine (an alkaloid which permanently opens voltage-sensitive Na^{2+} channels, thereby triggering the generation of a train of action potentials) markedly increased PENK mRNA levels (Siegel et al. 1985; Kley et al. 1986; Naranjo et al. 1986). This effect was inhibited by the Na^{2+} channel blocker tetrodotoxin (TTX) and the Ca^{2+} channel blockers D600 and Ca^{2+}, demonstrating that it is dependent on an increase in Na^{+} and Ca^{2+} flux (Kley et al. 1986). That Ca^{2+} entry alone may serve as a signal for gene induction is suggested by the findings that Ba^{2+}, which increases the Ca^{2+} permeability of Ca^{2+} channels (Hess and Tsien 1984), stimulates PENK mRNA expression by a TTX-resistant, but D600-sensitive, mechanism. This effect could be mimicked by the Ca^{2+} ionophore A23187, which increases intracellular Ca^{2+} levels (Ca_i^{2+}) in a depolarization-independent manner (Kley et al. 1986, 1987a). Washek et al. (1987) confirmed and extended these results by showing that the Ba^{2+} induction of EPNK mRNA strongly depends on extracellular Ca^{2+} (as opposed to secretory events), indicating that Ba^{2+} acts by promoting Ca^{2+} entry. In view of these data, the following model for nicotinic stimulation of PENK gene expression can be proposed. Binding of Ach to the nicotinic receptor causes an influx of Na^{+} ions, depolarization of the plasma membrane, and the subsequent opening of voltage-sensitive Ca^{2+} channels. The associated increase in Ca^{2+} then triggers activation of PENK gene expression.

Besides promoting Ca^{2+} entry, nicotinic stimulation has been reported to produce a marginal increase in intracellular cAMP levels (Eiden et al. 1984). Increasing intracellular cAMP levels by direct stimulation of adenylate cyclase with forskolin, or mimicking its increase by incubation with the stable cAMP analog 8brcAMP, has been shown to markedly stimulate PENK mRNA expression (Eiden and Hotchkiss 1983; Eiden et al. 1984; Quach et al. 1984; Kley et al. 1987a). This raises the question whether Ca^{2+} stimulates PENK gene expression indirectly through cAMP, or whether Ca^{2+} affects PENK mRNA levels independently of a rise in cAMP levels. Kley et al. (1987a) recently showed that agents which stimulate Ca^{2+} uptake (e.g. KCl, veratridine) do not increase cAMP levels in chromaffin cells. Furthermore, indirect evidence suggests that Ca^{2+} is unlikely to stimulate PENK gene expression primarily by raising intracellular cAMP levels. As compared to the dose-dependent effect of forskolin on intracellular cAMP levels and PENK mRNA levels, nicotinic receptor stimulation leads to only a small increase in cAMP levels, although the effects on PENK mRNA are comparable in magnitude (Kley et al. 1987a). In addition, nicotine has been reported to produce a transient elevation of cAMP levels (Kurosawa et al. 1976). However, persistent activation of the nicotinic receptor and a sustained entry of Ca^{2+} is required for efficient induction of PENK mRNA (N. Kley and C.-J. Farin unpubl. observ.). It is possible that an initial rise in cAMP may contribute to the nicotinic response. Indeed, Morita et al. (1987) recently reported that cAMP may promote the activation of Ca^{2+} channels in chromaffin cells. Thus, a rise in cAMP levels may activate PENK gene expression through a direct, Ca^{2+}-independent, and an indirect, Ca^{2+} mediated, mechanism.

Although Ca^{2+} appears to play a major role in the response to nicotinic stimulation, cAMP may also be important in the regulation of PENK gene expression by other neurotransmitters co-released with Ach during transsynaptic events: e.g. vasoactive intestinal peptide (VIP) has been shown to increase intracellular cAMP levels and opioid peptide synthesis (Wilson 1987, 1988).

An apparently different mechanism of PENK gene regulation has been reported for the rat adrenal medulla. In organ explants of this tissue, depolarizing stimuli as well as elevated cAMP have been reported to *decrease* the level of PENK mRNA (La Gamma et al. 1985, 1988). However, in primary cultures of rat adrenal medullary cells, depolarizing stimuli, as well as 8brcAMP, *increase* the cellular levels of PENK mRNA (Höllt et al. 1989). Thus, in dissociated rat and bovine chromaffin cells, PENK gene expression appears to be regulated in a similar manner and suggests that tissue dissociation markedly influences the regulation of the PENK gene in chromaffin cells.

3.2 Second Messenger Targets

In the previous section we discussed current evidence demonstrating that both cAMP and Ca^{2+} may serve as intracellular second messengers mediating effects of neuotransmitters on PENK gene expression. We now consider the intracellular targets of these messengers.

Induction of gene expression by cAMP has been analyzed in detail. Recent reports demonstrate that it is the catalytic subunit of protein kinase A that induces transcription of cAMP-responsive genes by activating proteins involved in gene transcription. This contention is supported by several observations: (1) the catalytic subunit (c-subunit) migrates from the cytoplasm into the nucleus after treatment with agents that raise intracellular cAMP levels (Nigg et al. 1985); (2) transfection of cells with a construct encoding an active protein kinase inhibitor peptide decreases cAMP-induced expression of a human PENK bacterial chloramphenicolacetyl-transferase (CAT) fusion gene (Grove et al. 1987); and (3) microinjection of purified regulatory and catalytic subunits of protein kinase A showed that the C-subunit induces expression of a fusion gene containing the VIP cAMP-responsive promoter/enhancer region joined to a bacterial reporter gene (LacZ) or the endogenous c-fos protooncogene, in the absence of an increase in cAMP levels (Riabowol et al. 1988).

The intracellular factors through which Ca^{2+} produces its effects are less well understood. Based on the sensitivity of secretion to calmodulin inhibitors, intracellular injection of calmodulin antibodies and protein kinase C activators (Kenigsberg and Trifaro 1985; TerBush et al. 1988), Ca^{2+}/calmodulin- and Ca^{2+}/phospholipid-dependent protein kinases have been suggested to mediate secretory responses to changes in Ca_i^{2+} in chromaffin cells. Some recent evidence points to a possible role for the Ca^{2+}/phospholipid-dependent protein kinase C in regulating PENK gene expression; phorbol esters (which activate protein kinase C) have been reported to induce PENK and mRNA expression in a dose- and time-dependent manner (Kley 1988); co-treatment with the Ca^{2+} ionophore A23187 and the phorbol ester PMA resulted in a synergistic activation of both protein kinase C (Brocklehurst and Pollard 1987) and PENK mRNA expression (Kley 1988). Also, it was recently reported that enhanced Ca^{2+} entry upon nicotine-receptor stimulation and membrane depolarization stimulates phosphoinositide hydrolysis in chromaffin cells (for review, see Eberhard and Holz 1988). This leads to the production of at least two second messengers, DAG and IP_3. DAG, like PMA, directly activates protein kinase C (Nishizuka 1984), whereas IP_3 promotes the release of Ca^{2+} from internal Ca^{2+} stores (Berridge and Irvine 1984; Stoehr et al. 1986). The findings that activation of protein kinase C by PMA and A23187 may promote PENK gene expression and that treatment with nicotine stimulates protein kinase C activity (TerBush et al. 1988) suggest that protein kinase C may be a Ca^{2+}-sensitive target which mediates the effects of membrane depolarization and nicotine on PENK gene expression.

A role for protein kinase C in regulating PENK gene expression is also supported by studies on the effects on the neurotransmitter histamine. Activation of histamine H1-receptors in chromaffin cells results in a pronounced breakdown of phosphoinositides (Noble et al. 1986; Kley 1988) and PENK gene expression (Kley et al. 1987b; Kley 1988) in chromaffin cells. The effect on PENK mRNA is inhibited by the protein kinase C inhibitors H7 and staurosporine (N. Kley et al., unpubl. observ.) clearly indicating a role for protein kinase C in mediating this effect. The calmodulin inhibitor W7 had no effect, emphasizing the differential role for Ca^{2+}/calmodulin-dependent protein kinases in regulating PENK secretory and biosynthetic events in chromaffin cells.

Recent reports suggest that regulation of PENK gene expression by protein kinase C is somewhat more complex and indicate that protein kinase C may also have inhibitory effects on PENK mRNA expression. The stimulating or inhibitory effects were shown to depend on, and vary with, the state of cellular activity, and to be mediated by an inhibitory action of protein kinase C on voltage-sensitive Ca^{2+} channels and phospholipase C. Thus, treatment of chromaffin cells with PMA was shown to inhibit Ca^{2+} influx enhanced by KCl stimulation and the dihydropyridine Ca^{2+} channel agonist BayK8644 (Kley 1988; Pruss and Stauderman 1988). This was associated with the ability of PMA to inhibit the KCl- and BayK8644-induced rise in PENK mRNA levels. The inhibition of voltage-sensitive Ca^{2+} channels (L-type) by PMA was observed after 10-min pretreatment and was sustained for at least 16–24 h in the continued presence of PMA. In contrast to other cell types, e.g. melanotrophs (see Sect. 2), incubation of chromaffin cells with PMA in the nanomolar range appears to produce only a gradual and partial down-regulation of protein kinase C over 32 h (Wilson 1989). This strongly suggests that it is the chronic activation of protein kinase C, rather than the loss of enzyme activity, which leads to long-term inhibition of Ca^{2+} channels and thus, PENK gene expression. Several observations also suggest that this inhibitory effect is not due to a direct blockade of Ca^{2+} channels by the phorbol ester. At very high concentrations, PMA (10^{-5} M) has been reported to exert a direct inhibitory effect (15%) on Ca^{2+} currents (Hockberger et al. 1989). However, the inhibitory effects of PMA on Ca^{2+} flux and PENK gene expression in chromaffin cells are significant in the nanomolar range. Furthermore, the inactive phorbol ester 4αPDD, which was also reported to exhibit an inhibitory effect on Ca^{2+} currents at 10^{-5} M (Hockberger et al. 1989), had no effect on either basal or stimulated Ca^{2+} flux and PENK mRNA expression in chromaffin cells (Kley 1988; Pruss and Stauderman 1988).

As well as inhibiting Ca^{2+} channels, PMA treatment has been reported to abolish the histamine-induced breakdown of phosphoinositides and the associated increase in PENK mRNA levels (Kley 1988). It thus appears that protein kinase C may not only play a role as an intracellular mediator in receptor-stimulated PENK gene expression (see above), but that it may also be involved in a feedback response by inhibiting Ca^{2+} entry through voltage-sensitive Ca^{2+} channels and the hydrolysis of inositol phospholipids and, thus, the generation of DAG and IP_3 which, in turn, would promote protein kinase C stimulation.

It remains to be investigated whether these different effects of the phorbol ester are mediated by different protein kinase C isozymes (Parker et al. 1986; Knopf et al. 1987; Ohno et al. 1987).

3.3 Regulation of Transcriptional Activity

Two common processes are involved in regulating cellular mRNA levels: changes in the rate of specific gene transcription and an alteration in the rate of degradation of the specific mRNA. Affolter et al. (1984) showed that membrane depolarization and increased cAMP levels lead to an elevation of PENK mRNA precursor levels in chromaffin cells. Nuclear run-on experiments indicate that this is, at least in part, mediated by enhanced PENK gene transcription (C.-J. Farin and N. Kley unpubl. observ.). However, stimulation of protein kinases does not appear to be sufficient to lead to a direct stimulation of PENK gene transcription by the activation of preexisting factors. Recent observations show that pretreatment with cycloheximide (a protein synthesis blocker) prevents the induction of PENK mRNA triggered by membrane depolarization, cAMP, and histamine (C.-J. Farin and N. Kley unpubl. results). A similar dependence was observed in nuclear run-on experiments. Thus, PENK gene induction requires on-going protein synthesis and de novo synthesis of proteins with rapid turnover rates. Thus, it appears that the molecular events subsequent to protein kinase activation may be complex, and a coordinated change in the expression of a set of genes encoding intracellular regulatory factors may be required to establish PENK gene induction in chromaffin cells. This contrasts to the PENK gene induction in C6 glioma cells by noradrenaline and glucocorticoids, which does not require on-going protein synthesis (Yoshikawa and Sabol 1986). It is likely that in the latter case, activation of glucocorticoid receptors directly mediates transcriptional activation.

In recent years, a group of oncoproteins have been identified which act as nuclear transcription factors and may be responsible for transducing short-term signaling events into long-term changes in specific gene expression. The c-fos protooncogene is of particular interest as it has been shown to be rapidly induced by a variety of external stimuli in neuronal (Greenberg et al. 1986; Curran and Morgan 1986; Morgan and Curran 1986; Hunt et al. 1987; Morgan et al. 1987) and chromaffin cells (N. Kley et al. unpubl. observ.) and to be involved in the transcriptional regulation of various genes (Distel et al. 1987; Sassone-Corsi et al. 1988). c-fos appears to regulate transcriptional events by virtue of its interaction with yet another nuclear protein, the transcription factor AP_1, which is responsible for the DNA-binding activity of the dimerized complex (for review, see Curran and Franza 1988).

Comb et al. (1986, 1988a,b) recently characterized a cAMP- and PMA-inducible enhancer in the 5′-flanking sequences of the human PENK gene, and showed that it consists of two functionally distinct elements, ENKCRE1 and ENKCRE2, to which at least four distinct proteins bind in vitro: AP_1, AP_2, AP_4, and a novel factor called ENKTF-1 (Fig. 2). Site-specific mutational analysis showed that inactivation of either of these elements eliminates cAMP- or PMA-inducible enhancer activity. Thus, the findings that: (1) the PENK gene encompasses an AP_1 binding site; (2) the c-fos protooncogene is expressed in chromaffin cells (N. Kley et al., unpubl. observ.); and (3) that induction of PENK gene expression in chromaffin cells requires on-going protein synthesis (N. Kley et al., unpubl. observ.) raise the question whether c-fos plays a role in the transcriptional regulation of the PENK gene. The recent observation that basic fibroblast growth factor (bFGF) induces, within the

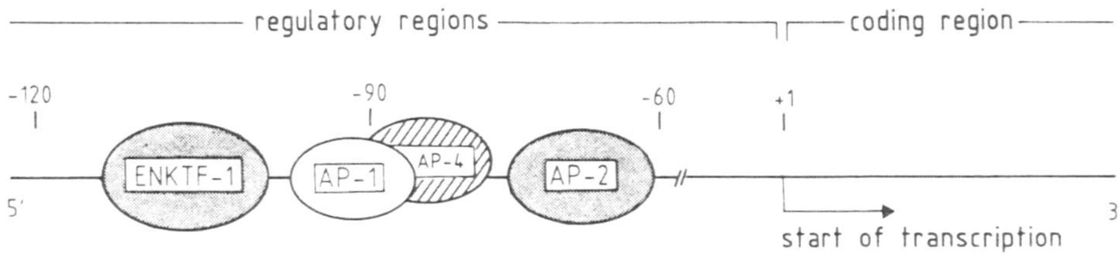

Fig. 2. DNA binding sites for nuclear transcription factors in the 5′-noncoding region of the proenkephalin gene, essentially as described by Comb et al. (1988b): AP_1, AP_2, AP_4: proteins of the c-jun-protooncogene family; *ENKTF-1*: unknown transcription factor

first 24 h of stimulation, a rapid induction of c-fos gene expression but not PENK gene expression (N. Kley et al., unpubl. observ.), suggests that an increase in c-Fos expression alone is not sufficient to promote PENK gene induction. This, however, does not preclude that c-Fos may act in concert with other induced factors to activate PENK gene transcription. Introducing c-fos antisense RNA would be an appropriate way to investigate this. Behr et al. (1989) recently described a novel gene-transfer technique applicable to primary cultured cells. An efficient transfection with expression plasmids, directing the synthesis of sense and antisense RNAs to a variety of nuclear transacting factors, should thus provide a powerful means in the future to study in detail the molecular mechanisms underlying the regulation of specific endogenous gene expression.

4 Summary

In the course of recent studies on signal transduction it has become clear that the specificity of the regulation of a particular gene may not only be determined by the presence or absence of tissue-specific factors involved in transcriptional regulation, but also by the cell-specific interaction and cross talk between second messenger systems which leads to a complex modulation of gene expression at a pretranscriptional level. Thus, for instance, the inhibitory action of PMA on PENK mRNA expression is only observed in activated chromaffin cells; this does not appear to be mediated by the ability of PKC to directly inhibit gene transcription but rather through its effects on voltage-sensitive Ca^{2+} channels and phospholipase C. Equally, one must distinguish between mechanisms responsible for the regulation of specific gene expression by one and the same second messenger. cAMP regulates PENK gene transcription via a cAMP-responsive element, suggesting that this site in the gene is responsible for the interaction with a cAMP-induced regulatory factor. In contrast, in intermediate and anterior pituitary cells cAMP induces POMC gene expression, although the POMC gene appears to lack a classical cAMP-responsive element, at least in the 5′ -1 kb sequence flanking the gene (see Eberwine et al. 1988). Thus, in this case it appears that cAMP may regulate the POMC gene in pituitary cells via its interaction with other second messenger generating systems. This is supported by raising intracellular Ca^{2+} levels through activation of Ca^{2+} channels.

It remains to be determined what type of Ca^{2+} responsive DNA element would mediate such effects. A combined effort in studies of cellular and molecular mechanisms controlling specific gene expression will be needed to further understand the basis of tissue-specific opioid gene expression. To this aim, the recent progress in gene transfer techniques applicable to primary cells should provide a powerful means to approach this issue.

References

Affolter HU, Giraud P, Hotchkiss AJ, Eiden LE (1984) Stimulus-secretion-synthesis coupling: a model for cholinergic regulation of enkephalin secretion and gene transcription in adrenomedullary chromaffin cells. In: Fraioli F (ed) Opiate peptides in the periphery. Elsevier, Amsterdam, pp 23–30

Behr JP, Demeneix BA, Loeffler JP, Perez T (1989) Efficient gene transfer into mammalian primary endocrine cells with lipopolyamine-coated DNA. Proc Natl Acad Sci USA (in press)

Berridge MJ, Irvine RF (1984) Inositol triphosphate, a novel second messenger in cellular signal transduction. Nature (London) 312:315–321

Biales B, Dichler M, Tischler A (1976) Electrical exitability of cultured adrenal chromaffin cells. J Physiol 262:743–753

Brocklehurst KW, Morita K, Pollard HB (1985) Characterisation of protein kinase C and its role in catecholamine secretion from bovine adrenal-medullary cells. Biochem J 228:35–42

Comb M, Birnberg NC, Seasholtz A, Herbert E, Goodman HM (1986) A cyclic AMP- and phorbol ester-inducible DNA element. Nature (London) 323:353–356

Comb M, Hyman SE, Goodman HM (1988a) Mechanism of transsynaptic regulation of gene expression. Trends Neurosci 10:473–478

Comb M, Mermod N, Hyman SE, Pearlberg J, Ross ME, Goodman HM (1988b) Proteins bound at adjacent DNA elements act synergistically to regulate human proenkephalin cAMP inducible transcription. EMBO J 7:3793–3805

Cote TE, Grewe CW, Tsuruta K, Stoot GC, Eskay RL, Kebabian GW (1982) D-2 dopamine receptor-mediated inhibition of adenylate cyclase activity in the intermediate lobe of the rat pituitary gland requires GTP. Endocrinology 110:812–816

Cote TE, Felder R, Kebabian JW, Sekura RD, Reisine T, Affolter HU (1986) D-2 dopamine receptor-mediated inhibition of proopiomelanocortin synthesis in rat intermediate lobe. J Biol Chem 261:4555–4561

Curran T, Franza BR (1988) Fos and jun: the API connection. Cell 55:395–397

Curran T, Morgan JI (1986) Barium modulates c-fos expression and post-translational modification. Proc. Natl Acad Sci USA 83:8521–8524

Dave JR, Eiden LE, Lozovsky D, Washek JA, Eskay RL (1987) Calcium-independent and calcium-dependent mechanisms regulate corticotropin-releasing factor-stimulated proopiomelanocortin peptide secretion and messenger ribonucleic acid production. Endocrinology 120:305–310

Distel RJ, Ro H-S, Rosen BS, Grovers DL, Spiegelman BM (1987) Nucleoprotein complexes that regulate gene expression in adipocyte differentiation: direct participation of c-fos. Cell 49:1835–1844

Douglas WW, Taraskevich PS (1978) Action potentials in gland cells of rat pituitary pars intermedia: inhibition by dopamine, an inhibitor of ASH secretion. J Physiol 285:171–178

Douglas WW, Kanno T, Sampson SR (1967) Influence of the ionic environment on the membrane potential of adrenal chromaffin cells and on the depolarising effect of acetylcholine. J Physiol 262:743–753

Eberhard DA, Holz RW (1988) Intracellular Ca^{2+} activates phospholipase C. Trends Neurosci 11:517–520

Eberwine JH, Jonassen JA, Evinger MJR, Roberts JL (1988) Complex transcriptional regulation by glucocorticoids and corticotropin-releasing hormone of proopiomelanocortin gene expression in rat pituitary cultures. DNA 6:483–492

Eiden LE, Hotchkiss AJ (1983) Cyclic adenosine monophosphate regulates vasoactive intestinal polypeptide and enkephalin biosynthesis in cultured bovine chromaffin cells. Neuropeptides 4:1–9

Eiden LE, Giraud P, Dave JR, Hotchkiss AJ, Affolter HU (1984) Nicotinic receptor stimulation activates enkephalin release and biosynthesis in adrenal chromaffin cells. Nature (London) 312:661–663

Eipper BA, Mains RE (1980) Structure and biosynthesis of pro-adrenocorticotropin/endorphin and related peptides. Endocrine Rev 1:1-27
Gagner J-P, Drouin J (1987) Opposite regulation of proopiomelanocortin gene transcription by glucocorticoids and CRH. Mol Cell Endocrinology 40:25-32
Gene CE, Chen CLC, Roberts JL, Thopson R, Watson SR (1983) Identification of proopiomelanocortin neurons in rat hypothalamus by in situ cDNA-mRNA hybridisation. Nature (London) 306:374-376
Greenberg ME, Ziff EB, Green LA (1986) Stimulation of neuronal acetylcholin receptors induces rapid gene transcription. Science 234:80-83
Grove JR, Price DJ, Goodman HM, Avruch J (1987) Recombinant fragment of protein kinase inhibitor blocks cyclic AMP-dependent gene transcription. Science 238:530-533
Hess P, Tsien RW (1984) Mechanism of ion permeation through calcium channels. Nature (London) 309:453-456
Hockberger P, Toselli M, Swandulla D, Lux D (1989) A diacylglycerol analogue reduces neuronal calcium currents independently of protein kinase C activation. Nature (London) 338:340-342
Höllt V (1983) Multiple endogenous opioid peptides. Trends Neurosci 1:24-26
Höllt V, Haarmann I (1984) Corticotropin-releasing factor differentially regulates proopiomelanocortin messenger ribonucleic acid levels in anterior as compared to intermediate pituitary lobes of rats. Biochem Biophys Res Commun 124:407-415
Höllt V, Haarmann I, Seizinger BR, Herz A (1981) Chronic haloperidol treatment increases the level of in vitro translatable messenger ribonucleic acid coding for the β-endorphin/adrenocorticotropin precursor proopiomelanocortin in the pars intermedia of the rat pituitary. Endocrinology 110:1885-1891
Höllt V, Kley N, Haarmann I, Reimer S (1989) Regulation of proenkephalin gene expression in the adrenal medulla: in vitro and in vivo studies. In: 5th Int Symp Chromaffin cell biology, the adrenal cell as a neuroendocrine model: from basis to clinical aspects. Jerusalem, April 1989, p 86 (Abstr)
Huang FL, Yoshida Y, Cunha-Melo JR, Beaven MA, Huang K-P (1989) Differential down-regulation of protein kinase C isozymes. J Biol Chem 264, 7:4238-4243
Hughes J, Smith TW, Kosterlitz HW, Fothergill LA, Morgan BA, Morris HR (1975) Identification of two related pentapeptides from the brain with potent opiate agonist activity. Nature (London) 258:577-579
Hunt SP, Pini A, Evan G (1987) Induction of c-fos-like protein in spinal cord neurons following sensory stimulation. Nature (London) 328:632-634
Jost JP, Moucharmont B, Jirincy J, Saluz H, Aertner T (1986) In vitro secondary activation (memory effect) of avian vitellogenin II gene in isolated liver nuclei. Proc Natl Acad Sci USA 83:43
Kakidani H, Furtani Y, Takahashi H, Noda M, Morimoto Y, Hirose T, Asai M, Inayama S, Nakanishi S, Numa S (1982) Cloning and sequence analysis of cDNA for porcine β-neoendorphin precursor. Nature (London) 298:245-249
Kazmarek LK (1987) The role of protein kinase C in the regulation of ion channels and neurotransmitter release. Trends Neurosci 10:30-34
Kenigsberg RL, Trifaro JB (1985) Microinjection of calmodulin antibodies into cultured chromaffin cells blocks catecholamine release in response to stimulation. Neuroscience 14:335-347
Kley N (1988) Multiple regulation of proenkephalin gene expression by protein kinase C. J Biol Chem 263:2003-2008
Kley N, Loeffler JP, Pittius CW, Höllt V (1986) Proenkephalin A gene expression in bovine adrenal chromaffin cells is regulated by changes in electrical activity. EMBO J 5:967-970
Kley N, Loeffler JP, Pittius CW, Höllt V (1987a) Involvement of ion channels in the induction of proenkephalin A gene expression by nicotine and cAMP in bovine chromaffin cells. J Biol Chem 262:4083-4089
Kley N, Loeffler JP, Höllt V (1987b) Ca^{2+}-dependent histaminergic regulation of proenkephalin mRNA levels in cultured adrenal chromaffin cells. Neuroendocrinology 46:89-92
Knopf JL, Lee MH, Shultzman LA, Kriz RW, Loomis CR, Hewick RM, Belc RM (1987) Cloning and expression of multiple protein kinase C cDNAs. Cell 46:491-502
Kurosawa A, Guidotti A, Costa E (1976) Induction of tyrosine 3-monooxygenase elicited by carbamycholin in intact and denervated adrenal medulla: role of protein kinase activation and translocation. Mol Pharmacol 15:420-430
La Gamma EF, White GD, Adler GE, Krause GE, McKelvy JF, Black IB (1985) Depolarization regulates adrenal preproenkephalin mRNA. Proc Natl Acad Sci USA 82:8252-8255
La Gamma EF, White GD, McKelvy GF, Black IB (1988) Increased cAMP or Ca^{2+} second messenger

reproduce effects of depolarization on adrenal enkephalin pathways. In: Johnson RG Jr (ed) The cellular and molecular biology of hormone and transmitter containing secretary vesicles. Ann NY Acad Sci, Washington DC, pp 26–32

Loeffler JP, Kley N, Pittius CW, Höllt V (1985) Corticotropin releasing factor and forskolin increase proopiomelanocortin messenger RNA levels in rat anterior and intermediate cells in vitro. Neurosci Lett 62:383–387

Loeffler JP, Demeneix DA, Pittius CW, Kley N, Haegele KD, Höllt V (1986a) GABA differentially regulates the gene expression of proopiomelanocortin in rat intermediate and anterior pituitary. Peptides 7:253–258

Loeffler JP, Kley N, Pittius CW, Höllt V (1986b) Calcium ion and cyclic adenosine 3′5′-monophosphate regulate proopiomelanocortin messenger ribonucleic acid levels in rat intermediate and anterior pituitary lobes. Endocrinology 119:2840–2847

Loeffler JP, Demeneix BA, Kley N, Höllt V (1988) Dopamine inhibition of proopiomelanocortin gene expression in the intermediate lobe of the pituitary. Neuroendocrinology 47:95–101

Loeffler JP, Kley N, Louis JC, Demeneix BA (1989) Ca^{2+} regulates hormone secretion and proopiomelanocortin gene expression in mealanotrope cells via the calmodulin and protein kinase C pathway. J Neurochem 152:1279–1283

Miyazaki K, Reisine T, Kebabian JW (1984a) Adenosine 3′5′-monophosphate (cAMP)-dependent protein kinase activity in rodent pituitary tissue: possible role in cAMP-dependent hormone secretion. Endocrinology 115:1933–1938

Miyazaki K, Goldman MW, Kebabian JW (1984b) Forskolin stimulates adenylate cyclase activity, adenosine 3′5′-monophosphate production and peptide release from the intermediate lobe of the rat pituitary gland. Endocrinology 114:761

Morgan JL, Curran T (1986) Role of ion flux in the control of c-fos expression. Nature (London) 322:552

Morgan JL, Cohen DR, Hempstead JL, Curran T (1987) Mapping patterns of c-fos expression in the central nervous system after seizure. Science 237:192–197

Morita K, Dohi T, Kitayama S, Koyama Y, Tsajimoto A (1987) Stimulation-evoked Ca^{2+}-fluxes in cultured bovine adrenal chromafin cells are enhanced by forskolin. J Neurochem 48:248–252

Munemura M, Eskay RL, Kebabian GW, Long RW (1980) Release of α-melanocyte-stimulating hormone from dispersed cells of the intermediate lobe of the rat pituitary gland: involvement of catecholamines and adenosine 3′5′-monophosphate. Endocrinology 106:1795

Nakanishi S, Inoue A, Kita T, Nakamura M, Chang ACY, Cohen SW, Numa S (1979) Nucleotide sequence of cloned cDNA for bovine corticotropin-β-lipotropin precursor. Nature (London) 278:423–427

Naranjo JR, Mochetti I, Schwartz JP, Costa E (1986) Permissive effect of dexamethasone on the increase of proenkephalin mRNA induced by depolarization of chromaffin cells. Proc Natl Acad Sci USA 83:1513–1517

Nigg EA, Hilz H, Eppenberger HM, Dutly F (1985) Rapid and reversible translocation of the catalytic subunit of cAMP-dependent protein kinase type II from the Golgi complex to the nucleus. EMBO J 4:2801–2806

Nishizuka Y (1984) The role of protein kinase C in cell surface signal transduction and tumour promotion. Nature (London) 308:693–699

Noble EP, Bommer M, Sincini E, Costa T, Herz A (1986) H_1-histaminergic activation stimulates Inositol-1-phosphate accumulation in chromaffin cells. Biochem Biophys Res Commun 135:566–573

Noda M, Furatani Y, Takahashi H, Toyosato M, Hirose T, Inayama S, Nakanishi S, Numa S (1982) Cloning and sequence analysis of cDNA for bovine adrenal preproenkephalin. Nature (London) 295:202–206

Oertel WH, Mugnani E, Tappaz ML, Weise VK, Dahl AL, Schmeckel DE, Kopin JJ (1982) Central GABAergic innervation of neurointermediate pituitary lobe: biochemical and immunocytochemical study in the rat. Proc Natl Acad Sci USA 79:675–679

Ohno S, Kawasaki H, Imajoh S, Suzuki K, Inagaki M, Yokokura H, Sakoh T, Hidaka H (1987) Tissue-specific expression of three distinct types of rabbit protein kinase C. Nature (London) 325:161–166

Parker PJ, Coussens L, Totty N, Rhee L, Young S, Chen E, Stabel S, Waterfield MD, Ullrich A (1986) The complete primary structure of protein kinase C – the major phorbol ester receptor. Science 233:853–859

Pittius CW, Kley N, Loeffler JP, Höllt V (1985) Quantification of proenkephalin A messenger RNA in

bovine brain, pituitary and adrenal medulla: correlation between mRNA and peptide levels. EMBO J 4:1257–1260
Pritchett DB, Roberts JL (1987) Dopamine regulates expression of the glandular-type kallikrein gene at the transcriptional level in the pituitary. Proc Natl Acad Sci USA 84:5545–5549
Pruss RM, Stauderman KA (1988) Voltage-regulated calcium channels involved in the regulation of enkephalin synthesis are blocked by phorbol ester treatment. J Biol Chem 263:13173–13178
Quach TT, Tang F, Kageyama H, Mocchetti I, Guidotti A, Meek JL, Costa E, Schwartz JP (1984) Enkephalin biosynthesis in adrenal medulla: modulation of proenkephalin mRNA content of cultured chromaffin cells by 8-bromo-adenosine 3′5′-monophosphate. Mol Pharmacol 26:255–260
Reisine T, Rongon G, Barbet J, Affolter HU (1985) Corticotropin releasing factor-induced adrenocorticotropin hormone secretion and synthesis is blocked by incorporation of the inhibitor of the cyclic AMP-dependent protein kinase into anterior pituitary cells by liposomes. Proc Natl Acad Sci USA 82:8261–8265
Riabowol KT, Fink JS, Gilman MZ, Walsh DA, Goodman RH, Feramisco JR (1988) The catalytic subunit of cAMP-dependent protein kinase induces expression of genes containing cAMP-responsive enhancer elements. Nature (London) 336:83–86
Sassone-Corsi P, Sisson JC, Verma IM (1988) Transcriptional autoregulation of the proto-oncogene fos. Nature (London) 334:314–319
Siegel RE, Eiden LE, Affolter HU (1985) Elevated potassium stimulates enkephalin biosynthesis in bovine chromaffin cells. Neuropeptides 6:543–552
Stoehr SJ, Smolen GE, Holz RW, Agranoff BW (1986) Inositol triphosphate mobilizes intracellular calcium in permeabilized adrenal chromaffin cells. J Neurochem 46:637–640
Taleb O, Loeffler JP, Trouslard J, Demeneix BA, Kley N, Höllt V, Feltz P (1986) Ionic conductances related to GABA action on secretory and biosynthetic activity of pars intermedia cells. Brain Res Bull 17:725–730
TerBush DR, Bittner MA, Holz RW (1988) Ca^{2+}-influx causes rapid translocation of protein kinase C to membranes. J Biol Chem 263:18873–18879
Tomiko SA, Tarakevich PS, Douglas WW (1983) GABA acts directly on cells of pituitary pars intermedia to alter hormone output. Nature (London) 301:706–707
Tomiko SA, Tarakevich PS, Douglas WW (1984) Effect of veratridine, tetrodotoxin and other drugs that alter electrical behaviour and secretion of melanocyte-stimulating hormone from melanotrophs of the pituitary pars intermedia. Neuroscience 12:1223
Tsuruta K, Grewe CW, Cote TE, Eshay RL, Kebabian JW (1982) Coordinate action of calcium ion and adenosine 3′5′-monophosphate upon the release of α-melanocyte-stimulating hormone from the intermediate lobe of the rat pituitary gland. Endocrinology 110:1133
Vallar L, Meldolesi J (1989) Mechanisms of signal transduction at the dopamine D2-receptor. Trends Pharmacol Sci 10:74–77
Viveros OH, Diliberto EJ, Hazum E, Chany KJ (1979) Opiate-like materials in the adrenal medulla: evidence for storage and secretion with catecholamines. Mol Pharmacol 16:1101–1108
Vuillez P, Perez SC, Stoeckel ME (1987) Colocalization of GABA and tyrosine hydroxylase immunoreactivities in the axons innervating the neurointermediate lobe of the rat pituitary: an ultrastructural immunogold study. Neurosci Lett 79:53–58
Washek JA, Dave JR, Eskay RL, Eiden LE (1987) Barium distinguishes separate calcium targets for synthesis and secretion of peptides in neuroendocrine cells. Biochem Biophys Res Commun 146:495–501
Wilson SP (1987) Vasoactive intestinal peptide and substance P increase levels of enkephalin-containing peptides in adrenal chromaffin cells. Life Sci 40:623–628
Wilson SP (1988) Vasoactive intestinal peptide elevates cAMP levels and potentiates secretion in bovine adrenal chromaffin cells. Neuropeptides 11:17–21
Wilson SP (1989) Chronic phorbol ester treatment inhibits secretion from bovine adrenal chromaffin cells. In: 5th Int Symp Chromaffin cell biology, the adrenal cell as a neuroendocrine model: from basic to clinical aspects. Jerusalem, April 1989, Abstr, p 84
Yoshikawa K, Sabol SL (1986) Expression of the enkephalin precursor gene in C6 rat glioma cells: regulation by β-adrenergic agonists and glucocorticoids. Mol Brain Res 1:75–83

Section III

CNS Opioidergic Systems: Distribution and Modulation

CHAPTER 10

Central Distribution of Opioid Receptors: A Cross-Species Comparison of the Multiple Opioid Systems of the Basal Ganglia

A. Mansour, M.K.H. Schafer, S.W. Newman, and S.J. Watson

1 Introduction

A necessary requirement in studying the function and regulation of the neurotransmitters and receptors of the brain is the careful delineation of their anatomical distributions. This detailed neurochemical mapping quickly becomes an overwhelming task, however, when the goal is to survey the entire brain. In view of the multiple receptor and neurotransmitter systems, a more meaningful and revealing approach is to examine the receptors and neurotransmitters within functionally and anatomically defined circuits. The basal ganglia provide a neuroanatomical system that is especially suited to this type of analysis. This system is well defined anatomically, in terms of the basic units which comprise it, demonstrates complex neurotransmitter-receptor interactions, and has been functionally implicated in sensory-motor integration and motivated behaviors.

This chapter focuses on the distribution of the opioid receptor types, μ, δ and κ, in the striatum, pallidum and substantia nigra of the rat, guinea pig, hamster and monkey. Species differences have been a hallmark of opioid receptor research and have added tremendously to our knowledge of multiple receptor types. It is hoped that this comparative perspective will provide more general principles concerning the distribution of opioid receptors, their interaction with opioid peptides and their possible function in the central nervous system.

Given the anatomical complexity of the basal ganglia (e.g., Penney and Young 1986; Alheid and Heimer 1988) and the differential distribution of peptides and receptors in different species, more questions may be raised than answered. However, the basic aim is to make comparisons within the framework of principles of organization that seem to pertain across species. As the emphasis of this chapter is to present concepts, the similarities and differences described will largely be qualitative and based on studies that have been done in the laboratories of the senior authors (Khachaturian et al. 1984, 1985a,b; Mansour et al. 1987, 1988; Neal and Newman 1989). Numerous laboratories have greatly contributed to this field (e.g., Atweh and Kuhar 1977; Bowen et al. 1981; Herkenham and Pert 1980, 1981, 1982; Quirion et al. 1983; Palacios and Maurer 1984; Morris and Herz 1986; Tempel and Zukin 1987; Zukin et al. 1988), making it difficult to cite all the papers that may be relevant. We have decided therefore to keep the citations to a minimum, choosing just one or two to illustrate a point. Prior to delineating the distribution of the opioid receptor types, an abbreviated description of the anatomy of these neural systems is essential.

2 Basal Ganglia Anatomy

Based upon close anatomical and functional interrelationships, a number of nuclei extending from the rostral telencephalon into the midbrain are now considered to constitute the basal ganglia. Intimately connected to the classically recognized deep telencephalic nuclei (the caudate nucleus, putamen, and globus pallidus) are the substantia nigra, ventral tegmental area (VTA) and subthalamic nucleus at the diencephalic/mesencephalic border. The thalamus, while not part of the classic basal ganglia, serves as much more than a relay station. The intralaminar nuclei, in particular, establish extensive interconnections with basal ganglia structures, especially with the striatum.

Recent studies reveal a previously unappreciated complexity of interactions between these structures. In particular, there is increasing evidence for reciprocity of connections between various nuclei within the system, and collateralization of these connections. In addition, the identity of the basal ganglia has been expanded with the recognition of the parallel, but separate, circuits connecting nuclei of the classical dorsal system (the striatum, or caudate nucleus and putamen, the globus pallidus, entopeduncular nucleus, and substantia nigra) and the more recently described projections linking the ventral striatum (nucleus accumbens and olfactory tubercle) with the ventral pallidum and VTA.

In spite of the proliferation of new anatomical information, however, we have not abandoned the basic concept of a flow of information funneled from the cerebral cortex through the striatum to the pallidum and midbrain tegmentum. In fact, the description of the parallel circuitry in the recently recognized ventral system has reinforced this concept. This basic pattern of connections is illustrated in Fig. 1, and which represents an organization common to all mammals in which these pathways have been studied. Although additional connections between these areas have been described, in this figure we illustrate the structures in which we have analyzed opioid receptors by autoradiographic techniques and the pathways connecting these structures. These fundamental similarities in the anatomical substrate are in striking contrast to the differences in receptor types and peptides within these systems which will be described.

Before describing these differences, however, several significant details of the anatomy of the basal ganglia should be mentioned. Of these, the most notable is the

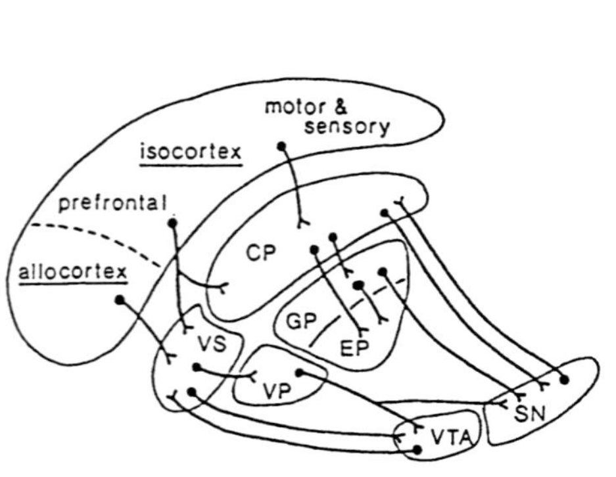

Fig. 1. Schematic diagram of the anatomical connections common to the basal ganglia of mammals. Note the two major systems connecting the cortex-ventral striatum-ventral pallidum-ventral tegmental area, on the one hand, and the cortex-caudate-globus pallidus-substantia nigra, on the other. Abbreviations: *CP* caudate-putamen; *EP* entopeduncular nucleus; *GP* globus pallidus; *SN* substantia nigra; *VP* ventral pallidum; *VTA* ventral tegmental area; *VS* ventral striatum

heterogeneity of the anatomical organization of the striatum, in which islands (or patches) and matrix regions can be differentiated on the basis of cell clustering and the distribution of neurotransmitters and transmitter receptors. In addition, recent investigations demonstrate that striatal afferents from the cerebral cortex, intralaminar thalamus, and substantia nigra/VTA all terminate in a nonhomogeneous fashion, although the interrelationships between these various mosaic patterns and the cell patches is largely unknown.

The anatomy and connections of the substantia nigra also deserve further comment. Although most of the nigrostriatal projections are known to arise from dopaminergic cells of the pars compacta, recent studies suggest that not all neurons in this system are dopaminergic and that neurons in the pars reticulata and the VTA also contribute to these connections. As indicated in Fig. 1, the majority of striatal projections from the VTA go to the nucleus accumbens and the olfactory tubercle, or the ventral striatum, rather than the dorsal striatum.

Finally, the structures illustrated in Fig. 1 are topographically organized. The details of this topography are beyond the scope of this review, but at the same time it is likely that they contribute directly to the distribution of receptors described below. For example, the tendency of the μ-receptor patches to be concentrated rostrally, medially, and ventrally within the striatum across species correlates with the differential distribution of cortical striatal inputs from prefrontal, limbic, and cortical areas.

3 Opioid Receptor Distributions

3.1 δ-Receptors

Of the three opioid receptor types, the distribution of δ-opioid receptors appears to be best conserved across mammalian species. δ-Receptors are densest in forebrain structures and sparse to nonexistent in most midbrain and brainstem areas. This relationship is particularly evident in the nigrostriatal system where moderate to dense δ-Receptor binding is observed in the caudate-putamen of the rat, guinea pig, hamster, and monkey (Fig. 2). The receptor binding is fairly uniform throughout caudate and putamen with the densest labeling in the more lateral aspects of these nuclei. δ-Receptor binding extends to the ventral striatum, with these species showing moderate to dense levels in the nucleus accumbens and olfactory tubercle. More caudally, in the pallidum, the levels of δ-Receptor binding are markedly reduced compared to striatum and sparse to nonexistent in the substantia nigra and VTA. Within the pallidum, little distinction in terms of receptor density can be made between the binding observed in the more dorsal globus and ventral pallidum. Small species differences are observed in the globus pallidus, where δ-binding in the guinea pig seems to be relatively greater than that observed in the rat, hamster, and monkey. Similarly, hamster substantia nigra, pars reticulata, seems to be devoid of δ-receptors, while low levels are observed in rat, monkey, and guinea pig. As δ-Receptor binding is extremely low in these mesencephalic structures, these latter differences may represent technical limitations rather than anatomical differences.

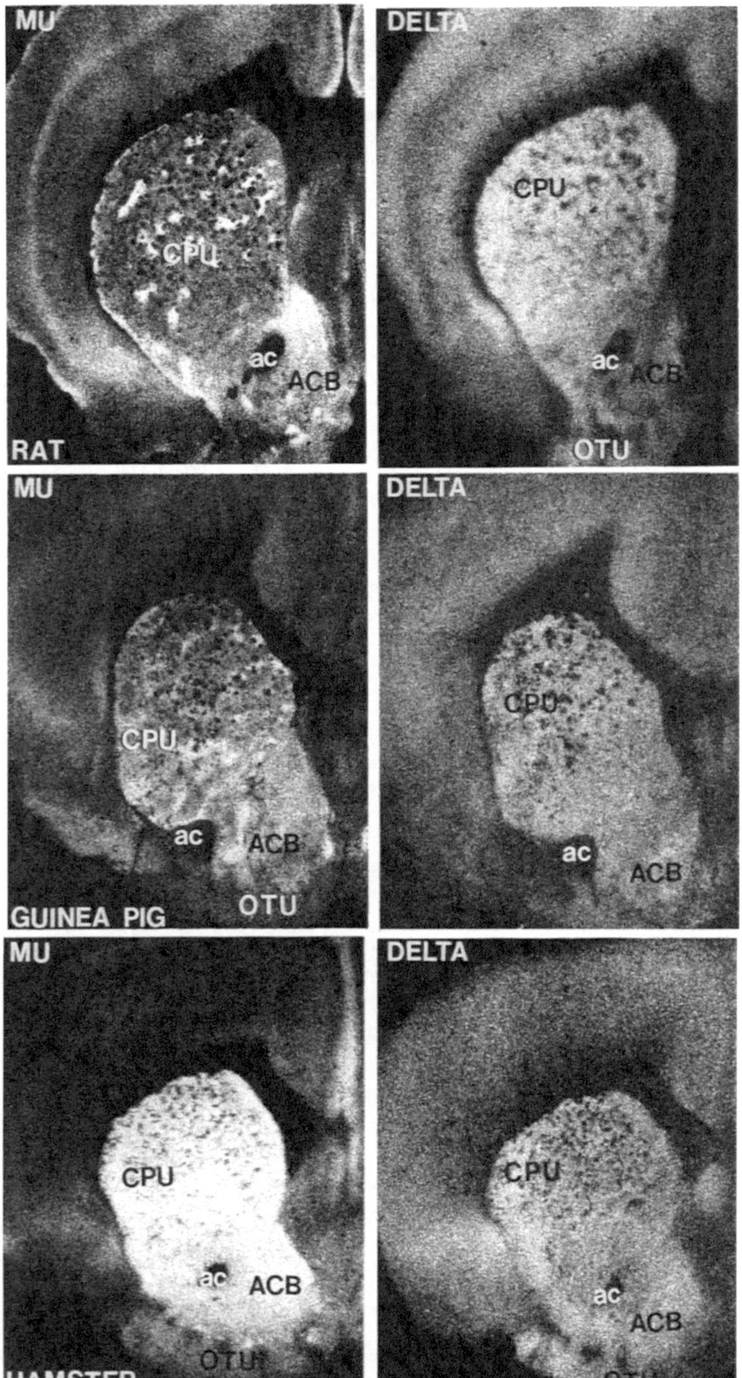
MU
CPU
ac
ACB
RAT
DELTA
CPU
ac
ACB
OTU
MU
CPU
ac
ACB
OTU
GUINEA PIG
DELTA
CPU
ac
ACB
MU
CPU
ac
ACB
OTU
HAMSTER
DELTA
CPU
ac
ACB
OTU

3.2 μ-Opioid Receptors

The distribution of μ-receptors is perhaps the next best conserved across mammals. In the rat, there are moderate to dense μ-receptors throughout the basal ganglia. μ-Binding is particularly dense in the caudate-putamen of each species examined (Fig. 2). Differences are observed, however, in the precise receptor distribution. Within the caudate-putamen, μ-binding can be separated into two compartments referred to as "patches" and "matrix." In the rat, guinea pig, and monkey the densest binding is observed in the patches with moderate levels seen in the matrix, the nonpatchy diffuse labeling found in the caudate-putamen. This distinction is most marked in the rat, and ventromedially in the monkey caudate; the guinea pig shows fairly dense labeling in both patches and matrix. In contrast to the other species, μ-binding in the hamster is diffusely and densely distributed throughout the striatum in an apparently homogeneous manner. The μ-receptor patches described in the rat, guinea pig, and monkey do not correspond to the patches or clustering of enkephalin cells described in the striatum and appear to more closely resemble the distribution of dynorphin peptides in this nucleus.

The μ-patches have a distinct gradient, being most prominent in the rostral regions across species and, in the case of the monkey and guinea pig, virtually disappearing at the level of the globus pallidus. μ-Binding appears to continue into the ventral striatum, with all species examined displaying moderate to dense labeling in the nucleus accumbens and olfactory tubercle. As seen with δ-receptor distribution, the μ-labeling in globus and ventral pallidum appears to be equivalent, with moderate to low levels observed in the hamster, rat, and monkey. Distinctly less μ-labeling is observed, however, in the ventral pallidum as compared to the globus pallidus of the guinea pig.

More caudally, species differences appear to be more striking in the entopeduncular nucleus, where dense μ-binding is observed in the rat and hamster (Fig. 3), while little to no labeling is seen in the monkey and guinea pig. While all four species show moderate to dense μ-labeling in the substantia nigra, they vary substantially with respect to its anatomical divisions. In the guinea pig and hamster, μ-labeling occurs predominantly in the pars reticulata with little or no binding observed in the pars compacta. The case is somewhat reversed in the rat, where fairly dense μ-binding is seen in the pars compacta with only moderate levels occurring in the pars reticulata. These anatomical boundaries do not appear as relevant for the monkey, where moderate μ-binding is seen in the medial portions of both pars compacta and reticulata (Fig. 4). μ-Receptors diminish in the lateral portions of this

Fig. 2. Dark-field autoradiograms of μ- and δ-receptor binding the striatum of rat, guinea pig, and hamster. The distribution of δ-receptors appears well-conserved across mammals in the striatum, while μ-receptor distribution varies markedly with species. μ-Binding in the caudate-putamen of the rat demonstrates a distinct patchy distribution, while the same binding in the hamster fails to show patches and is diffusely distributed. The guinea pig shows both dense binding in the form of patches and diffuse labeling. Abbreviations: *ac* anterior commissure; *ACB* nucleus accumbens; *CPU* caudate-putamen; *OTU* olfactory tubercle

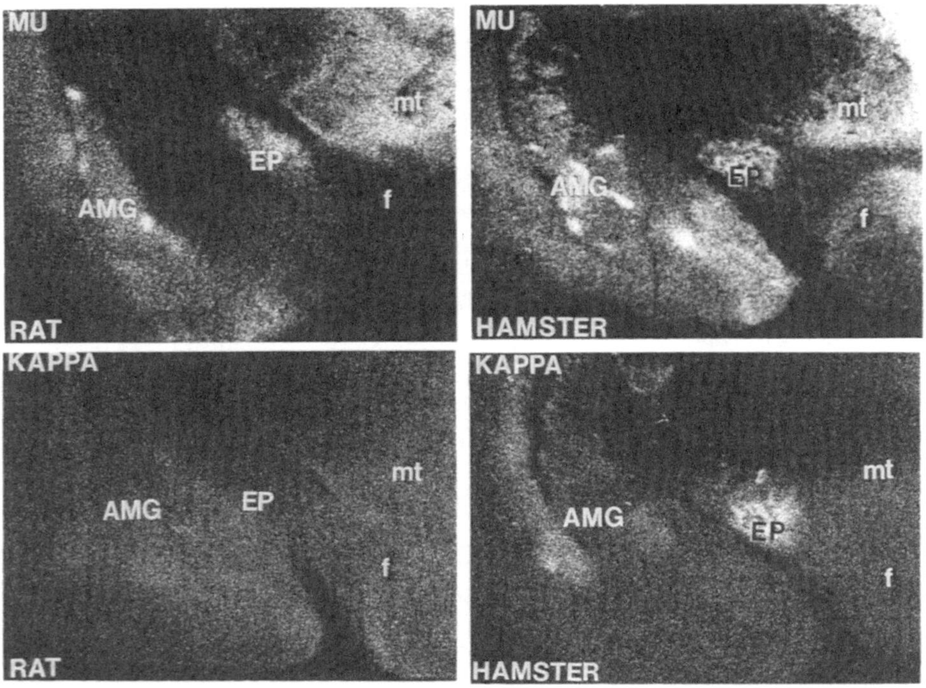

Fig. 3. Dark-field autoradiograms of μ- and κ-binding in the rat and hamster. Note that the hamster entopeduncular nucleus is densely endowed with μ- and κ-receptors, while the rat shows predominantly μ-binding in this region. Abbreviations: *AMG* amygdala; *EP* entopeduncular nucleus; *f* fornix; *mt* mammaliothalamic tract

nucleus and match the mediolateral gradient observed rostrally in the caudate-putamen. The VTA fails to show any significant species differences with low to moderate levels of μ-binding observed across the mammals examined.

3.3 κ-Opioid Receptors

The distribution of κ-opioid receptors demonstrates the largest species differences. On the simple level of relative receptor density, marked differences are seen in the rat as compared to the other mammals examined. Of the total opioid receptors in the rat forebrain, only 10% can be classified as κ. In contrast, κ-receptors in the guinea pig, monkey, and hamster comprise approximately one-third of the total opioid receptor population.

Because of these differences, and due to the fact that these are relative measures, the description of the κ-receptors in the rat do not reflect absolute amounts, but a relative measure within the rat κ-distribution. Therefore, areas described as having dense κ-labeling in the rat may be equivalent to areas of low binding in another species. The emphasis here is to point out qualitative differences in the distribution since the quantitative variations are inherent.

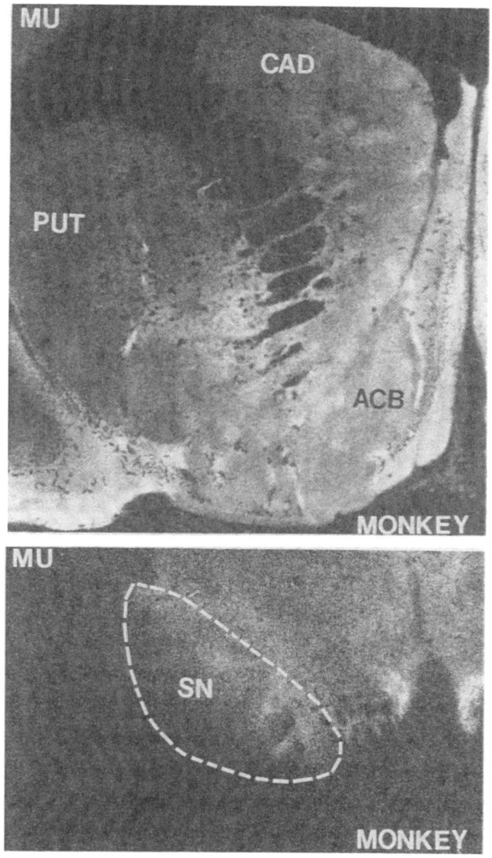

Fig. 4. Dark-field autoradiogram of μ-binding in the monkey striatum (*above*) and midbrain (*below*). μ-Binding appears denser in the medial portion of the caudate and putamen which shows good topographical correspondence to the μ-binding observed in the medioventral regions of the substantia nigra. Abbreviations: *ACB* nucleus accumbens; *CAD* caudate; *PUT* putamen; *SN* substantia nigra

Species differences are evident at several levels of the nigrostriatal and ventrostriatal-tegmental systems. The caudate-putamen of the guinea pig, rat, and hamster, for example, has moderate amounts of fairly diffuse receptor binding, while the same regions in the monkey have both a matrix and patchy pattern and a rostral-caudal gradient reminiscent of the μ-distribution found in this species. In constrast to the μ-receptor patches, however, the κ-patches appear to extend throughout the caudate and putamen. As seen with the other opioid receptor distributions, the κ-receptors extend into the ventral striatum, with moderate levels observed in the nucleus accumbens and olfactory tubercle. In the case of the rat, higher levels of κ-binding are observed in the nucleus accumbens as compared to the caudate-putamen.

In contrast to the μ- and δ-distribution in the pallidum, the density of κ-binding in this structure varies dorsoventrally with markedly higher levels observed in globus compared to ventral pallidum of the hamster, monkey, and guinea pig. In fact, the globus pallidus of the hamster contains one of the highest densities of κ-receptors found in this species. As can be seen in Fig. 5, κ-binding in the globus pallidus appears denser than the caudate of the hamster and monkey. This is not the case in

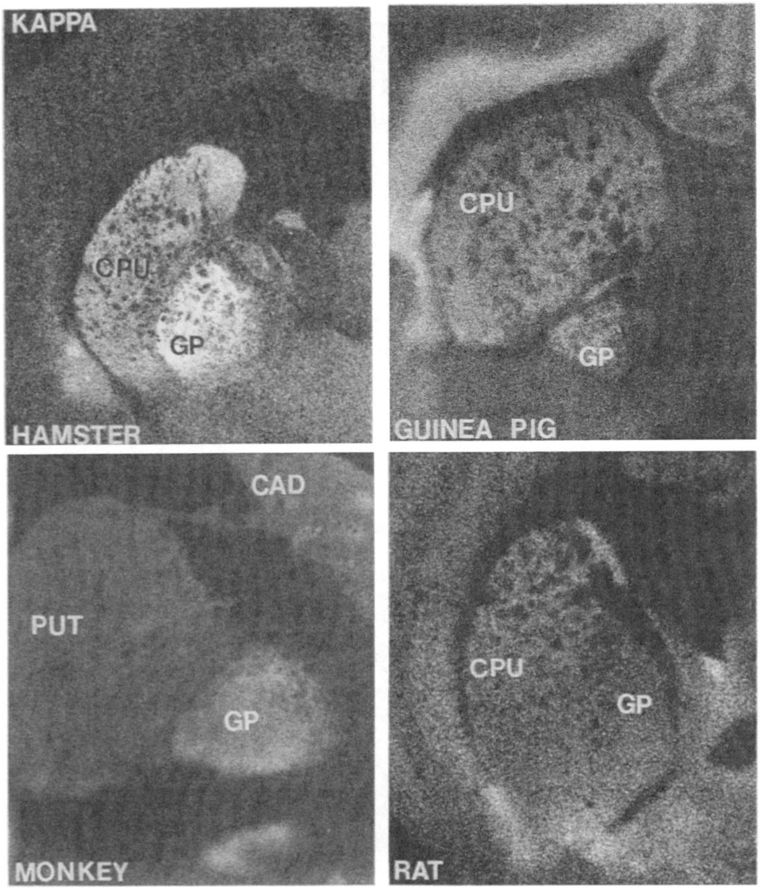

Fig. 5. Dark-field autoradiogram of κ-receptor binding in the hamster, guinea pig, monkey, and rat. Note that in the hamster and monkey, κ-receptors appear more densely in the globus pallidus as compared to the caudate and putamen. This relative difference in receptor density is not seen in the guinea pig and rat, where the density of κ-binding appears comparable in the caudate-putamen and globus pallidus. Abbreviations: *CAD* caudate; *CPU* caudate-putamen; *GP* globus pallidus; *PUT* putamen

the rat and guinea pig, where comparable levels of κ-binding are seen in the caudate-putamen and globus pallidus. Little difference can also be seen between levels of κ-binding of the globus and ventral pallidum of the rat.

Within the entopeduncular nucleus, a major projection area from the globus pallidus, marked species differences are noticed. A range of κ-receptor densities are observed that vary with species, with no κ-binding observed in the guinea pig, light labeling in the monkey and rat, and dense-binding in the hamster (Fig. 3).

At the level of the mesencephalon, further differences are seen in the substantia nigra. In the guinea pig and hamster, dense κ-labeling is seen in the pars reticulata while low to moderate levels are seen in the more dorsal pars compacta. These cellular boundaries are maintained to a lesser degree in the substantia nigra of the monkey and rat with moderate to low levels observed in both the pars compacta and reticulata. κ-Receptors in the monkey substantia nigra appear to be most dense in the lateroventral region as compared to its mediodorsal aspect. In the VTA, low levels of κ-binding are observed in the monkey and rat, with somewhat higher levels being found in the guinea pig and hamster.

4 Opioid Peptide Distribution

Comparing the distribution of the opioid peptides, proopiomelanocortin (POMC), enkephalin (ENK), and dynorphin (DYN) to the opioid receptors in the basal ganglia is a difficult task, as each of the four species examined here has not been adequately studied immunohistochemically. The distribution of the opioid peptides has been most intensely examined in the rat (e.g., Elde et al. 1976; Watson et al. 1982; Khachaturian et al. 1985b), followed by the cat (Graybiel et al. 1981) and monkey (Haber and Elde 1982; Khachaturian et al. 1984, 1985a), with little information available concerning the guinea pig and hamster (Neal and Newman 1989). In addition, variations in the primary amino acid structures of the opioid peptides across species may contribute to differences in immunohistochemical staining and localization. The following brief overview provides some of the salient features concerning the distribution of the opioids upon which most investigators agree. A more complete analysis of the distribution of the opioid peptides in the rat is presented earlier in this volume. As there is little published information on the distribution of the opioid peptides in the guinea pig, this species is not included.

The distribution of POMC peptides is well conserved across mammalian species. Cells originating in the arcuate nucleus and the medial basal hypothalamus project rostrally into the telencephalon and terminate in several brain regions including the nucleus accumbens. Some subtle species differences are observed in the striatum; whereas scattered fibers are seen in the monkey putamen (particularly in the rostral ventromedial aspect). POMC fibers are not detected in the caudate-putamen of the rat and hamster. Similarly, the nucleus accumbens of the monkey is more prominently innervated by POMC fibers as compared to the scattered fibers seen in the medial regions of the rat and hamster nucleus accumbens. Occasional POMC fibers are seen in the olfactory tubercle of each species examined with no fibers observed in the globus pallidus.

Caudal POMC projections extend into the mesencephalon with fibers coursing through the VTA of each species. Some scattered fibers have been observed in the pars compacta of the monkey and rat substantia nigra, with no immunoreactivity found in the entopeduncular nucleus and pars reticulata of the substantia nigra.

ENKergic neurons and processes are widely distributed throughout the basal ganglia. Of the three opioid peptide families, ENK is the most abundant in mammals, and demonstrates the most marked species differences. In the monkey, small cells and fibers appear scattered throughout the caudate and putamen and form distinct patches, being particularly concentrated in its rostral extent. ENK neurons also form patches in the rat caudate-putamen, but they are less dense and less widely distributed with the majority of cells being adjacent to the globus pallidus. The immunoreactive patches extend into the nucleus accumbens of the monkey, particularly its ventrolateral extent, and continue into the olfactory tubercle making the border between these structures difficult to discern. Distinct patches are not observed, however, in the rat and hamster nucleus accumbens. In the rat nucleus accumbens, dense fibers are seen with few cells observed medial to the anterior commissure, while the hamster accumbens shows lightly labeled scattered cells that are more pronounced in the shell portion of the accumbens. The staining in the olfactory tubercle demonstrates further species differences where, in the rat, it appears to be separate from the nucleus accumbens and forms a continuous band of cells that extend laterally and dorsally to the rhinal fissure. In contrast to both monkey and rat, no immunohistochemical ENK staining could be observed in the olfactory tubercle of the hamster.

More caudally in the globus pallidus, dramatic ENK staining, extending to the ventral pallidum, is observed across species. A high density of fine ENK fibers (woolly fibers) from the striatum form thick plexuses that envelope nonimmunoreactive dendrites of pallidal neurons, presumably modulating their transmission. The ENK fiber staining observed in the globus pallidus is the densest seen in any region of the brain and is one of the clearest examples of an ENK terminal field. ENK immunoreactivity in the monkey is substantially weaker in the internal division of the globus (which is equivalent to the entopeduncular nucleus in rodents) and corresponds well to the lack of ENK immunoreactivity in this nucleus in the rat and hamster.

In the mesencephalon, large species differences are seen in the substantia nigra. In the monkey, dense ENK fibers are seen in the substantia nigra, pars reticulata, particularly in its medial aspect, with moderate levels more laterally. In contrast, only a light labeling of fibers or an occasional fiber is seen in the hamster pars reticulata or the rat pars compacta, respectively. Light ENK fiber and cell staining is observed in the VTA of each animal examined.

The distribution of DYN neurons parallels that of ENK and appears to be densely distributed in the basal ganglia. While species differences are seen, they do not appear to be as marked as those found in the ENK system. In the rat, scattered DYN-positive cells and fibers are seen throughout the caudate-putamen. A similar pattern is found in the hamster, but the cells and fibers do not appear to be as numerous. DYN cells in the monkey, however, are concentrated rostrally in the ventromedial aspect of the caudate. While no positive cells are seen in the monkey putamen, scattered fibers are observed in the caudate and putamen. Within the ventral striatum, the DYN distribution appears consistent across species, with a moderate number of fibers and cells throughout the nucleus accumbens. The olfactory tubercle similarly shows scattered fibers and cells, especially in the Islands of Calleja.

More caudally in the globus pallidus, light fiber staining and scattered cells are seen in the hamster, with more numerous DYN cells and fibers found in the same region of the rat. The monkey demonstrates more intense DYN fiber staining in the form of woolly fibers in the globus pallidus which extends into the ventral pallidum. Moderate amounts of DYN fiber staining and scattered cells are seen in the ventral pallidum of the rat and hamster.

Few species differences are seen in the entopeduncular nucleus, where densely stained DYN fibers are seen in the monkey; moderate to light fiber staining is found in the rat and hamster, respectively. Similarly, in the substantia nigra, moderate DYN immunoreactivity is observed in the pars reticulata of the rat and hamster and dense fiber staining is seen in the monkey, particularly in the medial aspects of the pars reticulata. No DYN immunoreactivity is detected in the pars compacta of the hamster and monkey with an occasional DYN fiber observed in the rat. More dorsally in the VTA, scattered fibers and terminals are seen in the rat and hamster.

5 Conclusions

It is obvious from this review that despite a common mammalian anatomy, large species differences are seen in the distribution of the opioid receptors and peptides even within these well-delineated anatomical systems. The largest differences are seen in μ- and κ-receptor systems, while the distribution of δ-receptors is well conserved across species within the basal ganglia. In general, δ-sites are concentrated in the forebrain, while midbrain structures show comparatively few receptors. μ- and κ-sites, on the other hand, are more widely distributed throughout the basal ganglia, arguing that they may be involved in more complex forms of regulation than δ-sites. While it is difficult to know the physiological mechanisms involved, μ- and κ-receptors can potentially modulate these neuronal systems at each anatomical level that has been described.

A somewhat similar overall pattern is seen with the opioid peptides. The distribution of POMC fibers is well conserved across mammals, while the largest species differences are seen in the ENK and DYN systems of the basal ganglia. Their widespread distribution also allows for more complex forms of regulation, but compared to the opioid receptors, they demonstrate fewer, less dramatic species differences.

By focusing the review on the basal ganglia, we hoped to provide not only insights into the organization of these neural systems, but also a framework for examining the interaction of opioid peptides and receptors in the brain. These two themes are discussed below. While it is beyond the scope of this chapter to comprehensively discuss these issues, it provides a starting point for framing more intelligent questions concerning the distribution of opioid peptides and receptors and their functional role in the regulation of the basal ganglia.

One concept that has emerged from anatomical studies is that the basal ganglia is composed of two parallel neural systems; a dorsal one connecting cortex, caudate-putamen, globus pallidus, entopeduncular nucleus, and substantia nigra, and a more ventral system composed of connections between allocortex, ventral striatum (nucleus accumbens and olfactory tubercle), ventral pallidum, and VTA. The distribution of opioid receptors and peptides suggests that these distinct systems may be more neurochemically related than previously appreciated. μ- and δ-receptor binding, for example, appears to be homogeneous in the globus and ventral pallidum, suggesting that these structures may function in a coordinate fashion. This notion is further reinforced when examining the opioid peptides in these regions. ENK and DYN fibers extend in dense plexuses from the globus to the ventral pallidum, again suggesting that while these areas have distinct anatomical connections, they may function coordinately with regard to opioid transmission. κ-Receptors, on the other hand, fail to show this feature, with most species showing denser κ-binding in the globus pallidus as compared to the ventral pallidum. While this appears to be an interesting differentiation of κ-systems, the functional implications of these results are presently unclear.

A second example of an interaction between the dorsostriatal and ventrostriatal systems can be seen more rostrally in the caudate-putamen and nucleus accumbens. For each receptor type and species, the dorsal and ventral striata appear to be a unit, with the binding in the caudate and putamen appearing continuous with that seen in the nucleus accumbens and olfactory tubercle.

This apparent integration between the dorsal and ventral basal ganglia systems may be of functional interest as these circuits have been suggested to underlie distinct behaviors: the dorsal system has been linked to sensory-motor integration, while the ventrostriatal tegmental system is critical for reward and motivation (Fibiger and Phillips 1986). The intimate involvement of opioid systems in these neural circuits not only suggests that they may play a role in the behaviors, but based on their anatomical distribution, may aid in the coordination of sensory-motor inputs and their rewarding associations.

Using the basal ganglia as a model system to study peptide-receptor interactions, it is clear from the anatomical review presented above that there is no simple and consistent relationship between the three opioid receptors and peptides. Receptor binding studies had suggested that ENKs are selective for δ-receptors (Lord et al. 1977), while DYN peptides are selective for κ-sites (Chavkin et al 1982; Corbett et al. 1982). Such simple relationships are not seen anatomically. ENK fibers are not consistently localized with either μ- or δ-receptors, nor are DYN fibers consistently found with either μ- or κ-sites. These interactions are largely dependent on species and anatomical substrate. While these apparent inconsistencies may be due to technical limitations, such as the insensitivity of immunohistochemistry in visualizing finely scattered terminals or the inability of receptor autoradiographic techniques to detect regions with a low density of sites, these cannot be the entire explanation for these findings. The globus pallidus is a striking example where the relative abundance of opioid receptor types varies markedly with species, while the opioid peptide pattern remains consistent with dense ENK and DYN terminals observed across species. Based on abundance, ENK- and DYN-derived peptides are likely to interact with κ-receptors in the monkey and hamster, while in the rat they are more likely to interact with μ-receptors.

The entopeduncular nucleus and substantia nigra further demonstrate that the codistribution of opioid peptides and receptors varies substantially with species. The rat entopeduncular nucleus contains primarily μ-sites, while in the hamster both dense μ-and κ-receptors are observed. Despite these differences, both species have a similar opioid peptide distribution, with only DYN fibers being observed in the entopeduncular nucleus of each species. A similar observation can be made in the substantia nigra (pars reticulata) where, in the monkey, dense ENK and DYN fibers are codistributed with μ- and κ-receptors, while the same region in the hamster has only DYN fibers distributed with μ- and κ-sites. The pars reticulata of the rat similarly demonstrates a moderate density of μ-sites and a dense distribution of DYN fibers.

Such apparent inconsistency between specific opioid peptides and receptors is perhaps not surprising given that the affinity of these peptides for their receptors varies with posttranslational processing. DYN_{1-8}, for example, has good affinity for μ- and κ-receptors, while DYN_{1-17} appears more selective for κ-receptors (Corbett et al. 1982). Similarly, Quirion and Weiss (1983) report that proENK-derived peptides, such as Peptide E, bind potently to μ- and κ-sites in the guinea pig brain. Therefore, before precise statements can be made concerning the codistribution of opioid peptides and receptors, a better understanding of their posttranslational processing in each brain region is necessary. The presence of both μ- and κ-receptors in the guinea pig pars reticulata, for example, might suggest that DYN peptides are differentially processed in this species as compared to the rat; the latter contains primarily μ-sites in the pars reticulata. This is consistent with the findings of Lewis et al. (1985) who found that the conversion of DYN_{1-17} to DYN_{1-8} is more complete in the rat as compared to the guinea pig, producing a peptide more likely to bind to μ-receptors. Alternatively, the presence of different receptor populations within the same anatomical regions of several species may suggest different receptor functions across species. In other words, different opioid receptor types may subserve the same function in different species. Neither of these possibilities are mutually exclusive, and may both be involved in the organization of these neural systems.

While distribution studies such as the ones we have described in this chapter are helpful in providing a general framework for studying receptor function and regulation, more integrative studies combining electron microscopy, electrophysiology, and autoradiography are necessary to more precisely determine the exact interactions between opioid peptides and receptors. It is clear even at this anatomical level, however, that species differences are of primary importance in discussing the functions of the opioid receptors and provide insights into their interactions with the opioid peptides. Such species differences may be indicative of differential peptide processing or varied receptor function within anatomically equivalent structures. The basal ganglia with their well-delineated circuits and distinct opioid peptide and receptor distributions have provided the anatomical context within which to address these issues. The dense ENK and DYN terminal fields of the globus pallidus and substantia nigra are excellent neuroanatomical regions to compare the opioid receptor types. It is only with similar analyses in other functional and anatomical circuits, in conjunction with more integrative techniques, that a better understanding of opioid systems will be achieved.

Acknowledgments. We are grateful for the hamster opioid immunohistochemical data provided by Charles Neal of the University of Michigan and the following sources of grant funding: NIDA DA02265, NIMH MH422251, Lucille P. Markey Charitable Trust #88-46 and the Theophile Raphael Fund.

References

Alheid GF, Heimer L (1988) New perspectives in basal forebrain organization of special relevance for neuropsychiatric disorders: the striatatopallidal, amygdaloid, and corticopetal components of substantia innominata. Neuroscience 27:1-39

Atweh SF, Kuhar MJ (1977) Autoradiographic localization of opiate receptors in rat brain. III. The telencephalon. Brain Res 134:393-405

Bowen WD, Gentleman S, Herkenham M, Pert CB (1981) Interconverting mu and delta forms of the opiate receptor of rat striatal patches. Proc Natl Acad Sci USA 78:4818-4822

Charkin C, James I, Goldstein A (1982) Dynorphin is a specific endogenous ligand of the kappa opioid receptor. Science 215:413-415

Corbett AD, Paterson SJ, McKnight AT, Magnan J, Kosterlitz H (1982) Dynorphin (1-8) and dynorphin (1-9) are ligands for the kappa subtype of opiate receptor. Nature (London) 299:79-81

Elde R, Hokfelt T, Johansson O, Terenius L (1976) Immunohistochemical studies using antibodies to leucine-enkephalin: initial observations on the nervous system of the rat. Neuroscience 1:349-351

Fibiger HC, Phillips AG (1986) Reward, motivation, cognition: psychobiology of mesotelencephalic dopamine systems. In: Mountcastle VB, Bloom FE, Geiger SR (eds) Handbook of physiology. American Physiological Society, Bethesda, Maryland, pp 647-675

Graybiel AM, Ragsdale CW Jr, Yoneoka ES, Elde RP (1981) An immunohistochemical study of enkephalins and other neuropeptides in the striatum of the rat with evidence that the opiate peptides are arranged to form mosaic patterns in register with the striosomal compartments visible by acetylcholinertase staining. Neuroscience 6:377-397

Haber S, Elde R (1982) The distribution of enkephalin immunoreactive fibers and terminals in the monkey central nervous system: an immunohistochemical study. Neuroscience 7:1049-1095

Herkenham M, Pert CB (1980) In vitro autoradiography of opiate receptors in rat brain suggests loci of opiatergic pathways. Proc Natl Acad Sci USA 77:5532-5536

Herkenham M, Pert CB (1981) Mosaic distribution of opiate receptors, parafasicular projections and acetylcholinesterase in rat striatum. Nature (London) 291:415-418

Herkenham M, Pert CB (1982) Light microscopic localization of brain opiate receptors: a general autoradiographic method which preserves tissue quality. J Neurosci 2:1129-1149

Khachaturian H, Lewis ME, Haber SN, Akil H, Watson SJ (1984) Proopiomelanocortin peptide immunocytochemistry in the rhesus monkey brain. Brain Res Bull 13:785-800

Khachaturian H, Lewis ME, Haber SN, Houghten RA, Akil H, Watson SJ (1985a) Prodynorphin peptide immunohistochemistry in rhesus monkey brain. Peptides 6:155-166

Khachaturian H, Lewis ME, Schafer MK-H, Watson SJ (1985b) Anatomy of the CNS opioid systems. Trends Neurosci 8:111-119

Lewis ME, Lewis MS, Dores RM, Lewis JW, Khachaturian H, Watson SJ, Akil H (1985) Characterization of multiple opioid receptors and peptides in rat and guinea pig substantia nigra. J Biophys 47:54a

Lord JAH, Waterfield AA, Hughes J, Kosterlitz HW (1977) Endogenous opioid peptides: multiple agonists and receptors. Nature (London) 267:495-499

Mansour A, Khachaturian H, Lewis ME, Akil H, Watson SJ (1987) Autoradiographic differentiation of mu, delta, and kappa opioid receptors in the rat forebrain and midbrain. J Neurosci 7:2445-2464

Mansour A, Khachaturian H, Lewis ME, Akil H, Watson SJ (1988) Anatomy of CNS opioid receptors. Trends Neurosci 11:308-314

Morris BJ, Herz A (1986) Autoradiographic localization in rat brain of kappa opiate binding sites labelled by [^{3}H]bremazocine. Neuroscience 19:839-846

Neal CR, Newman SW (1989) Prodynorphin peptide distribution in the forebrain of the Syrian hamster and rat: a comparative study with antisera against dynorphin A, dynorphin B and the C-terminus of the prodynorphin precursor molecule. J Comp Neurol 288:353-386

Palacios JM, Maurer R (1984) Autoradiographic localization of drug and neurotransmitter receptors: focus on the opiate receptor. Acta Histochem 39:41-50

Penney JB, Young AB (1986) Striatal inhomogeneities and basal ganglia function. Move Disorders 1:3-15

Quirion R, Weiss AS (1983) Peptide E and other proenkephalin-derived peptides are potent kappa opiate receptor agonists. Peptides 4:445-449

Quirion R, Zajac JM, Morgat JL, Roques BP (1983) Autoradiographic distribution of mu and delta receptors in rat brain using highly selective ligands. Life Sci (Suppl 1) 33:227-230

Tempel A, Zukin RS (1987) Neuroanatomical patterns of mu, delta and kappa opioid receptors of rat brain as determined by quantitative in vitro autoradiography. Proc Natl Acad Sci USA 84:4308–4312
Watson SJ, Khachaturian H, Akil H, Coy DH, Goldstein A (1982) Comparison of the distribution of dynorphin systems and enkephalin systems in brain. Science 218:1134–1136
Zukin RS, Eghbali M, Olive D, Unterwald EM, Tempel A (1988) Characterization and visualization of rat and guinea pig kappa opioid receptors: evidence for K_1 and K_2 opioid receptors. Proc Natl Acad Sci USA 85:4061–4065

CHAPTER 11

Modulation of Central Opioid Receptors

B.J. Morris

1 Introduction

It appears to be a general rule of pharmacology that the number of receptors for a hormone or neurotransmitter in a given tissue is not fixed, but may fluctuate in response to a change in the level of receptor activation, or a change in the requirement of the tissue for receptor activation. The reported examples of alterations in receptor number following experimental manipulation cover more or less the entire range of hormone and neurotransmitter systems, and opioid receptors have been found to be as subject to modulation as any other receptor type.

This chapter attempts to summarize what has been learned about opioid receptors, and the cells that express them, from studies on the modulation of receptor number. There is also some evidence for the modulation of receptor affinity, and this will be mentioned where appropriate. It is also possible that other aspects of receptor function are subject to modulation, for example the rate of receptor desensitization. However, this chapter concentrates on changes in receptor number, where the evidence is most complete. It should be noted that in general it is the number of opioid binding sites that has been determined, rather than the number of functional receptors, and the question as to whether the changes detected have any functional significance has usually been left unanswered. Where it is used here, the term "receptor" should therefore be viewed as something assumed rather than proven.

There are three broad categories of experimental manipulation to which binding sites are susceptible: physiological, pharmacological and surgical. The first category generally represents an imposed alteration in the physiological condition of the animal, and a resulting change in the number of binding sites is used to infer a role for the receptor in that physiological process. The second category looks for changes in binding site number in response to administered drugs, and has been most frequently used to study the general principles governing binding site density. The final category investigates the effect of destroying particular neuronal pathways. An alteration in binding site number is then interpreted in terms of the anatomical location of the sites.

The amount of receptor binding can be determined either in tissue homogenates or in cryostat sections. The latter method has the advantage that it can be coupled with quantitative autoradiography to give a high level of anatomical resolution, and this is obviously the method of choice when any potential changes in binding sites may be limited to a particular brain region.

Three types of opiate receptor have been studied in detail: μ, δ, and κ. In this chapter, κ is used to denote the conventional kappa site which binds dynorphin$_{1\text{-}13}$ (DYN$_{1\text{-}13}$) and U50488H. The other proposed opioid binding sites are less well-defined, and have not yet been studied with respect to modulation. Some caution is required in interpreting the earlier reports, since highly selective ligands were not always available, and binding to other sites may have occurred.

2 Physiological Modulation

There is surprisingly little evidence for alterations in opioid binding after changes in the physiological condition of experimental animals. Endogenous and exogenous opioids have been implicated in the regulation of all of the physiological processes listed below. It should be remembered, however, that many of the experimental treatments which have been used involve a certain amount of stress for the animals, and the effects of the stress itself on opioid binding are not clear at the present time.

2.1 Nociception

In the rat models which use experimentally induced inflammation to simulate a condition of chronic pain, the animals become supersensitive to the antinociceptive effects of μ-agonists and subsensitive to the antinociceptive effects of κ-agonists (Millan et al. 1986, 1988a). However, there appears to be no alteration in the levels of μ, δ or κ binding sites (Millan et al. 1988a; Iadarola et al. 1988) in the segments of the spinal cord which are innervated by primary afferent fibres from the affected region. Equally, no change in μ-receptor binding has been found anywhere in the brain to explain the increased sensitivity to morphine. However, in the Freund's adjuvant-arthritis model of chronic pain, there is an increase in the level of κ-binding which is restricted to the periaqueductal grey matter (Millan et al. 1987), a region involved in the processing of nociceptive information. This would appear to be a functional response to the chronic discomfort, but it is not clear whether this has any relationship to the altered nociceptive sensitivity seen with this model.

2.2 Endocrine State

Little evidence has emerged to indicate that a changing endocrine condition is associated with an alteration in central opiate receptors. The only clear data relate to the regulation of μ-binding sites by gonadal steroids.

In contrast to earlier reports, a recent study failed to detect any alteration in ^{3}H-naloxone (^{3}H-NAL) binding in homogenates of whole brain or mediobasal hypothalamus following castration of male rats (Cicero et al. 1987). In the female rat, there are suggestions that long-term ovariectomy increases, while oestrogen treatment decreases, the B_{max} of μ-opioid binding in homogenates of hypothalamus (Vertes et al. 1986; Wilkinson et al. 1985). However, quantitative autoradiographic studies have revealed that the levels of ^{3}H-NAL binding in the sexually dimorphic region of the hypothalamic medial preoptic area are higher in dioestrous female rats than in male rats, although there was no significant change during the oestrous cycle. Postnatal treatment of female rats from birth with testosterone resulted in a pattern of binding similar to the males, while castration of neonatal male rats resulted in increased binding similar to that seen in the females (Hammer 1985 and references therein). These changes may reflect alterations in either binding affinity or capacity.

2.3 *Food and Water Deprivation*

Food deprivation has been suggested to increase the binding capacity of μ-sites in the midbrain (Tsujii et al. 1986).

Deprivation of water for 5 days, or administration of 2% saline for 5 days, is reported to decrease the binding to an opioid site in the posterior pituitary in rats by 35–50% (Brady and Herkenham 1987). The binding conditions used in this study (^{3}H-bremazocine binding at low temperature and high salt concentrations) label an unknown site in rat brain that has a distribution similar to μ-receptors, but has a pharmacological profile similar to, but different from, traditional κ-sites. Other workers have observed a 65% increase in δ, but not μ, binding in the nucleus tractus solitarius and spinal trigeminal nucleus following water deprivation for 5 days (Hwang et al. 1986). This may be an alteration in either binding affinity or capacity. One day of water deprivation is reported to increase total opioid binding (labelled by ^{3}H-diprenorphine) in the claustrum, lateral hypothalamus and ventral tegmental area, while decreasing binding in the cortex, lateral septum and periaqueductal grey matter (Blake et al. 1987). Since these changes were reversed by 15 min of drinking, they may reflect changes in the occupation of receptors by endogenous opioids, rather than changes in receptor density. One problem will all such studies is the difficulty in attributing the cause of any changes that are observed to the primary stimulus, and not the considerable degree of stress with which it is probably associated.

2.4 *Convulsive Stimuli*

Crain et al. (1987) have reported a decrease in μ- and δ-binding in the hippocampus 1 day after a kindled seizure. Since these changes were short-lived, and did not persist in parallel with the susceptibility to further seizures, this was seen as a consequence, rather than a cause, of the convulsion. Total (^{3}H-diprenorphine) opioid binding capacity is reportedly increased in many regions of rat brain, including hippocampus and striatum, after repeated electroconvulsive shocks (Hitzemann et al. 1987). Conversely, repeated electroconvulsive shock is reported to decrease the μ- and δ-binding capacity in the hypothalamus, hippocampus and striatum (Nakata et al. 1985). It is therefore not possible to draw any conclusions at present about the effect of seizure activity on opioid binding.

2.5 *Stress*

The effects of stress on opioid binding have not been extensively investigated, largely because of the difficulty of determining whether any changes observed are due to the primary stimulus causing the stress, or to the stress itself. The effects seen after some stressful stimuli have been mentioned above. In addition, Hnatowich et al. (1986) have observed a decrease in the number of total opioid (^{3}H-etorphine) binding sites following cold restraint stress, while Seeger et al. (1984) have suggested that total opioid binding decreases after both cold water swim stress and prolonged intermittent foot shock. Increases in μ- and δ-binding in various regions of rat brain have been reported following a single session of aversive foot shock (Stein et al. 1988).

2.6 *Miscellanea*

Beckman et al. (1986) have studied total (^{3}H-diprenorphine) opioid binding in the brain of the ground squirrel, and noted that hibernating squirrels show a decrease in many brain areas, relative to non-hibernating squirrels. However, since the changes are widespread, this could be a general effect of the decreased metabolic rate.

Haemorrhagic shock is reported to increase δ- and κ-binding in rat brainstem 24 h, but not 2 h, after the stimulus (Feuerstein et al. 1984).

3 Pharmacological Modulation

3.1 *Agonists*

Although it has been shown for many receptor types that chronic agonist treatment results in a decrease in receptor number, in the case of opioids it seems clear that the development of tolerance to morphine, and other μ- or δ-agonists, does not result from an alteration in the number of μ or δ receptors. However, it has proved difficult to determine unequivocally whether a change in opioid binding site number is an additional response to chronic activation.

Experiments using cultured cells (neuroblastoma × glioma) have shown that the application of full δ-receptor agonists will cause a decrease in the number of δ-binding sites (Blanchard et al. 1982; Law et al. 1983). Morphine appears to be a partial agonist on δ-receptors, and is ineffective in causing down-regulation. However, it has recently been reported that chronic morphine treatment will down-regulate μ-receptors in cell culture (Puttfarcken et al. 1988). A loss of responsiveness to μ- and δ-agonists can also be observed under these conditions, suggesting that the decreased binding may represent functional receptors. In both cases, the decrease in binding capacity occurs after a reduction in binding affinity, apparently caused by a decrease in receptor-effector coupling.

It might therefore be expected that opiate administration in vivo would lead to a decrease in the number of opiate binding sites. Since morphine may only be a partial agonist (discussed in Morris and Herz 1989), studies on the effect of chronic morphine treatment are difficult to interpret. Holaday and colleagues have consistently observed an increase in opioid binding, a finding replicated by other groups (Brandt et al. 1989; see also Rogers and El-Fakahany 1986; Rothman et al. 1987 and references therein). Other workers have seen clear evidence of decreased opioid binding after chronic morphine administration (Rogers and El-Fakahany 1986), which does not appear to be an effect of residual morphine from the chronic treatment. There is also evidence that opioid binding is unchanged (Klee and Streaty 1974; Geary and Wooten 1985). Part of the confusion undoubtedly derives from the multiplicity of opioid binding sites, the use of different doses of morphine, and the use of different ligands and experimental conditions in the binding assays. Chronic morphine administration to rats is thought to be a stressful stimulus, so there is the possibility that some of the changes in binding observed represent an indirect result of the stress rather than a direct action on opioid receptors.

Experiments in which ligands with high intrinsic activity were administered chronically have produced clearer results. Chronic etorphine treatment has been shown to down-regulate the number of μ- and δ-binding sites (Tao et al. 1987), and chronic bremazocine treatment has been demonstrated to decrease the number of κ-binding sites (Morris and Herz 1989). Thus opioid binding sites appear to respond in a similar way to other receptor types to chronic activation by a ligand with high intrinsic activity. Interestingly, the magnitude of the down-regulation varies in different brain regions (Tao et al. 1987; Morris and Herz 1989), and this may be indicative of differences in the properties of the receptors in different regions, or to regional differences in the basal level of receptor activation.

3.2 Antagonists

The phenomenon of supersensitivity, or increased tissue response to an agonist, can be observed in many tissues and with many receptor types, following blockade of the action of the endogenous agonist. Chronic administration of an opiate antagonist leads to a supersensitivity to opiate agonists and to an increase in opiate receptor binding (Zukin and Tempel 1986 and references therein; Millan et al. 1988b; Morris et al. 1988). It seems clear that up-regulation can be observed for μ-, δ- and κ-binding sites in rat brain, providing that a dose of antagonist which is active on the relevant receptor type is used (Zukin and Tempel 1986; Millan et al. 1988b; Morris et al. 1988). An example of the increases in κ-binding caused by chronic antagonist treatment is shown in Fig. 1. Increased receptor numbers are typically seen after a few days of antagonist administration. The time course of the development of behavioural supersensitivity parallels that of the up-regulation, suggesting that the increased binding represents functional receptors.

The increased binding is generally assumed to be a compensatory response resulting from the blockade of action of endogenous agonists, although there is also

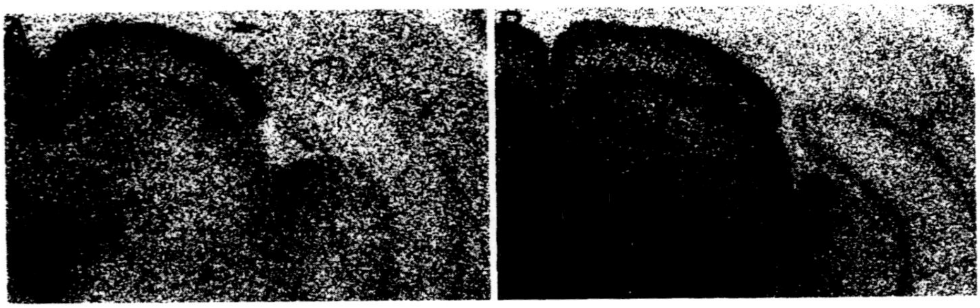

Fig. 1. Increase in κ-opioid binding following chronic administration of a high dose of NAL. The κ-binding sites were visualized by in vitro autoradiography following chronic in vivo infusion of either vehicle (**A**) or NAL (**B**). Note that binding was increased following NAL treatment, particularly in the periaqueductal grey matter (*arrowheads*) and the molecular layer of the hippocampus (*arrows*). Details as in Morris et al. (1988)

evidence that antagonists can themselves exert direct effects on cell function (discussed in Morris et al. 1988). Thus, apparently in the absence of any agonist, antagonists can induce an alteration in the basal level of receptor coupling to second-messenger systems and ion channels. However, it seems unlikely that the binding site up-regulation observed after chronic antagonist administration is caused by this intrinsic action. The observation, for each opioid receptor type, that the magnitude of the up-regulation varies in different brain regions (Zukin and Tempel 1986; Morris et al. 1988) argues against a direct action of the antagonist.

When NAL is administered chronically in a dose sufficient to block μ-, δ- and κ-receptors (Millan et al. 1988), a different neuroanatomical pattern of up-regulation is seen for each receptor type (Morris et al. 1988). The greatest increases in μ-binding are seen in the hippocampus, thalamus and hypothalamus, the greatest increase in δ-receptor binding is observed in the amygdala, while κ-receptor binding increases most in the hippocampus and periaqueductal grey matter. At present, the changes in receptor binding following chronic antagonist treatment appear to provide a better index of the level of receptor-mediated activity in a particular region than the absolute levels of the receptor itself. Studies of antagonist-induced up-regulation may therefore prove of great use in determining the sites where opioids act to produce their effects.

3.3 Drugs Not Acting on Opioid Receptors

Many drugs bind to opioid sites in vitro at high concentrations, but such interactions are unlikely to be physiologically or pharmacologically important. However, there are indications that some pharmacological agents can affect opioid binding at more relevant concentrations, either in vitro or in vivo. Cannabinoids will reduce the number of μ-, δ-, but not κ-binding sites in vitro, without affecting binding affinity (Vaysse et al. 1987). Ethanol can inhibit binding to δ-sites in vitro (Hiller et al. 1984) and, after chronic in vivo administration to rats and mice, is reported either to increase the affinity of δ-sites (Pfeiffer et al. 1981), or to decrease the affinity of μ-sites (Hoffman et al. 1982). Prenatal treatment of rats with diazepam will decrease the number of total opioid binding sites (Watanabe et al. 1983), while prolonged exposure to nitrous oxide decreases (μ) binding sites in rat brainstem (Ngai and Finck 1982). Chronic antidepressant administration is reported to increase the number of μ- and δ-sites in the spinal cord, and to decrease the number of these sites in the hypothalamus (Hamon et al. 1987, and references therein).

4 Surgical Modulation

4.1 Striatum

The striatum receives a unilateral dopaminergic projection from the substantia nigra which is known to exert a profound influence on striatal activity. Destruction of the dopaminergic projection using the selective neurotoxin 6-hydroxydopamine (6-OHDA) causes a reduction of between 30 and 40% in the levels of μ-opioid sites in the denervated striatum, compared to the contralateral side (Pollard et al. 1978; Reisine et al. 1979; Murrin et al. 1980; Gardner et al. 1980; Eghbali et al. 1987). This has invariably been interpreted as evidence that these μ-receptors are located on the dopamine (DA) terminals. However, it appears that there is a 50% loss of μ-opioid receptors on the contralateral side of the brain, where there is no decrease in DA levels and no disappearance of the terminals (Eghbali et al. 1987). This seems to indicate that the changes in striatal activity are sufficient in themselves to cause a marked alteration in receptor levels, and that such alterations cannot be interpreted in terms of a presynaptic location of the sites.

This is supported by the observation that a slightly more lateral placement of the nigral lesions, causing an equivalent decrease in ipsilateral DA content, results in, not a decrease, but a 25 to 30% increase in striatal μ-binding sites compared to the contralateral side (Gardner et al. 1980; B.J. Morris unpubl.). Thus, the alterations in striatal opioid binding in response to DA denervation are likely to be a compensatory effect due to a dramatic alteration in tissue function. It should also be remembered that destruction of a population of afferent fibres is likely to cause damage to postsynaptic structures. It has been shown that, following 6-OHDA lesions of the nigrostriatal tract, degenerating dendritic spines are found in the rat striatum postsynaptic to the degenerating boutons (Hattori and Fibiger 1982). Such trans-synaptic degenerative effects further preclude the interpretation of lesion-induced changes in receptor binding in terms of the anatomical location of the sites.

A study using electron microscope autoradiography has found that only 18% of μ-binding sites in the rat striatum are associated with axo-axonic interfaces (Hamel and Beaudet 1984), implying that only a small proportion of the striatal opioid receptors may be located on the terminals of afferent fibres. Moreover, it might be predicted on theoretical grounds that presynaptic receptors would constitute only a small percentage of the total receptor population (Stryer 1985). The large decreases in opioid binding observed after DA denervation are therefore unlikely simply to reflect the loss of presynaptic receptors. In fact, a recent report has provided evidence that the DA neurones of the substantia nigra, pars compacta, are insensitive to μ- and δ-agonists (Lacey et al. 1989). While it remains possible that opioid receptors could be expressed on the axon terminals in the striatum, and not on the cell bodies in the substantia nigra, it seems more likely that there are no functional μ- or δ-receptors on the DA terminals in the striatum.

Nigral 6-OHDA lesions resulting in an ipsilateral decrease in striatal μ-opioid binding also cause an ipsilateral loss of δ-binding, but reportedly no change in κ-binding (Eghbali et al. 1987). It has also been demonstrated that 6-OHDA lesions of the guinea pig substantia nigra do not affect the total number of opioid binding sites in the striatum (Foote and Maurer 1986). This could be due to opposite and approximately equal effects on two different types of opioid receptor, to the exact placement of the lesions within the nigra, bearing in mind the conflicting data obtained in rats, or to marked species differences.

Various workers have also examined the effect of intrastriatal injections of excitatory amino acids on the levels of opioid binding in rat striatum. These excitotoxins are thought to cause the destruction of perikarya, while sparing nerve terminals and axons of passage. Reductions of between 20 and 55% in striatal opioid (μ) binding have been reported, along with an 80% loss of δ-binding, and interpreted as evidence that the remaining sites are normally present presynaptically (Murrin et al. 1980; Abou-Khalil et al. 1984; Eghbali et al. 1987). However, the presence of opioid binding sites on glia remains a possibility, and the profound alteration in striatal function caused by this type of lesion may preclude such a straightforward interpretation.

Perhaps the best evidence available concerning the synaptic location of striatal opioid receptors comes from a study in which the nigrostriatal and striatonigral pathways were surgically sectioned (Van der Kooy et al. 1986). This appeared to cause an accumulation of ^{3}H-etorphine binding on both sides of the cut. The amount of binding in the vicinity of the cut was small in relation to total striatal opioid binding, but the evidence clearly suggests that at least some opioid receptors are transported in the mesostriatal tract. There would seem to be good reason to use this approach in parallel with labelling conditions selective for μ-, δ- and κ-binding sites.

Parenti et al. (1983) have reported a loss of striatal (μ) opioid binding capacity following destruction of the serotonergic nerve terminals in the striatum. The effects on opioid binding of lesioning the other afferent projections to the striatum remain to be investigated.

Overall, it appears that few definitive conclusions can be drawn concerning the anatomical location of striatal opioid receptors from lesion experiments. However, it seems that the levels of these opioid receptors are sensitive to changes in striatal activity. The alterations in opioid binding following DA denervation may be related to the dramatic changes in the activity of striatal enkephalinergic neurones seen with this treatment (Morris et al. 1989).

4.2 Substantia Nigra

An early study (Pollard et al. 1978) reported a 33% loss of μ-sites following 6-OHDA lesions of the nigrostriatal pathway which destroy the DA cells in the pars compacta. However, another report (Reisine et al. 1979) did not reproduce this finding. There is behavioural evidence showing that, after 6-OHDA lesions of the nigrostriatal tract, there is a decreased sensitivity to μ- and δ-agonists injected into the nigra, but a supersensitivity to the locomotor effects of microinjected κ-agonists (Matsumoto et al. 1988). It would be interesting to know whether these changes in sensitivity are receptor-related.

After kainic acid lesions of the striatum, which destroy the striatonigral projection neurones, Abou-Khalil et al. (1984) observed a 40% decrease in μ-binding and a 26% decrease in δ-binding in the substantia nigra. This was interpreted as evidence for a presynaptic location of the binding sites, although possible transsynaptic effects have not been investigated. Van der Kooy et al. (1986) observed an accumulation of total opioid (^{3}H-etorphine) binding sites on the striatal side of a knife cut interrupting the striatonigral pathway. This implies that some opioid receptors are transported to a presynaptic location in the pars reticulata. Interestingly, Herrera-Marschitz et al. (1984) have shown a supersensitivity to the locomotor effects of κ-agonists injected into the nigra following excitotoxin lesions of the striatum. This would appear to be denervation supersensitivity, and could conceivably be caused by an increase in the levels of κ-binding sites in the nigra. It would be of interest to determine whether this is the case, as the phenomena of supersensitivity and receptor up-regulation have until now not been shown to be separate.

4.3 Olfactory Bulb

Removal of the olfactory bulbs in mice causes both decreases and increases in μ-opioid binding in areas receiving efferent fibres (Hirsch 1980). Following unilateral bulbectomy, there is a gradual decrease in μ-opioid binding in the contralateral olfactory bulb, although there is no known direct projection from one olfactory bulb to the other (Hirsch and Margolis 1980). The results suggest a role for μ-receptors in olfactory function.

4.4 Cortex and Amygdala

There is a report that 6-OHDA lesions of the nigrostriatal pathway induce an increase in μ-receptor binding in the frontal cortex, possibly as a result of the loss of the mesocortical DA projection (Reisine et al. 1979). However, a more recent report describes a loss of μ-sites from the frontoparietal cortex following similar lesions (Eghbali et al. 1987). In the rat amyglada, μ-binding sites reportedly increase or decrease following destruction of the substantia nigra, according to the exact placement of the lesion (Gardner et al. 1980).

4.5 Pituitary

Sectioning of the pituitary stalk, with resultant degeneration of the neurohypophysial nerve terminals and proliferation of the pituicytes, induces an increase in the normally low levels of κ-binding in the rat posterior pituitary (Lightman et al. 1983; Bunn et al. 1985). Since the pituicytes are likely to be the only cell type present in large numbers in the neurohypophysis after stalk section, they represent the probable location of these sites. Immunoreactive leu-enkephalin (Leu-ENK), which may represent Leu-ENK itself or any of the dynorphins, has been detected in nerve terminals making apparently synaptoid contacts with pituicytes (Van Leeuwen et al. 1983). The κ-binding sites in the posterior pituitary are therefore probably functional receptors.

4.6 Spinal Cord

Attempts have been made to determine whether the high levels of opiate binding in the superficial laminae of the spinal cord are localized presynaptically on the nerve endings of the primary afferent fibres, or postsynaptically on the dorsal horn neurones. After the primary afferent fibres have been destroyed, there is a decrease in the levels of both μ- and δ-sites (LaMotte et al. 1976; Fields et al. 1980; Ninkovic et al. 1981), and this can be interpreted as evidence for a presynaptic location for some of the sites. The levels of conventional κ-binding sites in the rat lumbar spinal cord are low, although high levels of a related binding site are present (Morris and Herz 1987).

4.7 Other Regions

A presynaptic location for opioid receptors in the vagal and accessory optic systems has been proposed (Atweh et al. 1978).

5 General Considerations

The paucity, and frequently contradictory nature, of evidence showing altered opioid binding after physiological manipulations suggests that the number of binding sites is not a very sensitive index of the level of receptor-mediated activity. This might also be inferred from the fact that morphine tolerance, during which there is a profound change in opioid sensitivity, can be observed without any change in μ-binding (vide supra). However, since neurones show a degree of plasticity in the number of opioid receptors that they express, receptor number presumably does play a role in the regulation of cellular sensitivity to opioids. Further information is needed to determine under what circumstances this level of control is important.

Studies on the surgical manipulation of binding sites have frequently suffered from overinterpretation. Nevertheless, they can provide unique information on the function of opioid receptors in discrete anatomical circuits. Overall, there seems to be considerable scope for further investigations into the modulation of central opioid receptors.

References

Abou-Khalil A, Young AB, Penney JP (1984) Evidence for the presynaptic localization of opiate binding sites on striatal efferent fibres. Brain Res 323:21-29

Atweh SF, Murrin LC, Kuhar MJ (1978) Presynaptic localization of opiate receptors in the vagal and accessory optic systems: an autoradiographic study. Neuropharmacology 17:65-71

Beckman AL, Dean RR, Wamsley JK (1986) Hippocampal and cortical opioid receptor binding: changes related to the hibernation state. Brain Res 386:223-231

Blake MJ, Stein EA, Czech DA (1987) Drinking-induced alterations in reward pathways: an autoradiographic analysis. Brain Res 413:111-119

Blanchard SG, Chang K-J, Cuatrecasas P (1982) Characterization of the association of ^{3}H-enkephalin with neuroblastoma cells under conditions optimal for receptor down-regulation. J Biol Chem 258:1092-1097

Brady LS, Herkenham M (1987) Dehydration reduces κ-opiate binding in the neurohypophysis of the rat. Brain Res 425:212-217

Brandt SA, Livingston A, Roper M (1989) The effect of morphine tolerance on brain and spinal cord opioid and alpha$_2$-adrenergic receptors in the rat. Br J Pharmacol 96:314P

Bunn SJ, Hanley MR, Wilkin GP (1985) Evidence for a kappa-opioid receptor on pituitary astrocytes. Neurosci Lett 55:317-323

Cicero TJ, O'Connor LH, Bell RD (1987) Reevaluation of the effects of castration on naloxone-sensitive opiate-receptors in the male rat. Neuroendocrinology 46:176-184

Crain BJ, Chang K-J, McNamara JO (1987) An in-vitro autoradiographic analysis of mu and delta opioid binding in the hippocampal formation of kindled rats. Brain Res 412:343-351

Eghbali M, Santoro C, Paredes W, Gardner EL, Zukin RS (1987) Visualization of mutiple opioid receptor types in rat striatum after specific mesencephalic lesions. Proc Natl Acad Sci USA 84:6582-6586

Feuerstein G, Faden AI, Krumins SA (1984) Alteration in opiate receptor binding after haemorrhagic shock. Eur J Pharmacol 100:245-246

Fields HL, Emson PC, Leigh BK, Gilbert RFT, Iversen LL (1980) Multiple opiate receptor sites on primary afferent fibres. Nature (London) 284:351-353

Foote RW, Maurer R (1986) Opioid binding sites in the nigrostriatal system of the guinea pig are not located on nigral dopaminergic neurons. Neurosci Lett 65:341-345

Gardner EL, Zukin RS, Makman MH (1980) Modulation of opiate receptor binding in striatum and amygdala by selective mesencephalic lesions. Brain Res 194:232-239

Geary WA, Wooten GF (1985) Regional saturation studies of ^{3}H-naloxone binding in the naive, dependent and withdrawal states. Brain Res 360:214-223

Hamel E, Beaudet A (1984) Electron microscopic autoradiographic location of opioid receptors in rat striatum. Nature (London) 312:155-157

Hammer RP (1985) The sex hormone-dependent development of opiate receptors in the rat medial preoptic area. Brain Res 360:65-74

Hamon M, Gozlan H, Bourgoin S, Benoliel J, Mauborgne A, Taquet H, Cesselin F, Mico JA (1987) Opioid receptors and neuropeptides in the CNS in rats treated chronically with amoxapine or amitriptyline. Neuropharmacology 26:531-540

Hattori T, Fibiger HC (1982) On the use of lesions to localize neurotransmitter receptor sites in striatum. Brain Res 238:245-250

Herrera-Marschitz M, Christensson-Nylander L, Sharp T, Stines W, Reid M, Hokfelt T, Ungerstedt U (1984) Striatonigral dynorphin and substance P pathways in the rat. Exp Brain Res 64:193-207

Hiller JM, Angel LM, Simon EJ (1984) Characterization of the selective inhibition of the delta subclass of opioid binding sites by alcohols. Mol Pharmacol 25:249-255

Hirsch JD (1980) Opiate and muscarinic ligand binding in five limbic areas after bilateral olfactory bulbectomy. Brain Res 198:271-283

Hirsch JD, Margolis FL (1980) Influence of unilateral olfactory bulbectomy on opiate and other binding sites in the contralateral bulb. Brain Res 199:39–47
Hitzemann RJ, Hitzemann A, Blatt S, Meyerhoff JL, Tortella FC, Kenner JR, Belenky GL, Holaday JW (1987) Repeated electroconvulsive shock: effect on sodium dependency and regional distribution of opioid binding sites. Mol Pharmacol 31:562–566
Hnatowich MR, Labella FS, Kiernan K, Glavin GB (1986) Cold restraint stress reduces ^{3}H-etorphine binding to rat brain membranes. Brain Res 380:107–113
Hoffman PL, Urwyler S, Tabakoff B (1982) Alterations in opiate receptor function after chronic ethanol exposure. J Pharmacol Exp Ther 222:182–187
Hwang BH, Chang K-J, Severs WB (1986) Increased δ-, but not μ, opiate receptor binding in the medulla oblongata of Long-Evans rats following 5 day water deprivation. Brain Res 371:345–349
Iadarola MJ, Brady LS, Draisci G, Dubner R (1988) Enhacement of dynorphin gene expression in spinal cord following experimental inflammation. Pain 35:266–267
Klee WA, Streaty RA (1974) Narcotic receptor sites in morphine-dependent rats. Nature (London) 248:61–64
Lacey MG, Mercuri NB, North RA (1989) Two cell types in rat substantia nigra zona compacta distinguished by membrane properties and the actions of dopamine and opioids. J Neurosci 9:1233–1241
LaMotte C, Pert CB, Snyder SH (1976) Opiate receptor binding in primate spinal cord. Brain Res 112:407–412
Law PJ, Hom DS, Loh HH (1983) Opiate receptor down-regulation and desensitisation in neuroblastoma × glioma hybrid cells are separable cellular adaptation processes. Mol Pharmacol 24:413–424
Lightman SL, Ninkovic M, Hunt SP, Iversen LL (1983) Evidence for opiate receptors on pituicytes. Nature (London) 305:235–237
Matsumoto RR, Brinsfield KH, Patrick RL, Walker JM (1988) Rotational behaviour mediated by dopaminergic and non-dopaminergic mechanisms after intranigral microinjection of specific mu, delta and kappa opioid agonists. J Pharmacol Exp Ther 246:196–203
Millan MJ, Millan MH, Członkowski A, Höllt V, Pilcher CWT, Colpaert FC, Herz A (1986) A model of chronic pain in the rat: response of multiple opioid systems to adjuvant-induced arthritis. J Neurosci 6:899–906
Millan MJ, Morris BJ, Colpaert FC, Herz A (1987) A model of chronic pain in the rat: high resolution approach identifies alterations in multiple opioid systems in the periaqueductal grey. Brain Res 416:349–353
Millan MJ, Czlonkowski A, Morris B, Stein C, Arendt R, Huber A, Höllt V, Herz A (1988a) Inflammation of the hind limb as a model of unilateral, localized poin. Pain 35:299–312
Millan MJ, Morris BJ, Herz A (1988b) Antagonist-induced opioid receptor up-regulation: characterization of supersensitivity to selective mu and delta agonists. J Pharmacol Exp Ther 247:721–727
Morris BJ, Herz A (1987) Distinct distribution of opioid receptor types in rat lumbar spinal cord. Naunyn Schmiedeberg's Arch Pharm 336:240–243
Morris BJ, Herz A (1989) Control of opiate receptor number in vivo: simultaneous κ-receptor down-regulation and μ-receptor up-regulation following chronic agonist-antagonist treatment. Neuroscience 29:433–442
Morris BJ, Millan MJ, Herz A (1988) Antagonist-induced opioid receptor up-regulation: regionally-specific modulation of mu, delta and kappa binding sites revealed by quantitative autoradiography. J Pharmacol Exp Ther 247:729–736
Morris BJ, Herz A, Höllt V (1989) Localisation of striatal opioid gene expression, and its modulation by the mesostriatal dopamine pathway. J Mol Neurosci 1:9–18
Murrin LC, Coyle JT, Kuhar MJ (1980) Striatal opiate receptors: pre- and postsynaptic location. Life Sci 27:1175–1183
Nakata Y, Chang K-J, Mitchell CL, Hong JS (1985) Repeated electroconvulsive shock downregulates the opioid receptors in rat brain. Brain Res 346:160–163
Ngai SH, Finck AD (1982) Prolonged exposure to nitrous oxide decreases opiate receptor density in rat brainstem. Anesthesiology 57:26–30
Ninkovic M, Hunt SP, Gleave JRW (1981) Effect of dorsal rhizotomy on the autoradiographic distribution of opiate and neurotensin receptors in the rat spinal cord. Brain Res 230:111–119
Parenti M, Tirone F, Olgiati VR, Groppetti A (1983) Presence of opiate receptors on striatal serotonergic nerve terminals. Brain Res 280:317–322

Pfeiffer A, Seizinger BR, Herz A (1981) Chronic ethanol imbibition interferes with delta but not mu receptors. Neuropharmacology 20:1229–1232
Pollard H, Llorens C, Schwartz JC, Gros C, Dray F (1978) Localization of opiate receptors in the rat striatum in relationship to the nigrostriatal dopamine system. Brain Res 177:241–252
Puttfarcken PS, Werling LL, Cox BM (1988) Effects of chronic morphine exposure on opioid inhibition of adenyl cyclase in 7315c cell membranes. Mol Pharmacol 33:520–527
Reisine TD, Nagy JI, Beaumont K, Fibiger HC, Yamamura HI (1979) The localization of receptor binding sites in the substantia nigra and striatum of the rat. Brain Res 177:241–252
Rogers NF, El-Fakahany EE (1986) Morphine-induced opioid receptor down-regulation detected in intact adult rat brain cells. Eur J Pharmacol 124:221–227
Rothman RB, McLean S, Bykov V, Lessor RA, Jacobson AE, Rice KC, Holaday JW (1987) Chronic morphine upregulates a μ-opiate binding site labelled be cyclofoxy. Eur J Pharmacol 142:73–81
Seeger TF, Sforzo GA, Pert CB, Pert A (1984) In vivo autoradiography: visualisation of stress-induced changes in opiate receptor occupancy. Brain Res 305:303–311
Stein EA, Hiller JM, Simon EJ (1988) Alteration in opiate receptor binding following stressful stimuli in the rat. Adv Biosci 75:639–642
Stryer D (1985) Mismatch problem in receptor mapping studies. Trends Neurosci 8:522
Tao P-L, Law P-Y, Loh HH (1987) Decrease in delta and mu opioid receptor binding capacity in rat brain after chronic etorphine treatment. J Pharmacol Exp Ther 240:809–817
Tsujii S, Nakai Y, Fukata J, Koh T, Takahashi H, Usui T, Imura H (1986) Effects of food deprivation and high fat diet on opioid receptor binding in rat brain. Neurosci Lett 72:169–173
Van der Kooy D, Weinreich P, Nagy JI (1986) Dopamine and opiate receptors: localisation in the striatum and evidence for their axopiasmic transport in the nigrostriatal and striatonigral pathways. Neuroscience 19:139–146
Van Leeuwen FW, Pool CW, Sluiter AA (1983) Enkephalin immunoreactivity in synaptoid-elements on glial cells in the rat neural lobe. Neuroscience 8:229–241
Vaysse PJ-J, Gardner EL, Zukin RS (1987) Modulation of rat brain receptors by cannabinoids. J Pharmacol Exp Ther 241:534–560
Vertes M, Pamer Z, Garai J (1986) On the mechanism of opioid-oestradiol interactions. J Steroid Biochem 24:235–238
Watanabe Y, Shibuya T, Salafsky B, Hill HF (1983) Prenatal and postnatal exposure to diazepam: effects on opioid receptor binding in rat cortex. Eur J Pharmacol 96:141–144
Wilkinson M, Brawer JR, Wilkinson DA (1985) Gonadal steroid-induced modification of opiate binding sites in anterior hypothalamus of female rats. Biol Reprod 32:501–506
Zukin RS, Tempel A (1986) Neurochemical correlates of opiate receptor regulation. Biochem Pharmacol 35:1623–1627

CHAPTER 12

The Control of Hypothalamic Opioid Peptide Release

G. Burns and K.E. Nikolarakis

1 Introduction

It is generally agreed that one of the major factors involved in the response to stress is the 41 amino acid peptide corticotropin-releasing hormone (CRH) isolated by Vale and colleagues (Vale et al. 1981). This peptide is primarily known for its effect in stimulating the release of pituitary ACTH of which it appears to be the major modulator (Labrie et al. 1987). However, it has now become clear that CRH has a more general integrative role in the stress response. Thus, central administration of CRH produces a broad spectrum of autonomic and behavioral effects associated with stress. These include increased plasma levels of catecholamines and glucose (Brown et al. 1982), tachycardia and increased mean arterial pressure (Fisher et al. 1983), inhibited gastric acid secretion (Taché et al. 1983) and a variety of, generally locomotory, behavioral effects (Sutton et al. 1982; Vale et al. 1983). Furthermore, following acute or chronic stress, changes have been found in the levels of CRH-like immunoreactivity present in brain areas associated with stress responses (Chappell et al. 1986), coupled with increased synthesis and release of hypothalamic CRH (Haas and George 1988). At least one central response associated with stress (freezing behavior) can also be attenuated by the administration of a CRH receptor antagonist (Kalin et al. 1988). The mechanisms by which CRH elicits these effects are generally uncertain. However, since their discovery, the endogenous opioid peptides (EOPs) have been thought to play a major role in the reaction to stress (see Przewłocki et al., this Vol.) and a possible CRH-opioid interaction may therefore be envisaged. In support of this suggestion is the finding of a simultaneous co-expression of CRH and pro-opiomelanocortin (POMC) mRNA in specific regions of the brain (Thompson et al. 1987), implying an interaction between these systems. This would appear especially likely to occur in the hypothalamus, a region suggested to be the primary site of action for integrating the endocrine, autonomic and behavioral effects of CRH. The close neuroanatomical relationship between CRH and opioid neuronal systems would allow such a functional interaction between the two systems to occur. β-Endorphin (β-END) and CRH nerve terminals are present in close proximity in the medial basal hypothalamus and median eminence (Bloom et al. 1982; Swanson et al. 1983), while dynorphin$_{1-8}$ (DYN$_{1-8}$) and met-enkephalin (Met-ENK) have been reported to coexist with CRH in neurons in the paraventricular nucleus (Hökfelt et al. 1987; Roth et al. 1983); this is also the case for Met-ENK and CRH in the median eminence (Hisano et al. 1986).

In this chapter we describe the systems controlling the release of opioids from the brain, particularly the hypothalamus. Particular attention is paid to CRH, which, as will be discussed, appears to be the major factor as yet implicated in controlling EOP release in this tissue.

2 Control of Hypothalamic Opioid Peptide Release by CRH in the Rat

As mentioned above, it appears likely that the stress effects of CRH are mediated, at least partially, via an effect on brain opioid systems. In fact, investigation of the role of CRH in modulating the release of brain opioids supported this supposition, indicating CRH to be a potent modulator of their release. Application of CRH to hypothalamic slices in an in vitro perifusion system stimulated the release of β-END, DYN and Met-ENK (Nikolarakis et al. 1986, 1989a; Fig. 1). CRH was a very potent secretagogue in this system, being effective at concentrations as low as 10^{-12} M and, in all three cases was found to have a dose-dependent effect, maximal release occurring between 10^{-8} and 10^{-6} M CRH. The observed effects of CRH would appear to be mediated via specific receptors since the CRH receptor antagonist α-helical CRH_{9-41} was found to completely block the effect of CRH on the release of all three peptides. The time courses of the responses differed for each peptide with the release of β-END and Met-ENK being rapidly stimulated and that of DYN being somewhat delayed in comparison. However, an interesting finding in all cases was that, following maximal stimulation, the release rate declined, eventually reaching basal

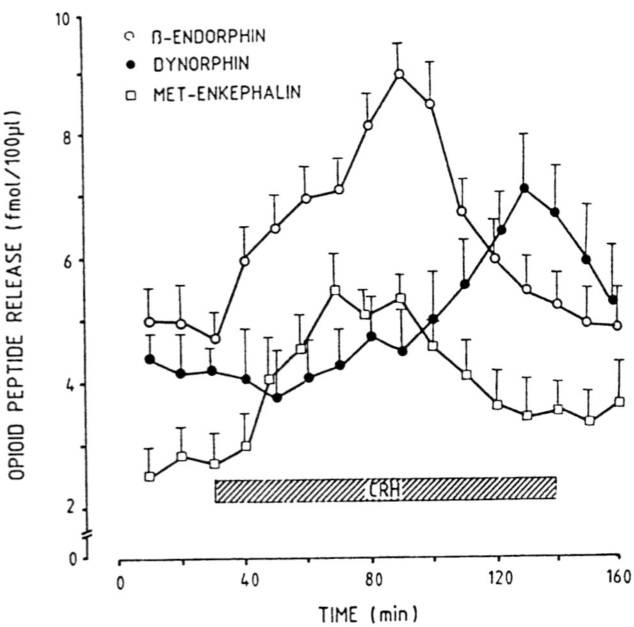

Fig. 1. Stimulation by CRH (10^{-8} M) of the basal release of β-endorphin, dynorphin and Met-enkephalin from perifused rat hypothalamic slices (After Nikolarakis et al. 1986, 1989a)

levels, despite the continued presence of CRH. A possible reason for this observation is presented later in this chapter.

CRH would appear to exert a tonic control over the release of β-END and Met-ENK since application of the CRH receptor antagonist in vitro lowered the basal release rate of both opioids (the basal level of DYN was also reduced, although not significantly; Nikolarakis et al. 1988a). This observation was also made in vivo when the antagonist was perfused into the arcuate-median eminence region of the medial basal hypothalamus using a push-pull cannula technique, the basal release rates of Met-ENK and β-END being again reduced (as was the basal release of DYN, again to an insignificant degree; Nikolarakis et al. 1988a). Preliminary results have also indicated that application of CRH via the push-pull technique stimulates opioid release from the hypothalamus. Thus, the CRH stimulation of EOP release may be physiologically relevant in maintaining the basal release rate of β-END and Met-ENK.

If a simple feedback loop was occurring, it would be expected that since CRH stimulates opioid release from the hypothalamus then in the converse situation opioids would be found to inhibit the release of CRH. However, the results from such studies are contradictory with in vitro application of opioids being found to be stimulatory or inhibitory to the release of CRH (Buckingham 1986; Yajima et al. 1986). Similarly conflicting results were found in vivo (Suemaru et al. 1985). The effect observed appears dependent upon the dose applied, method of application and the assay technique utilized to measure levels of CRH present. Thus, a more complex situation may exist with other neuronal systems involved in the effects observed. It is possible that CRH does not act directly to release opioid peptides but indirectly via further systems. Neurotransmitter control of opioid release from the brain has been little studied with the majority of work having been carried out on β-END. The release of this peptide from the hypothalamus has been clearly demonstrated to be stimulated by K^+ depolarization in a Ca^{2+}-dependent fashion (Osborne et al. 1979; Vermes et al. 1981). However, there is little strong evidence for any neurotransmitter systems affecting its release. A dopaminergic (DAergic) inhibition of hypothalamic release has been reported (Vermes et al. 1985; Sweep and Wiegant 1989). This was not, however, found to be the case for human β-END, the release of which was stimulated by DA (Rasmussen et al. 1987) or for αMSH, a major POMC-derived peptide in the hypothalamus, which was unaffected by DA (Warberg et al. 1979; Delbende et al. 1987). GABA was found to weakly stimulate β-END release from the rat hypothalamus (Nikolarakis et al. 1988b) while a possible role may also exist for the adrenergic system since clonidine, an α_2 agonist, stimulated the release of β-END from the brain-stem (Kunos et al. 1981). The β-receptor antagonist isoproterenol was, however, found to be without effect (Vermes et al. 1985; Sweep and Wiegant 1989). Serotonin (5-HT) levels in the brain also appear to affect the levels of β-END since reduction of brain 5-HT content by the use of the 5-HT depleters 5,7-dihydroxytryptamine or p-chlorophenylalanine reduced the amount of β-END present in a number of brain regions including the hypothalamus (Fukata et al. 1984). The mechanism underlying this effect is uncertain but it should perhaps be borne in mind that the tissue content does not necessarily reflect the release rate. Thus, the idea of these neurotransmitter systems having a role in the CRH-stimulated release of β-END is at best a tenuous one. Recent data from our laboratory does, however, suggest a role for arginine vasopressin (AVP), another hypothalamic peptide, in the CRH stimulation of β-END release from the rat hypothalamus. AVP is itself a potent secretagogue for hypothalamic β-END in vitro, stimulating its release with a similar magnitude to that found previously for CRH (Burns et al. 1989; Sweep and Wiegant 1989). Furthermore, blockade of AVP receptors using the AVP receptor antagonist $d(CH_2)_5[Tyr(OEt)^2,Val^4]$-AVP was found to reduce the in vitro stimulation of β-END release by CRH to a barely significant level, while in the presence of a high concentration of AVP no further stimulation of β-END release by CRH was observed (Burns et al. 1989). AVP would therefore appear to be a central factor in the CRH-stimulated release of β-END, acting as an intermediary factor, with CRH releasing AVP and thus, β-END. This neuronal interplay is postulated to occur in the median eminence where terminals containing all three peptides are known to exist.

This simple model is, however, complicated by the observation that the AVP receptor antagonist had no effect on the basal release of β-END. As we have previously suggested, CRH may be involved in maintaining the basal level of β-END via a tonic effect and, if this occurs through the intermediacy of AVP, then a lowering of the basal release rate would be expected with the AVP antagonist. One may therefore suggest the existence of two pathways, a direct effect of CRH upon the β-END neuron and a pathway routed via AVP. The former arrangement might be involved in maintaining basal levels and the latter, in the stimulated release of β-END. The intriguing suggestion that different stressors or physiological situations may activate differing pathways to release opioids is worthy of future investigation, possibly using an animal model such as the Brattleboro rat which is deficient in AVP.

The opioid-releasing action of CRH does not appear to be confined to the hypothalamus. Studies in other brain regions, such as the neostriatum and globus pallidus, showed that CRH stimulated the release of DYN and Met-ENK from these regions also. This was demonstrated both in vitro utilizing perifused rat neostriatal slices and in vivo perfusing the caudate nucleus and globus pallidus via a push-pull cannulae (Sirinathsinghji et al. 1989). In all studies CRH caused a dose-dependent release of both DYN and Met-ENK which could be blocked using the CRH-receptor antagonist α-helical CRH_{9-41}. Several points, however, indicate the mechanisms governing the response to differ between the hypothalamus and the neostriatum/globus pallidus: (1) the dynamics of the response differed, with slower responses being seen in the hypothalamus; (2) the stimulated release of DYN/Met-ENK in the neostriatal area did not reach a peak and then decline as in the hypothalamus but remained at a constant level; (3) application of the CRH antagonist did not lower basal release rates in vitro. Thus, the mechanisms governing EOP release may differ in the hypothalamus and neostriatum and require further investigation.

3 Second Messenger Systems Involved in the CRH Stimulation of β-END Release

A number of reviews have clearly described the nature and action of cellular second messenger systems (e.g., Drummond 1983; Guy and Kirk 1988) to which the reader is referred. We will instead concentrate on the second messenger systems involved in the CRH stimulation of β-END release from the rat hypothalamus. A useful starting point for this discussion is the pituitary gland where CRH also stimulates the release of ACTH/β-END. In the pituitary gland, CRH is the major modulator of release, acting directly upon the pituitary corticotrope. A number of lines of evidence have indicated that this effect is mediated exclusively via a stimulation of the adenylate cyclase complex which leads to a rise in intracellular cAMP levels. In support of this finding are (1) a CRH stimulation of intracellular cAMP levels, the dose-response curve for which mirrors that found for the CRH stimulation of ACTH release (Aguilera et al. 1983); (2) a direct stimulation of adenylate cyclase activity in pituitary homogenates (Abou-Samra et al. 1987); and (3) the activators of the adenylate cyclase complex, forskolin and cholera toxin, which "switch on" the enzymatic and stimulatory G-protein components of adenylate cyclase, respectively, both stimulate ACTH secretion in a nonadditive manner with CRH (Aguilera et al. 1983).

In the hypothalamus adenylate cyclase also appears to play a role in the action of CRH (Burns et al. 1989). Forskolin was found to stimulate the release of β-END with similar dynamics and magnitude to that found for CRH previously. This effect was nonadditive with CRH. The presence of a stimulatory G-protein (G_s) linking adenylate cyclase to the receptor was indicated by the slow, but highly significant, stimulation of β-END release by cholera toxin. Pretreatment with cholera toxin also prevented any further stimulation of release by CRH. However, it was further found that pertussis toxin (PTX), which has absolutely no effect on the stimulatory components of the adenylate cyclase complex, blocked the CRH stimulation of β-END release. This implied the presence of a further second messenger system connected to a PTX-sensitive G-protein. Since a complete blockade of stimulated release was observed with both cholera toxin and PTX, the results suggest that more than one cell type is involved, these being activated in a serial fashion. This correlates well with the concept described previously of a sequential series of events releasing β-END (i.e. CRH → AVP → β-END) and indicates that a different second messenger pathway is involved at each stage. The observation that PTX also completely blocks the effects of forskolin on β-END release would suggest that the PTX-sensitive event occurs subsequent to adenylate cyclase stimulation; thus, the CRH stimulation of AVP release would appear to be connected to the adenylate cyclase system and the AVP release of β-END to a PTX-sensitive G-protein-mediated event. In fact, PTX was also found to block the action of AVP on β-END release. The connection of the CRH receptor to the adenylate cyclase complex accords well with the known mechanisms of action of CRH in other tissue types including not only the pituitary but also the adrenal gland (Udelsman et al. 1986), and is in line with the finding of pharmacologically and kinetically similar CRH receptors in the brain and pituitary (De Souza and Kuhar 1986) (although the apparent molecular weight differs; Grigoriadis and De Souza 1988).

The messenger system connected to the PTX-sensitive G-protein is uncertain. Such G-proteins have been found to be connected to a number of signaling systems including phosphoinositide hydrolysis, receptor-gated Ca^{2+} and K^+ channels and the inhibitory control of adenylate cyclase (Gilman 1987); thus, any of these systems may be involved. However, AVP is also involved in the control of ACTH/β-END release from the pituitary gland. As with CRH, AVP acts directly upon the corticotrope to release ACTH/β-END and, although a weak secretagogue when used alone, AVP can interact with CRH to potentiate the effect of the latter, thus producing a greater than additive response (Gillies et al. 1982). The intracellular mechanism of action of AVP in this case is less clearly defined than that of CRH. Experimental evidence does, however, favor an involvement of the phosphoinositide cycle: AVP enhances phosphatidylinositol bisphosphate (PIP_2) hydrolysis in dispersed anterior pituitary cells (Bilezikjian and Vale 1987). The most important messenger for the effect of AVP generated by this action appears to be diacylglycerol (DAG). Substances mimicking the action of DAG, including phorbol esters and dioctanoylglycerol, have been demonstrated to stimulate ACTH release in a nonadditive manner with AVP (Abou-Samra et al. 1985, 1987). The possibility that this system also occurs in the hypothalamus is supported by several experimental observations. These include (1) the finding that lithium, which inhibits an enzyme in the phosphoinositide cycle leading to a buildup of the products of PIP_2 hydrolysis, stimulated the secretion of β-END in vitro; (2) the phorbol ester, phorbol-12 myristate-13-acetate stimulates β-END release in a nonadditive manner with AVP; no stimulation was seen with the inactive methylester form; and (3) raising endogenous DAG levels by the use of the DAG kinase inhibitor R59022 causes an increased release of β-END. The possibility that this is the second messenger system involved is also suggested by the observation that in the brain the majority of the receptors present appear to be of the V_1-subtype which are known to be connected to the phosphoinositide cycle (Cornett and Dorsa 1985). Interestingly, the V_1-receptor is believed to be PTX-insensitive (Fain et al. 1988); it is possible, therefore, that a different vasopressin receptor is involved in the system described here.

In conclusion, two second messenger systems appear to be involved in the β-END-releasing action of CRH. A cAMP-dependent system connected to a cholera toxin-sensitive G-protein and a second system, probably the phosphoinositide cycle, connected to a PTX-sensitive G-protein. We propose the schematic model presented in Fig. 2 as the possible system underlying the CRH stimulation of β-END release from the rat hypothalamus.

4 Feedback Inhibition of Opioid Peptide Release Via Presynaptically Located Opioid Receptors

Feedback inhibition of "classical" neurotransmitter release via presynaptic autoreceptors has been clearly demonstrated (Starke 1981; Chesselet 1984). Early speculation on the similar regulation of opioid peptide release (Kosterlitz and

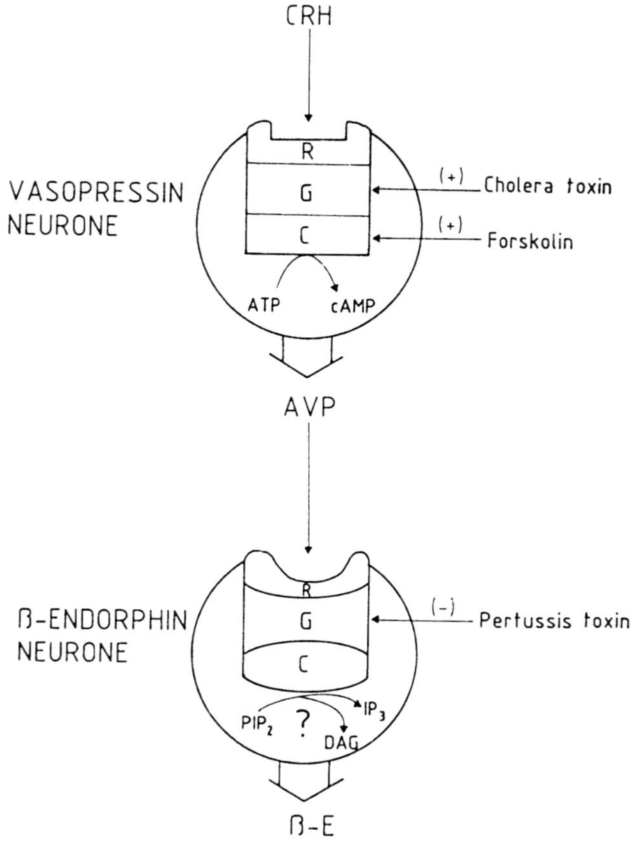

Fig. 2. A schematic representation of the postulated two-step mechanism mediating the CRH-stimulated release of β-endorphin (Burns et al. 1989)

Hughes 1975) was supported by several pharmacological and behavioral studies. Opioid agonists and antagonists have been shown to respectively inhibit and stimulate the release of Met-ENK from brain, spinal and peripheral neuronal tissue preparations (Yaksh and Elde 1981; Bruno et al. 1983; Jhamandas et al. 1984; Glass et al. 1986; Ueda et al. 1986, 1987). In addition, morphine inhibits β-END release into the hypophysial-portal circulation (Koenig et al. 1986). Opioid-induced analgesia has been extensively examined (see Millan et al., this Vol.), however, studies indicating that under some conditions opioid receptor antagonists can also have similar analgesic effects both in humans and in experimental animals have received little attention (Buchsbaum et al. 1977; Levine et al. 1978; Ueda et al. 1986; Vaccarino et al. 1989). This has been suggested to occur by blockade of presynaptic receptors mediating negative feedback effects on opioid peptide release.

The observation that the initial CRH-induced stimulation of β-END, DYN and Met-ENK release was followed by a decline in the release despite the continued presence of CRH in the superfusion medium (Fig. 1) led us to investigate the possibility that opioid receptors, stimulated by the increased levels of the released opioid, gradually inhibited their own release. To test this hypothesis, the opioid receptor antagonist naloxone (NAL) was applied to the superfusion medium together with CRH. The addition of NAL (10^{-6} M) not only prevented the drop in opioid release after the initial stimulation, but also resulted in a further increase of the CRH-induced (10^{-8} M) release (Fig. 3). Similar patterns of responses were observed after KCl (60 mM) stimulation of opioid peptide release with the fall in the release rate after the initial stimulation being abolished, and a further increase in opioid release occurring if NAL (10^{-6} M) was simultaneously applied with KCl. All these results support the initial hypothesis that opioid receptors mediate inhibitory feedback effects upon the secretory activity of β-END, Met-ENK and DYN neurons. Interestingly, when NAL alone was applied to the perifusion medium a significant increase in the release of the opioid peptides was observed. The involvement of opioid receptors was further confirmed by the lack of effect of (+) NAL, the inactive stereoisomer form of naloxone. The action of NAL was found to be Ca^{2+}-dependent while tetrodotoxin, which blocks Na^{+}-dependent nerve activity, blocking axonal transmission, was without effect. These findings indicate that presynaptically located opioid receptors tonically inhibit the release of opioid peptides from the hypothalamic neurons (Nikolarakis et al. 1987).

Since NAL is a universal opioid antagonist, the identification of the specific receptors involved was not possible in these earlier studies. In subsequent experiments specific opioid receptor antagonists were therefore used and their efficacy in modifying β-END, DYN and Met-ENK release was examined (Nikolarakis et al. 1989b).

Three antagonists were utilized, CTAP, a μ-receptor antagonist (Kazmierski et al. 1988); ICI 174864, specific for the δ-receptor (Cowan et al. 1985; Hirning et al. 1985) and nor-binaltorphimine (nor-BNI) acting upon the κ-receptor (Portoghese et al. 1987; Takemori et al. 1988). The EOPs were found to be differentially released by

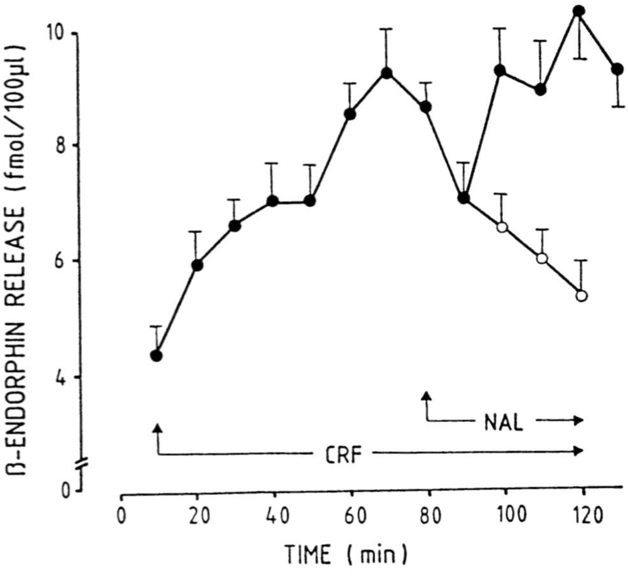

Fig. 3. Reversal of the decline in the CRH-stimulated release of β-endorphin from perifused rat hypothalami slices by the addition of naloxone

application of these antagonists. DYN (acting on κ-sites) was released by all three blockers; β-END (predominantly a μ-ligand) by CTAP and ICI 174864, and Met-ENK (a δ-ligand) by only ICI 174864. All changes described were again Ca^{2+}-dependent and tetrodotoxin-resistant (Nikolarakis et al. 1989b).

These data indicate a complex interaction of opioid peptide release, the release of each peptide being negatively controlled by homologous receptors (autoreceptors). In addition to this autoregulation, β-END release is also controlled by δ-type receptors, while DYN neurons bear not only κ- but also δ- and μ-receptors which regulate its release. These neurons therefore possess what might be termed "alleloreceptors". This name reflects the fact that the peptides belong to a single family and may often serve similar or synergistic roles. (A schematic model of this is presented in Fig. 4).

These opioid-opioid interactions at the terminal level explain the seemingly paradoxical findings that opioid agonists and antagonists have, in some cases, similar behavioral and pharmacological effects. Since other studies failed to demonstrate similar mechanisms in other tissue preparations (Richter et al. 1979; Osborne and Herz 1980; Sawynok et al. 1980), it is difficult to judge whether this phenomenon, which has been described in the hypothalamus, is a general or a region-specific phenomenon.

5 Conclusions

It is apparent from the discussion presented here that a number of systems control the release of brain opioid peptides.

1. CRH is clearly a major modulator of the release of all three opioid peptide "families" from hypothalamic and extrahypothalamic areas.

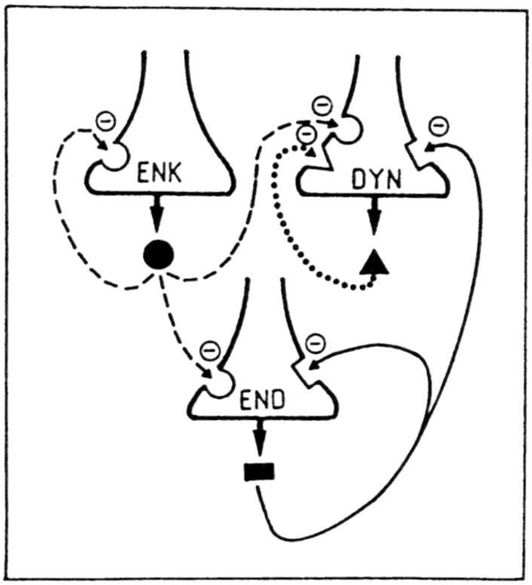

Fig. 4. A proposed model of the auto- and alleloreceptor control of opioid peptide release from the rat hypothalamus (Nikolarakis et al. 1989b)

2. Further control of release is afforded by a complex negative feedback system involving both auto- and, so-called, alleloreceptors.
3. Initial evidence has indicated AVP to have a role in controlling the release of hypothalamic β-END (its role in the control of release of other opioid peptides is as yet undefined). While AVP might itself be an important modulator of β-END release, it is also an essential intermediate in the β-END-releasing action of CRH.
4. Second messenger systems postulated to be involved in the CRH-stimulated release of β-END include an adenylate cyclase system involving a cholera toxin-sensitive G-protein connected to the CRH receptor and a second system, possibly the PI cycle, involving a PTX-sensitive G-protein linked to the AVP receptor.

CRH is the only neuropeptide so far characterized to have a definite role in controlling the release of brain opioid peptides. However, its functional significance is uncertain. As suggested in Section 1, an interaction between these systems is likely in mediating the central responses to stress. Experimental data also strongly support a CRH-opioid effect in the control of reproductive function (see Almeida and Pfeiffer, this Vol.). Direct evidence of opioidergic mediation of the effects of CRH, particularly in stress, is as yet unavailable; however, the close anatomical relationship between hypothalamic CRH and opioid neuronal systems, the simultaneous co-expression of CRH mRNA and POMC mRNA in the brain, and the data reported here of a CRH stimulation of brain opioid release are supportive of such an interaction.

Acknowledgments. Thanks must go to Ursula Bäuerle and Heide Roth for their help in the preparation of this manuscript.

References

Abou-Samra A-B, Catt KJ, Aguilera G (1985) Involvement of protein kinase C in the regulation of adrenocorticotropin release from rat anterior pituitary cells. Endocrinology 118:212-217

Abou-Samra A-B, Harwood JP, Catt KJ, Aguilera G (1987) Mechanism of action of CRF and other regulators of ACTH release in pituitary corticotrophs. Ann N Y Acad Sci 512:67-84

Aguilera G, Harwood JP, Wilson JX, Morell J, Brown JH, Catt KJ (1983) Mechanism of action of corticotropin releasing factor and other regulators of corticotropin release in rat pituitary cells. J Biol Chem 258:8039-8045

Bilezikjian LM, Vale WW (1987) Regulation of ACTH secretion from corticotrophs: the interaction of vasopressin and CRF. Ann N Y Acad Sci 512:85-96

Bloom FE, Battenberg ELF, Rivier J, Vale W (1982) Corticotropin-releasing factor (CRF): immunoreactive neurons and fibres in rat hypothalamus. Regul Peptide 4:43-48

Brown MR, Fisher LA, Spiess J, Rivier C, Rivier J, Vale W (1982) Corticotropin releasing factor: actions on the sympathetic nervous system and metabolism. Endocrinology 111:928-931

Bruno G De, Baune S de la, Gros C, Schwartz JC (1983) Inhibition of depolarisation-induced release of (met^5) enkephalin from slices of globus pallidus in the presence of thiorphan and bestatin. Cr hebd Scéanc Acad Sci, Paris 297:609-612

Buchsbaum M, Davis G, Bunney W (1977) Naloxone alters pain perception and somatosensory evoked potentials in normal subjects. Nature (London) 270:620-622

Buckingham JC (1986) Stimulation and inhibition of corticotropin-releasing factor secretion by β-endorphin. Neuroendocrinology 42:148-152

Burns G, Almeida OFX, Passarelli F, Herz A (1989) A two-step mechanism by which corticotropin-releasing hormone releases hypothalamic β-endorphin: the role of vasopressin and G-proteins. Endocrinology 125:1365–1372

Chappell PB, Smith MA, Kilts CD, Bissette G, Ritchie J, Anderson C, Nemeroff CB (1986) Alterations in corticotropin-releasing factor-like immunoreactivity in discrete rat brain regions after acute and chronic stress. J Neurosci 6:2908–2914

Chesselet MF (1984) Presynaptic regulation of neurotransmitter release in the brain. Neuroscience 12:347–375

Cornett LE, Dorsa DM (1985) Vasopressin receptor subtypes in dorsal hindbrain and renal medulla. Peptides 6:85–89

Cowan A, Zhu XZ, Porreca F (1985) Studies in vivo with ICI 174864 and [D-Pen²-D-Pen⁵]enkephalin. Neuropeptides 5:311–314

Delbende C, Jégou S, Transhand-Bunel D, Pelletier G, Vaudry H (1987) Hypothalamic α-melanocyte-stimulating hormone (α-MSH) is not under dopaminergic control. Brain Res 423:203–212

De Souza EB, Kuhar MJ (1986) Corticotropin-releasing factor receptors in the pituitary gland and central nervous system: methods and overview. Meth Enzymol 124:560–590

Drummond GI (1983) Cyclic nucleotides in the nervous system. In: Greengard P, Robinson GA (eds) Advances in cyclic nucleotide research, vol 15. Raven, New York, pp 373–494

Fain JN, Wallace MA, Wojcikiewicz RJH (1988) Evidence for involvement of guanine nucleotide-binding regulatory proteins in the activation of phospholipases by hormones. FASEB J 2:2569–2574

Fisher LA, Jessen G, Brown MR (1983) Corticotropin-releasing factor (CRF): mechanisms to evaluate mean arterial pressure and heart rate. Regul Peptide 5:153–161

Fukata J, Nakai Y, Imura H, Takeuchi J (1984) Effects of 5-HT depleters on the contents of β-endorphin, α-melanotropin and adrenocorticotropin in rat brain and pituitary. Brain Res 324:289–293

Gillies GE, Linton EA, Lowry PJ (1982) Corticotropin-releasing activity of the new CRF is potentiated several times by vasopressin. Nature (London) 299:355–357

Gilman AG (1987) G proteins: transducers of receptor-generated signals. Annu Rev Biochem 56:615–649

Glass J, Chan WC, Gintzler AR (1986) Direct analysis of the release of methionine-enkephalin from guinea pig myenteric plexus: modulation by endogenous opioids and exogenous morphine. J Pharmacol Exp Ther 239:742–747

Grigoriadis GE, De Souza EB (1988) The brain corticotropin-releasing factor (CRF) receptor is of lower apparent molecular weight than the CRF receptor in anterior pituitary. J Biol Chem 263:10927–10931

Guy GR, Kirk CJ (1988) Inositol phospholipids and cellular signalling. In: Cooke BA, King RJB, van der Molen HJ (eds) Hormones and their actions, pt 2. Elsevier, Amsterdam, pp 47–62

Haas DA, George SR (1988) Single or repeated mild stress increases synthesis and release of hypothalamic corticotropin-releasing factor. Brain Res 461:230–237

Hirning LD, Mosberg HI, Hurst R, Hruby VJ, Burkes TF, Porreca F (1985) Studies in vitro with ICI 174864, [D-Pen²-D-Pen⁵]enkephalin, [DPDPE] and [D-Ala²-NMe-Phe⁴, Gly-ol]-enkephalin (DAGO). Neuropeptides 5:383–386

Hisano S, Daikoku S, Yanaihara N, Shibasaki T (1986) Intracellular localization of CRF and Met-enk-8 in nerve terminals in the rat median eminence. Brain Res 370:321–326

Hökfelt T, Fahrenkrug J, Ju G, Ceccatelli S, Tsuruo Y, Meister B, Mutt V, Rundgren M, Brodin E, Terenius L, Hulting A-L, Werner S, Bjorklund H, Vale W (1987) Analysis of peptide histidine-isoleucine/vasoactive intestinal polypeptide-immunoreactive neurons in the central nervous system with special reference to their relation to corticotropin-releasing factor and enkephalin-like immunoreactivities in the paraventricular hypothalamic nucleus. Neuroscience 23:827–857

Jhamandas K, Yaksh TL, Go VL (1984) Acute and chronic morphine modifies the in vivo release of methionine enkephalin-like immunoreactivity from the cat spinal cord and brain. Brain Res 297:91–103

Kalin NH, Sherman JE, Takahashi LK (1988) Antagonism of endogenous CRH systems attenuates stress-induced freezing behavior in rats. Brain Res 457:130–135

Kazmierski W, Wire WS, Lui GK, Knapp RJ, Shook JE, Burks TF, Yamamura HI, Hruby VJ (1988) Design and synthesis of somatostatin analogues with topographical properties that lead to highly potent and specific μ opioid receptor antagonists with greatly reduced binding at somatostatin receptors. J Med Chem 31:2170–2177

Koenig J, Meltzer H, Gudelsky G (1986) Morphine or Capsaicin administration alters the secretion of β-endorphin into the hypophyseal portal vasculature of the rat. Neuroendocrinology 43:611–617
Kosterlitz HW, Hughes J (1975) Some thoughts on the significance of enkephalin, the endogenous ligand. Life Sci 17:91–96
Kunos G, Farsang C, Ramirez-Gonzales MD (1981) β-Endorphin: possible involvement in the antihypertensive effect of central α-receptor activation. Science 211:82–84
Labrie F, Giguère V, Meunier H, Simard J, Gossard F, Raymond V (1987) Multiple factors controlling ACTH secretion at the anterior pituitary level. Ann N Y Acad Sci 512:97–114
Levine JD, Gordon NC, Fields HL (1978) Naloxone dose dependently produces analgesia and hyperalgesia in postoperative pain. Nature (London) 278:740–741
Nikolarakis KE, Almeida OFX, Herz A (1986) Stimulation of hypothalamic β-endorphin and dynorphin release by corticotropin-releasing factor (in vitro). Brain Res 399:152–155
Nikolarakis KE, Almeida OFX, Herz A (1987) Feedback inhibition of opioid peptide release in the hypothalamus of the rat. Neuroscience 23:143–148
Nikolarakis KE, Almeida OFX, Sirinathsinghji DJS, Herz A (1988a) Concomitant changes in the in vitro and in vivo release of opioid peptides and luteinizing hormone-releasing hormone from the hypothalamus following blockade of receptors for corticotropin-releasing factor. Neuroendocrinology 47:545–550
Nikolarakis KE, Loeffler JPh, Almeida OFX, Herz A (1988b) Pre- and postsynaptic actions of GABA on the release of hypothalamus gonadotropin-releasing hormone (GnRH). Brain Res Bull 21:677–683
Nikolarakis KE, Almeida OFX, Herz A (1989a) Multiple factors influencing the in vitro release of [Met^5]-enkephalin from rat hypothalamic slices. J Neurochem 52:428–432
Nikolarakis KE, Almeida OFX, Yassouridis A, Herz A (1989b) Presynaptic auto- and allelo-receptor regulation of hypothalamic opioid peptide release. Neuroscience 31:269–273
Osborne H, Herz A (1980) K^+-evoked release of Met-enkephalin from rat striatum in vitro. Naunyn-Schmiedebergs Arch Exp Path Pharmak 310:203–209
Osborne H, Przewłocki R, Höllt V, Herz A (1979) Release of β-endorphin from rat hypothalamus in vitro. Eur J Pharmacol 55:425–428
Portoghese PS, Lipkowski AW, Takemori AE (1987) Binaltorphimine and nor-binaltorphimine, potent and selective κ-opioid receptor antagonists. Life Sci 40:1287–1292
Rasmussen DD, Liu JH, Wolf PL, Yen SSC (1987) Neurosecretion of human hypothalamic immunoreactive β-endorphin: in vitro regulation by dopamine. Neuroendocrinology 45:197–200
Richter JA, Wesche DL, Frederickson RCA (1979) K^+-stimulated release of Leu- and Met-enkephalin from rat striatal slices: lack of effect of morphine and naloxone. Eur J Pharmacol 56:105–113
Roth KA, Weber E, Barchas JD, Chang JK (1983) Immunoreactive dynorphin (1–8) and corticotropin releasing factor in subpopulation of hypothalamic neurons. Science 219:189–191
Sawynok J, Labella FS, Pinsky C (1980) Effects of morphine and naloxone on the K^+ stimulated release of methionine-enkephalin from slices of rat corpus striatum. Brain Res 189:483–493
Sirinathsinghji DJS, Nikolarakis KE, Herz A (1989) Corticotropin-releasing factor stimulates the release of methionine-enkephalin and dynorphin from the neostriatum and globus pallidus of the rat: in vitro and in vivo studies. Brain Res 490:276–291
Starke K (1981) Presynaptic receptors. Annu Rev Pharmacol Toxicol 21:7–30
Suemaru SK, Hashimoto K, Ota Z (1985) Effects of morphine on hypothalamic corticotropin-releasing factor (CRF), norepinephrine and dopamine in non-stressed and stressed rats. Acta Med Okayama 39:463–470
Sutton RE, Koob GF, Moal ML, Rivier J, Vale W (1982) Corticotropin releasing factor produces behavioural activation in rats. Nature (London) 297:331–333
Swanson LW, Sawchenko PE, Rivier J, Vale WW (1983) Organization of ovine corticotropin-releasing factor immunoreactive cells and fibres in the rat brain: an immunohistochemical study. Neuroendocrinology 36:164–186
Sweep CGJ, Wiegant VM (1989) Release of β-endorphin-immunoreactivity from rat pituitary and hypothalamus in vitro: effects of isoproterenol, dopamine, corticotropin-releasing factor and $arginine^8$-vasopressin. Biochem Biophys Res Commun 161:221–228
Taché Y, Goto Y, Gunion MW, Vale W, Rivier J, Brown M (1983) Inhibition of gastric acid secretion in rats by intracerebral injection of corticotropin-releasing factor. Science 222:935–937
Takemori A, Ho B, Naeseth J, Portoghese P (1988) Nor-binaltorphimine, a highly selective kappa-opioid antagonist in analgesic and receptor binding assays. J Pharmacol Exp Ther 246:255–258

Thompson RC, Seasholtz AF, Douglass JO, Herbert E (1987) The rat corticotropin-releasing hormone gene. Ann NY Acad Sci 512:1-11

Udelsman R, Harwood JP, Millan MA, Chrousos GP, Goldstein DS, Zimlichman R, Catt KJ, Aguilera G (1986) Functional corticotropin releasing factor receptors in the primate peripheral sympathetic nervous system. Nature (London) 319:147-150

Ueda H, Fukushima N, Kitao T, Ge M, Tagaki H (1986) Low doses of naloxone produce analgesia in the mouse brain by blocking presynaptic autoinhibition of enkephalin release. Neurosci Lett 65:247-252

Ueda H, Fukushima N, Ge M, Tagaki H, Satoh M (1987) Presynaptic opioid κ-receptor and regulation of the release of Met-enkephalin in the rat brainstem. Neurosci Lett 81:309-313

Vaccarino AL, Tasker RAR, Melzack R (1989) Analgesia produced by normal doses of opioid antagonists alone and in combination with morphine. Pain 36:103-109

Vale W, Spiess J, Rivier C, Rivier J (1981) Characterization of a 41-residue ovine hypothalamic peptide that stimulates secretion of corticotropin and β-endorphin. Science 213:1394-1397

Vale W, Rivier C, Brown MR, Spiess J, Koob G, Swanson L, Bilezikjian L, Bloom F, Rivier J (1983) Chemical and biological characterization of corticotropin-releasing factor. Rec Prog Hormone Res 39:245-270

Vermes I, Mulder GH, Berkenbosch F, Tilders FJH (1981) Release of β-lipotropin and β-endorphin from rat hypothalami in vitro. Brain Res 211:248-254

Vermes I, Tilders FJH, Stoof JC (1985) Dopamine inhibits the release of immunoreactive β-endorphin from rat hypothalamus in vitro. Brain Res 326:42-46

Warberg J, Oliver C, Barnea A, Parker CR, Porter JC (1979) Release of immunoreactive α-MSH by synaptosome-enriched fractions of homogenates of hypothalami. Brain Res 175:247-257

Yajima F, Suda T, Tomori N, Sumitomo T, Nakagami Y, Ushiyama T, Demura H, Shizame K (1986) Effects of opioid peptides on immunoreactive corticotropin-releasing factor release from the rat hypothalamus in vitro. Life Sci 39:181-186

Yaksh TL, Elde RP (1981) Factors governing release of methionine enkephalin-like immunoreactivity from mesencephalon and spinal cord of the rat in vivo. J Neurophysiol 46:1056-1075

CHAPTER 13

Modulation of Catecholamine Release in the Central Nervous System by Multiple Opioid Receptors

P. Illes and R. Jackisch

Introduction

Exogenous and endogenous opioids elicit a number of their effects in the organism by modulating the release of transmitters in the peripheral and central nervous system (CNS). This is due to the activation of receptors located at either axon terminals (presynaptic receptors) or at cell somata and/or dendrites (somato-dendritic receptors). Indirect effects via neighboring neurons are also possible. In recent years, it has become apparent that opioids are able to inhibit the action potential-induced secretion of norepinephrine (NE) and dopamine (DA) in the CNS (Starke 1977; Westfall 1977; Henderson et al. 1979; Vizi 1979; Langer 1981; Szekely and Ronai 1982a,b; Jackisch et al. 1988; Illes 1989). This chapter will review research concerning this topic and focus specifically upon the idea of multiple opioid receptor types in modulating catecholamine release.

Transmitter secretion may be directly measured in brain slices or synaptosomes. It may also be determined in anaesthetized or conscious animals by superfusing or dialyzing certain areas of the CNS. In addition, the firing rate of NE or DA neurons (in vivo and in vitro) gives indirect information on the release of catecholamines. Results obtained by these mutually complementary approaches will be discussed in the following sections.

1.1 Multiple Opioid Receptors

The original postulate of a homogeneous opioid receptor population was challenged by experiments in the chronic spinal dog (Martin et al. 1976). The three receptor types (μ, κ, σ) initially described were later complemented by the discovery of δ- (Lord et al. 1977) and ε-receptors (Schulz et al. 1981). The latter receptor types were found in isolated organs, namely, in the vasa deferentia of mice (δ) and rats (ε). The prototypic agonists for the μ-, δ-, κ-, and ε-receptors are morphine, [Leu^5]enkephalin (Leu-ENK), ethylketocyclazocine (EKC), and β-endorphin (β-END), respectively. The effects of all four substances are antagonized by naloxone (NAL). However, higher concentrations of NAL are required to antagonize the effects of EKC and Leu-ENK as compared to morphine and β-END. Certain benzomorphans, such as N-allylnormetazocine (SKF10047), produce their psychotomimetic effects at the σ-receptor. Originally, it was suggested that phencyclidine (PCP) also acts here, but separate σ- and PCP-recognition sites have recently been described (Manallack et al.

1986; Sonders et al. 1988). Because of its NAL insensitivity, the σ-receptor is no longer considered an opioid receptor (Zukin and Zukin 1984). Therefore, the effects of σ-ligands will not be discussed.

1.2 Receptor-Type Selective Agonists and Antagonists

In addition to the prototypic opioids enumerated above, other agonists are routinely used for receptor characterization (Kosterlitz 1985; Leslie 1987; Illes 1989). Normorphine and [D-Ala^2, Me, Phe^4, Gly-ol^5] enkephalin (DAGO) are both μ-agonists. Met-ENK, [D-Ala^2, D-Leu^5]enkephalin (DADLE), and [D-Pen^2, D-Pen^5] enkephalin (DPDPE) are preferential for δ-receptors, while U-50488H, dynorphin (DYN) 1–13, and DYN 1–17 are selective κ-receptor agonists. However, it should be emphasized that the preference of a given agonist for one type of receptor is not absolute; only DAGO and normorphine (μ), DPDPE (δ), and U-50488H (κ) are highly selective. Unfortunately, the intrinsic activity of normorphine is low and, therefore, in most studies, DAGO is a more useful μ-receptor ligand. All other agonists mentioned primarily activate the preferred receptor, but they also interact with other opioid receptor types.

As already discussed, NAL is a preferential μ-antagonist at low concentrations, whereas at higher concentrations it acts at all types of opioid receptors. ICI 154129 and ICI 174864 are δ-selective; MR-2266 binds with a slightly higher affinity to κ- than to μ-receptors. In contrast to these antagonists, β-funaltrexamine (β-FNA) and β-chlornaltrexamine (β-CNA) act irreversibly. β-FNA is μ-selective, whereas β-CNA alkylates all opioid receptor types.

2 Modulation of Catecholamine Release

2.1 Norepinephrine (NE)

Cortex

The ability of narcotic analgesics to depress the electrically induced release of [^{3}H]NE from cortical slices of rat brain was demonstrated more than a decade ago (Montel et al. 1974a,b). It was later shown that the naturally occurring ENKs (Taube et al. 1977) as well as their stable synthetic analogs (Hagan and Hughes 1984) also produced this effect. However, the low dissociation constant (K_D) of the μ-preferential NAL in antagonizing the effects of selective μ-agonists (normorphine, DAGO) as well as DADLE, a δ-agonist which also has some μ-activity, favored the exclusive involvement of μ-receptors in this process (Table 1; Hagan and Hughes 1984). This conclusion was recently confirmed by studies comparing the effects of highly selective μ- (DAGO) and δ- (DPDPE) agonists (Schoffelmeer et al. 1988). The failure of bremazocine to alter the release of [^{3}H]NE argued against the presence of κ-receptors.

When high K^+ was used as a stimulus of transmitter secretion in brain slices, the preferential μ-receptor agonist morphine (Schoffelmeer and Mulder 1983, 1984;

Table 1. Opioid receptor types involved in the modulation of catecholamine release from brain neurons

Species	Area	Trans-mitter	Stimu-lation	Receptor type		References
Rat	Cortex	NE	EFS	μ↓	(κ)	Hagan and Hughes (1984)
			EFS	μ↓		Schoffelmeer et al. (1988)
			K^+	μ↓	(κ)	Mulder et al. (1987)[a]
			K^+	μ↓		Werling et al. (1987)
		DA	K^+	κ↓		Werling et al. (1988a)
	Hippocampus	NE	EFS	μ↓		Jackisch et al. (1986)
	Striatum	DA	K^+	κ↓		Mulder et al. (1984)
						Schoffelmeer et al. (1988)
						Werling et al. (1988a)
			No	κ↓ [μ↑]	(δ)	Di Chiara and Imperato (1988)[b]
			No	δ↑	(κ)	Lubetzki et al. (1982)
						Petit et al. (1986)
	Nucleus accumbens	DA	No	κ↓ [μ↑]	(δ)	Di Chiara and Imperato (1988)[b]
	Hypothalamus	DA	No	μ↓	(δ, κ)	Gudelsky and Porter (1979)[c]
Guinea pig	Cortex	NE	K^+	μ, δ, κ↓		Werling et al. (1987, 1989)
		DA	K^+	κ↓		Werling et al. (1988a)
	Hippocampus	NE	K^+	μ, δ, κ↓		Werling et al. (1987)
	Cerebellum	NE	K^+	μ, δ, κ↓		Werling et al. (1987)
	Striatum	DA	K^+	κ↓		Werling et al. (1988a)
Rabbit	Hippocampus	NE	EFS	κ↓		Jackisch et al. (1986)
	Colliculus superior	NE	EFS	κ↓		Wichmann and Starke (1988)

Slices were dissected from the indicated area and the release of the endogenous or radiolabeled transmitter was measured. Exceptions were the following:
[a] Synaptosomes were prepared from cortical tissue.
[b] Dopamine release was measured by microdialysis in conscious animals.
[c] Dopamine concentration was determined in pituitary stalk plasma during anaesthesia.
Stimulation of release was by high potassium (K^+) or electrical field stimulation (EFS). The activation of opioid receptors mostly depressed (↓), but sometimes enhanced (↑), transmitter release. Receptors in square brackets indicate that they are probably situated at nondopaminergic cells and only indirectly modulate catecholamine release. Only those publications, in which a detailed characterization of receptors was performed, are listed. NE, norepinephrine; DA, dopamine.

Schoffelmeer et al. 1986a), as well as the δ-preferential agonists Met-ENK (Göthert et al. 1979), DADLE (Schoffelmeer and Mulder 1984; Schoffelmeer et al. 1986b), and [D-Ala^2, Met^5]enkephalinamide (Jones and Marchbanks 1982), were all active. The inability, however, of the more selective δ-agonist DPDPE to impair the K^+-induced secretion of [^{3}H]NE from cortical synaptosomes, in spite of a pronounced depression by DAGO and morphine, suggests a dominant role for the μ-receptor (Table 1; Mulder et al. 1987). The effects of μ-agonists in the synaptosomal preparation prove that, in the intact tissue, the site of action is at the nerve terminals.

When [^{3}H]NE release was evoked by K^+ in cortical slices of guinea pigs, all three types of receptor-selective opioids, namely, DAGO (μ), DPDPE (δ), and U-50488H (κ) were inhibitory (Table 1; Werling et al. 1987, 1988a). Although the inhibitory effects of these compounds were NAL-reversible, higher concentrations of NAL were required to antagonize the effects of δ- and κ-agonists (Werling et al. 1987). The δ-selective antagonist ICI 174864 only counteracted the effect of DPDPE, while the κ-selective antagonist nor-binaltorphimine (nor-BNI) only antagonized that of U-50488H (Werling et al. 1989). The action of DAGO on [^{3}H]NE secretion from cortical slices was greatly reduced, when guinea pigs were rendered tolerant to morphine by its long-term in vivo infusion (Werling et al. 1988b). However, U-50488H continued to act even under these conditions. Converse changes were observed subsequent to pretreatment of guinea pigs with U-50488H; tolerance to inhibition of transmitter release by the κ-agonist was apparent, but the sensitivity to DAGO was unaltered. These results indicate the absence of cross-tolerance to opioids and also suggest that all three types of opioid receptors are present at terminals of noradrenergic axons of the guinea pig cortex.

Noncortical Areas

In cerebellar (Montel et al. 1975) and hypothalamic slices (Diez-Guerra et al. 1986; Taube et al. 1977) of rats, morphine depressed the electrically evoked secretion of [^{3}H]NE, suggesting the existence of μ-receptors. A similar conclusion was reached on the basis of experiments made on rat hippocampal slices; among μ- (DAGO), δ- (DADLE), and κ-agonists (EKC), only DAGO had a considerable potency (Table 1; Jackisch et al. 1986). By contrast, in the rabbit hippocampus, the presynaptic opioid receptors mediating the inhibition of NE release are mainly of the κ-type (Table 1; Jackisch et al. 1986, 1988). The κ-preferential agonists U-50488H, EKC, DYN 1–13, and DYN 1–17 were more active than some prototypic μ- and δ-agonists. The effects of all three classes of opioids were antagonized by the preferential κ-receptor antagonist MR-2266, but not by the δ-antagonist ICI 174864. The K_D values of MR-2266 and the μ-preferential NAL against EKC were compatible with the presence of a κ-receptor. A similar analysis of opioid effects in slices of the rabbit superior colliculus documented a κ-, but not μ- or δ-inhibition of [^{3}H]NE release (Wichmann and Starke 1988). The possibility of considerable species variability is further highlighted by studies in which receptor-selective agonists were applied to hippocampal and cerebellar slices of guinea pigs. In these tissues, the three classic opioid receptor types were shown to modulate NE release (Table 1; Werling et al. 1987).

Cell Bodies in the Locus Coeruleus

Almost half of the NE terminals in the CNS originate from cell bodies in the locus coeruleus. This nucleus contains a homogeneous group of neurons, which discharge in rats, both under in vivo and in vitro conditions, with a frequency of 0.5–2 Hz (Williams et al. 1986). The iontophoretic application of the μ-agonist morphine (Korf et al. 1974) and levorphanol (Bird and Kuhar 1977), as well as the δ-agonist Met-ENK (Guyenet and Aghajanian 1979) caused a marked and long-lasting depression of the extracellularly recorded spontaneous firing in anaesthetized animals. This effect was NAL-antagonizable. Intracellular recordings in locus coeruleus slices confirmed the inhibition of neuronal activity by opioids, and showed that the blockade of spike discharge is due to an increase in K^+ conductance and subsequent hyperpolarization (Pepper and Henderson 1980; Williams et al. 1982; North and Williams 1985; Aghajanian and Wang 1986).

It was concluded that these receptors are of the μ-type (Williams and North 1984; North 1986a,b). Thus, although both μ- (normorphine) and δ-agonists (Met-ENK, DADLE) produced a hyperpolarization which was NAL-reversible, the low K_D values obtained for such antagonism are characteristic of μ-receptor mediation (Williams and North 1984). The δ-antagonist ICI 154129 had high K_D values against both normorphine and Met-ENK; a finding which would also exclude the presence of δ-receptors. EKC was active, whereas other κ-agonists, such as DYN 1–13 and the highly selective U-50488H, were not. Since the irreversible μ-antagonist β-FNA blocked the effects of EKC, normorphine, and DADLE, a role for κ-receptors can be discounted. The involvement of μ-, but not δ-receptors was confirmed in other studies. Under the same conditions in which DAGO hyperpolarized the cells (Christie et al. 1987; North et al. 1987), the δ-selective agonist DPDPE had no effect (North et al. 1987). Preincubation with the non-preferential, irreversible antagonist β-CNA decreased the hyperpolarization produced by normorphine to a larger extent than that produced by DAGO, indicating less receptor reserve, i.e., a lower intrinsic activity of normorphine (Christie et al. 1987).

The κ-agonist U-50488H depressed the excitatory postsynaptic potentials (EPSPs) evoked by stimulation of afferent inputs to the locus coeruleus, without influencing the depolarization produced by local application of glutamic acid (McFadzean et al. 1987). Both the high selectivity of this opioid and the high K_D required for antagonism by the μ-preferential NAL, strongly suggest the existence of κ-receptors which modulate transmitter release from the excitatory neuronal projections. Thus, opioid regulation takes place at multiple sites, both at the terminals and somata of locus coeruleus cells, and, in addition, at the fibers innervating them.

An interesting suggestion is that the reaction chains coupled to both somatic and presynaptic receptors may be identical (Illes 1986; North 1986a,b). For example, μ-agonists may enhance K^+ conductance in the terminal axons and produce hyperpolarization. This will either hinder the propagation of action potentials to the varicosity, or prevent the activation of Ca^{2+} entry, since depolarization is then insufficient to reach a threshold value (Bug et al. 1986).

2.2 *Dopamine (DA)*

Striatum

Several early studies indicated that the μ-agonist morphine reduces the K^+-evoked release of [^{3}H]DA from rat striatal slices (Celsen and Kuschinsky 1974; Loh et al. 1976; Subramanian et al. 1977). However, subsequent investigations failed to show an effect of morphine on [^{3}H]DA secretion (Arbilla and Langer 1978; Westfall et al. 1983; Mulder et al. 1984; Dewar et al. 1987), except at very high concentrations (Starr 1978), when nonreceptor-mediated effects may appear.

In rat striatal slices, various κ-agonists, including DYN 1–13, inhibited both the spontaneous outflow of [^{3}H]DA and that increased by a high K^+ concentration in the perfusing medium (Table 1; Mulder et al. 1984, 1988). The finding that the K^+-induced release of [^{3}H]DA is sensitive to κ-agonists only was recently confirmed (Table 1; Schoffelmeer et al. 1988; Werling et al. 1988a) and extended to the guinea-pig striatum (Table 1; Werling et al. 1988a). In synaptosomes prepared from this brain area, K^+ stimulated the release of [^{3}H]DA; U-50488H (κ), but not DAGO (μ) or DPDPE (δ), was inhibitory. Thus the nigrostriatal cells seem to possess presynaptic κ-, but not μ-receptors. In agreement with the previous results it was shown by means of brain dialysis in freely moving rats that κ-agonists decrease the secretion of DA in the striatum (Di Chiara and Imperato 1988). Under the same conditions μ-agonists increased the release, presumably due to modulation of neighboring non-DAergic neurons.

In the rat striatum, the basal outflow of the transmitter continuously formed from [^{3}H]tyrosine was raised by the δ-selective agonists [D-Thr^2, Leu^5]enkephalyl-Thr and [D-Ser^2, Leu^5]enkephalyl-Thr, whereas the μ-agonists morphine and DAGO were without effect (Table 1; Lubetzki et al. 1982; Petit et al. 1986). NAL and ICI 154129 prevented the actions of the δ-agonists. The K^+-induced secretion was facilitated by concentrations of [D-Thr^2, Leu^5]enkephalyl-Thr which were higher than those needed to elevate basal outflow (Petit et al. 1986). Unexpectedly, the δ-agonist only enhanced the outflow of [^{3}H]DA, freshly synthesized after preincubation of the tissue with [^{3}H]tyrosine, but had no effect, when [^{3}H]DA itself was incorporated into the transmitter stores (Westfall et al. 1983; Mulder et al. 1984; Petit et al. 1986). Thus, in addition to κ-receptors, δ-receptors may also exist at the terminals of the nigrostriatal cells. In cats, injection of δ-agonists into the caudate nucleus produced an enhancement followed by a depression of the spontaneous outflow of [^{3}H]DA, after previous application of the precursor [^{3}H]tyrosine (Chesselet et al. 1981; 1982).

The action of opioids upon the release of endogenous and [^{3}H]DA evoked by the ganglionic stimulant dimethylphenylpiperazinium (DMPP) or glutamate has also been studied in rat striatal slices. The DMPP-, but not the electrically or K^+-induced transmitter release was depressed by the μ-agonist morphine; the δ-preferential agonists Leu-ENK and DADLE were inactive (Westfall et al. 1983). However, even high concentrations of morphine produced only moderate inhibition; no data regarding the effect of NAL were presented. Therefore, the suggestion of the authors about an interaction of μ-type opioids with the nicotinic receptor situated at DAergic nerve terminals needs additional support. The secretion of endogenous DA in striatal slices was enhanced by L-glutamic acid; it was reduced both by the N-methyl-D-aspartate antagonist DL-2-amino-7-phosphonoheptanoic acid and, in the presence of the DA uptake blocker nomifensine, by DADLE (Jhamandas and Marien 1987). In view of the low concentrations of the ENK tested, and of the antagonism by NAL, there is no doubt that the inhibition is opioid receptor-mediated. The evidence for its δ-nature is less strong, since no highly selective δ-agonists (DPDPE) or antagonists (ICI 174864) were used.

Cell Bodies in the Substantia Nigra

In other experiments, an electrophysiological approach was utilized to localize the site of opioid action. Systematic application of morphine enhanced the extracellularly recorded firing rate of neurons in the nigral pars com pacta (Iwatsubo and Clouet 1977; Finnerty and Chan 1981; Jurna 1981). However, iontophoresis of morphine to this area was without effect (Hommer and Pert 1983). These results, and the finding that both systemic (Finnerty and Chan 1981) and iontophoretically applied morphine (Hommer and Pert 1983) depress activity in the nigral pars reticulata suggest that the site of action is at inhibitory interneurons modulating the DAergic cells of the pars compacta. By contrast, intravenous injection of the κ-selective agonist U-50488H reduced the firing rate of compacta neurons (Walker et al. 1987). The facilitation by morphine was antagonized by lower doses of NAL than was the κ-receptor-mediated depression. Microinfusion of U-50488H into the caudate nucleus also inhibited the discharge of DAergic neurons in the pars compacta. Moreover, iontophoresis of another κ-agonist DYN 1–13, into the reticulata, depressed neuronal activity, whereas iontophoresis into the compacta was without effect (Lavin and Garcia-Munoz 1985). Thus, κ-receptors may be situated at the axon terminals and dendrites of DAergic neurons, but are probably missing from cell somata.

Nucleus Accumbens

Brain dialysis in freely moving rats has shown that μ-agonists stimulate DA release more effectively, and at lower doses, in the nucleus accumbens as compared to the striatum (Di Chiara and Imperato 1988). By contrast, κ-agonists produce a similar depression in both areas.

Cortex

In slices prepared from the cortex of rats and guinea pigs, the K^+-induced secretion of [^{3}H]DA was inhibited by U-50488H (κ), but not by DAGO (μ) or DPDPE (δ) (Table 1; Werling et al. 1988a). The larger maximum inhibition produced by U-50488H in the cortex than in the striatum of guinea pigs suggests that, in this species, the mesocortical DA system is more sensitive to κ-agonists than the nigrostriatal system.

Hypothalamus

Although, in the rat striatum, the electrically or K^+-induced secretion of [^{3}H]DA was insensitive to μ-agonists, it was depressed in other regions of the brain. Both somatodendritic receptors and classic presynaptic receptors, localized at the axon terminals, may mediate the actions of opioids in the hypothalamus. When morphine was iontophoretically applied to cell bodies of the tuberoinfundibular DA neurons in the arcuate nucleus of the hypothalamus, both their firing rate (Haskins and Moss 1983) and the concentration of DA released into the hypophysial portal blood were markedly decreased (Haskins et al. 1981). The systemic administration of NAL prevented this effect. Morphine, introduced subcutaneously, or by the intracerebroventricular route, produced a similar reduction in the plasma concentration of DA (Table 1; Gudelsky and Porter 1979), whereas the concentrations of NE and epinephrine (E) were not altered (Reymond et al. 1983). The absence of a change in NE levels of the portal blood contrasts with results from experiments performed on slices of the medial preoptic area; in this tissue morphine depressed the electrically induced secretion of [^{3}H]NE (Diez-Guerra et al. 1986). In mediobasal hypothalamic slices (containing the arcuate nucleus), β-END inhibited the spontaneous release of DA (Wilkes and Yen 1980). NAL antagonized these opioid effects and, when given alone, enhanced the spontaneous release of DA, NE, and E (Leadem et al. 1985). Thus, morphine injected subcutaneously or intracerebroventricularly may not reach the presynaptic μ-receptors of NE neurons terminating at the portal capillaries. However, under in vitro conditions these receptors can be exposed to, and activated by μ-agonists. An inhibitory control of catecholamine release by endogenous opioids may also be operative in the hypothalamus.

Retina

In the rabbit retina, [^{3}H]DA release elicited by electrical stimulation was depressed by both μ- (morphine) and δ-preferential (DADLE) agonists (Dubocovich and Weiner 1983). (−)NAL antagonized the inhibition; the nonopioid stereoisomer (+)NAL was inactive. These results suggest that this tissue contains μ-receptors. Since there are no data available regarding the effects of highly selective δ-agonists and antagonists, it is not clear, whether this receptor type is also present in the retina.

2.3 Interaction Between Transmitter Systems

In rat cortical slices, α_2-adrenoceptors and μ-receptors situated at the terminals of NE neurons were suggested to interact with each other (Schoffelmeer et al. 1986b). The application of the α_2-agonist clonidine reduced the inhibitory effect of morphine on the K^+-induced secretion of [^{3}H]NE, whereas the blockade of α_2-adrenoceptors by phentolamine enhanced the potency of the μ-agonist. Such a phenomenon was also demonstrated in rabbit hippocampal and cortical tissue and with electrical stimulation of [^{3}H]NE release; in this case, the opioid used was the κ-agonist EKC (Jackisch et al. 1986; Limberger et al. 1986). The α_2-antagonist yohimbine potentiated the effect of EKC, while clonidine attenuated it. Subsequent publications showed that α_2- and κ-receptors reciprocally influence not only each other, but also adenosine A_1-receptors (Allgaier et al. 1987, 1989; Limberger et al. 1988a,b).

The two possible mechanisms of interaction are firstly, the mutual relation between the receptors themselves in the neuronal membrane and secondly, the utilization of common signal transduction systems. As far as presynaptic receptors are concerned, it has been shown in a most elegant series of experiments that α_2-adrenoceptors must be activated in order to diminish the effects of opioid κ- and adenosine A_1-receptor agonists (Limberger et al. 1988b). In this study two different kinds of electrical stimulation were used, single pulses and trains of stimuli. α_2-Adrenoceptor antagonists, such as yohimbine increased [^{3}H]NE release only when trains were applied and, thereby, the α_2-adrenoceptor-mediated autoinhibitory feedback was functional. Similarly, yohimbine only potentiated the opioid κ- and adenosine A_1-agonist effects under these conditions.

There appears to be a different situation in the case of somatodendritic α_2-adrenoceptors, which interfere with opioid μ-receptors even when they are not activated by endogenous NE or exogenous α_2-agonists (Nörenberg and Illes 1989).

In the rat locus coeruleus, the α_2-antagonist rauwolscine did not change the spike discharge on its own, but it increased the inhibitory action of Met-ENK. The failure of rauwolscine to influence the firing rate indicates that the α_2-adrenoceptors were not stimulated by endogenous NE. Thus, α_2- and μ-ligands interfered with each other even under conditions when the α_2-adrenoceptors, and consequently their signal transduction mechanisms, were not activated. This finding suggests that a mere occupation of the α_2-adrenoceptor by the antagonist is sufficient to initiate the reported changes. Hence, the interaction may take place in the plasma membrane between the receptors themselves.

3 Conclusions

In the rat, both the cell somata and terminals of NE locus coeruleus neurons are endowed with inhibitory μ-receptors. μ-Agonists enhance K^+ conductance and, thereby, may depress cell firing and transmitter release. Thus, regardless of their location on a particular neuron (e.g., cell body or axon terminal), opioid receptors in the locus coeruleus are of the same receptor type and have similar transduction mechanisms. Experiments in the rabbit hippocampus indicate that presynaptic opioid receptors on NE neurons are exclusively of the κ-type. By contrast, in slices from various areas of the guinea-pig brain (cortex, hippocampus, cerebellum), NE release is inhibited by all three classes of opioid agonists (μ, δ, κ).

The nigrostriatal DA neurons of rats appear to possess presynaptic and dendritic, but not somatic κ-receptors. κ-Agonists reduce both the spike discharge and transmitter release. Stimulatory δ- and μ-effects have also been demonstrated, although the latter ones were suggested to involve non-DA cells which, in turn, modulate catecholamine release. The receptor population of the tuberoinfundibular DA neurons of rats belongs to the μ-type and is inhibitory. In guinea pigs, the nigrostriatal system is depressed by κ-agonists only.

Presynaptic receptors of NE neurons interact with each other. The blockade of α_2-adrenoceptors potentiates the inhibitory action of opioid agonists; α_2-activation has the opposite effect. Somatodendritic receptors interact in a similar manner, although in this case the activation of α_2-adrenoceptors does not attenuate the opioid effect.

References

Aghajanian GK, Wang YY (1986) Pertussis toxin blocks the outward currents evoked by opiate and α_2-agonists in locus coeruleus neurons. Brain Res 371:390–394

Allgaier C, Hertting G, Kügelgen OV (1987) The adenosine receptor-mediated inhibition of noradrenaline release possibly involves a N-protein and is increased by α_2-autoreceptor blockade. Br J Pharmacol 90:403–412

Allgaier C, Daschmann B, Sieverling J, Hertting G (1989) Presynaptic κ-opioid receptors on noradrenergic nerve terminals couple to G proteins and interact with the α_2-adrenoceptors. J Neurochem 53:1629–1635

Arbilla S, Langer SZ (1978) Morphine and β-endorphin inhibit release of noradrenaline from cerebral cortex but not of dopamine from rat striatum. Nature (London) 271:559–561

Bird SJ, Kuhar MJ (1977) Iontophoretic application of opiates to the locus coeruleus. Brain Res 122:523–533

Bug W, Williams JT, North RA (1986) Membrane potential measured during potassium-evoked release of noradrenaline from rat brain neurons: effects of normorphine. J Neurochem 47:652–655

Celsen B, Kuschinsky K (1974) Effects of morphine on kinetics of ^{14}C-dopamine in rat striatal slices. Naunyn-Schmiedeberg's Arch Pharmacol 284:159–165

Chesselet MF, Cheramy A, Reisine TD, Glowinski J (1981) Morphine and δ-opiate agonists locally stimulate in vivo dopamine release in cat caudate nucleus. Nature (London) 291:320–322

Chesselet MF, Cheramy A, Reisine TD, Lubetzki C, Glowinski J, Fournie-Zaluski MC, Roques B (1982) Effects of various opiates including specific delta and mu agonists on dopamine release from nigrostriatal dopaminergic neurons in vitro in the rat and in vivo in the cat. Life Sci 31:2291–2294

Christie MJ, Williams JT, North RA (1987) Cellular mechanisms of opioid tolerance: studies in single brain neurons. J Pharmacol Exp Ther 32:633–638

Dewar D, Jenner P, Marsden CD (1987) Effects of opioid agonist drugs on the in vitro release of ^{3}H-GABA, ^{3}H-dopamine and ^{3}H-5HT from slices of rat globus pallidus. Biochem Pharmacol 36:1738–1741

Di Chiara G, Imperato A (1988) Opposite effects of mu and kappa opiate agonists on dopamine release in the nucleus accumbens and in the dorsal caudate of freely moving rats. J Pharmacol Exp Ther 244:1067–1080

Diez-Guerra FJ, Augood S, Emson PC, Dyer RG (1986) Morphine inhibits electrically stimulated noradrenaline release from slices of rat medial preoptic area. Neuroendocrinology 43:89–91

Dubocovich ML, Weiner N (1983) Enkephalins modulate [^{3}H]dopamine release from rabbit retina in vitro. J Pharmacol Exp Ther 224:634–639

Finnerty EP, Chan SHH (1981) The participation of substantia nigra zona compacta and zona reticulata neurons in morphine supression of caudate spontaneous neuronal activities in the rat. Neuropharmacology 20:241–246

Göthert M, Pohl IM, Wehking E (1979) Effects of presynaptic modulators on Ca^{2+}-induced noradrenaline release from central noradrenergic neurons. Noradrenaline and enkephalin inhibit release by decreasing depolarization-induced Ca^{2+} influx Naunyn-Schmiedeberg's Arch Pharmacol 307:21–27

Gudelsky GA, Porter JC (1979) Morphine- and opioid peptide-induced inhibition of the release of dopamine from tuberoinfundibular neurons. Life Sci 25:1697–1702

Guyenet PG, Aghajanian GK (1979) ACh, substance P and Met-enkephalin in the locus coeruleus: pharmacological evidence for independent sites of action. Eur J Pharmacol 53:319–328

Hagan RM, Hughes IE (1984) Opioid receptor sub-types involved in the control of transmitter release in cortex of the brain of the rat. Neuropharmacology 23:491–495

Haskins JT, Moss RL (1983) Differential effects of morphine, dopamine and prolactin administered iontophoretically on arcuate hypothalamic neurones. Brain Res 268:185–188

Haskins JT, Gudelsky GA, Moss RL, Porter JC (1981) Iontophoresis of morphine into the arcuate nucleus: effects on dopamine concentrations in hypophysial portal plasma and serum prolactin concentrations. Endocrinology 108:767–771

Henderson G, Hughes J, Kosterlitz HW (1979) Modification of catecholamine release by narcotic analgesics and opioid peptides. In: Paton DM (ed) The release of catecholamines from adrenergic neurons. Pergamon, Oxford, pp 217–228

Hommer DW, Pert A (1983) The actions of opiates in the rat substantia nigra: an electrophysiological analysis. Peptides 4:603–607

Illes P (1986) Mechanisms of receptor-mediated modulation of transmitter release in noradrenergic, cholinergic and sensory neurons. Neuroscience 17:909–928

Illes P (1989) Modulation of transmitter and hormone release by multiple neuronal opioid receptors. Rev Physiol Biochem Pharmacol 112:139–233

Iwatsubo K, Clouet DH (1977) Effects of morphine and haloperidol on the electrical activity of rat nigrostriatal neurons. J Pharmacol Exp Ther 202:429–436

Jackisch R, Geppert M, Illes P (1986) Characterization of opioid receptors modulating noradrenaline release in the hippocampus of the rabbit. J Neurochem 46:1802–1810

Jackisch R, Geppert M, Lupp A, Huang HY, Illes P (1988) Types of opioid receptors modulating neurotransmitter release in discrete brain regions. In: Illes P, Farsang C (eds) Regulatory roles of opioid peptides. VCH, Weinheim, pp 240–258

Jhamandas K, Marien M (1987) Glutamate-evoked release of endogenous brain dopamine: inhibition by an excitatory amino acid antagonist and an enkephalin analogue. Br J Pharmacol 90:641–650
Jones CA, Marchbanks RM (1982) Effects of (D-alanine^2, methione^5)enkephalinamide on the release of acetylcholine and noradrenaline from brain slices and isolated nerve terminals. Biochem Pharmacol 31:455–458
Jurna I (1981) Changes in the activity of nigral neurones induced by morphine and other opiates in rats with an intact brain and after prenigral decerebration. Naunyn-Schmiedeberg's Arch Pharmacol 316:149–154
Korf J, Bunney BS, Aghajanian GK (1974) Noradrenergic neurons: morphine inhibition of spontaneous activity. Eur J Pharmacol 25:165–169
Kosterlitz HW (1985) Opioid peptides and their receptors. Proc R Soc London Ser B 225:27–40
Langer SZ (1981) Presynaptic regulation of the release of catecholamines. Pharmacol Rev 32:337–361
Lavin A, Garcia-Munoz M (1985) Electrophysiological changes in substantia nigra after dynorphin administration. Brain Res 369:298–302
Leadem CA, Crowley WR, Simpkins JW, Kalra SP (1985) Effects of naloxone on catecholamine and LHRH release from the perifused hypothalamus of the steroid primed rat. Neuroendocrinology 40:497–500
Leslie FM (1987) Methods used for the study of opioid receptors. Pharmacol Rev 39:197–249
Limberger N, Späth L, Hölting T, Starke K (1986) Mutual interaction between presynaptic α_2-adrenoceptors and opioid κ-receptors at the noradrenergic axons of rabbit brain cortex. Naunyn-Schmiedeberg's Arch Pharmacol 334:166–171
Limberger N, Singer EA, Starke K (1988a) Only activated but not non-activated presynaptic α^2-autoreceptors interfere with neighbouring presynaptic receptor mechanisms. Naunyn-Schmiedeberg's Arch Pharmacol 338:62–67
Limberger N, Späth L, Starke K (1988b) Presynaptic α_2-adrenoceptor, opioid κ-receptor and adenosine A_1-receptor interactions on noradrenaline release in rabbit brain cortex. Naunyn-Schmiedeberg's Arch Pharmacol 338:53–61
Loh HH, Brase DA, Sampath-Khanna S, Mar JB, Way EL (1976) β-Endorphin in vitro inhibition of striatal dopamine release. Nature (London) 264:567–568
Lord JAH, Waterfield AA, Hughes J, Kosterlitz HW (1977) Endogenous opioid peptides: multiple agonists and receptors. Nature (London) 267:495–499
Lubetzki C, Chesselet MF, Glowinski J (1982) Modulation of dopamine release in rat striatal slices by delta opiate agonists. J Pharmacol Exp Ther 222:435–440
Manallack DT, Beart PM, Gundlach AL (1986) Psychotomimetic σ-opiates and PCP. Trends Pharmacol Sci 7:448–451
Martin WR, Eades CG, Thompson JA, Huppler RE, Gilbert PE (1976) The effects of morphine- and nalorphine-like drugs in the nondependent and morphine-dependent chronic spinal dog. J Pharmacol Exp Ther 197:517–532
McFadzean I, Lacey MG, Hill RG, Henderson G (1987) Kappa opioid receptor activation depresses excitatory synaptic input to rat locus coeruleus neurons in vitro. Neuroscience 20:231–239
Montel H, Starke K, Weber F (1974a) Influence of morphine and naloxone on the release of noradrenaline from rat brain cortex slices. Naunyn-Schmiedeberg's Arch Pharmacol 283:357–369
Montel H, Starke K, Weber F (1974b) Influence of fentanyl, levorphanol and pethidine on the release of noradrenaline from rat brain cortex slices. Naunyn-Schmiedeberg's Arch Pharmacol 283:371–377
Montel H, Starke K, Taube HD (1975) Influence of morphine and naloxone on the release of noradrenaline from rat cerebellar cortex slices. Naunyn-Schmiedeberg's Arch Pharmacol 288:427–433
Mulder AH, Wardeh G, Hogenboom F, Frankhuyzen AL (1984) κ- and δ-opioid receptor agonists differentially inhibit striatal dopamine and acetylcholine release. Nature (London) 308:278–280
Mulder AH, Hogenboom F, Wardeh G, Schoffelmeer ANM (1987) Morphine and enkephalins potently inhibit [^{3}H]noradrenaline release from rat brain cortex synaptosomes: further evidence for a presynaptic localization of μ-opioid receptors. J Neurochem 48:1043–1047
Mulder AH, Frankhuyzen AL, Schoffelmeer ANM (1988) Modulation by opioid peptides of dopaminergic neurotransmission at the pre- and postsynaptic level. In: Illes P, Farsang C (eds) Regulatory roles of opioid peptides. VCH, Weinheim, pp 268–281
Nörenberg W, Illes P (1989) Blockade of somato-dendritic α_2-adrenoceptors increases the inhibitory

effect of μ-opioid agonists on the firing rate of rat locus coeruleus neurons. Naunyn-Schmiedeberg's Arch Pharmacol 339:R103
North RA (1986a) Receptors on individual neurons. Neuroscience 17:899–907
North RA (1986b) Opioid receptor types and membrane ion channels. Trends Neurosci 9:114–117
North RA, Williams JT (1985) On the potassium conductance increased by opioids in rat locus coeruleus neurons. J Physiol 364:265–280
North RA, Williams JT, Surprenant A, Christie MJ (1987) μ and δ receptors belong to a family of receptors that are coupled to potassium channels. Proc Natl Acad Sci USA 84:5487–5491
Pepper CM, Henderson G (1980) Opiates and opioid peptides hyperpolarize locus coeruleus neurons in vitro. Science 209:394–396
Petit F, Hamon M, Fournie-Zaluski MC, Roques BP, Glowinski J (1986) Further evidence for a role of δ-opiate receptors in the presynaptic regulation of newly synthesized dopamine release. Eur J Pharmacol 126:1–9
Reymond MJ, Kaur CK, Porter JC (1983) An inhibitory role for morphine on the release of dopamine into hypophysial portal blood and on the synthesis of dopamine in tuberoinfundibular neurons. Brain Res 262:253–258
Schoffelmeer ANM, Mulder AH (1983) Differential control of Ca^{2+}-dependent [^{3}H]noradrenaline release from rat brain slices through presynaptic opiate receptors and α-adrenoceptors. Eur J Pharmacol 87:449–458
Schoffelmeer ANM, Mulder AH (1984) Presynaptic opiate receptor- and α_2-adrenoceptor-mediated inhibition of noradrenaline release in the rat brain: role of hyperpolarization? Eur J Pharmacol 105:129–135
Schoffelmeer ANM, Putters J, Mulder AH (1986a) Activation of presynaptic α_2-adrenoceptors attenuates the inhibitory effect of μ-opioid receptor agonists on noradrenaline release from brain slices. Naunyn-Schmiedeberg's Arch Pharmacol 333:377–380
Schoffelmeer ANM, Wierenga EA, Mulder AH (1986b) Role of adenylate cyclase in presynaptic α_2-adrenoceptor- and μ-opioid receptor-mediated inhibition of [^{3}H]noradrenaline release from rat brain cortex slices. J Neurochem 46:1711–1717
Schoffelmeer ANM, Rice KC, Jacobson AE, Van Gelderen JG, Hogenboom F, Heijna MH, Mulder AH (1988) μ-, δ- and κ-opioid receptor-mediated inhibition of neurotransmitter release and adenylate cyclase activity in rat brain slices: studies with fentanyl isothiocyanate. Eur J Pharmacol 154:169–178
Schulz R, Wüster M, Herz A (1981) Pharmacological characterization of the ε-opiate receptor. J Pharmacol Exp Ther 216:604–606
Sonders MS, Keana JFW, Weber E (1988) Phencyclidine and psychotomimetic sigma opiates: recent insights into their biochemical and physiological sites of action. Trends Neurosci 11:37–40
Starke K (1977) Regulation of noradrenaline release by presynaptic receptor systems. Rev Physiol Biochem Pharmacol 7:1–124
Starr MS (1978) Investigation of possible interactions between substance P and transmitter mechanisms in the substantia nigra and corpus striatum of the rat. J Pharm Pharmacol 30:359–363
Subramanian N, Mitznegg P, Sprügel W, Domschke W, Domschke S, Wünsch E, Demling L (1977) Influence of enkephalin on K^{+}-evoked efflux of putative neurotransmitters in rat brain. Naunyn-Schmiedeberg's Arch Pharmacol 299:163–165
Szekely JI, Ronai AZ (1982a) Opioid peptides, vol 1. Research methods. CRC, Boca Raton
Szekely JI, Ronai AZ (1982b) Opioid peptides, vol 2. Pharmacology. CRC, Boca Raton
Taube HD, Starke K, Borowski E (1977) Presynaptic receptor systems on the noradrenergic neurons of rat brain. Nauyny-Schmiedeberg's Arch Pharmacol 299:123–141
Vizi ES (1979) Presynaptic modulation of neurochemical transmission. Prog Neurobiol 12:181–290
Walker JM, Thompson LA, Frascella J, Friederich MW (1987) Opposite effects of μ and κ opiates on the firing-rate of dopamine cells in the substantia nigra of the rat. Eur J Pharmacol 134:53–59
Werling LL, Brown SR, Cox BM (1987) Opioid receptor regulation of the release of norepinephrine in brain. Neuropharmacology 26:987–996
Werling LL, Frattali A, Portoghese PS, Takemori AE, Cox BM (1988a) Kappa receptor regulation of dopamine release from striatum and cortex of rats and guinea pigs. J Pharmacol Exp Ther 246:282–286
Werling LL, McMahon PN, Cox BM (1988b) Selective tolerance at mu and kappa opioid receptors modulating norepinephrine release in guinea pig cortex. J Pharmacol Exp Ther 247:1103–1106
Werling LL, McMahon PN, Portoghese PS, Takemori AE, Cox BM (1989) Selective opioid antagonist

effects on opioid-induced inhibition of release of norepinephrine in guinea-pig cortex. Neuropharmacology 28:103–107
Westfall TC (1977) Local regulation of adrenergic neurotransmission. Physiol Rev 57:659–728
Westfall TC, Grant H, Naes L, Meldrum M (1983) The effect of opioid drugs on the release of dopamine and 5-hydroxytryptamine from rat striatum following activation of nicotinic-cholinergic receptors. Eur J Pharmacol 92:35–42
Wichmann T, Starke K (1988) Uptake, release and modulation of release of noradrenaline in rabbit superior colliculus. Neuroscience 26:621–634
Wilkes MM, Yen SSC (1980) Reduction by β-endorphin of the efflux of dopamine and DOPAC from superfused medial basal hypothalamus. Life Sci 27:1387–1391
Williams JT, North RA (1984) Opiate receptor interactions on single locus coeruleus neurons. Mol Pharmacol 26:489–497
Williams JT, Egan TM, North RA (1982) Enkephalin opens potassium channels on mammalian central neurons. Nature (London) 299:74–77
Williams J, Henderson G, North A (1986) Locus coeruleus neurons. In: Dingledine R (ed) Brain slices. Plenum, New York, pp 297–311
Zukin RS, Zukin SR (1984) The case for multiple opiate receptors. Trends Neurosci 7:160–164

Section IV

Functional Aspects

CHAPTER 14

Adaptation of Opioid Systems to Stress

R. Przewłocki, B. Przewłocka, W. Lasoń

1 Introduction

The reaction of biological systems to stressors has often been utilized for the elucidation of neuronal and endocrine control functions of the organism. Stress might be defined as a response to sufficiently intense or frequent environmental stimuli, and the consequence of the reaction. It results in a wide range of biological responses including changes in brain neuronal activity and behavior, as well as in functional alterations of the endocrine, cardiovascular, and immune systems. Furthermore, repeated stress may be involved in the pathophysiology and etiology of some functional disorders.

Endogenous opioids as well as catecholamines, ACTH, and glucocorticoids are important factors in the neurochemical mechanisms of stress. Indeed, considerable evidence demonstrates that a variety of stressors influence endogenous opioid systems. However, the biochemical basis of these processes is still poorly understood. Over the last few years there have been considerable advances in our understanding of the biogenesis of various opioid peptides (EOP), their anatomical distribution, and characteristics of the multiple receptors with which they interact. Recombinant DNA technology has shown that all known EOP are derived from three different precursor proteins: proopiomelanocortin (POMC), prodynorphin (PDYN), and proenkephalin (PENK) (see Höllt, this Vol.). These three propeptides are similar in number of amino acids and all contain the sequence Met- or Leu-enkephalin (ENK). Furthermore, they are derived from remarkably similar genes suggesting that, although they are not closely linked in the genome, these three genes might have arisen via a common evolutionary mechanism (Herbert et al. 1985).

All three endogenous opioid systems are widely represented in regions which are heavily involved in the stress response. They can be found in the hypothalamus, pituitary, and adrenals. Similarly, autonomic nervous system centers have been shown to be innervated by central and peripheral opioidergic neurons. A question arises concerning the conditions that result in the regulatory response of opioid systems to stress (Fig. 1). It is important to note that opioid systems are not usually tonically active and, hence, opioid antagonists have little or no effect in the state of homeostasis. On the contrary, they are activated by various stressful stimuli, thus allowing them to influence certain effects of such stimuli. In this chapter we have focused our attention on those alterations in EOP systems which occur as a result of the organism's reaction and adaptation to stressful and harmful stimuli, and the differential involvement of multiple EOP in this phenomenon.

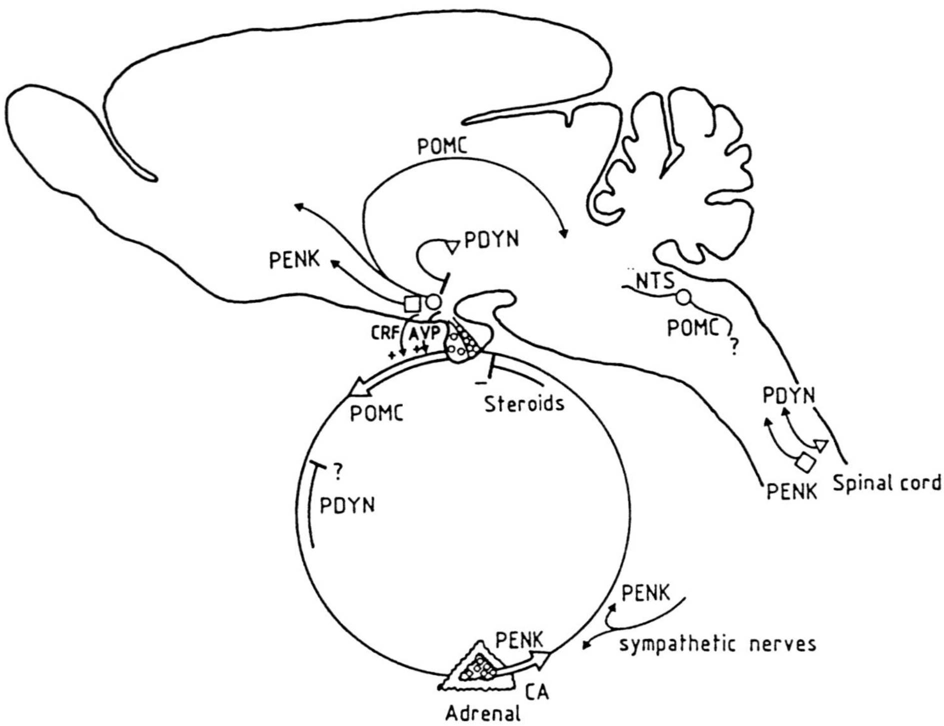

Fig. 1. Schematic representation of brain, spinal cord, and adrenal medulla opioid systems activated or inhibited in response to stress. Abbreviations: *AVP* arginine vasopressin; *CRF* corticotropin releasing factor; *CA* catecholamines; *NTS* nucleus tractus solitarius; *PDYN* prodynorphin; *PENK* proenkephalin; *POMC* proopiomelanocortin

2 Biochemical Alteration of Opioid Peptides in Stress

2.1 Proopiomelanocortin (POMC) System

Peptides derived from POMC are present in neurons of the nucleus arcuatus, which is localized in the mediobasal hypothalamus, nucleus tractus solitarius of the caudal medulla, endocrine cells, of both the anterior and intermediate lobes of pituitary, and in cells of some peripheral tissues (Bloom et al. 1978; Watson et al. 1978). An extensive fiber system, originating in the arcuate nucleus, terminates in many areas of the brain which have been implicated in the stress response, e.g., the hypothalamic nuclei, limbic and raphe nuclei, and some pontine nuclei. In addition, some of these structures might also be innervated by POMC neurons located in the nucleus tractus solitarius, which project laterally and ventrally, and which also enter the spinal cord (Schwartzberg and Nakane 1981).

In the hypothalamus, the major form of β-endorphin (β-END) appears to be β-END$_{1\text{-}31}$, which is the most potent analgesic among the endogenous opioids. In some terminals, C-terminal processing to shorter forms, such as β-END_{1-27}, which are less biologically active that the parent peptide, may also occur (Akil et al. 1981). The existence of N-acetylated forms in neuronal terminals has also been postulated (Akil et al. 1983). However, Weber et al. (1981) suggested the lack of acetylated forms in the brain. A minor quantity of the N-acetyl-β-END_{1-31} (less than 2%) detected in brain tissue disappeared following hypophysectomy, indicating its pituitary origin. Thus, although posttranslation processing of the POMC peptides in the brain is still a matter of controversy, it appears that N-acetylation, which may lead to nonopiate derivatives of β-END, occurs: it does so much less frequently in brain than in other tissue. Furthermore, it is clear that β-END_{1-31} is the main POMC-derived OP in the brain.

There is increasing evidence that β-END exists within releasable pools of the hypothalamus. It was previously demonstrated that β-END may be released from hypothalamic slices in vitro in response to high K^+ ion concentrations, in a Ca^{2+}-dependent manner (Osborne et al. 1979). Recent studies have shown that an important factor stimulating this release is corticotropin releasing hormone (CRF) (see Burns and Nikolarakis, this Vol.). In addition, β-END-containing cells appear to remain under tonic CRF control, since the CRF receptor antagonist α-helical CRF_{9-41} inhibits β-END release in vitro (Nikolarakis et al. 1986). Given the role of CRF in mediating various stress responses, the interaction between CRF- and POMC-containing neurons within the hypothalamus strongly suggests the involvement of hypothalamic β-ENDergic neurons in the regulation of the stress response.

Increasing evidence suggests that both acute and chronic stress enhances the activity of the POMC system in the brain. It was previously shown that a short-term noxious foot-shock stress causes the depletion of β-END in the hypothalamus, septum, and periaqueductal gray, indicating an enhanced release of the peptide therein (Millan et al. 1981a; Przewłocki et al. 1982). More recently, it was shown that acute swim stress decreases the β-END content in the nucleus accumbens septi, again suggesting an accelerated release of this peptide from nerve terminals (Przewłocki et al. 1989). However, more severe stress, i.e., acute prolonged intermittent foot shock induces no consistent alteration in hypothalamic and midbrain levels of the peptide (Akil et al. 1986; Przewłocki et al. 1987). Thus, the stress-induced decrease in the hypothalamic β-END level might only be detectable with mild, short-term stress. Various kinds of chronic stress (repeated foot shock, prolonged inflammatory pain, repeated electroconvulsive shock) do not induce reliable alterations in brain β-END levels (Millan et al. 1986; Lasoń et al. 1987; Przewłocki et al. 1984, 1987). It is important to recognize that tissue levels of a peptide are influenced by several processes: release, biosynthesis, and axonal transport. Thus, one likely explanation for the apparent unchanged β-END levels in brain is that high intensity or long-term stress induces a rapid stimulation of POMC biosynthesis in the brain, which, in turn, may prevent the depletion of POMC-derived peptides from this tissue. On the other hand, long-term food deprivation significantly lowered the β-END level in the nucleus accumbens septi (Przewłocki et al. 1989). Interestingly, a recent study in our laboratory has shown that conditioned fear-induced stress markedly decreases the hypothalamic content of β-END

(Przewłocka et al. 1989). This result is in line with the previous observation that there is a measurable decrease in midbrain β-END content when chronically stressed rats were restressed acutely (Akil et al. 1986). These data indicate that there is an enhanced releasability of β-END or an increase in the releasable pool of this peptide following repeated exposure to stress. Nevertheless, additional studies of POMC biosynthesis and release in the brain are needed before the ability of stress to activate brain POMC systems can be clarified.

A number of studies have described multiple changes in pituitary POMC systems in response to stress. In the anterior lobe of the pituitary, POMC peptides are present in corticotrophs. In these cells, unmodified β-END_{1-31} and $ACTH_{1-39}$ are primarily produced. In contrast, POMC in intermediate lobe cells is processed into β-END_{1-27} and β-END_{1-26} (Evans et al. 1981; Smyth 1983), as well as into peptides devoid of opioid receptor specificity, e.g., N-acetyl-β-END_{1-27}. Similarly, ACTH in the intermediate lobe is processed to $ACTH_{18-39}$ (CLIP) and to a peptide with an acetylated N-terminus and an amidated C-terminus $ACTH_{1-13}$-NH_2 (α-MSH).

Acute stress leads to a substantial elevation of β-END and ACTH in the plasma of rats. Corticotrophs in the anterior lobe react to acute stress by an enhanced release of β-END which results in a decrease in peptide level in the anterior pituitary, and an increase in the blood (Guillemin et al. 1977). In addition to affecting release, acute stress appears to increase the rate of both POMC synthesis and processing (Shiomi and Akil 1982). The intermediate lobe also responds to stress in a similar way (Akil et al. 1985; Przewłocki et al. 1982), leading to an elevation in the plasma POMC products, even though β-END levels in this lobe are only slightly decreased or remain unchanged. Young et al. (1986) have found that the predominant peptide released by stress was β-END_{1-31}, which appeared to be of anterior pituitary origin. On the other hand, there was also a slight but significant increase in N-acetyl-β-END in response to stress. Thus, it appears that, at least in rodents, both the anterior and intermediate lobe pools seem to contribute to the changes in the circulating levels of β-END.

Chronic stress such as foot shock, forced deprivation of food intake, or chronic pain leads to a substantial increase in β-END levels in both the anterior and intermediate lobes of the pituitary (Akil et al. 1986; Höllt et al. 1986; Przewłocki et al. 1987). Interestingly, the level of POMC mRNA increases in the anterior but not in the intermediate lobe after 1 week of repeated foot shock (Höllt et al. 1986). This finding suggests a selective activation of POMC synthesis only in the anterior lobe. In contrast, other studies have demonstrated that rats subjected to chronic foot shock or forced swimming stress also show an increase in POMC mRNA in the intermediate lobe of the pituitary (Kelsey et al. 1984). It is likely that prolonged treatment is necessary for increased biosynthesis of POMC in intermediate lobe cells. Chronic stress leads to an increase in the release of N-acetyl-β-END into plasma (Akil et al. 1986), whereas the release of pituitary β-END_{1-31} is lower in chronic than in acute stress. Thus, adaptation of pituitary POMC systems to stress results in an increase in biosynthetic activity in both pituitary lobes, while the profile of released peptides shifts from opioid forms of β-END in acute stress, to its nonopioid or less active forms in chronic stress.

In the adrenal medulla, the expression of POMC-derived peptides appears to be species-specific and has only been observed in humans and monkeys (Evans et al. 1985b). It appears that in other species, POMC message in the tissue is shorter than that in the brain and, as has been shown in bovine adrenal medulla cells, does not appear to be expressed (Jingami et al. 1984). Its role in stress awaits to be established.

2.2 Prodynorphin (PDYN) System

PDYN neurons are widely distributed in brain areas associated with stress responses (Khachaturian et al. 1982). These peptides are present in the magnocellular neurons of the paraventricular nucleus of the hypothalamus where they are costored with vasopressin. Vasopressin is a known secretogogue of ACTH and β-END in anterior lobe corticotrophs (Przewłocki et al. 1979). In addition, the PDYN peptides have been found in the nucleus tractus solitarius, an area classically associated with the regulation of vagal and other autonomic functions. Further, PDYN neurons occur in the limbic system and in the area of the spinal cord which is involved in transmission of nociceptive stimuli. Cells expressing PDYN mRNA are also present in at least a subpopulation of anterior lobe gonadotrophs (Khachaturian et al. 1982), posterior pituitary neuronal terminals, and the adrenals, most predominantly in the adrenal cortex (Day et al. 1988), a structure which is extremely sensitive to POMC products. PDYN mRNA levels in adrenals were found to be higher than those in the brain but might not be translated (Civelli et al. 1985).

The processing of PDYN to its active opioid and nonopioid products has not been fully elucidated. Several sizes of DYNs appear to exist. In the brain, PDYN yields DYN_{1-17}, DYN_{1-32}, DYN B and α-Neo-END (Weber et al. 1982). DYN_{1-17} may be further processed into DYN_{1-8}, which is the major product in the rat brain and posterior pituitary, while DYN_{1-32} may be further converted to DYN B. In the anterior lobe gonadotrophs, PDYN seems to be processed into higher molecular weight forms and into α-Neo-END. The processing of PDYN appears to be relatively simple compared with that of POMC since there is no evidence for α-N-acetylation of PDYN peptides. In addition, the production of Leu-ENK from PDYN peptides does not appear to predominate in the brain since a comparative immunohistochemical study has shown that the PENK and PDYN systems are, in fact, anatomically distinct in this tissue (Watson et al. 1982).

DYN levels in the hypothalamus remain unchanged (Przewłocki et al. 1987) or even slightly increased in response to acute foot-shock stress (Millan et al. 1981b). Interestingly, the level of α-Neo-END significantly increases after acute swimming stress, indicating an inhibition of peptide release (Przewłocki et al. 1989). In contrast, a pronounced fall in hypothalamic DYN levels is observed in response to single electroconvulsive shock (Lasoń et al. 1987), indicating an enhancement of PDYN-derived peptide release during seizures. There is no evidence that chronic stress, such as recurrent foot shock (Przewłocki et al. 1987), chronic pain (Millan et al. 1986; Przewłocki et al. 1985), and food deprivation (Przewłocki et al. 1983b) or conditioned fear-induced stress, influence the PDYN system in the hypothalamus

(Przewłocka et al. 1989). On the contrary, repeated electroconvulsive shocks markedly increase DYN content and PDYN mRNA levels in the hypothalamus (Hong et al. 1985; Lasoń et al. 1987). Thus, hypothalamic PDYN neurons appear to be particularly sensitive to seizures. Furthermore, these neurons also respond to dehydration as evidenced by an increase in PDYN-derived peptides (Höllt et al. 1981; Majeed et al. 1986a) and PDYN mRNA levels (Sherman et al. 1986). Intermittent foot shock produced no alteration of PDYN peptides in the anterior and intermediate lobes of the pituitary (Przewłocki et al. 1987), although it was also reported that acute, short-term foot shock produced a decrease in DYN levels in the anterior lobe (Millan et al. 1981b). If PDYN-derived peptides from this lobe are released under acute stress, an increase in the plasma PDYN peptides should be observed. However, plasma α-Neo-END levels slightly decrease after either acute ether stress or electroconvulsive shock (Przewłocki, unpubl. observ.). This observation indicates that the release of PDYN immunoreactive peptides may, in fact, be inhibited by stress. This finding is in line with other data which have shown that DYN levels in the posterior lobe of the pituitary (containing DYNergic neuronal terminals) remain stable after acute foot shock (Przewłocki et al. 1987). Taken together, these results suggest that stress inhibits, rather than activates, PDYN systems in the brain and pituitary.

Several investigations have focused upon alterations in the spinal cord PDYN system, which occur in response to stress and chronic arthritic pain. We first demonstrated that levels of DYN were dramatically increased in the lumbar part of the spinal cord of rats with local inflammation of the hind limb (Przewłocki et al. 1985). Other studies have confirmed and extended these observations (Iadarola et al. 1986; Millan et al. 1986). Furthermore, an increase of PDYN mRNA levels induced by chronic pain has also been demonstrated (Höllt et al. 1986; Iadarola et al. 1986). An increase of PDYN peptide levels was also demonstrated in the spinal cord of rats subjected to repeated foot shock (Przewłocki et al. 1987). Thus, such results indicate that a chronic exposure to painful stimuli enhances the biosynthetic activity of PDYN neurons. Further, the levels of PDYN peptides are elevated. Interestingly, this system seems to adapt to chronic painful stimuli by a decrease in PDYN peptide release (Przewłocki et al. 1986). Consequently, this leads to a supersensitivity of spinal opioid receptors (Przewłocki et al. 1985). Recent biochemical studies indicate that an experimental spinal shock also increases the level of PDYN in the spinal cord of rats (Faden et al. 1985; Przewłocki et al. 1988). Furthermore, a recent mRNA hybridization study revealed that PDYN mRNA levels increased in rat spinal cord after a traumatic injury (Przewłocki et al. 1988). The above results suggest that chronic pain and injury activates PDYN neurons in the spinal cord; however, they also indicate that adaptation to permanent pain results in a decrease, rather than an increase, in PDYN peptide release.

2.3 *Proenkephalin (PENK) System*

PENK neurons have a wide distribution throughout the central and peripheral nervous system. They are localized predominantly in short interneurons, some forming local circuits and others forming longer tract projections. PENK neurons are abundant in the paraventricular nucleus and the nucleus arcuatus of the hypothalamus. In these regions they can interact with CRF-containing hypothalamic cell bodies and processes located in the paraventricular nucleus and the median eminence, respectively. There is also evidence for the coexistence of ENK and CRF in the paraventricular nucleus and in terminals in the median eminence. A number of PENK neuronal systems exist in limbic system structures, e.g., hippocampus, septum, and the bed nucleus of stria terminalis, where they may participate in the modulation of emotion and memory. Septal PENK neurons project directly to the amygdala. The fibers extend throughout the bed nucleus of stria terminalis and neurons from this nucleus project to the paraventricular nucleus and the median eminence. Neurons containing PENK peptides can be seen in the spinal cord cranial sensory systems and in the major pain signaling network. These neurons are likely to be involved in the modulation of somato- and viscerosensory information, which may influence CRF synthesis and release. A variety of ENK-containing cells are present in the adrenal medulla (Viveros et al. 1979). These cells also produce catecholamines. Both PENK peptides and catecholamines may be released simultaneously in response to stressful events (Lewis et al. 1982).

Processing of PENK is tissue-specific as in the case of other endogenous opioids. In the brain, one mainly finds the smaller PENK peptides (Weber et al. 1983). Of the hypothalamic ENKs, octa- and heptapeptides appear to be the major products of PENK processing. On the other hand, in the adrenals processing appears to be less complete, resulting in high molecular weight species such as peptide F, peptide E, and BAM-peptides, in addition to Met-ENK, amidorphin, and metorphamide which are also present in the brain. Foot shock stress has been shown to decrease the hypothalamic content of Leu-ENK (Rossier et al. 1978). Furthermore, an augmented release of Met-ENK into the cerebrospinal fluid has been observed in response to acute noxious stimuli (Cesselin et al. 1982; Kurumaji et al. 1987) and it has also been demonstrated that 30 min of immobilization stress decreased Met-ENK levels in the hypothalamus. This decrease most probably results from an increase in peptide release. In contrast, other studies have shown no visible influence of foot shock stress upon Met-ENK in the rat hypothalamus (Millan et al. 1981a). Recent studies indicate, however, that stressful conditions such as opiate withdrawal and intraperitoneal injection of hypertonic saline enhance PENK gene expression in cells localized in the parvocellular part of the paraventricular nucleus (Lightman and Young 1989). An increase in the biosynthetic activity of the hypothalamic PENK system has also been demonstrated in rats subjected to repeated electroconvulsive shock (Hong et al. 1985). Thus, activation of hypothalamic neurons may be of considerable importance in the neuroendocrine response to stress. PENK neurons in the paraventricular nucleus may, for example, modulate the response to stress by acting on vasopressin- or CRF-containing neurons in this nucleus.

Lewis and colleagues (1982) have implicated adrenal medulla opioids in the response to stress. Various manipulations, which affect the function of adrenals, e.g., adrenalectomy, adrenal demedullation, or adrenal medullary denervation, at least partly diminish stress-induced analgesia. Exposure of rats to short, intermittent foot shock causes a decrease in the adrenal medullary content of ENKs and peptide F (Alessi et al. 1982; Lewis et al. 1982), suggesting enhanced release of these opioids from medullary cells. The levels of EOP in the adrenal medulla are decreased after acute stress but return to control levels in chronically stressed rats. This result suggests that the biosynthetic activity of PENK cells increases upon prolonged stress in order to compensate for enhanced release of peptides into the blood. Van Loon and his collaborators (1988b) have suggested that plasma Met-ENK is derived from sympathetic peripheral neurons and not from the adrenals. They observed that exposure of rats to restraint stress results in the release of Met-ENK into the plasma and that peak release first occurs after 30 s and then again 30 min after the onset of the stressor. Repeated, daily exposure of rats to restraint stress resulted in an adaptive loss of the plasma Met-ENK response. The authors concluded that plasma Met-ENK is derived largely from sympathetic nerves and not from the adrenal medulla, since chemical sympathectomy, but not adrenal demedullation, prevented the response to restraint stress. Thus, it would appear that in the periphery both adrenal opioid cells and sympathetic neurons contribute to the opioid component of the stress response. Additionally, it may be the case that different types of stress selectively activate different pools of PENK peptides.

3 Opioid Receptor Response to Stress

There is considerable evidence for the existence of at least three main opioid receptor types, μ, κ, and δ (see Simon, this Vol.). μ- and κ-Receptor subtypes have also been postulated. Several attempts have been made to assess stress-induced alterations of opioid receptors. Conditioned fear-induced stress was shown to cause a decrease in ^{3}H-Leu-ENK binding to rat brain (Chance et al. 1978). Similar results have been obtained in rats after exposure to a forced swim (Pert and Bowie 1979; Christie et al. 1981). More recently, Seeger et al. (1984) reported that either prolonged, intermittent foot shock or forced swimming caused a significant reduction in ^{3}H-diprenorphine binding in the hypothalamus and other brain structures, as measured by autoradiography. A decrease in high affinity ^{3}H-etorphine binding after restraint stress in rats has also been documented (Hnatowich et al. 1986). Thus, acute stress appears to decrease the binding of opioid receptor ligands, suggesting that there is a persistent activation of the opioid receptor due to an enhanced release of endogenous opioids.

Chronic, recurrent stress induced by repetitive electroconvulsive shocks has been shown to cause downregulation of both μ- and δ-receptors in some structures of the rat brain (Nakata et al. 1985). It is reasonable to assume that the effect may be an adaptive response to the enhanced release of EOP during stress. In contrast, Holaday et al. (1982) reported an upregulation of δ-opioid receptors in response to repeated electroconvulsive shock treatments. Furthermore, a recent study of Stein et al. (1988) showed that stressors such as intermittent, 20-min foot shock and 4-day water deprivation induced an increase in μ- and δ-binding sites in the rat limbic system. In contrast, Lewis et al. (1987) were unsuccessful in demonstrating any changes in the number or affinity of μ-, δ-, and κ-sites in various brain and spinal cord regions of rats exposed to chronic stress. Evidently, the results of the studies are variable, sometimes contradictory, and rely predominantly on the measurement of total receptor content. Clearly, more studies are needed before the influence of stress on specific opioid receptors can be evaluated.

4 Involvement of Opioids in Some Behavioral and Physiological Adaptations to Stress

4.1 Stress-Induced Analgesia

Akil et al. (1976) were the first to report that exposure to stress caused potent antinociception in rats. Furthermore, they found that naloxone (NAL) antagonized this analgesia, suggesting an involvement of endogenous opioid systems. Other studies, however, have shown that NAL only slightly decreases (Przewłocki et al. 1979) or has no influence on the stress-induced elevation of tail-flick latency; the results apparently depend on the length of exposure to the stressor (Hayes et al. 1978). Lewis et al. (1980) later clarified this issue by finding that analgesia induced by prolonged (but not short-term) foot shock is blocked by NAL. Przewłocki et al. (1979) and Millan et al. (1981a) have shown that electrolytic lesions of the arcuate nucleus (e.g., a structure containing β-END neurons) augmented foot shock, stress-induced analgesia. Such results indicate that the β-END-containing cells may, at least partly, be involved in this phenomenon. A more recent study has suggested the potential role of β-END cells within the nucleus tractus solitarius in the response; electrical stimulation of this structure induces opioid-mediated analgesia in the rat (Lewis et al. 1985).

An involvement of pituitary β-END has also been postulated. Millan et al. (1980) showed that adenohypophysectomy abolished stress-induced antinociception. The study suggested that the anterior, but not the intermediate, lobe plays a critical role. On the other hand, it is not clear whether pituitary pools of β-END play a role in the phenomena since other forms of stress induce peptide release into the blood without affecting nociception. Adenohypophysectomy also decreases the content of β-END in the hypothalamus (Przewłocki et al. 1982), which may also contribute to the analgesic response to stress. Furthermore, peripheral steroids may also be involved since the effect of hypophysectomy is blocked by corticosterone treatment (MacLennan et al. 1982). There is evidence suggesting a role of adrenal medulla PENK peptides in stress-induced analgesia. Thus, it has been shown that adrenal demedullation abolishes the analgesic response (Lewis et al. 1982). Evidently, integration of the hypothalamo-pituitary-adrenal-axis is essential for maintaining the analgesic response to stress. Furthermore, central and peripheral POMC and PENK systems may respond to stress at different levels of the nociceptive system.

The involvement of the PDYN system in the mediation of stress-induced analgesia remains unclear. It appears, however, that some kinds of stress may inhibit the PDYN system. Thus, it has been shown that the in vitro hypothalamic release of α-Neo-END as well as the level of this peptide in plasma (in contrast to β-END) are lowered in rats upon conditioned fear-induced stress. This effect is also accompanied by NAL-reversible analgesia (Przewłocka et al. 1989).

Various stresses, such as tail-pinch, restraint, food and water deprivation, are known to augment consumer behavior. Interestingly, the effect appears to be mediated by EOP systems since opioid antagonists abolished the eating response (Holtzman 1974; Majeed et al 1986b, Morley et al. 1983). Furthermore, food and water deprivation results in alterations of EOP levels in the brain and pituitary similar to those induced by chronic stress (Majeed et al. 1986; Przewłocki et al. 1983b, 1989). Thus, although EOP do not regulate feeding behavior alone, they play an important role in the mediation of its stress-evoked component.

Cardiovascular responses to stress include increased catecholamine secretion from the adrenals, peripheral vasoconstriction in certain vascular beds, tachycardia due to increased cardiac sympathetic and reduced vagal efferent activity, arrhythmia, and hypotension. Little is known about the involvement of particular opioid systems in these processes. It is known, however, that the pituitary POMC peptides regulate the biosynthesis and release of corticoids in the adrenal cortex, and the substances are known to have pressor activity. In the brain, the POMC system may control the function of the nucleus tractus solitarius, a structure which is known to participate in the control of cardiac function. Intravenous or intracerebroventricular administration of β-END, as well as injection into the nucleus tractus solitarius, have been shown to decrease blood pressure (Hassen et al. 1982; Sitsen et al. 1982).

PDYN peptides appear to modulate the release of vasopressin from the posterior lobe and may in this way regulate diuresis. κ-Opioid agonists are powerful diuretics (Leander 1982). Furthermore, DYN decreases blood pressure and produces bradycardia, when applied intravenously or into the cisterna magna (Laurent and Schmitt 1983). In contrast, application of DYN into the nucleus tractus solitarius or cerebral ventricles did not alter cardiovascular function (Hassen et al. 1982; Glatt et al. 1987).

Cardiovascular responses to centrally administered PENK products are contradictory and provide little insight into the physiological role of central PENK endogenous systems in cardiovascular functions. However, adrenal PENK peptides released by stimulation of the splanchnic nerve may induce bradycardia and hypotension as shown in reserpinized dogs (Hanbauer et al. 1982). If this is the case, it is likely that EOP may, in some circumstances, counteract the cardiovascular effects of moderate stress such as tachycardia and increased blood pressure. In contrast, EOP appear to mediate the cardiovascular depression which occurs in responses to severe stress. In fact, a number of studies have demonstrated that NAL reverses the hypotension induced by most shock states. Subsequent studies have shown analogous effects of other opioid antagonists, a lack of effect of inactive stereoisomers, as well as enhanced release of EOP in shock (Holaday 1983).

A variety of stressors have been found to alter immune functions in animals and affect pathological processes in humans. Endogenous opioids released in stress may interact with the immune system by modulating immune responses to various factors. The mechanisms of this interaction are still not completely clarified and the findings to date are contradictory. Thus, in vitro studies suggest that EOP enhance immune responses (Plotnikoff and Miller 1983; Wybran 1985). In contrast, in vivo studies appear to lead to the opposite conclusion. In fact, stress can suppress immune functions in rats and decrease their resistance to tumor challenge. Greenberg et al. (1984) have found that stress induced by tail shock suppressed natural killer cell cytotoxicity, an effect which was blocked by naltrexone (NLTRX). Furthermore, in a subsequent study Shavit et al. (1986) showed that inescapable foot shock stress decreased the response of T-lymphocytes to mitogens, decreased natural killer cell activity, and finally reduced the resistance of rats to a mammary ascites tumor. These effects were blocked by NLTRX, indicating that they were mediated by EOP mobilized during stress. Evidently, the mechanisms of the interaction of the immune system with endogenous opioids in response to stress is not, as yet, fully understood. It is, however, apparent that particular forms of stress can suppress immune functions.

5 Summary and Conclusions

The papers discussed suggest an involvement of endogenous opioids in the modulation of stress. Activation of certain opioid systems depends on the type of stressor, the duration of exposure, its intensity, and biological significance. The acute, mild, short-lasting stressors appear to mobilize opioids, which may act to oppose stress-induced physiological reactions and, in concord with other factors, function to counteract the initial response. In acute severe stress, such as traumatic injury, circulatory shock and hypoxia, opioids appear to mediate pathological responses and to facilitate death, when all compensatory reactions fail. Giuffre et al. (1988) found that specific immune neutralization of β-END prolonged survival following severe surgical stress. That study suggested that circulating β-END might have deleterious effects during severe stress. Interestingly, the PDYN system seems to be strongly activated by severe destructive stimuli, such as neuronal injury (Faden et al. 1985; Przewłocki et al. 1988), seizures (Lasoń et al. 1983; Hong et al. 1985), and dehydration (Höllt et al. 1981; Majeed et al. 1986a; Sherman et al. 1986). At present, it is not clear whether the PDYN peptides mediate deleterious or adaptive effects, or both. In the spinal cord PDYN peptides induce flaccid paralysis of the hind limbs (Faden et al. 1985; Przewłocki et al. 1983c). This effect is also associated with neuronal cell death. In contrast, PDYN derivatives may have protective actions in seizures and cerebral injury (Przewłocka et al. 1983; Lee and Smith 1984; Tortella 1988).

Adaptation to chronic stress should enable an organism to cope with environmental demands. Opioids appear to be involved in this process. However, due to the fragmentary and sometimes inconsistent data presently available, their exact role remains unclear. Understanding of these processes may be of great importance for human pathology. Many stressors encountered by humans are, in fact, chronic in nature, and the elucidation of the mechanisms of adaptation of opioid systems to stress may have important therapeutic implications for humans who cope poorly with stress.

References

Akil H, Mayer DJ, Liebeskind JC (1976) Antagonism of stimulation-produced analgesia by naloxone, a narcotic antagonist. Science 191:961-962

Akil H, Ueda Y, Lin HL, Lewis JW, Walker JM, Shiomi H, Liebeskind JC, Watson SJ (1981) Multiple forms of β-endorphin in pituitary and brain: effect of stress. In: Takagi H, Simon E (eds) Advances in endogenous and exogenous opioids. Kodansha, Tokyo, pp 116-118

Akil H, Lin HL, Veda Y, Knoblock M, Watson SJ, Coy D (1983) Some of the α-NH_2-acetylated β-endorphin-like material in rat and monkey pituitary and brain is acetylated α- and β-endorphin. Life Sci 33 (Suppl):9-12

Akil H, Watson SJ, Young E, Lewis ME, Khachaturian H, Walker JM (1984) Endogenous opioids: biology and function. Annu Rev Neurosci 7:223-255

Akil H, Shiomi H, Matthews J (1985) Induction of the intermediate pituitary by stress. Synthesis and release of a non-opioid form of β-endorphin. Science 227:424-426

Akil H, Young E, Walker JM, Watson SJ (1986) The many possible roles of opioids and related peptides in stress-induced analgesia. Ann NYA Sci 467:140-153

Alessi N, Taylor L, Akil H (1982) Peptide F (pro-enkephalin fragment): radioimmunoassay and stress-induced changes in adrenal. Life Sci 31:1875-1878

Bloom FE, Battenberg E, Rossier J, Ling N, Guillemin R (1978) Neurons containing β-endorphin in rat brain exist separately from those containing enkephalin: immunocytochemical studies. Proc Natl Acad Sci USA 75:1591-1595

Cesselin F, Oliveras JL, Bourgoin S, Sieralta F, Michelot R, Besson JM, Hamon M (1982) Increased levels of Met-enkephalin-like material in the CSF of anesthetized cats after tooth pulp stimulation. Brain Res 237:325-338

Chance WT, White AC, Krynock GM, Rosecrans JA (1978) Conditional fear-induced decreases in the binding of (^{3}H)-Leu-enkephalin to rat brain. Brain Res 141:371-374

Christie MJ, Cheser GB, Bird KD (1981) The correlation between swim-stress induced antinociception and [^{3}H]leu-enkephalin binding to brain homogenates in mice. Pharmacol Biochem Behav 15:853-857

Civelli O, Douglass J, Goldstein A, Herbert E (1985) Sequence and expression of the rat prodynorphin gene. Proc Natl Acad Sci USA 82:4291-4295

Day R, Schafer MKH, Douglass I, Ortega MR, Watson SJ, Akil H (1988) The localization of prodynorphin and proenkephalin mRNAs in the rat adrenal gland by in situ hybridization. Soc Neurosci Abstr 14:544

Evans CJ, Weber E, Barchas JD (1981) Isolation and characterization of α-N-acetyl β-endorphin$_{1-26}$ from the rat posterior/intermediate pituitary lobe. Biophys Res Commun 102:897-904

Evans CJ, Barchas JD, Esch FS, Bohlen P, Weber E (1985a) Isolation and characterization of an endogenous C-terminal fragment of the α-neoendorphin/dynorphin precursor from bovine caudate nucleus. J Neurosci 5:1803-1807

Evans CJ, Erdelyi E, Barchas JD (1985b) Opioid peptides in the adrenal pituitary axis. Psychopharm Bull 21:466-471

Faden AJ, Molineaux CJ, Rosenberg JG, Jacobs TP, Cox BM (1985) Endogenous opioid immunoreactivity in rat spinal cord following traumatic injury. Ann Neurol 17:368-390

Giuffre KA, Udelsman R, Listwak S, Chrousos GP (1988) Effects of immune neutralization of corticotropin-releasing hormone, adrenocorticotropin, and β-endorphin in the surgically stressed rat. Endocrinology 122:306-310

Glatt ChE, Kenner JR, Long JB, Holaday JW (1987) Cardiovascular effects of dynorphin A_{1-13} in conscious rats and its modulation of morphine bradycardia over time. Peptides 8:1089-1092

Greenberg A, Dyck D, Sandler L (1984) Opponent processes, neurohormones and natural resistance. In: Fox B, Newberry B (eds) Impact of psychoendocrine systems in cancer and immunity. Hogrefe, Toronto, pp 225-258

Guillemin R, Vargo T, Rossier J, Minick S, Ling N, Rivier C, Vale W, Bloom F (1977) β-Endorphin and adrenocorticotropin are secreted concomitantly by the pituitary gland. Science 197:1367-1369

Hanbauer I, Govoni F, Majane E, Yang H-T, Costa E (1982) In vivo regulation of the release of Met-enkephalin-like peptides from dog adrenal medulla. Adv Biochem Psychopharmacol 33:63-69

Hassen AH, Feuerstein GZ, Faden AI (1982) Cardiovascular responses to opioid agonists injected into the nucleus of the tractus solitarius of anaesthetized cats. Life Sci 31:2193-2196

Hayes RL, Bennett GJ, Newlon PG, Mayer DJ (1978) Behavioral and physiological studies of non-narcotic analgesia in the rat elicited by certain environmental stimuli. Brain Res 166:69–90

Herbert E, Civelli O, Douglass J, Martens G, Rosen H (1985) Generation of diversity of opioid peptides. In: Litwack G (ed) Biochemical actions of hormones, vol 12. Academic Press, New York London, pp 1–36

Hnatowich MR, Labella FS, Kiernan K, Glavin GB (1986) Cold-restraint stress reduces [^{3}H] etorphine binding to rat brain membranes: influence of acute and chronic morphine and naloxone. Brain Res 380:107–113

Holaday JW, Hitzeman RJ, Curell I, Tortella FC, Belenky GI (1982) Repeated electroconvulsive shock or chronic morphine treatment increases the number of ^{3}H-D-Ala^2-D-Leu^5-enkephalin binding sites in rat brain membranes. Life Sci 31:2359–2362

Holaday JW (1983) Cardiovascular effects of endogenous opiate systems. Annu Rev Pharmacol Toxicol 23:541–594

Höllt V, Haarmann I, Seizinger BR, Herz A (1981) Levels of dynorphin (1–13) immunoreactivity in rat neurointermediate pituitaries are concomitantly altered with those of leucine-enkephalin and vasopressin in response to various endocrine manipulations. Neuroendocrinology 33:333–359

Höllt V, Przewłocki R, Haarmann I, Almeida OFX, Kley N, Millan MJ, Herz A (1986) Stress-induced alterations in the levels of messenger RNA coding for proopiomelanocortin and prolactin in rat pituitary. Neuroendocrinology 43:277–282

Höllt V, Haarmann MJ, Millan MJ, Herz A (1987) Prodynorphin gene expression is enhanced in the spinal cord of chronic arthritic rats. Neurosci Lett 73:90–94

Holtzman SG (1974) Behavioural effects of separate and combined administration of naloxone and d-amphetamine. J Pharmacol Exp Ther 189:51–60

Hong JS, Yoshikawa K, Kanamatsu T, McGinty JF, Sabol SL (1985) Effects of repeated electroconvulsive shock on the biosynthesis of enkephalin and concentration of dynorphin in the rat brain. Neuropeptides 5:557–560

Iadarola MJ, Civelli O, Douglass J, Naranjo JR (1986) Increased spinal cord dynorphin mRNA during peripheral inflammation. In: Holaday JW, Law PY, Herz A (eds) Progress in opioid research. NIDA Res Monogr 75:406–409

Jingami H, Nakanishi S, Imura H, Numa S (1984) Tissue distribution of messenger RNA coding for opioid peptide precursors and related RNA. Eur J Biochem 142:441–447

Kelsey SJ, Watson SJ, Akil H (1984) Changes in pituitary POMC mRNA levels. Soc Neurosci Abstr 10:359

Khachaturian H, Watson SJ, Lewis ME, Coy D, Goldstein A, Akil H (1982) Dynorphin immunocytochemistry in the rat central nervous system. Peptides 3:941–954

Kurumaji A, Takashima M, Shibuya H (1987) Cold and immobilization stress-induced changes in pain responsiveness and brain Met-enkephalin-like immunoreactivity in the rat. Peptides 8:355–359

Lasoń W, Przewłocka B, Stala L, Przewłocki R (1983) Changes in hippocampal immunoreactive dynorphin and α-neoendorphin content following intraamygladar kainic acid induced seizures. Neuropeptides 3:399–404

Lasoń W, Przewłocka B, Przewłocki R (1987) Single and repeated electroconvulsive shock differentially affects the prodynorphin and proopiomelanocortin system in the rat. Brain Res 403:301–307

Laurent S, Schmitt H (1983) Central cardiovascular effects of kappa agonists dynorphin$_{1-13}$ and ethylketocyclazocine in the anaesthetized rat. Eur J Pharmacol 96:165–169

Leander JD (1982) A kappa opioid effects increased urination in the rat. J Pharmacol Exp Ther 224:89–94

Lee NM, Smith AP (1984) Possible regulatory function of dynorphin and its clinical implications. Trends Pharm Sci 5:108–110

Lewis JW, Cannon JT, Liebeskind JC (1980) Opioid and non-opioid mechanisms of stress analgesia. Science 208:623–625

Lewis JW, Tordoff MG, Sherman JE, Liebeskind JD (1982) Adrenal medullary enkephalin-like peptides may mediate opioid stress analgesia. Science 217:557–559

Lewis JW, Baldrighi G, Watson SJ, Akil H (1985) Electrical stimulation of the nucleus tractus solitarius (NTS) causes opioid mediated analgesia in the rat. Soc Neurosci Abstr 11:637

Lewis JW, Mansour A, Khachaturian H, Watson SJ, Akil H (1987) Opioids and pain regulation. In: Akil H, Lewis JW (eds) Neurotransmitters and pain control. Karger, Basel, pp 129–159

Lightman S, Young III WS (1989) Influence of steroids on the hypothalamic corticotropin-releasing factor and proenkephalin mRNA responses to stress. Proc Natl Acad Sci 86:4306–4310

MacLennan AJ, Drugan RC, Hyson RL, Maier SF, Madden J, Barchas JD (1982) Corticosterone. A critical factor in an opioid form of stress-induced analgesia. Science 215:1530–1532

Majeed NH, Lasoń W, Przewłocka B, Przewłocki R (1986a) Brain and peripheral opioids after changes in ingestive behaviour. Neuroendocrinology 42:267–272

Majeed NH, Przewłocka B, Wedzony K, Przewłocki R (1986b) Stimulation of food intake following opioid microinjection into the nucleus accumbens septi in rats. Peptides 7:711–716

Millan MJ, Przewłocki R, Herz A (1980) A non-β-endorphinergic adenohypophyseal mechanism is essential for an analgetic response to stress. Pain 8:343–353

Millan MJ, Przewłocki R, Jerlicz MH, Gramsch C, Höllt V, Herz A (1981a) Stress-induced release of brain and pituitary β-endorphin: major role of endorphin in generation of hyperthermia not analgesia. Brain Res 208:325–328

Millan MJ, Tsang YF, Przewłocki R, Höllt V, Herz A (1981b) The influence of foot shock stress upon brain, pituitary and spinal cord pools of immunoreactive dynorphin in rats. Neurosci Lett 24:75–79

Millan MJ, Millan MH, Członkowski A, Höllt V, Pilcher CWT, Herz A, Colpaert FC (1986) A model of chronic pain in the rat: response of multiple opioid systems to adjuvant-induced arthritis. J Neurosci 6:899–906

Morley JE, Levine AS, Yim GK (1983) Opioid modulation of appetite. Neurosci Biobehav Rev 7:281–305

Nakata Y, Chang KJ, Mitchell CL, Hong JS (1985) Repeated electroconvulsive shock downregulates the opioid receptors in rat brain. Brain Res 346:160–163

Nikolarakis KE, Almeida OFX, Herz A (1986) Stimulation of hypothalamic β-endorphin and dynorphin release by corticotropin-releasing factor (in vitro). Brain Res 399:152–155

Osborne H, Przewłocki R, Höllt V, Herz A (1979) Release of β-endorphin from rat hypothalamus in vitro. Eur J Pharmacol 55:425–428

Pert CB, Bowie DL (1979) Behavioral manipulations of rats causes alterations in opiate receptor occupancy. In: Usdin E, Bunney WE, Kline NS (eds) Endorphins in mental health research. MacMillan, New York, pp 93–104

Plotnikoff NP, Miller GC (1983) Enkephalins as immunomodulators. Int J Immunopharmacol 5:437–441

Przewłocka B, Stala L, Lasoń W, Przewłocki R (1983) The effect of various opiate receptor agonists on seizure threshold in the rat. Is dynorphin an endogenous anticonvulsant? Life Sci 33 (Suppl):595–598

Przewłocka B, Lasoń W, Sumova A, Przewłocki R (1989) Differential regulation of opioid system in response to conditioned stress in rats. In: Quirion R, Jhamandas K, Gianoulakis Ch (eds) The international narcotic research conference (INRC) '89. Alan R Liss, New York, pp 319–323

Przewłocki R, Höllt V, Voight KH, Herz A (1979) Modulation of in vitro release of β-endorphin from the separate lobes of the rat pituitary. Life Sci 24:1601–1608

Przewłocki R, Millan MJ, Herz A (1980) Is β-endorphin involved in the analgesia generated by stress? In: Way EL (ed) Endogenous and exogenous opiate agonists and antagonists. Pergamon, Oxford, pp 391–394

Przewłocki R, Millan MJ, Gramsch C, Millan MH, Herz A (1982) The influence of selective adeno- and neurointermedio-hypophysectomy upon plasma and brain levels of β-endorphin and their response to stress in rats. Brain Res 242:107–117

Przewłocki R, Gramsch C, Pasi A, Herz A (1983a) Characterization and localization of immunoreactive dynorphin, β-neo-endorphin, met-enkephalin and substance P in human spinal cord. Brain Res 280:95–103

Przewłocki R, Lasoń W, Konecka A, Gramsch C, Herz A, Reid L (1983b) The opioid peptide dynorphin, circadian rhythms and starvation. Science 219:71–72

Przewłocki R, Shearman GT, Herz A (1983c) Mixed opioid/non-opioid effects of dynorphin and dynorphin-related peptides after their intrathecal injection in rats. Neuropeptides 3:233–239

Przewłocki R, Przewłocka B, Lasoń W, Garzon J, Stala L, Herz A (1985) Opioid peptides, particularly dynorphin and chronic pain. In: Besson JM, Lazorthes Y (eds) Spinal opioids and the relief of pain. INSERM, Paris, pp 159–170

Przewłocki R, Lasoń W, Silberring J, Herz A, Przewłocka B (1986) Release of opioid peptides from spinal cord of rats subjected to chronic pain. In: Holaday JW, Law P-Y, Herz A (eds) Progress in opioid research. NIDA Res Monogr 75, Washington, pp 422–425

Przewłocki R, Lasoń W, Höllt V, Silberring J, Herz A (1987) The influence of chronic stress on multiple opioid peptide systems in the rat: pronounced effects upon dynorphin in spinal cord. Brain Res 413:219

Przewłocki R, Haarman I, Nikolarakis K, Herz A, Höllt V (1988) Prodynorphin gene expression in spinal cord is enhanced after traumatic injury in the rat. Mol Brain Res 4:37–41
Przewłocki R, Majeed NH, Wedzony K, Przewłocka B (1989) The effect of stress on the opioid peptide systems in the rat nucleus accumbens. In: Van Loon GR, Kvetnansky R, McCarty R, Axelrod J (eds) Stress: neurochemical and humoral mechanisms. Gordon and Breach, New York, pp 155–161
Rossier J, Guillemin R, Bloom FE (1978) Foot-shock induced stress decreases Leu^5-enkephalin immunoreactivity in rat hypothalamus. Eur J Pharmacol 48:465–466
Schwartzberg DG, Nakane PK (1981) Pro-ACTH/endorphin antigenicities in medullary neurons of the rat. Soc Neurosci Abst 7:224
Seeger TF, Sforzo GA, Pert CB, Pert A (1984) In vivo autoradiography: visualizaton of stress-induced changes in opiate receptor occupancy in the rat brain. Brain Res 305:303–311
Shavit Y, Lewis JW, Terman G, Gale RP, Liebeskind JC (1986) Stress, opioid peptides and immune function. In: Frederickson RCA, Hendrie JN, Hingtgen HC, Aprison MH (eds) Neuroregulation of autonomic endocrine and immune systems. Martinus Nijhoff, Boston, pp 343–366
Sherman TG, Civelli O, Douglass J, Herbert E, Watson SJ (1986) Coordinate expression of hypothalamic pro-dynorphin and pro-vasopressin mRNA with osmotic stimulation. Neuroendocrinology 44:222–228
Shiomi H, Akil H (1982) Pulse-chase studies of the POMC/β-endorphin system in the pituitary of acutely and chronically stressed rats. Life Sci 31:2271–2273
Sitsen JMA, Van Ree JM, De Jong W (1982) Cardiovascular and respiratory effects of beta-endorphin in anaesthetized and conscious rats. J Cardiovasc Pharmacol 4:883–888
Smyth DG (1983) β-Endorphin and related peptides in pituitary, brain, pancreas and atrium. Br Med Bull 39:25–30
Stein EA, Hiller JM, Simon EJ (1988) Alteration in opiate receptor binding following stressful stimuli in the rat. Int Narcotics Res Conf Albi, France, Abstr, p 161
Tortella FC (1988) Endogenous opioid peptides and epilepsy: quieting the seizing brain? Trends Pharm Sci 9:366–372
Van Loon GR, Kiritsy-Roy JA, Marson L (1988a) Physiologic stress produces hypotension in opioid-pretreated rats. In: Int Narcotics Res Conf, Albi, France, Abstr, p 117
Van Loon GR, Pierzchała K, Brown LV, DR (1988b) Plasma met-enkephalin and cardiovascular responses to stress. In: Stumpe KO, Kraft K, Faden AI (eds) Opioid peptides and blood pressure control. Springer, Berlin Heildelberg New York, pp 117–126
Viveros OH, Diliberto EJJ, Hazum E, Chang KJ (1979) Opiate-like materials in the adrenal medulla: evidence for storage and secretion with catecholamines. Mol Pharmacol 16:1101–1108
Watson SJ, Richard CW, Barchas JD (1978) Adrenocorticotropin in rat brain: immunocytochemical localization in cells and axons. Science 200:1180–1182
Watson SJ, Khachaturian H, Akil H, Coy DH, Goldstein A (1982) Comparison of the distribution of dynorphin systems and enkephalin in brain. Science 218:1134–1136
Weber E, Evans CJ, Barchas JD (1981) Acetylated and nonacetylated forms of endorphin in pituitary and brain. Biochem Biophys Res Commun 103:982–989
Weber E, Roth KA, Barchas JD (1982) Immunohistochemical distribution of α-neo-endorphin/dynorphin neuronal systems in rat brain: evidence for colocalization. Proc Natl Acad Sci USA 79:3062–3066
Weber E, Evans CJ, Barchas JD (1983) Multiple endogenous ligands for opioid receptors. Trends Neurosci 6:333–336
Wybran E (1985) Enkephalins and endorphins as modifiers of the immune system: present and future. Fed Proc 44:92–96
Young EA, Akil H (1985) Corticotropin-releasing factor stimulation of adrenocorticotropin and β-endorphin release: effects of acute and chronic stress. Endocrinology 117:23–30
Young EA, Lewis J, Akil H (1986) The preferential release of β-endorphin from the anterior pituitary lobe by corticotropin releasing factor (CRF). Peptides 7:603–607

CHAPTER 15

Endogenous Opioid Systems in the Control of Pain

M.J. Millan, E. Weihe, and A.C. Czlonkowski

1 Introduction

Our current inability to satisfactorily manage pain, in particular chronic pain, represents a serious clinical problem and a continuing challenge to basic research and drug development. Morphine and similar opioids are ubiquitously employed for the treatment of severe and/or long-term pain. However, certain types of pain are refractory to the application of opioids and there are significant drawbacks associated with their usage, e.g. respiratory depression, emesis, development of tolerance, and their abuse potential. It is arguable that the more judicious and intelligent administration of opioids by thoroughly trained physicians (specializing in pain management) would substantially contribute to the overcoming of such problems. Nevertheless, the conviction remains that there is a need for alternative modes of pain therapy and that an enhanced understanding of the operation of systems physiologically involved in the response to pain should facilitate the development of novel analgetic agents.

The discovery of a multiplicity of opioid receptors and the endogenous ligands thereof, the endorphins (or endogenous opioid peptides), naturally raised questions as to their individual roles in the response to pain (Millan 1986; Zieglgänsberger 1986). In this respect, two key questions arise. Firstly, what is the role of individual endorphins in the response to pain? Secondly, what are the roles of the various opioid receptor types? The question as to the role of individual receptor types is of potentially profound therapeutic importance. Thus, morphine is a preferential ligand of μ-receptors which are believed to mediate both its desired antinociceptive and undesired secondary actions. Certain authors have advanced arguments for the existence of μ-receptor subtypes (Clark et al. 1988) and, on this basis, attempted to separate the desired analgesic properties from untoward side effects. However, in recent years, a major effort has been made to develop selective agonists (and antagonists) at non-μ-receptor sites in order to evaluate their utility as antinociceptive agents. In particular, the κ-receptor type, which does not appear to mediate either respiratory depression or positive reward (and may, therefore, lack abuse potential) and for which subtypes may exist (Traynor 1989), has been the focus of much attention.

This chapter briefly summarizes the approaches which we have employed over the past several years in elucidating the role of multiple opioid systems in the response to pain.

2 Approaches Adopted

In general, a complementary biochemical and functional approach has been adopted in order to evaluate the significance of the various opioid peptides (Table 1) and opioid receptor types (Table 2) which exist in the CNS.

The models employed were designed to examine the response of opioid systems to both acute and chronic noxious stimulation: indeed, our primary objective in recent work has been to characterize the nature of opioid systems uniquely or predominantly responding to chronic pain. As briefly summarized below, foot shock (FS) was selected as a paradigm of acute pain and polyarthritis (PA) of chronic pain. Unilateral inflammation (UI) was originally conceptualized as a less severe (and more ethically acceptable) model of chronic pain but, as shall be pointed out below, the response of opioid systems thereto is apparently very rapid following its induction. Finally, stimulation-produced analgesia was elicited from the periaqueductal grey (PAG) in a study aimed at ascertaining

Table 1. Multiple opioid peptides

Endogenous Opioid Peptides *Precursor*	 *Major processing products*
Pro-opiomelanocortin (POMC)	β-Endorphin ACTH[a] /α-MSH[a] /β-lipotropin[a]
Pro-enkephalin (PENK)	Met-enkephalin Leu-enkephalin Octapeptide (MERGL) Heptapeptide
Pro-dynorphin (PDYN)	Dynorphin A1-17, 1-8 Dynorphin B α-Neo-endorphin
"*Exogenous*" opioids	
Source	*Opioid*
Milk	Casomorphins
Frog skin	Dermorphin

[a] Non-opioids.

Table 2. Multiple opioid receptors and their ligands

	Prototypical agonist	Selective agonists	Protypical antagonists	Selective antagonists	Endogenous ligand
μ	Morphine	DAGO Morphiceptin	Naloxone Naltrexone	β-Funaltrexamine CTOP	β-Endorphin ???
δ	Enkephalin	DPDPE	Naloxone Naltrexone	ICI 174,154 Naltrindole	Enkephalin ????
κ	Ethylketocyclazocine	U50, 488H U69,593 ICI 197,067 PD 117,302	Naloxone Naltrexone	Nor-binaltorphimine	Dynorphin?

Note: Subtypes of μ- and κ-receptors may exist. The number of question marks is inversely proportional to the degree of confidence in the making of these assertions.

the identity of the opioid system responsible for its mediation. This is of particular relevance in view of the use of brain stimulation for the relief of chronic pain in man. The models were as follows:

Acute (5 min) intermittent FS as a model of 'environmentally'-induced antinociception or stress-induced antinociception (SIA).

Polyarthritis: chronic, generalized, systemic disease state involving painful, arthritic inflammation of all limbs: slow onset of symptoms (2 weeks) and peak (3–4 weeks).

Unilateral inflammation: localized, painful inflammation of a single limb of rapid onset (< 24 h) and long duration (many weeks).

Acute (5 min) electrical stimulation of the periaqueductal grey (PAG) of freely moving animals. This is termed stimulation-produced antinociception (SPA).

The data presented concerning chronic FS were generated by Przewłocki et al. (1987), employing a twice-daily 30 min FS schedule for 1 week: rats were killed 24 h following the last shock.

3 Patterns of Response of Endogenous Opioid Peptides

In Tables 3 to 5 the influence of each manipulation upon levels of opioid peptides representing each of the three families known (see Table 1) is summarized. From these data, some general patterns may be derived which we may integrate with the findings of functional studies. These data provide a framework for a discussion of the roles of various opioid peptide networks, in relation to the various opioid receptor types, in the response to and control of pain.

3.1 β-Endorphin (β-END)

Acute FS elicits a rapid rise in levels of immunoreactive (ir)- β-END in the systemic circulation (Millan and Herz 1985). The intermediate pituitary provides a variable contribution to this elevation which largely originates in the anterior lobe from where β-END is co-secreted with β-lipotropin and ACTH (Millan et al. 1981a). Indeed, synthesis and processing of pro-opioimelanocortin (POMC) is accelerated therein under acute stress (Shiomi and Akil 1982). Since levels in plasma are elevated only in PA, but not Ul or chronic FS, it appears that chronic, continuous and rather intense stimulation may be required for such a long-term rise (Millan M.J. et al. 1986, 1988a; Przewłocki 1987). This increase in PA is accompanied by a rise in mRNA levels encoding POMC in the anterior (but not intermediate) lobe which reflects a sustained enhancement in synthetic activity (Millan et al. 1986). The relationship of changes in hypophysial β-END to alterations in pain threshold is nevertheless questionable. In general, no compelling evidence for a role of hypophysial β-END in antinociceptive processes is available: indeed, blockade (by hypophysectomy) of β-END secretion into the blood neither affects the pain of PA or UI nor this model of SIA (Millan, unpublished). β-END in the circulation still awaits a functional role.

Table 3. Influence of various noxious stimuli and analgetic stimulation of the periaqueductal grey upon levels of β-endorphin, a processing product of pro-opiomelanocortin[a]

	Polyarthritis (3 weeks)	Unilateral inflammation (1 week)	Foot shock (1 week)	Foot shock (acute)	PAG Stim. (acute)
Plasma	↑	—	—	↓	—
Ant. pituitary	↑	—	↑	↓	—
Int. pituitary	—	—	↑	↓	—
Hypothalamus	—	—	—	↓	—
PAG	↓	—	?	↓	↓
Septum	—	—	?	↓	—
Thalamus	—	—	—	?	?

[a] No change (—); increase (↑); decrease (↓); not determined (?).

Notably, ir-β-END levels in plasma are not affected by PAG stimulation which effects a naloxone (NAL)-sensitive antinociception comparable to that elicited by an acute stress such as FS (Millan M.H. et al. 1986; Millan et al. 1987b). (This observation, incidentally, indicates the non-stressful nature of this model of SPA.) In contrast, both SIA and SPA are associated with a rapid depletion of pools of ir-β-END in the PAG (Millan et al. 1981a, 1987b). Further, radiofrequency destruction of brain (but not pituitary) pools of β-END attenuates both SIA and SPA, indicating a role of brain β-END in their mediation (Millan et al. 1980; Millan M.H. et al. 1986). The contention of a role of PAG pools of β-END in antinociceptive processes is reinforced by reports of (1) analgesia induced by PAG application of β-END (Luttinger et al. 1984) and (2) the inhibition of acupuncture-induced antinociception by neutralization of β-END by administration of antibodies against β-END into the PAG (Xie et al. 1983).

SPA was further shown to develop tolerance and exhibit cross-tolerance to the μ-ligand, morphine, but not the κ-ligand, U50–488H (Millan et al. 1987a). Moreover, perfusion of NAL at a low dose sufficient to block μ- but not δ- or κ-receptors could prevent SPA (Millan et al. 1987a). Thus, μ-receptors seem to mediate SPA. Evidently, one may reasonably infer that it is β-END which acts directly via μ-receptors, but definitive proof of this contention is still lacking.

A further observation focusing on PAG pools of ir-β-END was the reduction in these seen in polyarthritic rats (Millan et al. 1987d). This is of unclear functional significance but of interest in view of clinical reports of a reduction in CSF levels of ir-β-END in chronic pain patients (see Millan 1986 for review). The above observations on SPA are of clinical relevance to the findings that electrical stimulation of this periventricular region in man may elevate CSF levels of β-END (Akil et al. 1978; but see Fessler et al. 1984), and that SPA can NAL-reversibly alleviate chronic pain (Adam 1976). This mode of therapeutic pain relief was reported to develop tolerance and show cross-tolerance to morphine (Adams 1976). Such data are consistent with a role of μ-receptors in this alleviation of chronic pain in man. There is a clear parallel between these clinical reports and our experimental findings on SPA: in each case, the data may indicate a role of β-END and μ-receptors.

3.2 Met-Enkephalin (Met-ENK)

No conclusive biochemical or functional evidence has been forthcoming from either our studies (Table 4) or those of others for a role of hypophysial pools of Met-ENK in pain control (Millan 1986). Table 4 also reveals a conspicuous absence of alterations in brain pools of Met-ENK or other pro-enkephalin A (PENK) derived opioids in response to the manipulations employed. However, it is necessary to qualify the above statements as follows. (1) The major origin of circulating pools of ir-Met-ENK is probably the adrenal medulla (plus sympathetic nerve endings) which has been reported to be activated by FS in other studies (Millan and Herz 1985). The above manipulations have not been evaluated for their influence on adrenal medullary secretion of Met-ENK and other PENK-derived opioids. In fact, a role of medullary pools in the mediation of SIA has been suggested (Lewis et al. 1982) and an interaction of circulating Met-ENK with receptors external to the CNS (see Stein et al. 1988 and 1989) remains a possibility. (2) With respect to the brain, perikarya for β-END are confined to the arcuate hypothalamus and nucleus tractus solitarius, whereas those for Met-ENK are very widely distributed. Thus, the "fractional releasable pool" of β-END in such tissue studies is very much greater for β-END than that for Met-ENK. Further, studies of in vivo perfusates suggest a release of Met-ENK in response to noxious stimulation (Yaksh and Elde 1981; Cesselin et al. 1982; Kuraishi et al. 1984). (3) The broad distribution of Met-ENK neurons renders their selective lesioning virtually impossible. However, antibody studies have suggested a role of PAG pools of PENK-derived opioids in the mediation of antinociception elicited by acupuncture (Han et al. 1984). Further, inhibitors of ENK degradation have been reported to elicit antinociception via a prevention of ENK degradation in the brain (Schwarz 1981): however, in view of their lack of selectivity such data should be viewed with extreme caution. (4) Studies of gene

Table 4. Influence of various stimuli upon levels of met-enkephalin/MERGL, processing products of pro-enkephalin A, in comparison to those of leu-enkephalin[a]

	Polyarthritis (3 weeks)	Unilateral inflammation (1 week)	Foot shock (1 week)	Foot shock (acute)
		ME/MERGL		
Ant. pituitary	–	–	?	–
Int. pituitary	–	–	?	–
Hypothalamus	–	–	–	–
PAG	–	–	?	–
Thalamus	–	–	?	–
Dorsal horn	↑	–	↑[1]	↓[1]
Ventral horn	–	–	↑[1]	↓[1]
		LE		
Dorsal horn	↑	–	?[1]	?[1]
Ventral horn	–	–	?[1]	?[1]

[a] No change (–); increase (↑); decrease (↓); not determined (?); 1 signifies that no attempt was made to separate dorsal horn from ventral horn.

expression may be expected to yield novel insights into the response of discrete brain pools of PENK-derived (and other) pools of opioid peptides to pain.

In the spinal cord, noxious FS elicits a rapid depletion of Met-ENK pools (Millan 1986), suggestive of their release. More convincingly, noxious inputs such as sciatic nerve stimulation evokes an elevation in levels of ir-Met-ENK and ir-MERGL in spinal cord perfusates (Cesselin et al. 1984, 1989; Yaksh and Elde 1981) and enhances expression of the gene for PENK (Noguchi et al. 1989). Chronic noxious stimulation by FS or PA elicits a minor elevation in spinal cord pools of ir-Met-ENK, apparently confined to the dorsal horns and to the segments receiving noxious information (Cesselin et al. 1980; Millan M.J. et al. 1986). However, in UI, little effect is seen (Millan et al. 1988a), and there is considerable variability in the response of PA animals. The low magnitude and variability of this response perhaps questions its significance. In fact, it has reported that the spontaneous release of ir-Met-ENK in the spinal cord of polyarthritic animals is decreased (Bourgoin et al. 1988). Further, we have been unable to detet elevated levels of mRNA encoding PENK in the spinal cord of polyarthritic animals and only a minor rise in these have been detected in rats with UI (Millan et al. 1988a). Nevertheless, there are several points which indicate that a role of spinal pools of Met-ENK in response to long-term pain should *not* be dismissed. First, it is unwise to assume that the magnitude of changes in peptide levels correlates with their functional importance. Second, levels of ir-Met-ENK released into the spinal perfusate of PA in response to K^+ depolarization are equal to (or greater than) those of control animals: indeed, the response to noxious joint stimulation is enhanced (Bourgoin et al. 1988). Third, others have seen a minor, but significant, response of mRNA encoding PENK in UI (Iadorola et al. 1988a,b; Ruda et al. 1988). Fourth, there is a clear explanation for the low magnitude and variability in the response of pools of ir-Met-ENK. Immunohistochemical studies have revealed that only a subpopulation of dorsal horn Met-ENK containing neurons located in deep laminae IV/V show an enhanced staining in PA and UI: these co-stain for dynorphin (DYN) (Weihe et al. 1988). Met-ENK neurons in other laminae are unaffected. Fifth, no direct functional evidence for or against a role of Met-ENK in the control of nociception under conditions of chronic pain are available. Thus, under these conditions, the role of spinal pools of Met-ENK remains to be clarified. Finally, where examined, spinal cord pools of leu-enkephalin (Leu-ENK) are co-modulated with those of Met-ENK but not of DYN (Millan et al. 1988a). This is of significance in view of the presence of Leu-ENK in the sequence of both pro-DYN (PDYN) and PENK A (Millan and Herz 1985). Indeed, we have observed the co-occurrence of Leu-ENK and PENK A peptides in laminae IV neurons of polyarthritic rats (Weihe et al. 1988). Thus, there may be a coordinated response of spinal pools of Met-ENK and Leu-ENK to pain.

3.3 Dynorphin (DYN)

Whereas the receptor type via which β-END and Met-ENK exert their actions remains a matter of discussion, there is reasonable evidence that DYN may act via κ-receptors. In view of this apparent correspondence between DYN (and other

PDYN derivatives) and κ-receptors, the effects of activation and deactivation of κ-receptors can be related to findings acquired with DYN itself (Table 5).

Comparatively few effects of acute or chronic noxious stimulation upon DYN levels in the pituitary and brain have been seen. The opposite modulation of adenohypophysial DYN by acute FS and PA is intriguing but cannot, as yet, be related to antinociceptive processes (Millan et al. 1981b; Millan 1986). Similarly, the response of hypothalamic pools of DYN to acute FS is of unclear significance (Millan et al. 1981b). The response of thalamic polls of ir-DYN is of interest in light of the role of this structure in the processing of nociceptive information ascending from the spinal cord (Millan M.J. et al. 1986). Further, neurons therein showed altered responses to opioid agonists and antagonists in polyarthritic rats (Guilbaud et al. 1982). However, in neither UI nor chronic (or acute) FS are pools of ir-DYN modified in the thalamus.

In contrast to the brain and pituitary, spinal cord pools of ir-DYN are consistently modified both by acute and chronic noxious stimulation (Millan et al. 1981b; Millan M.J. et al. 1986, 1988a; Przewłocki et al. 1987). These data focus on a role of spinal, rather than brain, pools of DYN in antinociceptive processes. Correspondingly, introduction of antibodies into the spinal cord may inhibit the induction of antinociception by acupuncture (Han and Xie 1984). The ability of intrathecal injection of antibodies against DYN to modify the antinociception associated with pregnancy in rats also focuses on an antinociceptive role of DYN in the spinal cord (Sander et al. 1989). In this model, the similar action of the κ-antagonist, nor-binaltorphimine indicates that this action may be mediated by κ-receptors (Sander et al. 1988). Spinal κ-receptors and DYN have also been proposed to mediate at least certain forms of SIA based on the actions of preferential κ-antagonists, such as MR 2266, in comparison to those of NAL (Panerai et al. 1984).

However, it is the possible role of spinal DYN in the response to inflammatory conditions which has aroused most interest. In our original work, we found that polyarthritic rats display a pronounced elevation in spinal cord levels of ir-DYN, in addition to ir-α-neo-END and ir-$DYN_{1\text{-}8}$ (Millan M.J. et al. 1986). Subsequently, it proved possible to demonstrate a parallel rise in the content of mRNA expressing

Table 5. Influence of various noxious stimuli upon levels of dynorphin, a processing product of pro-dynorphin[a]

	Polyarthritis (3 weeks)	Unilateral inflammation (1 week)	Foot shock (1 week)	Foot shock (acute)
Ant. pituitary	↑	–	–	↓/–
Int. pituitary	–	–	↑	–
Hypothalamus	–	–	–	↓/–
PAG	–	–	?	–
Thalamus	↑	–	–	–
Dorsal horn	↑	↑[1]	↑[2]	↓[2]
Ventral horn	↑	–	↑[2]	↓[2]

[a] No change (-); increase (↑); decrease (↓); not determined (?); 1 signifies ipsilateral dorsal horn; 2 signifies that no attempt was made to separate dorsal and ventral horn.

PDYN, indicative of an accelerated formation of DYN (Höllt et al. 1987). Unfortunately, largely for technical reasons, no data is available to directly demonstrate a (presumed) augmentation in the release of ir-DYN. Nevertheless, the following observations were consistent with an increased activity of DYN: (1) the density of κ-receptors was slightly reduced in spinal cord; (2) U50,488H, a selective κ-agonist, had a diminished antinociceptive efficacy against noxious pressure applied to the inflamed paws; and (3) MR 2266, which possesses greater κ-antagonist activity than NAL at equivalent doses, potentiated the hyperalgesia shown by the inflamed paws to noxious pressure (Millan et al. 1987c). These data may be interpreted as follows: A sustained increase in the release of DYN would, in analogy to the effects of chronic treatment with exogenous κ-agonists, lead to a downregulation of κ-receptors, and a cross-tolerance to the κ-ligand, U50,488H. The blockade of κ-receptors evoked by MR 2266 potentiates the hyperalgesia of PA. Thus, we concluded that DYN acts via κ-receptors in the spinal cord to control pain, a conclusion of potential importance.

Perhaps inevitably, subsequent research has revealed that the situation is not so simple. The essential issue is the generality of these phenomena and whether they are characteristic of "chronic pain". In this respect, the following observations are of pertinence: (1) Noxious stimuli (e.g. FS or painful inflammation) may be sufficient to activate DYN neurons in the spinal cord. Noxious stimulation may *not*, however, be necessary. Comparable elevations in spinal cord levels of DYN (and mRNA encoding PDYN) accompany spinal trauma, intrathecal cannulation and deafferentation (Cho and Basbaum 1988; Przewłocki et al. 1988; Millan et al. 1989). These conditions might, despite our assumptions of the contrary, be painful to the animal. Nevertheless, the deafferentation data at least show that a sustained enhanced primary afferent (noxious) input is probably not essential for an increase in levels of DYN in neurons of the (dorsal horn) of the spinal cord, unless the very brief post-cut discharge is effective. (2) Studies at the single cell level, employing techniques of immunocytochemistry and in situ hybridization, have revealed that DYN neurons responding to painful inflammation are located in the superficial laminae I/II and deeper laminae IV/V known to be involved in the primary processing of noxious information (Ruda et al. 1988; Weihe et al. 1988, 1989). As mentioned above, the latter population co-stains for PENK-derived opioids. Some of these neurons may *not* be interneurons active in local processing, but projection neurons ascending to the brain. Further, there is a dramatic intensification of staining for DYN in terminals surrounding the motoneurons in the ventral horn (Weihe et al. 1989). This is suggestive of an additional role and may perhaps be related to observations of the flaccid paralysis effected by intrathecal administration of DYN (Przewłocki et al. 1983). One provocative, and as yet untested, hypothesis is that an action (not necessarily opioid-like) of DYN in the ventral horn to inhibit motoneurons would encourage immobility of the animal, thereby minimizing movement-induced pain and facilitating recovery from limb injury. One could envisage such a function as complementary and synergistic to DYN in suppressing the effects of noxious stimuli in the dorsal horn. (3) Animals with UI likewise show an elevation in the content of ir-DYN restricted to the dorsal horn ipsilateral to inflammation; this is also associated with an increase in the expression of the gene encoding PDYN (Iadorola et al. 1988a; Millan et al. 1988a). Moreover, in these animals, MR 2266 can further potentiate the hyperalgesia of the inflamed paw to noxious pressure (Millan et al. 1988a). However, UI animals show neither alterations in the density of κ-receptors nor a reduced antinociceptive potency of U50,488H, findings which differ from those obtained in the polyarthritic model. Moreover, a key observation is that the response of DYN neurons in UI animals, as regards the increase in expression of the PDYN gene and the rise in levels of DYN, is seen rapidly (within 24 h) (Iadorola et al. 1988a,b; Ruda et al. 1988). Indeed, other types of inflammatory agents (e.g. yeast, carrageenan) have now been shown to induce a very rapid (within 8 h) elevation in the spinal cord content of mRNA encoding PDYN: correspondingly, levels of ir-DYN also rapidly increase (Iadorola et al. 1988a,b). These data suggest that "chronicity" is not a critical precondition for the activation of DYN neurons in spinal cord. Upon reflection, this is not so surprising in view of the apparent ability of acute noxious FS to influence spinal pools of ir-DYN (Millan et al. 1981b). Nevertheless, these data may require a reappraisal of our concept of "chronic". (4) As mentioned above, blockade of (κ) opioid receptors enhances the hyperalgesia to noxious pressure shown by PA and UI animals to noxious pressure. A high dose of NAL (sufficient to block κ-receptors) has also been shown to enhance the spontaneous firing rate of dorsal horn neurons in PA animals (Lombard and Besson 1989). Such acute manipulations are consonant with a role of DYN in counteracting the hyperalgesia attributable to inflammation. However, the fundamental question is as to whether this role of DYN is essential. In the event of a permanent interruption of DYN activity, is the response to noxious pressure continuously enhanced? This question cannot be addressed by means of acute application of antagonists. In a recent study (Millan et al. unpublished), we examined this issue by employing a technique we have recently developed for the long-term blockade of various opioid receptor subtypes (Millan et al. 1988b; Morris et al. 1988). UI rats were infused via osmotic minipumps for 1 week with either a low or high dose of NAL sufficient to block only μ- or both μ- and κ-receptors, respectively. NAL (at both doses) led to a reduction in both food and water intake which was apparent in the former case throughout the duration of infusion. In contrast to the results in control rats, the high dose of NAL reduced paw-pressure thresholds in UI animals 48 h following implantation of the minipumps. This provided novel evidence for the role of κ-, rather than μ-, receptors in combatting inflammatory hyperalgesia. (Interestingly, as shown in Table 6, other functional actions of κ- as compared to μ-blockade were modified in UI rats.) However, by 6 days post-pump implantation the effect of NAL had subsided completely. Evidently, the animals can compensate for a long-term interruption in the activity of opioid systems. The most probable mechanism is the progressive engagement of a non-opioid antinociceptive mechanism: conceivably, it is *this* system which is the key component of the response to *chronic* pain. In confirmation of the above interpretation, following removal of pumps, rats which had been undergoing κ-receptor blockade showed normalized thresholds, that is thresholds on the inflamed paw were elevated to a level not different from those of the uninflamed paw. This change may reflect a supersensitivity to an endogenous opioid (presumably DYN). This last experiment suggests that the role of opioid systems, specifically DYN/κ-receptors, in combating chronic pain may *not* be indispensable.

Table 6. Alterations in the actions of naloxone in unilateral long-term inflammatory pain[a]

	Hypophagia	Hypodipsia	Loss of body weight	Hypothermia
0.16	↑	–	–	–
3.0	↑	–	↑	–

	Hypoactivity	Blockade of Morphine	Blockade of U 69593	Hyperalgesia of inflamed paw
0.16	–	–	–	–
3.0	↑	–	–	↑

[a] Unchanged potency (–); ↑ enhanced potency (↑).

4 Opioid Receptor Subtypes: κ-Agonists as Analgesics?

From the above data (summarized in Table 7), we may infer a role of β-END and μ-receptors (though not necessarily β-END via μ-receptors) in the mediation of SPA elicited from the PAG, and certain forms of SIA. There is no convincing evidence for a role in the response to longer-term pain. DYN, via κ-receptors, might mediate (other) forms of acute SIA but is not involved in SPA from the PAG. In fact, DYN and κ-receptors seem to play an important role in the response to inflammatory and, possibly, chronic pain: this function may not, however, be indispensable. Currently, no compelling evidence for a physiological role of δ-receptors in pain control is available. These conclusions concern CNS-localized opioid receptors and, as discussed by Stein (this Vol.), novel data provide evidence for additional roles of peripheral μ; δ- and κ-receptors in antinociceptive processes in inflamed tissue. The physiological significance of such mechanisms awaits clarification.

In view of the above data focusing on μ- and κ-receptors, it is not surprising that great efforts have been made to characterize the comparative pharmacology of μ- and κ-mediated antinociception.

For several years, two factors have been considered as decisive in regards to the role of the various receptor types. Firstly, the site of action in the brain or spinal cord and, secondly, the nature of the noxious stimulus (Millan 1986). Whereas μ-

Table 7. Role of endogenous opioid peptides and receptors in pain control[a]

	Polyarthritis	Unilateral inflammation	Acute foot shock	PAG Stimulation
β-Endorphin	?	N	Y[1]	Y
Met-enkephalin	?	?	?	N
Dynorphin	Y	Y	Y[2]	N
μ	N	N	Y[1]	Y
δ	N	N	?	N
κ	Y	Y	Y[2]	N

[a] Y, major role; N no role; ? uncertain. Note: for met-enkephalin/δ-receptors, data are still very unclear.
1,2 certain forms of stress-induced analgesia.

receptors certainly act against all types of noxious stimulus via actions exerted in the brain and spinal cord, it was claimed that κ-agonists act only at segmental sites and against non-thermal stimuli (Tyers 1980; Wood et al. 1981; von Voigtlander et al. 1983; Schmauss and Yaksh 1984). In retrospect, this distinction was too attractive and straightforward to have been correct. More recent work employing genuinely selective κ-agonists has shown that κ-receptors do mediate antinociception at both the segmental *and* suprasegmental level against noxious thermal and non-thermal stimuli (Porrecca et al. 1984; Belknap et al. 1987; Czlonkowski et al. 1987; Calcagnetti et al. 1988; Fleetwood-Walker et al. 1988; Millan et al. 1988c, 1989; Tiseo et al. 1988). Complementary electrophysiological (Headley et al. 1984; Parsons et al. 1986) and behavioural (Millan et al. 1988c, 1989) analyses indicate that the previous failure to detect actions of κ-agonists vs noxious heat, in rats, reflected the use of intense heat stimuli and the failure to employ algesiometric tests allowing for convincing comparisons. For example, it is somewhat absurd to automatically attribute differences between diverse tests [e.g. the abdominal writhing test (chemical), paw withdrawal to pressure test and the latency to jump in the hot-plate test] simply to stimulus quality. Employing moderate, carefully intensity-matched stimuli, κ- (as well as μ-) agonists can be shown to be equipotent against heat and pressure stimuli applied to the tail, as judged by withdrawal latency in concious animals (Millan 1989) and the response of dorsal horn neurons in extracellular recordings (Parsons et al. 1986). However, in behavioural studies, the action of κ-agonists against heat, in contrast to pressure, is ***intensity***-dependent; increasing and decreasing stimulus intensity is associated with, respectively, a decrease and increase in potency. This difference is ***not*** apparent with μ-ligands. The mechanism underlying this intensity dependence remains to be explored. However, this distinction between heat and pressure suggests that the population of κ-receptors controlling heat sensitivity, although revealing an identical pharmacology, may differ from that controlling pressure sensitivity. This possibility receives support from the following observations: (1) in PA animals, blockade of κ-receptors modifies (enhances) the response to noxious pressure but not heat (Millan 1986); (2) long-term blockade of κ-receptors in normal rats leads to a supersensitivity to the antinociceptive actions of κ-ligands against pressure but not heat (Millan et al. 1988b; Morris et al. 1988).

Thus, both μ- and κ-receptors can mediate antinociception against thermal and non-thermal stimuli via actions in the brain and spinal cord. Further, in inflamed tissue, additional actions (Stein et al. 1988) appear to be expressed in the periphery. In view of these observations, it is understandable that κ-ligands have attracted attention as potential novel analgesics. However, there are potential difficulties with their use.

In recent work, we observed a very clear qualitative distinction between μ- and κ-agonists. μ-Agonists potently raise the threshold to vocalize in response to electrical stimulation of the tail, whereas all κ-agonists are completely inactive (Millan 1989). We suggest this test may reflect the differences between μ- and to κ-ligands upon mood. μ-Receptors mediate positive reward, whereas an activation of κ-receptors is strongly aversive (Mucha and Herz 1985; Pfeiffer et al. 1987; Shippenberg et al. 1987). Evidently, κ-ligands would have the advantage of a low-abuse potential, but their aversiveness would severely limit their utility. Further, it is arguable that the "positive" influence on mood is the decisive component of the antinociceptive efficacy of μ-opioids. The absence of this action would correspondingly be reflected in a much lesser ability of κ-agonists to relieve pain. It is this distinction which may (in contrast to reflexive tests) be monitored in the vocalization test. Nevertheless, we reasoned that in subjects undergoing pain the rewarding properties of analgesics may be different: if κ-ligands *do* elicit antinociception, this may counteract any aversive properties. Indeed, we recently showed that in UI rats, κ-agonists are *not* aversive but motivationally neutral (Shippenberg et al. 1988). If a similar situation exists in patients suffering from chronic pain, this finding would have major implications for the future potential use of κ-analgetics in the management of chronic pain.

Further problems associated with κ-analgesics would be their sedative (and diuretic) actions. However, there are two possible avenues via which such difficulties might be circumvented. Firstly, since the aversive and sedative actions of κ-ligands are mediated centrally, a κ-agonist not passing the blood-brain barrier would not be expected to elicit such effects. The discovery of peripheral analgesia mediated by κ-ligands (Stein et al. 1989) suggests the possibility of useful antinociceptive actions exerted in inflamed tissue via a κ-ligand acting exclusively in the periphery. Secondly, very recent data suggest the existence of κ-receptor subtypes (Traynor 1989). It should be feasible to discover κ-agonists acting selectively at particular subtypes. Although their functional correlates are not known, it is possible that thereby the desirable (antinociceptive) actions of κ-receptors could be dissociated from their undesirable effects.

Finally, mention should briefly be made of δ-receptor-mediated antinociception. δ-Receptors apparently also act peripherally (Stein et al. 1989). In regard to their actions in the CNS, a role of brain-localized δ-receptors in the mediation of antinociception has long been a subject of controversy: however, the balance of evidence is definitely in favour of such an action. Spinal δ-receptors are active against both heat and pressure and, probably, chemical stimuli (Millan 1986; Vaught et al. 1988; Sanchez-Blanchez and Garzon 1989). The lack of availability of δ-ligands capable of penetrating the blood-brain barrier, together with an absence of data indicating a physiological role in pain control (see above), has unfortunately curtailed an interest in these ligands. Further, δ-ligands seem to share some of the secondary actions of μ-opioids such as respiratory depression and mediation of reward (Shippenberg et al. 1987). Thus, although intrathecal administration of δ-agonists has proven effective in patients tolerant to morphine (Onofrio and Yaksh 1983), regrettably little progress has been made to encourage development of novel δ-agonists as analgesics.

5 Concluding Comments

Over the past 10 years, much has been learned concerning the molecular biology, biochemistry, electrophysiology, anatomy, physiology and pharmacology of the opioids. As described herein, recent work has succeeded in identifying certain opioid systems involved in the response to pain and antinociceptive processes. Further, important advances in the pharmacology of the antinociception mediated by various opioid receptor subtypes have been achieved. It is somewhat disappointing to reflect that this research has yielded rather little in terms of tangible improvements in pain therapy. It is to be hoped that in the not too distant future our enhanced knowledge of the role of opioid and non-opioid systems in the response to pain may ultimately lead to more concrete improvements in the management of pain. Moreover, this may be an appropriate moment to extend our horizons and consider the potential significance of non-opioid networks implicated in pain transmission, modulation and suppression. Therein lies, perhaps, the future of pain research.

References

Adams JE (1976) Naloxone reversal of analgesia produced by brain stimulation in the human. Pain 2:161-168

Akil H, Richardson DE, Barchas JD, Li CH (1978) Appearance of β-endorphin-like immunoreactivity in human ventricular fluid upon analgesic electrical stimulation. Proc Natl Acad Sci USA 75:5170-5172

Belknap P, Danielson W, Lawsen SE, Nooedewier B (1987) Selective breeding for levorphanol-induced antinociception on the hot-plate assay: communalities in mechanism of action with morphine, pentazocine, ethylketocyclazocine, U-50488H and clonidine in mice. J Pharmacol Exp Ther 241:477-481

Bourgoin S, Le Bars D, Clot AM, Hamon M, Cesselin F (1988) Spontaneous and evoked release of met-enkephalin-like material from the spinal cord of arthritic rats in vivo. Pain 32:107-114

Calcagnetti DJ, Helmstetter FJ, Fanselow MS (1988) Analgesia produced by centrally administered DAGO, DPDPE and U50 488H in the formalin test. Eur J Pharmacol 153:117-122

Cesselin F, Montastruc JL, Gros C, Bourgoin S, Hamon M (1980) Met-enkephalin levels and opiate receptors in the spinal cord of chronic suffering rats. Brain Res 191:289-293

Cesselin F, Oliveras JL, Bourgoin S, Sierralta F, Michelot R, Besson J-M, Hamon M (1982) Increased levels of met-enkephalin-like material in the CSF of anaesthetized cats after tooth pulp stimulation. Brain Res 237:325-328

Cesselin F, Bourgoin S, Artaud F, Hamon M (1984) Basic and regulatory mechanisms of in-vitro release of met-enkephalin from the dorsal horn of the rat spinal cord. J Neurochem 43:763-773

Cesselin F, Bourgoin S, Clot AM, Hamon M, Le Bars D (1989) Segmental release of met-enkephalin-like material from the spinal cord of rats elicited by noxious thermal stimuli. Brain Res 484:71-77

Cho HJ, Basbaum AI (1988) Increased staining of immunoreactive dynorphin cell bodies in the deafferented spinal cord of the rat. Neurosci Lett 84:125-130

Clark JA, Houghten R, Pasternak GW (1988) Opiate binding in calf thalamic membranes: a selective μ_1 binding assay. Mol Pharmacol 34:309-317

Czlonkowski A, Millan MJ, Herz A (1987) The selective κ-opioid agonist, U50 488H, produces antinociception in the rat via a supraspinal action. Eur J Pharmacol 142:183-184

Fessler RG, Brown FD, Rachlin JR, Mullan S (1984) Levels of β-endorphin in cerebrospinal fluid after electrical brain stimulation: artifact of contrast infusion? Science 224:1017-1019

Fleetwood-Walker SM, Hope PJ, Mitchell R, El-Yassir N, Molony S (1988) The influence of opioid receptor subtypes on the processing of nociceptive inputs in the spinal dorsal horn of the cat. Brain Res 451:213-226

Guilbaud G, Benoist JM, Gautron M, Kayser V (1982) Effects of systemic naloxone upon ventrobasal thalamus neuronal responses in arthritic rats. Brain Res 59-66

Han JS, Xie GX (1984) Dynorphin: important mediator for electroacupuncture analgesia in the spinal cord of the rabbit. Pain 18:367-376

Han JS, Fei H, Zhou ZF (1984) Met-enkephalin-Arg^6-Phe^7-like immunoreactive substances mediate electroacupuncture analgesia in the periaqueductal gray of the rabbit. Brain Res 322:289-296

Headley PM, Parsons CG, West DC (1984) Comparison of mu, kappa and sigma preferring agonists for effects on spinal nociceptive and other responses in rats. Neuropeptides 5:249-252

Höllt V, Haarman I, Millan MJ, Herz A (1987) Prodynorphin gene expression is enhanced in the spinal cord of chronic arthritic rats. Neurosci Lett 73:90–94

Iadorola MJ, Brady LS, Draisci G, Dubner R (1988a) Enhancement of dynorphin gene expression in spinal cord following experimental inflammation: stimulus specificity, behavioural parameters and opioid receptor binding. Pain 35:313–326

Iadorola MJ, Douglas J, Civelli O, Naranjo JR (1988b) Differential activation of spinal cord dynorphin and enkephalin neurones during hyperalgesia: evidence using cDNA hybridization. Brain Res 455:205–212

Kuraishi Y, Sugimoto M, Hamada T, Kyanoki Y, Takagi H (1984) Noxious stimuli and met-enkephalin release from nucleus reticularis gigantocellularis. Brain Res Bull 12:123–127

Lewis JW, Tordoff MG, Sherman JR, Liebeskind JC (1982) Adrenal medullary enkephalin-like peptides may mediate opioid stress analgesia. Science 217:557–559

Lombard MC, Besson J-M (1989) Electrophysiological evidence for a tonic activity of the spinal cord intrinsic opioid systems in a chronic pain model. Brain Res 477:48–56

Luttinger D, Hernandez DE, Nemeroff CB, Prange AJ (1984) Peptides and nociception. Int Rev Neurobiol 25:185–272

Millan MH, Millan MJ, Herz A (1986) Depletion of central β-endorphin blocks midbrain stimulation produced analgesia in the rat. Neuroscience 18:641–649

Millan MJ (1986) Multiple opioid systems and pain. Pain 26:303–349

Millan MJ (1989) κ-Opioid receptor mediated antinociception in the rat: I comparative actions of μ- and κ-opioids against noxious thermal pressure and electrical stimuli. J Pharmacol Exp Ther 251:334–341

Millan MJ, Herz A (1985) The endocrinology of the opioids. Int Rev Neurobiol 26:1–84

Millan MJ, Gramsch C, Przewłocki R, Höllt V, Herz A (1980) Lesions of the hypothalamic arcuate nucleus produce a temporary hyperalgesia and attenuate the analgesia evoked by stress. Life Sci 27:1513–1523

Millan MJ, Przewłocki R, Jerlicz MH, Gramsch C, Höllt V, Herz A (1981a) Stress-induced release of brain and pituitary β-endorphin: major role of endorphins in generation of hyperthermia not analgesia. Brain Res 208:325–328

Millan MJ, Tsang Y, Przewłocki R, Höllt V, Herz A (1981b) The influence of foot shock stress upon brain, pituitary and spinal cord pools of immunoreactive dynorphin in rats. Neurosci Lett 24:75–79

Millan MJ, Millan MH, Czlonkowski A, Höllt V, Pilcher CWT, Herz A, Colpaert FC (1986) A model of chronic pain in the rat: response of multiple opioid systems to adjuvant induced arthritis. J Neurosci 6:899–906

Millan MJ, Czlonkowski A, Herz A (1987a) Evidence that μ-opioid receptors mediate stimulation-produced analgesia in the freely moving rat. Neuroscience 22:885–896

Millan MJ, Czlonkowski A, Millan MH, Herz A (1987b) Activation of periaqueductal grey pools of β-endorphin by analgetic electrical stimulation of the freely moving rat. Brain Res 407:199–203

Millan MJ, Czlonkowski A, Pilcher CWT, Almeida OFX, Millan MH, Colpaert FC, Herz A (1987c) A model of chronic pain in the rat: functional correlates of alterations in the activity of opioid systems. J Neurosci 7:77–87

Millan MJ, Morris B, Colpaert FC, Herz A (1987d) High resolution neuroanatomical approach reveals alterations in multiple opioid systems in the periaqueductal grey of polyarthritic rats. Brain Res 416:349–353

Millan MJ, Czlonkowski A, Morris B, Höllt V, Stein C, Arendt R, Huber A, Herz A (1988a) Inflammation of the hind-limb as a model of unilateral localized inflammatory pain: influence on multiple opioid systems in the spinal cord of the rat. Pain 35:299–312

Millan MJ, Morris B, Herz A (1988b) Antagonist-induced opioid upregulation I. Characterization of supersensitivity to selective μ- and κ-agonists. J Pharmacol Exp Ther 247:721–728

Millan MJ, Stein C, Weihe E, Nohr D, Höllt V, Czlonkowski A, Herz A (1988c) Dynorphin and κ-receptors in the control of nociception: response to peripheral inflammation and the pharmacology of κ-antinociception. In: Besson JM, Guilbaud G (eds) The arthritic rat as a model of clinical pain? Elsevier, Amsterdam, pp 153–171

Millan MJ, Czlonkowski A, Lipkowski A, Herz A (1989) κ-Opioid receptor mediated antinociception in the rat: II supraspinal in addition to spinal sites of action. J Pharmacol Exp Ther 251:342–350

Morris BJ, Millan MJ, Herz A (1988) Antagonist induced opioid-upregulation. II Regionally specific modulation of μ, δ and κ binding sites revealed by quantitative autoradiography. J Pharmacol Exp Ther 247:729–736

Mucha R, Herz A (1985) Motivational properties of kappa and mu opioid receptor agonists studied with place and taste preference conditioning procedures. Psychopharmacology 86:247–280
Noguchi K, Morita Y, Kiyami H, Sato M, Ono K, Tohyama M (1989) Proenkephalin gene expression in the rat spinal cord after noxious stimuli. Mol Brain Res 5:227–234
Onofrio BM, Yaksh TL (1983) Intrathecal delta-receptor produces analgesia in man. Lancet 1:1386–1387
Panerai AE, Martini A, Sacerdote P, Mantegazza P (1984) κ-Receptor antagonist reverses 'non-opioid' stress-induced analgesia. Brain Res 304:153–156
Parsons CG, West DC, Headley PM (1986) Similar actions of kappa and mu agonists on spinal nociceptive reflexes in rats and their reversibility by naloxone. NIDA Res Monogr 75:461–464
Pfeiffer A, Brantl V, Herz A, Emrich HM (1987) Psychotomimesis mediated by κ-opioid receptors. Science 233:774–776
Porrecca F, Mosberg HI, Hurst R, Hruby VJ, Burks TF (1984) Roles of mu, delta and kappa opioid receptors in spinal and supraspinal mediation of gastrointestinal transit effects and hot-plate analgesia in the mouse. J Pharmacol Exp Ther 230:341–348
Przewłocki R, Stala L, Greczek M, Shearman GT, Przewlocka B, Herz A (1983) Analgesic effects of μ-δ- and κ-opiate agonists at the spinal level. Life Sci 33 Suppl 1:649–652
Przewłocki R, Lason W, Höllt V, Silberring J, Herz A (1987) The influence of chronic stress on multiple opioid peptide systems in the rat: pronounced effects upon dynorphin in spinal cord. Brain Res 413:213–219
Przewłocki R, Haarman I, Nikolarakis K, Herz A (1988) Prodynorphin gene expression is enhcanced after traumatic injury in the cat. Mol Brain Res 4:37–41
Ruda MA, Iadorola MJ, Cohen LV, Young WS (1988) In situ hybridization histochemistry and immunocytochemistry reveal an increase in spinal dynorphin biosynthesis in a rat model of peripheral inflammation and hyperalgesia. Proc Natl Acad Sci USA 85:622–626
Sanchez-Blanchez P, Garzon J (1989) Evaluation of δ receptor mediation of supraspinal opioid analgesia by in vivo protection against the β-funaltrexamine antagonist effect. Eur J Pharmacol 159:9–23
Sander HW, Portoghese PS, Gintzler AR (1988) Spinal κ-opiate receptor involvement in the analgesia of pregnancy: effects of intrathecal norbinaltorphimine, a κ-selective antagonist. Brain Res 474:343–347
Sander HW, Kream RM, Gintzler AR (1989) Spinal dynorphin involvement in the analgesia of pregnancy: effects of intrathecal antibody against dynorphin. Eur J Pharmacol 159:205–209
Schmauss C, Yaksh TL (1984) In vivo studies on spinal opiate receptor systems mediating antinociception II: pharmacological profiles suggesting a differential association of mu, delta and kappa receptors with visceral chemical and cutaneous thermal stimuli in the rat. J Pharmacol Exp Ther 228:1–12
Schwarz JC (1981) Metabolism of enkephalins and the inactivating neuropeptides concept. Trends Neurosci 216:604–616
Shiomi H, Akil H (1982) Pulse-chase studies of the POMC/β-endorphin system in the pituitary of acutely and chronically stressed rat. Life Sci 31:2185–2188
Shippenberg TS, Bals-Kubik R, Herz A (1987) Motivational properties of opioids: evidence that an activation of δ-receptors mediates reinforcement processes. Brain Res 436:234–238
Shippenberg TS, Stein C, Huber A, Millan MJ, Herz A (1988) Motivational effects of opioids in an animal model of prolonged inflammatory pain: alterations in the effects of κ-but not μ-agonists. Pain 35:179–186
Stein C, Millan MJ, Yassouridis A, Herz A (1988) Antinociceptive effects of μ- and κ-agonists in inflammation are enhanced by a peripheral opioid receptor specific mechanism. Eur J Pharmacol 155:255–264
Stein C, Millan MJ, Shippenberg TS, Peter K, Herz A (1989) Peripheral opioid receptors mediating antinociception in inflammation: evidence for involvement of μ· δ-, κ-receptors. J Pharmaçol Exp Ther 248:1269–1275
Tiseo PJ, Geller EB, Adler MW (1988) Antinociceptive action of intracerebroventricularly administered dynorphin and other opioid peptides in the rat. J Pharmacol Exp Ther 246:449–453
Traynor J (1989) Subtypes of the κ-opioid receptor: fact or fiction. Trends Pharmacol 10:52–53
Tyers MB (1980) A classification of opiate receptors that mediate antinociception in animals. Br J Pharmacol 69:503–512
Vaught J, Mathiasen JR, Raffa RB (1988) Examination of the involvement of supraspinal and spinal mu and delta receptors in analgesia using the mu receptor deficient CXBK mouse. J Pharmacol Exp Ther 245:13–15

Von Voigtlander PF, Lahti RA, Ludens JH (1983) U50,488H: a selective and structurally novel non-mu (kappa) opioid agonist. J Pharmacol Exp Ther 224:7–12
Weihe E, Millan MJ, Leibold A, Nohr D, Herz A (1988) Co-localization of pro-enkephalin and pro-dynorphin derived opioid peptides in laminae IV/V neurones of the spinal cord revealed in arthritic rats. Neurosci Lett 85:187–192
Weihi E, Millan MJ, Höllt V, Nohr D, Herz A (1989) Induction of the gene encoding prodynorphin by experimentally induced arthritis enhances staining for dynorphin in the spinal cord of rats. Nueroscience 1:77–95
Wood PL, Rackham A, Richard J (1981) Spinal analgesia: comparison of the mu agonist morphine and the kappa agonist ethylketocyclazocine. Life Sci 28:2119–2225
Xie GX, Han JS, Höllt V (1983) Electroacupuncture analgesia blocked by microinjection of anti-beta-endorphin antiserum into periaqueductal grey of the rabbit. Int J Neurosci 18:287–292
Yaksh TL, Elde RP (1981) Factors governing release of methionine-enkephalin-like immunoreactivity from mesencephalon and spinal cord of the cat in vivo. J Neurophysiol 46:1056–1075
Zieglgänsberger W (1986) Central control of nociception: In: Mountcastle VB, Bloom FE, Geiger SR (eds) Handbook of physiology – the nervous system, vol 4. Williams & Wilkins, Baltimore, pp 581–645

CHAPTER 16

Opioid Interactions with Other Neuropeptides in the Spinal Cord: Relevance to Nociception

M. Satoh and Y. Kuraishi

1 Introduction

The spinal dorsal horn is endowed with neural structures rich in a variety of neurotransmitters and neuromodulators, including peptides, monoamines, and amino acids. Many of these are concentrated in the superficial dorsal horn.

The peptides substance P (SP), somatostatin (SS), calcitonin gene-related peptide (CGRP), cholecystokinin (CCK), and vasoactive intestinal polypeptide (VIP) are found in nerve terminals of small-diameter myelinated or nonmyelinated primary afferents of the dorsal roots. These small fibers are known to terminate in the superficial layers (laminae I, II, and III) of the dorsal horn from where they convey nociceptive information from the periphery. We proposed that SP and SS transmit mechanically and thermally induced nociceptive information, respectively, from the skin to the spinal dorsal horn (Kuraishi et al. 1985). This hypothesis was supported by recent data showing that the intrathecal administration of antisera against SP or SS produced, in rats, modality-specific antinociception in the paw-pressure and hot-plate tests, respectively (Nance et al. 1987; Ohno et al. 1988).

Many local spinal interneurons are also peptidergic, containing SP, SS, CCK, enkephalins (ENK), and dynorphins (DYN). However, the physiological roles of neuropeptides in the spinal dorsal horn interneurons in nociception and antinociception remain to be elucidated.

It is therefore fruitful to investigate interactions between neurotransmitters or neuromodulators such as amino acids, amines, and neuropeptides in the spinal dorsal horn for a better understanding of the mechanisms of transmission and regulation of nociceptive information at the gateway to the central nervous system. In this chapter, we describe several examples of interactions between neuropeptides (ENK-SP, DYN-SP, CGRP-SP, and SP-SS) in the spinal dorsal horn, which have been investigated in our laboratory.

2 Experimental Approaches

Three types of experimental approach were used in our studies of neuropeptidergic interactions in the spinal cord, a brief description of which follows:

2.1 In Situ Perfusion of the Dorsal Horn

Pentobarbital-anesthetized rabbits were decerebrated at the rostral end of the diencephalon, and a unilateral push-pull cannula was inserted into the dorsal horn (Kuraishi et al. 1983). The tips of the push-pull cannulae reached into the dorsolateral area where there are high concentrations of SP, SS, and CGRP. After basal conditions were established, the dorsal horn was perfused with artificial cerebrospinal fluid, and samples of perfusate collected for the assay of SP and SS at 20-min intervals. Samples were also collected during the intermittent application of noxious mechanical (pinch) and thermal (radiant heat) stimuli to the clipped skin of a hind leg, ipsilateral to the perfusion site.

2.2 In Vitro Superfusion of the Dorsal Horn

The dorsal halves of the cervical and lumbar enlargements of the spinal cords of rats were minced and superfused with Krebs Ringer bicarbonate containing peptidase inhibitors (bestatin 10 μM, captopril 5 μM, leupeptin 1 μM, and chymostatin 1 μM). Superfusates were collected at 6-min intervals and assayed for SP and Met-ENK. Stimulation of the tissues in vitro was achieved by addition of 50 mM K^+ or 0.5 μM capsaicin (K^+ nonselectively stimulates all neurons, whereas capsaicin selectively acts on unmyelinated primary afferents; both agents cause the release of neurotransmitters in a Ca^{2+}-dependent manner).

2.3 Behavioral Measurements of Nociception

Nociceptive responsiveness of the hind paws of male rats to thermal stimulation was measured using the hot-plate test; licking of the hind paw was used as the nociceptive response. Nociceptive thresholds of the hind paw for mechanical stimulation were measured by the paw-pressure method using an analgesimeter, in which struggling was considered as nociceptive response. The responses were measured following intrathecal injections of neuropeptides into the subarachnoid space through the foramen intervertebrale, between L3 and L4 (Satoh et al. 1983).

3 Interactions Between Met-ENK and SP

Considerable evidence has demonstrated opioid-induced modulation of the release of SP in the dorsal horn. In a pioneering study, Jessell and Iversen (1977) showed that opiates and opioid peptide analogs inhibited the in vitro K^+-evoked release of SP from slices of the spinal trigeminal nucleus but not from those of the substantia nigra. Inhibitory actions of morphine (10 μM) and [D-Ala^2]-Met-ENK (3 μM) were antagonized by naloxone (NAL; 1 μM). Subsequently, using the method of in vivo superfusion of the cat spinal cord, Yaksh et al. (1980) showed that superfusion with morphine (500 μM) produced a NAL-reversible suppression of sciatic nerve stimulation-induced SP release into the ventricular space.

Using the method of in situ perfusion of a localized area in the rabbit dorsal horn, we demonstrated that the noxious pinch-evoked release of SP was depressed by systemic administration of morphine (10 mg/kg); this was partially antagonized by NAL (50 μM) applied topically to the dorsal horn (Kuraishi et al. 1983). Further, the depressant effect of systemic morphine was also reduced by topical application of prazosin (50 μM), an α_1-adrenergic blocker. Therefore, the inhibition of systemic morphine (in a large dose of 10 mg/kg) of the noxious pinch-evoked release of SP was suggested to be mediated by opioid receptors and the descending noradrenergic system in the dorsal horn of the spinal cord. Indeed, local application of norepinephrine (10 μM) or morphine (10 μM) to the perfusion site inhibited the pinch-evoked release of SP; the actions of each of these drugs was reversed by yohimbine (an α_2-adrenergic blocker) and NAL, respectively (Hirota et al. 1985a; Kuraishi et al. 1985).

Met-ENK, which has a higher affinity for δ-receptors than for μ-receptors (McKnight et al. 1983; Paterson et al. 1984), is concentrated in the superficial layers of the dorsal horn (LaMotte and de Lanerolle 1983; Khachaturian et al. 1985), but morphological investigations, including immunocytochemistry, have not provided clear evidence for axo-axonic connections, i.e., for presynaptic SP and Met-ENK interactions (Hunt et al. 1980; Cuello et al. 1982).

Does endogenous Met-ENK inhibit SP release from primary afferent terminals? Noxious peripheral stimuli-evoked SP release into dorsal horn perfusates was greatly inhibited by topically applied Met-ENK (10 μM), an action that was antagonized by NAL (10 μM) (Hirota et al. 1985a). In addition, the release of SP in intrathecal superfusates evoked by sciatic nerve stimulation was suppressed by an ENK analog [D-Ala^2,D-Leu^5]-ENK (DADLE; 10 μM), and the inhibitory action was antagonized by NAL (100 μM; Yaksh 1986). These findings indicate that Met-ENK and its analog suppress the release of SP evoked by peripheral stimulation. However, two questions still remain: (1) whether the evoked release of SP occurs mainly from the primary afferent terminals, and (2) whether endogenous Met-ENK released from the dorsal horn can suppress SP release.

Capsaicin is thought to act selectively to evoke the release of SP from primary afferent terminals (Gamse et al. 1979; Helke et al. 1981). Such an assumption is supported by findings that in experiments on the in vitro release of SP from dorsal-half slices of the rat spinal cord, the capsaicin (0.5 μM)-evoked release of SP from slices prepared from rats with chronic (2 weeks) bilateral sciatic nerve sections was negligible. When the same dose of capsaicin (0.5 μM) was applied to dorsal-half slices of the spinal cord pretreated with 50 mM K^+, the release of SP was only 85% of that seen following the K^+-stimulation (Table 1). However, concurrent application of tetrodotoxin (TTX; 0.3 μM) and capsaicin significantly increased the capsaicin-induced release of SP. Since TTX blocks most axonal conduction, these data suggest that capsaicin acts indirectly, via the release of primary afferent transmitter(s), to activate inhibitory interneurons in the dorsal horn. In this context, it is interesting that Met-ENK (1 μM) abolished the potentiating effect of TTX; the

Table 1. Effects of various agents on the capsaicin-evoked release of immunoreactive substance P (SP) from dorsal-half slices of the rat spinal cord

Agent	Concentration (μM)	n	Control (%)
TTX	0.3	4	140.7 ± 10.7[a]
TTX + Met-ENK	1	4	88.8 ± 15.0[b]
TTX + DYN	1	4	123.7 ± 5.0
NAL	10	6	129.6 ± 7.9[a]
Mr2266	10	4	120.1 ± 10.1

[a] $P < 0.05$ compared with control.
[b] $P < 0.05$ compared with TTX alone.

simultaneous application of NAL (10 μM, but not 1 μM) produced a slight, but significant, increase in the capsaicin-induced release of SP (see Table 1); and capsaicin (10 μM) facilitated the in vitro release of Met-ENK from slices of the dorsal-half of the spinal cord, an increase which was blocked by the concurrent application of an SP antagonist, [D-Arg^1,D-Pro^2,D-Trp^{7,9},Leu^{11}]-SP (10 μM). Furthermore, SP releases Met-ENK into in vitro and in vivo superfusates of the rat spinal cord (Tang et al. 1983; Cesselin et al. 1984). Taken together, these observations suggest that SP released from primary afferent terminals evokes the release of Met-ENK from intrinsic spinal neurons; these neurons most probably exert a partially inhibitory influence on the SP-containing primary afferents (Fig. 1). If an SP-ENK inhibitory feedback interaction applies in vivo, NAL would be expected to increase peripheral stimuli-induced release of SP from the dorsal horn. Indeed, Yaksh (1986) reported just such an effect in in vivo superfusions of the spinal cord of anaesthetized cats.

Although a major source of dorsal horn SP may be the primary afferent axons, many SP-immunoreactive neurons are also found in the superficial dorsal horn

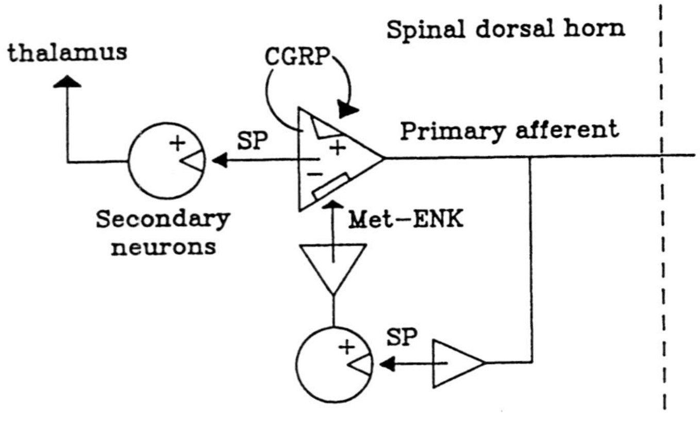

Fig. 1. Hypothetical interactions between Met-enkephalin (Met-ENK) or calcitonin gene-related peptide (*CGRP*) with substance P (*SP*) in the spinal dorsal horn. +, Excitatory; −, inhibitory

(Barber et al. 1979; Hunt et al. 1981). Adjuvant-induced inflammation enhances expression of the gene coding for preprotachykinin A (from which SP is derived) in the spinal cord (Minami et al. 1989). This finding suggests that SP-containing spinal neurons receive nociceptive information. In addition, about 95% of SP-immunoreactive neurons in the superficial layers of the rat spinal cord also show Met-ENK-Arg^6-Gly^7-Leu^8 immunoreactivity (Senba et al. 1988), and about 50% of these neurons stain positively for SP (Katoh et al. 1988; Senba et al. 1988). However, the functional significance of the coexistence of SP and ENK in the same neurons of the dorsal horn remains to be elucidated.

4 Interactions Between DYN and SP

DYN, which has a high affinity for κ-opioid receptors (Corbett et al. 1982), is present in the superficial layers of the dorsal horn (Khachaturian et al. 1982, 1985), where κ-receptors are also concentrated (Slater and Patel 1983). It is relevant to note that spinal DYN systems are activated during inflammation (Millan et al. 1985, 1986, 1988; Iadarola et al. 1988). Immunoreactive DYN has also been detected in guinea-pig primary afferents. About one-half of dorsal root ganglion cells contain both immunoreactive SP and DYN (Gibbins et al. 1987). The question thus arises as to whether DYN-κ-receptor systems modulate SP release from primary afferent terminals.

The topical application of 10 μM DYN (in contrast to Met-ENK) did not significantly inhibit the noxious pinch-evoked release of SP in the rabbit dorsal horn in in situ perfusion experiments (Hirota et al. 1985a). The κ-selective agonist U50488H (1000 μM) also did not significantly inhibit the in vivo evoked release of SP into spinal superfusates in studies of the cat (Yaksh 1986). Furthermore, in in vitro perfusion experiments with dorsal-half slices of the rat spinal cord, DYN (1 μM) did not significantly inhibit the facilitatory effect of TTX on the capsaicin-evoked release of SP, and the (relatively) κ-selective antagonist Mr2266 (1 and 10 μM) was also without effect (Table 1). Thus, spinal DYN systems do not seem to exert an inhibitory influence on the SP-containing primary afferents. This view is consistent with the electrophysiological observations of Fleetwood-Walker et al. (1988) who showed that the iontophoretic application of the κ-agonists, DYN_{1-13} and U50488H, onto the substantia gelatinosa does not inhibit the nociceptive responses of the cat spinocervical tract neurons, although SP-containing primary afferents terminate densely in the former area.

5 Interactions Between CGRP and SP

CGRP is present in many small and large dorsal root ganglion cells (Ju et al. 1987), and is concentrated in the superficial layers of the dorsal horn (Gibson et al. 1984; Wiesenfeld-Hallin et al. 1984; Skofitsch and Jocobowitz 1985). The major origin of CGRP in the dorsal horn is suggested to be the primary afferent neurons (Chung et al. 1988; Gibson et al. 1984).

The intrathecal injection of CGRP lowers the nociceptive threshold for mechanical stimulation in rats (Oku et al. 1987). In contrast, injection of anti-CGRP serum via the same route increases the nociceptive threshold (Kuraishi et al. 1988); the increase in the nociceptive threshold for both mechanical and thermal stimuli was more marked in adjuvant-inflamed animals than in noninflamed ones (Kawamura et al. 1989). In addition, adjuvant-induced inflammation has been shown to increase the content of CGRP in the dorsal root ganglia and the axonal flow of this peptide in primary afferents of the rat (Kuraishi et al. 1989). Adjuvant-induced inflammation also increases the capsaicin-evoked release of CGRP from dorsal-half slices of the lumbar cord in rats (Nanayama et al. 1989). Taken together, these findings suggest the involvement of primary afferent CGRP in the transmission or regulation of nociceptive information in the dorsal horn.

With regard to interactions between CGRP and SP, it is interesting that most immunoreactive SP primary afferent neurons also show CGRP immunoreactivity (Wiesenfeld-Hallin et al. 1984; Lee et al. 1985; Gibbins et al. 1987; Ju et al. 1987). In behavioral experiments on rats, Wiesenfeld-Hallin et al. (1984) showed that the intrathecal injection of CGRP (5.3 nmol) prolonged the duration of intrathecally applied SP-induced (7.4 nmol) aversive responses (e.g., caudally directed biting and scratching); however, CGRP alone had no behavioral effect. It was suggested that CGRP might inhibit the degradation of SP in the spinal cord (Greves et al. 1985). However, Oku et al. (1987) did not observe any potentiation of the SP-induced aversive responses by CGRP, but instead found that CGRP (5 nmol) alone produced significant hyperalgesia in response to mechanical noxious stimuli which activate the SP-containing primary afferents in the rat. The lack of an interaction between intrathecal CGRP and SP was also noted in mice (Gamse and Saria 1986). Thus, Oku et al. (1987) examined the effect of CGRP on the in vitro release of SP from slices of the dorsal-half of the spinal cord. CGRP (1 μM) significantly potentiated the capsaicin-evoked release of SP but did not alter its basal release. This data provided the first evidence that when two neuropeptides coexist, the release of one (SP) may be increased by the other (CGRP; see Fig. 1). In this context, Oku et al. (1988) demonstrated that CGRP promoted the influx of $^{45}Ca^{2+}$ into synaptosomes of the rat spinal dorsal horn in a concentration-dependent manner (from 1 nM to 1 μM). Further, CGRP (100 nM) inhibited the K^+-evoked efflux of $^{45}Ca^{2+}$ from synaptosomes of the spinal dorsal horn. This action of CGRP would seem to be specific to the spinal cord since CGRP does not alter the influx or efflux of $^{45}Ca^{2+}$ into, or from, synaptosomes prepared from rat cerebral cortex, a brain area with reportedly few CGRP binding sites. These findings make it plausible that CGRP increases Ca^{2+} influx into synaptosomes of the spinal dorsal horn by interacting with its own receptors; the release of SP from the primary afferent terminals is thus enhanced.

6 Interactions Between SP and SS

SS is distributed in the superficial layers of the spinal dorsal horn (Hökfelt et al. 1976; Gibson et al. 1981) and is present in small and intermediate dorsal root ganglion cells (Hökfelt et al. 1976; Price 1985). SS-immunoreactive dorsal root ganglion cells have been demonstrated to innervate the skin and muscle (Molandar et al. 1987), suggesting the involvement of SS-containing primary afferents in somatosensory processing.

The SS present in primary afferents appears to be involved in the transmission of information related to thermal, but not mechanical, noxious stimulation. Thus, peripheral noxious heat stimulation increases the in situ release of SS from the (rabbit) dorsal horn (Kuraishi et al. 1985), whereas noxious mechanical stimuli do not increase SS release (cat), as measured by the immunoprobe method (Morton et al. 1988). Furthermore, the intrathecal injection of anti-SS serum has been shown to inhibit thermal, but not mechanical, nociceptive responses in the rat (Ohno et al. 1988). Also supportive of the stimulus specificity of this response is the observation that the intrathecal injection of SS (0.006–1 nmol) facilitates the thermal, but not mechanical, nociceptive responses of rats (Hirota et al. 1985b; Wiesenfeld-Hallin 1986).

The intrathecal injection of SP inhibits the nociceptive responses to thermal stimulation (Doi and Jurna 1981; Hirota et al. 1985b; Rodriguez and Rodriguez 1987; Ryall and Pini 1987). As mentioned above, spinal ENK systems are implicated in such antinociception. However, when noxious mechanical stimuli are applied, intrathecal injections of SP produce hypernociception, but not antinociception (Hirota et al. 1985b; Matsumura et al. 1985; Wiesenfeld-Hallin 1986). Since the mode-specific antinociception of intrathecal SP cannot be explained by opioid mediation alone, we examined the effects of SP on SS release from the dorsal horn. As shown in Fig. 2, the topical application of SP (10 μM) abolished the peripheral heat-evoked release of SS without affecting basal release. This result suggests that intrathecal SP-induced thermal antinociception may involve an inhibition of SS release from primary afferents.

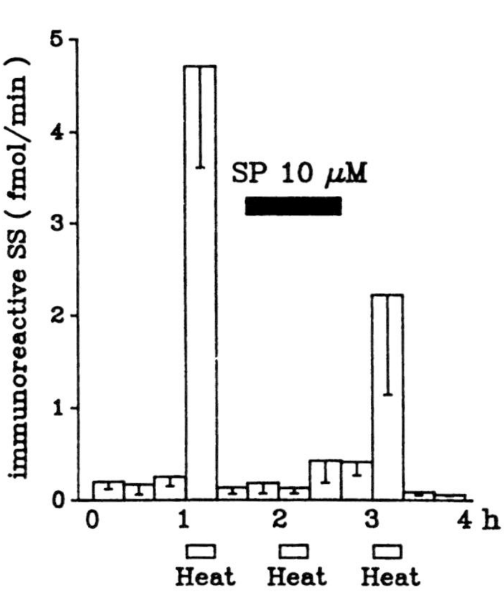

Fig. 2. Release of immunoreactive somatostatin (*SS*) from rabbit dorsal horn (perfused in situ) in response to heat stimulation of the skin, and its suppression by substance P (*SP*) applied to the perfusate. Radiant heat was applied to the shaved skin until the subcutaneous temperature was elevated to about 50°C; it was subsequently cooled (by air) until the temperature was reduced to 40°C. This procedure was repeated for the period (20 min) indicated by *open bars*. SP was applied to the dorsal horn for the period indicated by the *black bar*. Values represent mean and SEM from 3 experiments

7 Conclusions

There is evidence for several types of neuropeptidergic interactions in the dorsal horn of the spinal cord, e.g., between ENK and SP, DYN and SP, CGRP and SP, and SP and SS. The main focus of this chapter has been on the negative and positive regulation of neuropeptide (SP and SS) release from the primary afferent terminals which transmit nociceptive information from the periphery to the spinal cord. Opioid peptides like ENK and DYN produce the former at the spinal dorsal horn and play important roles in antinociception through their presynaptic actions. However, this does not exclude the possible involvement of postsynaptic mechanisms in their antinociceptive actions. On the other hand, as mentioned in the introduction, there are many neurotransmitters and neuromodulators which seem to be involved in nociception and its inhibition, and more investigations into the neurochemical bases of nociception and antinociception are warranted.

Acknowledgements. We thank Emeritus Prof. H. Takagi for encouragement, and our colleagues N. Hirota, R. Oku, S. Kawabata, and T. Nanayama for their valuable contributions to this work. Studies by the authors were partly supported by grants-in-aid for scientific research from the Japanese Ministry of Education, Sciences and Culture, and the Naito Foundation Subsidy for Promotion of Specific Research Projects.

References

Barber RP, Vaughn JE, Slemmon JR, Salvaterra PM, Roberts E, Leeman SE (1979) The origin, distribution and synaptic relationships of substance P axons in rat spinal cord. J Comp Neurol 184:331–351

Cesselin F, Bourgoin S, Artaud F, Hamon M (1984) Basic and regulatory mechanisms of in vitro release of Met-enkephalin from dorsal zone of the rat spinal cord. J Neurochem 43:763–773

Chung K, Lee WT, Carlton SM (1988) The effects of dorsal rhizotomy and spinal cord isolation on calcitonin gene-related peptide-labeled terminals in the rat lumbar dorsal horn. Neurosci Lett 90:27–32

Corbett AD, Paterson SJ, McKnight AT, Magnan J, Kosterlitz HW (1982) Dynorphin$_{1-8}$ and dynorphin$_{1-9}$ are ligands for the κ-subtype of opiate receptor. Nature (London) 299:79–81

Cuello AC, Priestley JV, Matthews MR (1982) Localization of substance P in neuronal pathways. In: Porter R, O'Connor M (eds) Substance P in the nervous system. Pitman, London, pp 55–83

Doi T, Jurna I (1981) Intrathecal substance P depresses the tail-flick response – antagonism by naloxone. Naunyn-Schmiedberg's Arch Pharmacol 317:135–139

Fleetwood-Walker SM, Hope PJ, Mitchell R, El-Yassir N, Molony V (1988) The influence of opioid receptor subtypes on the processing of nociceptive inputs in the spinal dorsal horn of the cat. Brain Res 451:213–226

Gamse R, Saria A (1986) Nociceptive behavior after intrathecal injections of substance P, neurokinin A and calcitonin gene-related peptide in mice. Neurosci Lett 70:143–147

Gamse R, Molnar A, Lembeck F (1979) Substance P release from spinal cord slices by capsaicin. Life Sci 25:629–636

Gibbins IL, Furness JB, Costa M (1987) Pathway-specific patterns of the co-existence of substance P, calcitonin gene-related peptide, cholecystokinin and dynorphin in neurons of the dorsal root ganglia of the guinea-pig. Cell Tissue Res 248:417–437

Gibson SJ, Polak JM, Bloom SR, Wall PD (1981) The distribution of nine peptides in rat spinal cord with special emphasis on the substantia gelatinosa and on the area around the central canal (lamina X). J Comp Neurol 201:65–79

Gibson SJ, Polak JM, Bloom SR, Sabate IM, Mulderry PM, Ghatei MA, McGregor GP, Morrison JFB,

Kelly JS, Evans RM, Rosenfeld MG (1984) Calcitonin gene-related peptide immunoreactivity in the spinal cord of man and of eight other species. J Neurosci 4:3101-3111

Greves PL, Nyberg F, Terenius L, Hökfelt T (1985) Calcitonin gene-related peptide is a potent inhibitor of substance P degradation. Eur J Pharmacol 115:309-311

Helke CJ, Jacobowitz DM, Thoa NB (1981) Capsaicin and potassium evoke substance P release from the nucleus tractus solitarius and spinal trigeminal nucleus in vitro. Life Sci 29:1779-1785

Hirota N, Kuraishi Y, Hino Sato Y, Satoh M, Takagi H (1985a) Met-enkephalin and morphine, but not dynorphin, inhibit noxious stimuli-induced release of substance P from rabbit dorsal horn in situ. Neuropharmacology 24:567-570

Hirota N, Ohara H, Kuraishi Y, Satoh M, Takagi H (1985b) The role of substance P and somatostatin on pain transmission in the spinal dorsal horn. Jpn J Pharmacol 39 Suppl:73P

Hökfelt T, Elde R, Johansson D, Luft R, Nilsson G, Arimura A (1976) Immunohistochemical evidence for separate populations of somatostatin-containing and substance P-containing primary afferent neurons in the rat. Neuroscience 1:131-136

Hunt SP, Kelly JS, Emson PC (1980) The electron microscopic localisation of methionine-enkephalin within the superficial layers (I and II) of the spinal cord. Neuroscience 5:1871-1890

Hunt SP, Kelly JS, Emson PC, Kimmel JR, Miller RJ, Wu J-Y (1981) An immunohistochemical study of neuronal populations containing neuropeptides or gamma-aminobutyrate within the superficial layers of the rat dorsal horn. Neuroscience 6:1883-1898

Iadarola MJ, Douglass J, Civelli O, Naranjo R (1988) Differential activation of spinal cord dynorphin and enkephalin neurons during hyperalgesia: evidence using cDNA hybridization. Brain Res 455:205-212

Jessell TM, Iversen LL (1977) Opiate analgesics inhibit substance P release from rat trigeminal nucleus. Nature (London) 268:549-551

Ju G, Hökfelt T, Brodin E, Fahrenkrug J, Fischer JA, Frey P, Elde RP, Brown JC (1987) Primary sensory neurons of the rat showing calcitonin gene-related peptide immunoreactivity and their relation to substance P-, somatostatin-, galanin-, vasoactive intestinal polypeptide- and cholecystokinin-immunoreactive ganglion cells. Cell Tissue Res 247:417-431

Katoh S, Hisano S, Kawano H, Kagotani Y, Daikoku S (1988) Light- and electron-microscopic evidence of costoring of immunoreactive enkephalins and substance P in dorsal horn neurons of rat. Cell Tissue Res 253:297-303

Kawamura M, Kuraishi Y, Minami M, Satoh M (1989) Antinociceptive effect of intrathecally administered antiserum against calcitonin gene-related peptide on thermal and mechanical noxious stimuli in experimental hyperalgesic rats. Brain Res 497:199-203

Khachaturian H, Watson SJ, Lewis ME, Coy D, Goldstein A, Akil H (1982) Dynorphin immunocytochemistry in rat central nervous system. Peptides 3:941-954

Khachaturian H, Lewis ME, Schafen MKH, Watson SJ (1985) Anatomy of the CNS opioid systems. Trends Neurosci 8:111-119

Kuraishi Y, Hirota N, Sugimoto M, Satoh M, Takagi H (1983) Effects of morphine on noxious stimuli-induced release of substance P from rabbit dorsal horn in vivo. Life Sci 33 Suppl 1:693-696

Kuraishi Y, Hirota N, Sato Y, Hino Y, Satoh M, Takagi H (1985) Evidence that substance P and somatostatin transmit separate information related to pain in the spinal dorsal horn. Brain Res 325:294-298

Kuraishi Y, Nanayama T, Ohno H, Minami M, Satoh M (1988) Antinociception induced in rats by intrathecal administration of antiserum against calcitonin gene-related peptide. Neurosci Lett 92:325-329

Kuraishi Y, Nanayama T, Ohno H, Fujii N, Otaka A, Yajima H, Satoh M (1989) Calcitonin gene-related peptide increases in the dorsal root ganglia of adjuvant arthritic rat. Peptides 10:447-452

LaMotte CC, de Lanerolle NC (1983) Ultrastructure of chemically defined neuron systems in the dorsal horn of the monkey. II. Methionine-enkephalin immunoreactivity. Brain Res 274:51-63

Lee Y, Kawai Y, Shiosaka S, Takami K, Kiyama H, Hillyard CJ, Girgis S, MacIntyre I, Emson PC, Tohyama M (1985) Coexistence of calcitonin gene-related peptide and substance P-like peptide in single cells of the trigeminal ganglion of the rat: immunohistochemical analysis. Brain Res 330:194-196

Matsumura H, Sakurada T, Hara A, Sakurada S, Kisara K (1985) Characterization of the hyperalgesic effect induced by intrathecal injection of substance P. Neuropharmacology 24:421-426

McKnight AT, Corbett AD, Kosterlitz HW (1983) Increase in potencies of opioid peptides after peptidase inhibition. Eur J Pharmacol 86:393–402
Millan MJ, Millan MH, Pilcher CWT, Czlonkowski A, Herz A, Colpaert FC (1985) Spinal cord dynorphin may modulate nociception via κ-opioid receptor in chronic arthritic rats. Brain Res 340:156–159
Millan MJ, Millan MH, Czlonkowski A, Höllt V, Pilcher CWT, Herz A, Colpaert FC (1986) A model of chronic pain in the rat: response of multiple opioid systems to adjuvant-induced arthritis. J Neurosci 6:899–906
Millan MJ, Czlonkowski A, Morris B, Stein C, Arendt R, Huber A, Höllt V, Herz A (1988) Inflammation of the hind limb as a model of unilateral, localized pain: influence on multiple opioid systems in the spinal cord of the rat. Pain 35:299–312
Minami M, Kuraishi Y, Kawamura M, Yamaguchi T, Masu Y, Nakanishi S, Satoh M (1989) Enhancement of preprotachykinin A gene expression by an adjuvant-induced inflammation in the rat spinal cord: possible involvement of substance P-containing spinal neurons in nociception. Neurosci Lett 98:105–110
Molandar C, Ygge J, Dalsgaard C-J (1987) Substance P-, somatostatin- and calcitonin gene-related peptide-like immunoreactivity and fluoride resistant acid phosphatase activity in relation to retrogradely labeled cutaneous, muscular and visceral primary sensory neurons in the rat. Neurosci Lett 74:37–42
Morton CR, Hutchison WD, Hendry IA (1988) Physiological stimuli releasing somatostatin and calcitonin gene-related peptide with substance P in the spinal cord. Regul Peptides 22 Suppl:129P
Nanayama T, Kuraishi Y, Ohno H, Satoh M (1989) Capsaicin-induced release of calcitonin gene-related peptide from dorsal horn slice is enhanced in adjuvant arthritic rats. Neurosci Res 6:569–572
Nance PW, Sawynok J, Nance DM (1987) Modality specific analgesia produced by intrathecal anti-substance P antibody. In: Henry JL, Couture R, Cuello AC, Pelletier G, Quirion R, Regoli D (eds) Substance P and neurokinins. Springer, Berlin Heidelberg New York, pp 282–284
Ohno H, Kuraishi Y, Minami M, Satoh M (1988) Modality-specific antinociception produced by intrathecal injection of anti-somatostatin antiserum in rats. Brain Res 474:197–200
Oku R, Satoh M, Fujii N, Otaka A, Yajima H, Takagi H (1987) Calcitonin gene-related peptide promotes mechanical nociception by potentiating release of substance P from the spinal dorsal horn in rats. Brain Res 403:350–354
Oku R, Nanayama T, Satoh M (1988) Calcitonin gene-related peptide modulates calcium mobilization in synaptosomes of rat spinal dorsal horn. Brain Res 475:356–360
Paterson SJ, Robson LE, Kosterlitz HW (1984) Opioid receptors. In: Udenfriend S, Meienhofer J (eds) The peptides: analysis, synthesis, biology, vol 6. Academic Press, New York London Orlando, pp 147–189
Price J (1985) An immunohistochemical and quantitative examination of dorsal root ganglion neuronal subpopulations. J Neurosci 5:2051–2059
Rodriguez RE, Rodriguez FD (1987) Spinal and supraspinal substance P antinociception: synergistic interaction. In: Henry JL, Couture R, Cuello AC, Pelletier G, Quirion R, Regoli D (eds) Substance P and neurokinins. Springer, Berlin Heidelberg New York, pp 331–333
Ryall RW, Pini AJ (1987) A depressant action of substance P on spinal nociceptive neurons in the rat and cat. In: Henry JL, Couture R, Cuello AC, Pelletier G, Quirion R, Regoli D (eds) Substance P and neurokinins. Springer, Berlin Heidelberg New York, pp 239–240
Sastry BR (1979) Presynaptic effects of morphine and methionine-enkephalin in the feline spinal cord. Neuropharmacology 18:367–375
Satoh M, Yasui M, Fujibayashi K, Takagi H (1983) Bestatin potentiates analgesic effect of intrathecally administered dynorphin in rats. IRCS Med Sci 11:965–966
Senba E, Yanaihara C, Yanaihara N, Tohyama M (1988) Co-localization of substance P and Met-enkephalin-Arg^6-Gly^7-Leu^8 in the intraspinal neurons of the rat, with special reference to the neurons in the substantia gelatinosa. Brain Res 453:110–116
Skofitsch G, Jocobowitz DM (1985) Calcitonin gene-related peptide: detailed immunohistochemical distribution in the central nervous system. Peptides 6:721–745
Slater P, Patel S (1983) Autoradiographic localization of opiate κ receptors in the rat spinal cord. Eur J Pharmacol 92:159–160
Tang J, Chou J, Yang HYT, Costa E (1983) Substance P stimulates the release of Met^5-enkephalin-Arg^6-Phe^7 and Met^5-enkephalin from rat spinal cord. Neuropharmacology 22:1147–1150

Wiesenfeld-Hallin Z (1986) Substance P and somatostatin modulate spinal cord excitability via physiologically different sensory pathways. Brain Res 372:172–175

Wiesenfeld-Hallin Z, Hökfelt T, Lundberg JM, Forssmann WG, Reinecke M, Tschopp FA, Fischer JA (1984) Immunoreactive calcitonin gene-related peptide and substance P coexist in sensory neurons to the spinal cord and interact in spinal behavioral responses of the rat. Neurosci Lett 52:199–204

Yaksh TL (1986) The central pharmacology of primary afferents with emphasis on the disposition and role of primary afferent substance P. In: Yaksh TL (ed) Spinal afferent processing. Plenum, New York, pp 165–195

Yaksh TL, Jessel TM, Gamse R, Mudge AW, Leeman SE (1980) Intrathecal morphine inhibits substance P release from mammalian spinal cord in vivo. Nature (London) 286:155–157

CHAPTER 17

Opioid Analgesia at Peripheral Sites

C. Stein

1 Introduction

Traditionally, opiates are considered to exert analgesic effects through actions within the central nervous system (CNS). Recently, however, the notion that opioid antinociception can be brought about by activation of opioid receptors located peripherally, i.e. outside the CNS, has gained more widespread recognition. One of the earliest reports was that of Wood in 1855 who showed that morphine elicited analgesic effects when applied topically to "painful areas" in the periphery. Numerous clinical and experimental reports of similar observations have occurred since. However, most of the former are merely anecdotal and many of the latter were discounted because of their lack of demonstration of the principal criterion for opioid receptor-mediated effects, namely naloxone (NAL) reversibility. Moreover, as already discussed in Wood's paper, the question as to whether these effects result from a truly peripheral rather than from a central site of action (e.g. via uptake of the agent into the circulation and transport to the CNS) has been raised repeatedly. In this chapter, studies examining peripheral antinociceptive effects of opioids in animals and humans will be reviewed.

2 Animal Studies

2.1 *Aspects to Consider*

The following aspects of each paper were evaluated in detail and comprise the headings in Table 1.

1. Exclusion of Central Effects. Several strategies can be pursued to demonstrate that an effect is mediated at a peripheral site as opposed to within the CNS. One approach is to use charged, lipophobic agonists or antagonists with minimal capability to cross the blood-brain barrier. Examples are quaternary alkaloids (e.g. N-methyl morphine, N-methyl NAL) or polar peptides. However, penetration of the blood-brain barrier by these compounds cannot be entirely excluded. Moreover, the introduction of polarizing residues almost invariably results in a decreased affinity for the receptor (Kosterlitz and Waterfield 1975). Consequently, to achieve effects comparable to their tertiary congeners, the dose of quaternary compounds will have to be increased, which, in turn, will increase the probability of penetration of the

Table 1. Animal studies

No.	Species	Assay	Agonist	RO	Antagonism		DR	ST
1	Rat	Paw pressure	Morphine Met-Enk Leu-Enk Pentazocine NAL Nalorphine	1 5 6 4 3 2	No,	NAL was agonist	Yes (not shown)	Ø
2	Mouse	Abdominal constriction	Normorphine Morphine Codeine Oxymorphine Pentazocine Ketocyclazocine Met-Enk Leu-Enk Levorphanol Dextrorphan	5 4 9 3 10 1 7 8 2 6	Yes, No,	NAL vs morphine Met-Enk NAL vs pentazocine ketocyclazocine	Yes (not shown)	Yes
3	Rat	Paw pressure	Morphine Met-Enk Levorphanol Dextrorphan	1 2 3 4	Yes,	NAL at low dose but agonist at higher dose	Yes morphine only	Yes
4	Mouse	Writhing (hot plate)	Morphine N-methyl-morphine	1 2	Yes,	NAL N-methyl-nalorphine	Yes	Ø
5	Rat	Paw pressure (hot plate)	NAL N-methyl-NAL MR 2266 MR 2267	1 1 2 3	No,	see agonists	Bell-shaped	Yes
6	Rat	Paw pressure (hot plate)	NAL N-methyl-NAL	2 1	No,	see agonists	Bell-shaped	Ø

Inflammatory agents and time of peak effect	Further characterization	Exclusion of systemic effects	Reference
PGE_2 (3–6h) Isoprenaline Ca-ionophore $BaCl_2$ Db-cAMP (PGI_2 Carrageenan, not shown)	Data incomplete on agents other than PGE_2	Morphine Met-Enk Ipsi- vs contra-lateral injection	Ferreira and Nakamura (1979)
Acetic acid (6 min) Acetylcholine (s)	Tolerance to morphine No cross-tolerance to Leu-Enk Insufficient NAL used vs pentazocine and ketocyclazocine	Morphine Met-Enk Leu-Enk iv. vs i.p.	Bentley et al. (1981)
PGE_2 (3–6 h)	Peripheral component of systemic morphine	Morphine Ipsi- vs contra-lateral injection	Ferreira et al. (1982)
Acetic acid (15 min)	Peripheral mechanism effective in writhing, but not in hot plate test	By analysis of CNS penetration of radioactive compounds	Smith et al. (1982)
Carrageenan (3–5 h)	No antiinflammatory effects (no data presented)	N-methyl-NAL vs morphine	Rios and Jacob (1982)
Carrageenan (4–6 h)		N-methyl-NAL vs morphine (i.pl. vs s.c. agonists)	Rios and Jacob (1983)

Table 1. (*Continued*)

No.	Species	Assay	Agonist	RO	Antagonism		DR	ST
7	Rat	Paw pressure	N-methyl-nalorphine N-methyl-morphine Morphine	NR		Ø	Yes	Ø
8	Rat	Paw heat	Fentanyl EKC Levorphanol Dextrorphan	NR		Ø	Yes, levor-phanol only	Yes
9	Mouse	Writhing (hot plate tail flick)	BW 443 C Morphine Pethidine D-propoxyphene	2 1 2 2	Yes,	NAL N-methyl-nalorphine	Yes, in writhing U-shaped in heat assays	Ø
10	Rat	Paw Formalin Tail flick	Morphine EKC	Ø	Yes,	NAL NAL-methyl-bromide	Yes	Ø
11	Rat	Paw pressure	Fentanyl		Yes,	NAL stereospec.	Yes	Yes
12	Rat	Paw pressure	Morphine U-50,488H	1 2	Yes,	NAL stereospec.	Yes	Yes
13	Rat	Paw pressure	DAMGO DPDPE U-50,488H Morphine (+)-Morphine Tifluadom (−)-Tifluadom	1 4 4 3 5 2 5	Yes,	NAL stereospec. CTAP ICI 174,864 Nor-BNI	Yes	Yes, mor-phine tiflu-adom

Inflammatory agents and time of peak effect	Further characterization	Exclusion of systemic effects	Reference
Carrageenan (3 h) PGE_2	Oral application No tolerance to peripheral effects No antiinflammatory effects No GI effects of N-methyl-nalorphine Peripheral component of systemic morphine	Quaternary compounds	Ferreira et al. (1984)
Carrageenan (2–4 h)		Local vs systemic application	Joris et al. (1987)
Phenyl-p-benzoquinone acetic acid (10–20 min)	Inhibition of GI motility Greater potency in writhing compared with heat assays	Polar peptide Quaternary antagonists applied i.p. and i.c.v.	Follenfant et al.(1988)
Formalin (20–25 min)	Peripheral component of systemic EKC but not morphine	Systemic vs i.c.v. application Quaternary antagonist	Abbott (1988)
Freund's adjuvant (4–6 days)		Local vs. systemic application	Stein et al. (1988a)
Freund's adjuvant (4–6 days)	Peripheral component of systemic morphine and U-50,488H	Local vs systemic application	Stein et al. (1988a)
Freund's adjuvant (4–6 days)	Three distinct receptor types (μ,δ,κ) active	Local vs systemic application	Stein et al. (1989)

Abbreviations: DR, dose-response relationship; ST, stereospecificity; NR, not reported; NAL: naloxone; Enk, enkephalin; Ø, not examined; RO, rank order of potency: 1 = highest, 10 = lowest; No., number of study; i.c.v., intracerebroventricular; i.v., intravenous; i.p., intraperitoneal; i.pl., intraplantar; s.c. subcutaneous; EKC, ethylketocyclazocine; CTAP, D-Phe-Cys-Tyr-D-Trp-Arg-Thr-Pen-Thr-NH_2; nor-BNI = nor-binaltorphimine.

blood-brain barrier. Alternatively, demonstration of effects after local, rather than systemic administration of equivalent doses of agonists or antagonists, provides evidence for a peripheral site of action.

2. Criteria for Opioid Receptor-Mediated Effects. Antagonism by NAL or related compounds (e.g. naltrexone) is the *conditio sine qua non* of opioid receptor-mediated effects. Dose dependency and stereospecificity of agonist and antagonist effects are equally well-established criteria.

3. Multiplicity of Receptor Types. Early studies have examined differences in the potency of different nonselective agonists and antagonists to identify the type of receptor involved in a given effect. In recent years, however, such identification has been facilitated considerably by the availability of highly selective agonists and antagonists for each of the different opioid receptor types (see Kosterlitz this Vol.).

4. Algesiometry. Traditional algesiometric tests measure an animal's latency of response to stimuli (e.g. heat, pressure) applied acutely to noninjured tissue (e.g. tail, paw). Recently, clinically more relevant models of tissue injury (e.g. inflammation) have become increasingly popular. In these models, either the animal's latency of response to superimposed stimuli (e.g. pressure) or behavioral alterations following application of the irritant alone (e.g. abdominal constrictions, writhing) are measured.

5. Induction of Tissue Injury. The intraplantar injection of various agents (e.g. carrageenan, different prostaglandins, Freund's adjuvant) into one or both hind-paws of rats produces a localized inflammatory response that is confined to the treated paw(s). This response is typically comprised of the classical symptoms: hyperalgesia, hyperemia, increased local temperature and swelling. Intraperitoneal administration of irritants (e.g. acetic acid) presumably also produces an inflammatory reaction, although this has not yet been characterized in terms of the symptoms outlined above.

6. Further Characterization. The development of tolerance to peripheral opioid effects, antiinflammatory and gastrointestinal motility effects of opioids have been addressed in some studies.

2.2 Review of Behavioral Studies in Animals

Table 1 summarizes 13 behavioral studies which have examined peripheral antinociceptive effects of opioids. To differentiate central and peripheral actions, polarized compounds were used in six and local vs systemic application was used in seven. The essential issue of antagonism by NAL was examined in 11 papers. NAL (or related compounds) functioned as an antagonist at peripheral receptors in the majority of cases. However, four studies (1,3,5,6) demonstrated agonist effects: In three of those (3,5,6) "bell-shaped" dose-response curves were found, i.e., antagonist effects at low doses and agonist effects at higher doses. Such phenomena have been observed with opioid mixed agonist-antagonists in other assays and have been ascribed to the presence of multiple receptors mediating opposite effects. Indeed, the presence of multiple opioid receptor types at peripheral sites has also been suggested by studies using various selective ligands (see below). Dose dependency of agonist effects was claimed in 11 papers, although the data were not shown in four (1,2,3,4) of these. Stereospecificity of agonist and/or antagonist effects was shown in seven papers. On the whole, the majority of studies support the occurrence of opioid receptor-specific effects at peripheral sites.

Most studies have used a panel of agonists with differing receptor affinities. Preferential μ-ligands were most potent in six (1,3,5,9,12,13) and κ-ligands in one (2). δ-Ligands were generally less potent than μ-agonists (1,2,3,13). These findings are in line with data obtained from central application of such agents (Porreca et al. 1984; Fang et al. 1986). Furthermore, different potencies of NAL to block different agonists were shown in three studies (2,12,13). Lastly, highly selective μ-, δ-, and κ-agonists and antagonists were shown to be effective in one paper (13). Taken together, these observations suggest that pharmacologically distinguishable populations of opioid receptors are involved in the mediation of antinociceptive effects in the periphery.

2.3 *Role of Inflammation*

All studies have employed models of inflammation (this is assuming that the intraperitoneal administration of irritants also causes an inflammatory reaction). Ten studies used intraplantar injections of one of the following agents: carrageenan (1,5–8), Freund's adjuvant (11–13), prostaglandin (1,3,7), and formalin (10). Three studies used intraperitoneal administration of acetic acid (2,4,9), acetylcholine (2), or phenylbenzoquinone (9). None provided a comprehensive documentation of the symptomatology of the inflammatory reaction, although four papers refer to earlier studies characterizing their model in this respect (8,11,12,13). Edema, one of the cardinal symptoms, was examined in two studies (5,7) and claimed to be unaffected by opioids. We (C. Stein et al. unpubl. observations) and others (K.M. Hargreaves, pers. commun.) have likewise found no dramatic effects on this parameter. Hyperalgesia and the time course of its development was documented in four studies (1,5,6,8) and mentioned in the rest. In all studies but one (5), opioid effects were examined at a time when hyperalgesia was well established. This time point varied from a few seconds (2) to several days (11,12,13) after administration of the respective inflammatory agent. Studies comparing opioids in inflamed vs noninflamed tissue have unequivocally reported greater antinociceptive effects in inflammation (1,3,5,11,12,13). Taken together, these observations suggest that peripheral opioid antinociceptive effects are recruited primarily under inflammatory conditions.

2.4 *Summary*

The case for peripheral opioid receptor-specific antinociceptive effects as well as for a multiplicity of receptors mediating these effects is convincing. The question arises as to where these receptors are located. On the one hand, the fact that so far, these antinociceptive effects have only been demonstrated under inflammatory conditions points toward an involvement of the immune system. Numerous opioid-mediated actions upon immune cells and their functions (e.g. chemotaxis, superoxide production, mast cell degranulation) have been reported (Sibinga and Goldstein 1988). However, at least two facts argue against the notion that local antinociceptive effects of opioids could be mediated indirectly through such mechanisms: Firstly, the rapid onset of antinociceptive effects (within minutes of administration of the opioid agoists) and secondly, the apparent lack of marked antiinflammatory actions. Therefore, one is more inclined to invoke a neural mechanism. In fact, there is abundant biochemical and electrophysiological evidence for the existence and functional significance of opioid receptors on primary afferent neurons (for references, see 13). Recently, we have been able to demonstrate such receptors immunocytochemically (Stein et al. 1990b; A.H.S. Hassan et al. unpubl. data) and functionally (Barthó et al. 1990). It is conceivable that, in inflammatory conditions, the de novo synthesis and peripherally directed axonal transport of opioid receptors are increased and/or that preexistent, but inactive, receptors on the terminals are rendered active. Thus, changes in tissue pH, as is the case under the acidic conditions in inflamed tissue, may result in an alteration of receptor conformation that may entail a transformation from a desensitized to an active state or vice versa. Following occupation of such receptors by an opioid agonist, an antinociceptive effect may be produced by at least two mechanisms. First, the excitability of the nociceptive input terminal may be attenuated. Second, the release of colocalized excitatory transmitters (e.g. substance P) from central and/or peripheral endings of primary afferents may be inhibited (for references, see 12, 13).

3 Human Studies

3.1 *Aspects to Consider*

The following aspects of clinical studies were evaluated in detail and comprise the headings in Table 2.

1. Condition and Type of Pain. According to a recent definition (International Association for the Study of Pain, IASP 1986), pain is "an unpleasant sensory and emotional experience associated with actual or potential tissue damage, or described in terms of such damage". Pain is always subjective and is an experience consisting of sensory and affective components. Acute pain has to be clearly differentiated from chronic pain. An extensive classification of chronic pain exists (IASP 1986) and should be used for standardization between studies.

Table 2. Human studies

No.	Condition	Pain assessment	Agonists	Dose (mg)	ANT	DR	ST	Exclusion of central effects	Locus of application	Reference
1	Post-operative pain	VAS VRS	Morphine	2–6	Ø	2 Doses	Ø	Comparing left vs right limb	Perineural	Bullingham et al. (1983)
2	Post-operative pain	VAS (not shown) VRS	Morphine Buprenorphine	1–4 0.12	Ø	Ø	Ø	Comparing left vs right limb	Perineural i.v. regional	Bullingham et al. (1984)
3	Not specified	VAS VRS	Morphine	0.04–6	Ø	Ø	Ø	In one group only Local vs i.m. application	Perineural	Mays et al. (1987)

Abbreviations: Ø, not examined; VAS, visual analogue scale; VRS, verbal rating scale.

2. Pain Assessment. Intensity of pain can be assessed by use of the "visual analogue scale (VAS)" (Scott and Huskisson 1976) and by numerical scales. A variety of verbal scales are available to evaluate affective components, sensory qualities, and intensity of pain (Karoly and Jensen 1987).

3. Exclusion of Central Effects. To differentiate between peripheral and central effects, again the effectiveness of local compared to systemic administration of equivalent doses of agonists can be examined. An alternative design (used in studies 1 and 2) is identical surgery on both extremities; in this case an opioid agonist can be applied to one side and a control substance to the other.

4. Criteria for Opioid-Mediated Effects. The same principles as outlined for animal studies apply.

5. Locus of Application. The locus of application of drugs is obviously of critical importance. If one assumes that peripheral opioid antinociceptive effects are brought about by an action upon primary afferent neurons, at least two issues must be taken into consideration: Firstly, the anatomical circumstances vary considerably between the axon and the nerve terminal and secondly, the functional state of receptors may be different in different regions along the nerve. The specific conditions of the tissue (e.g. pH) surrounding the nerve will influence both the penetration and pharmacological activity of drugs and possibly also the activation state of receptors.

3.2 Review of Clinical Studies

Numerous anecdotal case reports have appeared describing prolonged pain relief after peripheral application of opiates (e.g. Rynd 1845; Wood 1855; Mitchell 1872; Mocavero 1981; Sanchez et al. 1984). However, only three studies were carried out in a blind and controlled fashion and those are summarized in Table 2. Of these only one (3) found evidence for a local action of an opioid agonist whereas the other two (1,2) did not.

1. Condition and Assessment of Pain. Two studies (1,2) investigated conditions of acute, namely, postoperative pain. The third paper (3) merely described patients as suffering from "chronic intractable pain amenable to nerve block therapy". This is an insufficient specification. All three studies assessed intensity of pain by use of visual analogue scales and verbal rating scales. None evaluated the quality of pain.

2. Criteria for Opioid-Mediated Effects. None of these papers examined NAL reversibility or stereospecificity. An attempt to assess dose dependency was made in one (1), however, only two doses were tested.

3. Locus of Application. Perineural (i.e., in the vicinity of the nerve axon rather than the nerve terminal) application was used in all three studies; additionally, an intravenous regional technique was applied in one (2). The perineural administration of agents was carried out using standard nerve block techniques (i.e., the insertion of a needle near the target nerve by use of anatomical landmarks) in all studies. The application of opioids in the vicinity of peripheral nerve terminals has not been studied so far. Administration of a drug by an intravenous regional technique (2) requires exsanguination of the limb and, thus, can result in tissue hypoxia and acidosis. This, in turn, can influence drug activity by altering the relative fraction of the agent present in its charged or uncharged form (Savarese and Covino 1986). Moreover, as discussed by the authors, any drug applied by this technique will be diluted and washed out rapidly after removal of the tourniquet (see also Murphy 1986). In summary, the evidence from clinical studies either in favor or against the occurrence of peripheral analgesic actions of opioids is inconclusive for the following reasons: First, the number of controlled studies is too small. Second, all of the existing studies suffer from serious methodological shortcomings. Thus, more investigations carried out in a blind and controlled manner using not only perineural but also peripheral application of opioids are needed.

4 Conclusions

Strong evidence for peripheral antinociceptive actions of opioids has been provided by animal studies. So far, however, clinical investigations have been equivocal. Several issues have to be considered:

1. Perineural application is different from application at the nerve terminal. Assuming that opioids act upon receptors on primary afferent neurons, these may be present in different states of functional activity at different loci along the neuron. Furthermore, the anatomical circumstances are different. Axons are encased in Schwann cells and, in the case of $A\delta$ fibers, in several layers of lipoid myelin. These layers progressively disappear toward the peripheral terminals. Thus, peripherally applied agents may gain access to the nerve more readily.
2. From animal studies, inflammatory hyperalgesic conditions seem to be especially amenable to peripheral opioid antinociceptive actions. Thus, direct local application of agonists in clinical conditions of prolonged inflammatory pain may be particularly rewarding to study. On the other hand, the fact that opioid receptor activation can obviously occur very rapidly (see e.g. 2,4,9) suggests that peripheral application of opioids may be effective also in acute pain situations (e.g. intra- and postoperative pain). Polar compounds, unable to cross the blood-brain barrier or the placenta, might lack side effects such as respiratory depression, sedation, or neonatal depression.
3. Two important questions remain to be clarified: What are the endogenous ligands of these peripheral receptors and what is their origin? Three sources are conceivable: The pituitary-adrenal axis (Millan and Herz 1985), peripheral nerves (Weihe 1989), or the immune system (Sibinga and Goldstein 1988). Studies addressing these issues have commenced in our laboratory (Stein et al. 1990a,b; Parsons et al. 1990).

References

Abbott F (1988) Peripheral and central antinociceptive actions of ethylketocyclazocine in the formalin test. Eur J Pharmacol 152:93-100

Barthó L, Stein C, Herz A (1990) Involvement of capsaicin-sensitive neurones in hyperalgesia and enhanced opioid antinociception in inflammation. Naunyn Schmiedeberg's Arch Pharmacol (in press)

Bentley GA, Newton SH, Starr J (1981) Evidence for an action of morphine and the enkephalins on sensory nerve endings in the mouse peritoneum. Br J Pharmacol 73:325-332

Bullingham R, O'Sullivan G, McQuay H, Poppleton P, Rolfe M, Evans P, Moore A (1983) Perineural injection of morphine fails to relieve postoperative pain in humans. Anesth Analg 62:164-167

Bullingham RES, McQuay HJ, Moore RA (1984) Studies on the peripheral action of opioids in postoperative pain in man. Acta Anaesth Belg 35 Suppl:285-290

Fang GF, Fields HL, Lee NM (1986) Action at the mu receptor is sufficient to explain the supraspinal analgesic effect of opiates. J Pharmacol Exp Ther 238:1039-1044

Ferreira SH, Nakamura M (1979) Prostaglandin hyperalgesia II: the peripheral analgesic activity of morphine, enkephalins and opioid antagonists. Prostaglandins 18:191-200

Ferreira SH, Molina N, Vettore O (1982) Prostaglandin hyperalgesia, V: a peripheral analgesic receptor for opiates. Prostaglandins 23:53-60

Ferreira SH, Lorenzetti BB, Rae GA (1984) Is methylnalorphinium the prototype of an ideal peripheral analgesic. Eur J Pharmacol 99:23-29

Follenfant RL, Hardy GW, Lowe LA, Schneider C, Smith TW (1988) Antinociceptive effects of the novel opioid peptide BW443C compared with classical opiates; peripheral versus central actions. Br J Pharmacol 93:85-92

IASP – International Association for the Study of Pain (ed) (1986) Classification of chronic pain. Pain Suppl 3:S217

Joris JL, Dubner R, Hargreaves KM (1987) Opioid analgesia at peripheral sites: a target for opioids released during stress and inflammation. Anesth Analg 66:1277-1281

Karoly P, Jensen MP (eds) (1987) Multimethod assessment of chronic pain. Pergamon, New York, pp 42-57

Kosterlitz HW, Waterfield AA (1975) In vitro models in the study of structure activity relationships of narcotic analgesics. Annu Rev Pharmacol Toxicol 15:29-47

Mays KS, Lipman JJ, Schnapp M (1987) Local analgesia without anesthesia using peripheral perineural morphine injections. Anesth Analg 66:417-420

Millan MJ, Herz A (1985) The endocrinology of the opioids. Int Rev Neurobiol 26:1-83

Mitchell SW (1872) Injuries of the nerves and their consequences. Lippincott, Philadelphia, pp 122-123

Mocavero G (1981) Analgesia selettiva con morfina perinervosa. Incont Anest Rianimaz Sci Affini 16:1-3

Murphy TM (1986) Nerve blocks. In: Miller RD (ed) Anesthesia, 2nd edn. Livingstone, New York, p 1015

Parsons CG, Członkowski A, Stein C, Herz A (1990) Peripheral opioid receptors mediating antinociception in inflammation. Activation by endogenous opioids and involvement of pituitary-adrenal axis. Pain 41:81-93

Porreca F, Mosberg HJ, Hurst R, Hruby VJ, Burks TF (1984) Roles of mu, delta and kappa opioid receptors in spinal and supraspinal mediation of gastrointestinal transit effects and hot-plate analgesia in the mouse. J Pharmacol Exp Ther 230:341-348, 1984

Rios L, Jacob JJC (1982) Inhibition of inflammatory pain by naloxone and its N-methyl quaternary analogue. Life Sci 31:1209-1212

Rios L, Jacob JJC (1983) Local inhibition of inflammatory pain by naloxone and its N-methyl quaternary analogue. Eur J Pharmacol 96:277-283

Rynd F (1845) Neuralgia: introduction of fluid to the nerve, vol 13. Medical Press, Dublin, pp 167-168

Sanchez R, Nielsen H, Heslet L, Iversen AD (1984) Neural blockade with morphine. Anaesthesia 39:788-789

Savarese JJ, Covino BG (1986) Basic and clinical pharmacology of local anesthetic drugs. In: Miller RD (ed) Anesthesia, 2nd edn. Livingstone, New York, p 985

Scott J, Huskisson EC (1976) Graphic representation of pain. Pain 2:175-184

Sibinga NES, Goldstein A (1988) Opioid peptides and opioid receptors in cells of the immune system. Annu Rev Immunol 6:219-249

Smith TW, Buchan P, Parsons DN, Wilkinson S (1982) Peripheral antinociceptive effects of N-methyl morphine. Life Sci 31:1205–1208

Stein C, Millan MJ, Shippenberg TS, Herz A (1988a) Peripheral effect of fentanyl upon nociception in inflamed tissue of the rat. Neurosci Lett 84:225–228

Stein C, Millan MJ, Yassouridis A, Herz A (1988b) Antinociceptive effects of μ- and κ-agonists in inflammation are enhanced by a peripheral opioid receptor-specific mechanism of action. Eur J Pharmacol 155:255–264

Stein C, Millan MJ, Shippenberg TS, Peter K, Herz A (1989) Peripheral opioid receptors mediating antinociception in inflammation. Evidence for involvement of mu, delta and kappa receptors. J Pharmacol Exp Ther 248:1269–1275

Stein C, Gramsch C, Herz A (1990a) Intrinsic mechanisms of antinociception in inflammation. Local opioid receptors and β-endorphin. J Neurosci 10:1292–1298

Stein C, Hassan AHS, Przewłocki R, Gramsch C, Peter K, Herz A (1990b) Opioids from immunocytes interact with receptors on sensory nerves to inhibit nociception in inflammation. Proc Natl Acad Sci USA 87:5935–5939

Weihe E (1989) Neuropeptides in primary afferent neurons. In: Zenker W, Neuhuber W (eds) The primary afferent neuron. Plenum, New York, pp 127–159

Wood A (1855) New method of treating neuralgia by the direct application of opiates to the painful points. Edinburgh Med Surg J 82:265–281

CHAPTER 18

Regulation of Intestinal Motility by Peripheral Opioids: Facts and Hypotheses

W. Kromer

1 Introduction

A functional opioid system intrinsic to the gastrointestinal wall was recognized following a few key observations some 10 years ago (Table 1). The report by van Nueten et al. (1976) of an unexpected reversal of peristalsis in the "fatigued" guinea pig ileum (GPI) upon application of the opioid antagonist naloxone (NAL) in vitro raised the question as to the biological significance of intestinal opioids for "normal" peristalsis. A comprehensive review of gastrointestinal opioid functions has been published recently (Kromer 1988). The present chapter will therefore focus on intestinal mechanisms controlling motility. The following section is only intended to provide a background for integration of available observations on opioids; further details may be found in the references cited.

2 Myogenic and Neurogenic Mechanisms Underlying Intestinal Motility

Although coordinated peristaltic activity depends on the integrity of the myenteric plexus (Bortoff 1972), the smooth muscle layers are capable of myogenic electrical and mechanical activity in the absence of tonic neuronal inhibition (disinhibition of the smooth muscle) (Wood 1987). This should not be confused with expulsive peristalsis. Myogenic rhythmic electrical activity (slow waves; "slow potentials" in guinea pig) is based on oscillations in Na^+ conductance of the cell membrane and is augmented by acetylcholine (ACh) (Szurszewski 1987).

A hypothetical scheme of how complex neuronal reflex mechanisms may work is depicted in Fig. 1. Luminal distention increases the discharge rate of mechanoreceptor 1 leading to excitation of neurons 2, 3 and 4 (Fig. 1a), followed by activation of the cholinergic motor neuron 5 which then contracts the circular muscle proximal to the distention stimulus (*ascending excitation*). *Ascending disinhibition* (Fig. 1b) of the spontaneously active circular muscle may be achieved by inhibition of a spontaneously active neuron 7 via activation of inhibitory neuron 6. This switches off neuron 8 which, in the absence of a distension stimulus, inhibits smooth muscle contractility. Vasoactive intestinal peptide (VIP) may be the final inhibitory transmitter (Wood 1987). These proposed pathways are also thought to include nicotinic synapses since they can be interrupted by hexamethonium (Kosterlitz and Lees 1964; Costa and Furness 1976).

Table 1. Observations which led to the recognition of an opioid system intrinsic to the guinea pig intestine in vitro

Observation	Reference
1. Naloxone-induced release of ACh	Waterfield and Kosterlitz (1975)
2. Naloxone reversal of "fatigued" peristalsis	van Nueten et al. (1976)
3. Opioid immunoreactivity in intestinal wall	Smith et al. (1976)
4. Opioid biosynthesis in intestinal wall	Sosa et al. (1977)
5. Stereospecificity of the naloxone effect on normal peristalsis	Kromer and Pretzlaff (1979)
6. Inverse relationship between dynorphin-/enkephalin release and peristaltic activity	Kromer et al. (1981a) Clark and Smith (1981)

Preparation of the path ahead of the luminal contents is achieved by *descending inhibition* of myogenic activity (Fig. 1c) (Hirst and McKirdy 1974). First, neuron 2 may activate neuron 9 to release VIP, thus inhibiting circular smooth muscle contraction ahead of the luminal bolus. This pathway is probably comprised of nicotinic but not serotonin (5-HT)-synapses (Costa and Furness 1976; Wood 1987), although the latter point seems to be controversial (Hodgkiss and Lees 1986). Second, smooth muscle contraction might be prevented (see Hodgkiss and Less 1986), for example, by inhibition of spontaneously active neuron 11 which supposedly drives motor neuron 5. Since a distension stimulus from a distal site might also activate motor neuron 5, proximal activity has to suppress distal activity to allow propulsion of luminal contents in an oral-aboral (proximal-distal) direction. In fact, descending inhibition dominates ascending excitation (Costa and Furness 1976; Hodgkiss and Lees 1986) if the two are simultaneously induced. Thus, hypothe-

Fig. 1a-f. Schematic diagram of potential myenteric plexus reflex mechanisms controlling gut motility. The postulated scheme which is based on heterogeneous experimental data is incomplete and only represents a working hypothesis. It mainly considers mechanisms relating to opioids. In particular, details of neuronal connections, such as modulatory inputs onto soma vs processes, and morphological types of neurons, are arbitrary or have been altogether ignored in the scheme due to a paucity of specific information. The assembly of opioid neurons 6, 10, 15, and 16 indicate possible explanations for available experimental data. Several other alternatives may equally apply. The submucosal plexus has not been considered since it does not appear to be essential for the peristaltic reflex.

LM Longitudinal muscle; *CM* circular muscle. Neurons have been *numbered* for easy reference to the text; their potential transmitters are indicated at their cell bodies (Nō.), but this attribution is, in several cases, speculative: *ACh* acetylcholine (*1, 2, 5*); *5-HT* 5-hydroxytryptamine (serotonin) (*3, 7, 14*); *VIP* vasoactive intestinal peptide (*8, 9*); *SP* substance P (*4, 13*); *CCK* cholecystokinin (*12*); *OP* opioids (*6, 10, 13, 15, 16*); *OR* opioid receptor. *Squares* (◊) indicate supposed spontaneously-active neurons as opposed to *circles* (○, driven neurons). *Heavily outlined squares* (◊) and *circles* (○) indicate inhibitory neurons, as opposed to excitatory neurons.

The various neuronal pathways of the myenteric plexus are presented in **a-f** in order of increasing complexity due to the stepwise addition of one pathway to the other. *Pathways to circular muscle*: **a** ascending excitation; **b** ascending disinhibition; this may be of minor importance in the guinea pig since no myogenic contractile activity appears to be present in this species. (Note added in proof: A recent publication by Wang and Tson (1989) supports the hypothetical assumption in **b** of opioid-induced inhibition of opioid release within the myenteric plexus; see **b**, neurons *6* and *16*). **c** Descending inhibition;

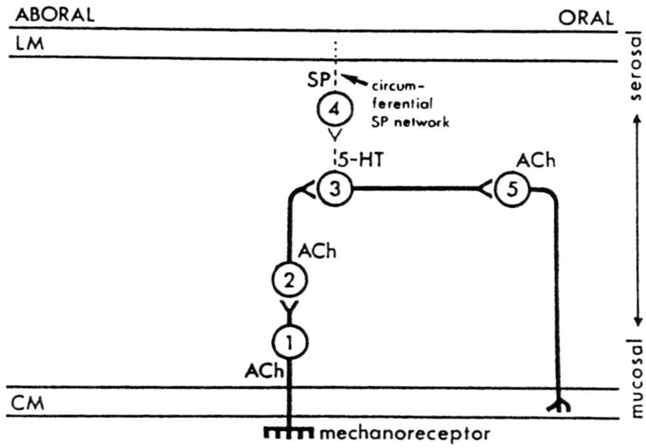

a) ASCENDING EXCITATION OF CIRC. MUSCLE

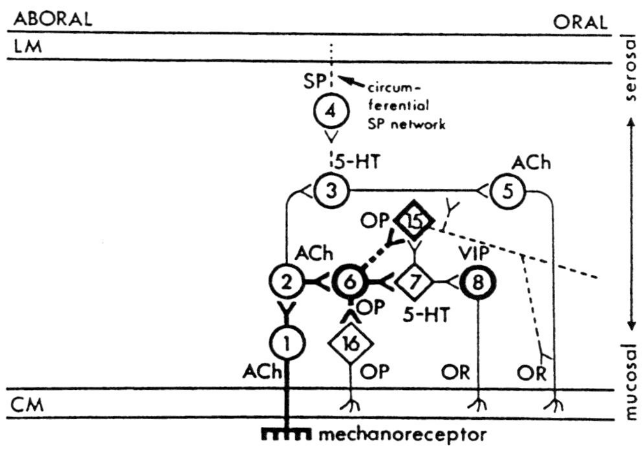

b) ASCENDING DISINHIBITION OF CIRC. MUSCLE

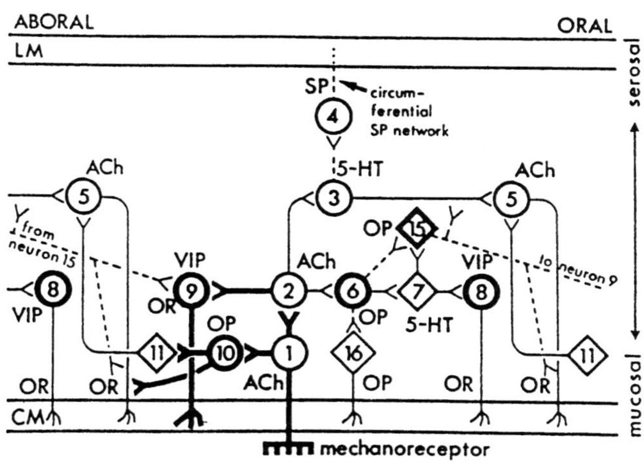

c) DESCENDING INHIBITION OF CIRC. MUSCLE

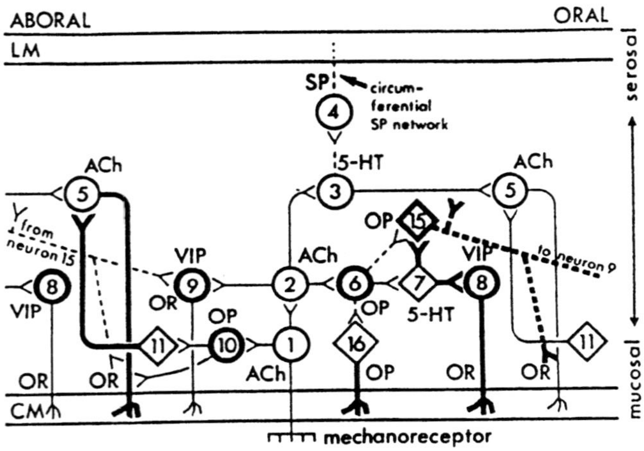

d) SEGMENTING MODE OF CIRC. MUSCLE

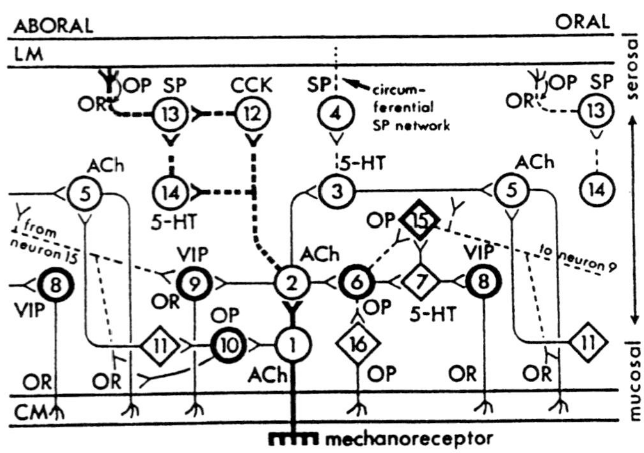

e) CCK-SP AND 5-HT-SP PATHWAYS TO LONG. MUSCLE

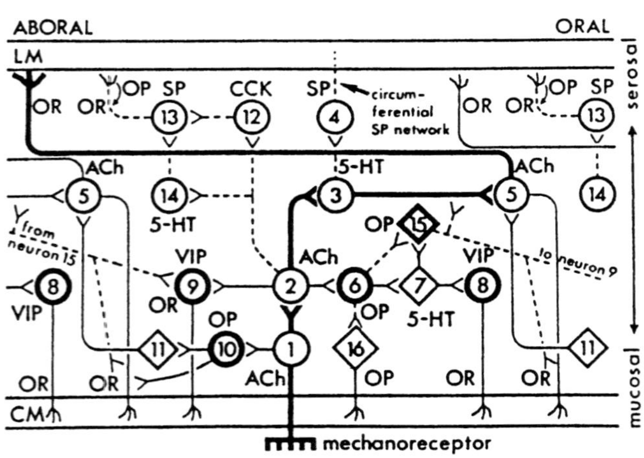

f) DESCENDING EXCITATION OF LONG. MUSCLE

Fig. 1. Continued (d-f). **d** segmenting mode. *Pathways to longitudinal muscle*: **e** CCK-SP and 5-HT-SP pathways; **f** descending excitation. The particular pathways involved are stressed by heavily outlined neuronal processes (as opposed to heavily outlined cell bodies which have a different meaning; see above). Broken lines do not have any special meaning except to make reading of the scheme easier

tically, neuron 10 may inhibit motor neuron 5. Descending inhibition may account for the observation of long synaptic pathways directed aborally which elicit inhibitory postsynaptic potentials (IPSPs) on the circular smooth muscle distal to the stimulus (Hirst and McKirdy 1974).

While in some species the migrating motor complex (MMC) develops under fasting conditions, a random motility pattern develops in other species under fed conditions (Weisbrodt 1987). Both patterns may achieve intestinal transit. In the latter, this may be due to a frequency gradient decreasing in an oral-aboral direction. In the neuronal circuit depicted in Fig. 1d, rhythmic spontaneous activity of neuron 7 (leading to inhibition of circular smooth muscle) and of neuron 11 (leading to excitation of adjacent circular smooth muscle) shows how ring contractions may develop (*segmenting mode*). These may move over short distances in either direction, depending on the state of activity of the proximal and distal segments.

One of the predominant transmitters in the enteric ganglia may be 5-HT (Mathias and Sninsky 1985), the release of which is enhanced during peristalsis (Kosterlitz and Lees 1964). 5-HT is known to release ACh from the myenteric plexus (Wood 1987) shown in the scheme (Fig. 1a) as a 5-HT neuron (3) synapsing on cholinergic motor neuron (5). Substance P (SP) neurons which receive cholinergic (Furness and Costa 1980) and possibly 5-HT (Gintzler 1980) inputs, excite other neurons (Wood 1987) and thereby enhance both peristalsis and segmentations (Kosterlitz and Lees 1964). SP neurons form a circumferential network which may synchronize ring contractions (Wood 1987). This brief and incomplete overview serves merely as a background for discussion of opioid functions and quotes, therefore, mainly reviews.

3 Potential Sites of Action of Opioids in the Intestinal Wall

Endogenous opioids of all three classes (enkephalins: ENK; endorphins: END, and dynorphins: DYN) have been detected in nerve fibers and neuronal cell bodies of the myenteric and submucosal plexi throughout the gastrointestinal tract of a variety of species, including man (for review, see Kromer 1988). Nerve fibers containing ENK immunoreactivity project both aborally to the circular smooth muscle and orally to other myenteric neurons (Furness et al. 1983). The former projection may relate to the spasmogenic effect of opioids, the latter to their inhibitory action on peristalsis (see below). This is represented in Fig. 1b where spontaneously active neuron 15 inhibits neuron 5 to prevent contractile activity. On the other hand, opioids, at nanomolar concentrations, cause NAL-reversible contraction of isolated circular smooth muscle cells devoid of neuronal elements (Makhlouf 1987). This effect is associated with an extracellular Ca^{2+}-independent increase in intracellular free

Ca^{2+} which is possibly released from intracellular stores by inositol triphosphate. A shallow concentration-response curve and the comparison between various opioids led the authors to suggest the presence of μ-, κ- and δ-type opioid receptors. All three of them lead to a maximum of 30–40% decrease in circular smooth muscle cell length, but no effect on longitudinal muscle cells was detected. This may contribute to the spasmogenic effect of opioids mentioned above (see neuron 16, Fig. 1d). Opioids did not affect the contraction of the circular muscle in response to myotropic stimulants (Harry 1963), a point needing further clarification.

A neuronal opioid action switches the motility pattern of the gut from peristaltic to segmenting activity (Weisbrodt et al. 1980). Endogenous intestinal opioids have a similar function as evidenced by the opposite effect of NAL (Kromer et al. 1980a, 1981b). Although Fig. 1d suggests independent discharge of spontaneously active neurons as the basis of the segmenting mode, it provides no ready explanation for the action of exogenous opioids as a command signal. It may be that interruption of coordinated peristalsis by opioids eventually results in random escape of pacemaker neurons which then induce local segmentations. Exogenous opioids inhibit ACh release from myenteric neurons (Paton 1957; Schaumann 1957) and intestinal opioids operate similarly (Waterfield and Kosterlitz 1975). This is represented in Fig. 1d as inhibitory opioid inputs to neurons 3 and 5. Indeed, it has been shown that morphine inhibits nicotine- and 5-HT-induced circular smooth muscle contractions within the myenteric plexus (Harry 1963) and normorphine impairs the in vitro peristalsis induced by exogenous ACh (Kromer and Schmidt 1982).

The depicted divergence of opioid fibers and their attribution to pre- vs post-synaptic sites on cholinergic and 5-HT neurons in Fig. 1d has been chosen arbitrarily to simply indicate modulation, without giving favor to one or the other site. However, opioid neurons may provide convergence of input rather than divergence of output. Bornstein et al. (1984) showed that ENK immunoreactivity in the guinea pig myenteric plexus is found almost exclusively in Dogiel type-1 cells which have one long and many short processes. The opioid neurons may be tonically driven (for example, by neuron 7 in Fig. 1d), or may themselves discharge spontaneously (for example, neurons 15 and 16). In fact, both DYN (Kromer et al. 1981a) and [Met^5]-ENK (Clark and Smith 1981) are spontaneously released (i.e., without an exogenous stimulus) from the GPI in vitro; their release is reduced upon distension-induced reflex peristalsis. This is shown in the scheme as an inhibition of neurons 15 and 16 by neuron 6 which is activated by a distention stimulus. Alternatively, inhibition of neuron 7 which drives the hypothetical opioid neuron 15 in the scheme would likewise result in decreased opioid release. These mechanisms are highly speculative.

Morphine has been shown to release 5-HT, and to thereby trigger intestinal motility in the dog (Burks and Long 1967). This could result from activation of opioid receptors on neuron 15, thus disinhibiting neuron 3; this assumes that neuron 15 in Fig. 1d is a nonopioid inhibitory neuron and that neuron 3 has no opioid receptors. On the other hand, there is indirect evidence which suggests release of DYN by 5-HT in the rat myenteric plexus (Majeed et al. 1985); if this is true, this would be consistent with a hypothetical input from neuron 7 to (in this case) opioid neuron 15 (Fig. 1d).

Bornstein et al. (1984) reported that ENK immunoreactive Dogiel type-1/S-neurons receive cholinergic inputs, i.e., they had fast EPSPs; interneuron 10 may account for this observation. Again, the attribution is entirely speculative.

Grider and Makhlouf (1987) observed a decrease in DYN release during relaxation distal to a radial stretch stimulus in the guinea pig and rat colon in vitro; conversely, an increase in DYN release was seen during contraction proximal to a radial stretch stimulus. Moreover, opioids inhibited relaxation distal to, and increased contraction proximal to, stretch stimuli in their experiments. These are just the opposite effects to those found by Kromer et al. (1980a, 1981b) when expulsive peristaltic contractions in the guinea pig small and large intestine were investigated in vitro. The observation by Grider and Makhlouf (1987) may possibly relate to the segmenting mode when variable spontaneous discharges of neurons 7, 11, 15, and 16 may cause random circular muscle contractions at alternating loci, depending on the relative temporal discharge pattern of these neurons (Fig. 1d). However, the functional relationship between their activity and the stretch stimulus remains obscure. A population of mechanoreceptors responding to contraction, as opposed to distension (Wood 1987), may not be operative in a longitudinally cut preparation under radial stretch, as used by Grider and Makhlouf (1987); however, the functional significance of this is not known. It should be noted that the assumption in Fig. 1d of a series of functionally coupled opioid neurons is entirely speculative and has been presented simply to illustrate the diversity of possible mechanisms. Other mechanisms are also likely; e.g., Grider and Makhlouf (1987) found that radial stretch also increased VIP release during circular muscle relaxation distal to the stimulus; opioids impaired this influence on VIP release, an observation that is consistent with the scheme showing opioid receptors on neuron 9.

Barthó et al. (1987) have recently reported that NAL antagonizes the hexamethonium-induced inhibition of the peristaltic reflex in the isolated GPI. Although this is in conflict with data by Kromer and Schmidt (1982), it may point to an endogenous opioidergic modulation of redundant pathways producing ascending contraction and descending inhibition, without the involvement of nicotinic synapses. It should also be noted that, in the dog, the neuropeptide motilin may trigger gut motility by releasing endogenous opioids with contractile activity (Fox and Daniel 1987). Conversely, an ENK analogue inhibited motilin release in man which was suggested to partially explain the antimotility effect of opioids (Sekiya et al. 1986). These results have not been considered in Fig. 1.

Garzón et al. (1987) demonstrated that, in the presence of atropine, the application of NAL shortly after treatment with the excitatory peptide CCK-8, resulted in contraction of the longitudinal muscle myenteric plexus preparation of the GPI. The response to NAL was abolished by desensitization of the preparation to SP, and an ENK analogue impaired the CCK-8-induced contraction. Therefore, CCK-8 may release both endogenous opioids and SP, possibly from common nerve terminals (Domoto et al. 1984). It is probable that opioids mediate the negative feedback of SP release, thus causing smooth muscle relaxation (Fig. 1e). In fact, an enhanced release of SP is seen upon NAL challenge of the intestine of morphine-treated guinea pigs (Gintzler 1980). 5-HT may play a role similar to that of CCK since it stimulates the release of a noncholinergic excitatory transmitter, the release of which is inhibited by an opioid (Gintzler 1979). Such *excitatory pathways to the longitudinal muscle* are shown as neurons 12 to 14 (Fig. 1e).

Descending cholinergic excitation to the longitudinal muscle, which helps to prepare the path ahead of the luminal bolus, is depicted in Fig. 1f. A situation in which the release of DYN from, for example, neurons 6, 10, or 13 into the vascular effluent of the intestinal segment is enhanced during peristalsis (Donnerer et al. 1984), rather than decreased (as measured in the serosal bathing solution; Kromer et al. 1981a) may, of course, exist. In fact, DYN-containing nerve terminals, in the rat, have close contact to arterioles (Wolter 1985). For further discussion of this aspect, see Kromer (1989).

4 The Action of Intestinal Opioids at the Cellular Level

Early extracellular recordings from myenteric neurons demonstrated that morphine inhibits spontaneous, nicotinic- and 5-HT-induced neuronal discharge (Sato et al. 1973; Dingledine et al. 1974). It later became apparent from intracellular recordings that opioids hyperpolarize and, thereby, decrease the excitability of a population of S/type-1 (mainly) and also AH/type-2 neurons (North and Tonini 1977). Iontophoretic application of the opioid identified the cell processes as an important site of action (Morita and North 1981). It was later shown that μ and δ agonists activate a Ca^{2+}-dependent K^+ conductance in myenteric (Morita and North 1982) and submucosal neurons (North et al. 1987). No evidence was found for the mediation of this action by cAMP-dependent protein kinase or protein kinase C, but the opioid effect persisted in the presence of nonhydrolyzable GTP[S], suggesting coupling between opioid receptors and K^+ channels by a G-protein. Since morphine inhibits spontaneously active neurons in the absence of extracellular Ca^{2+} (Dingledine and Goldstein 1976), it has been suggested that opioids release Ca^{2+} from intracellular stores to stimulate the Ca^{2+}-dependent K^+ conductance, thereby hyperpolarizing the cell membrane and blocking impulse propagation along the cell processes. As a consequence of high K^+ conductance, the duration of the action potential is shortened and the Ca^{2+} influx during the plateau phase reduced. Hence, high extracellular Ca^{2+} concentrations prevent the opioid effect (Morita and North 1982).

A decrease in synaptosomal Ca^{2+} content following the application of opioids (Cardenas and Ross 1976) is consistent with the above view. Regional Ca^{2+} depletion in brain tissue (Ross et al. 1974) cannot be explained solely by sequestration of Ca^{2+} within the cell, and displacement of Ca^{2+} by opioids from the plasma membrane (Cardenas and Ross 1976), which carries the K^+ channel, should decrease rather than increase K^+ conductance. Nevertheless, opioid-induced increase of in vitro binding of $^{45}Ca^{2+}$ to brain synaptic plasma membranes was interpreted by Yamamoto et al. (1978) as an indication of prior dissociation of endogenous Ca^{2+} from the membrane which, in turn, causes an increased number of binding sites to be available to $^{45}Ca^{2+}$. In fact, Cherubini and North (1985) reported on another primary mechanism of opioid action on myenteric neurons. They demonstrated a direct coupling of κ opioid receptors to a voltage-dependent Ca^{2+} channel resulting in decreased Ca^{2+} entry during the action potential, with a subsequent reduction in transmitter release. Both mechanisms, enhancement of Ca^{2+}-dependent K^+ conductance with a secondary decrease in Ca^{2+} entry, and primary inhibition of Ca^{2+} entry, have even been identified in the same cell and attributed to μ and κ receptors, respectively (Cherubini and North 1985).

Szerb (1980) found that morphine and low extracellular Ca^{2+} concentrations differentially affected the kinetics of ^{3}H-ACh efflux from the myenteric plexus of the GPI. Whereas morphine reduced the size of the releasable pool, low extracellular Ca^{2+} concentrations decreased the rate of efflux. This was taken as evidence for morphine-induced hyperpolarization which results in the complete inhibition of some neurons, while not affecting transmitter release from other more excitable neurons. However, another point must also be considered. The drug 4-aminopyridine (4-AP), which blocks K^+ channels (Dolly 1988) and dramatically increases Ca^{2+} entry during the action potential, prevents the inhibitory effect of morphine on intestinal peristalsis in vitro (Kromer et al. 1980b). Since 4-AP alone considerably enhances peristalsis, blockade of the opioid effect cannot be explained solely on the basis of Ca^{2+}-induced activation of K^+ conductance and, hence, hyperpolarization which may anticipate and thereby prevent the opioid effect. Rather, 4-AP causes hyperexcitability of the neurons by blocking potassium channels.

5 Opioid Receptor Types

Most in vitro studies examining the opioid receptor types involved in the control of gut motility have been done in the longitudinal muscle myenteric plexus preparation of the guinea pig ileum. The electrically stimulated preparation displays completely different responses compared to the intact intestinal segment in vitro which develops circular muscle reflex peristalsis. A degree of caution is necessary when extrapolating from one preparation to the other. This becomes obvious when a comparison between the effects of the opiate N-allyl-normetazocine (SKF 10,047) in both preparations is made. While SKF 10,047, like other opiates, inhibits electrically induced longitudinal muscle contractions in a NAL-blockable fashion, it enhances, in contrast to other opioids, both ACh-induced and reflex peristalsis; SKF 10,047 thus acts similarly to NAL in this respect (Kromer et al. 1982). The different actions have been attributed to the activation of supposed σ- and blockade of μ and κ receptors by SKF 10,047; these different receptor types are probably of different functional significance in the two preparations.

It may well be that a certain receptor population, though detectable in one assay, may be silent in the other. For example, Leslie et al. (1980) found δ binding sites to be selectively labeled by ^{3}H-[D-Ala^2-D-Leu^5]-ENK in homogenates of the GPI longitudinal muscle myenteric plexus, although functional studies revealed the presence of only μ and κ receptors (Lord et al. 1977). A similar finding was reported for longitudinal muscle myenteric plexus preparations when a cross-tolerance paradigm was employed (Schulz et al. 1981), and in circular muscle with the plexus removed, leaving only nerve terminals (Johnson et al. 1988). The latter study proved that the cholinergic terminals supplying the circular muscle bear both μ and κ opioid receptors, both of which inhibit transmitter release. These data agree with intracellular recordings from single myenteric neurons (Cherubini and North 1985) which demonstrated the presence of μ and κ opioid receptors. Submucosal neurons, in contrast, seem to express δ opioid receptors (Mihara and North 1986). Their significance for organ function is unknown but may relate to the antisecretory effect of opioids (for review, see Kromer 1988).

In vivo, peripheral opioid inhibition of gastrointestinal transit in the mouse was produced by μ, but not δ or κ opioid agonists (Shook et al. 1987). Other investigators have also shown the involvement of μ opioid receptors in the mouse (Ward and Takemori 1982) and guinea pig (Culpepper-Morgan et al. 1988). However, there are additional reports of the participation of κ opioid receptors in the peripheral inhibition of gastrointestinal transit in the mouse (Ward and Takemori 1982; Ramabadran et al. 1988) and guinea pig (Culpepper-Morgan et al. 1988). In vivo studies in the dog have found opioids to produce a premature MMC with subsequent inhibition of the fasted motility pattern; both these actions occur via μ opioid receptor activation (Sarna and Lang 1985; Vaught et al. 1985). In addition, however, δ opioid receptors may contribute to the contractile effects of opioids in the dog (Vaught et al. 1985). It is particularly interesting that κ opioid agonists have been found to have either no (Vaught et al. 1985), or an entirely inhibitory, effect in the dog, in terms of both the fasted (MMC) and fed motility pattern (Telford et al. 1988). Although completely speculative, the distinct cellular mechanisms of action of μ and δ agonists, on the one hand, and κ agonists, on the other (Cherubini and North 1985; see above), suggest a differential distribution of receptor types which may account for the functional differences observed.

It should be noted that the functional significance of phase III of the MMC for propulsion of luminal contents is still uncertain. Although Borody et al. (1985; M. Wienbeck, personal communication) found long-lasting enhancement of phase III activity of the MMC by opioids in man, there is no doubt that opioids, nevertheless, cause constipation. In fact, phase II activity in the fasted and irregular continuous activity in the fed state, for example in the dog, may be of greater significance for intestinal transit than phase III activity. Thus, though important advances in opioid research on gastrointestinal motility have been made over the past decade, the functional relationship between many of these findings has to be elaborated in the future.

References

Barthó L, Holzer P, Lembeck F (1987) Is ganglionic transmission through nicotinic receptors essential for the peristaltic reflex in the guinea-pig ileum? Neuropharmacology 26:1663–1666

Bornstein JC, Costa M, Furness JB, Lees GM (1984) Electrophysiology and enkephalin immunoreactivity of identified myenteric plexus neurones of guinea-pig small intestine. J Physiol 351:313–325

Borody TJ, Quigley EMM, Phillips SF, Wienbeck M, Tucker RL, Haddad A, Zinsmeister AR (1985) Effects of morphine and atropine on motility and transit in the human ileum. Gastroenterology 89:562–570

Bortoff A (1972) Digestion: motility. Annu Rev Physiol 34:261–290

Burks TF, Long JP (1967) Release of intestinal 5-hydroxytryptamine by morphine and related agents. J Pharmacol Exp Ther 156:267–276

Cardenas HL, Ross DH (1976) Calcium depletion of synaptosomes after morphine treatment. Br J Pharmacol 57:521–526

Cherubini E, North RA (1985) μ and κ opioids inhibit transmitter release by different mechanisms. Proc Natl Acad Sci USA 82:1860–1863

Clark SJ, Smith TW (1981) Peristalsis abolishes the release of methionine-enkephalin from guinea-pig ileum in vitro. Eur J Pharmacol 70:421–424

Costa M, Furness JB (1976) The peristaltic reflex: an analysis of the nerve pathways and their pharmacology. Naunyn-Schmiedeberg's Arch Pharmacol 294:47–60

Culpepper-Morgan J, Kreek MJ, Holt PR, Laroche D, Zhang J, O'Bryan L (1988) Orally administered kappa as well as mu opiate agonists delay gastrointestinal transit time in the guinea pig. Life Sci 42:2073–2077

Dingledine R, Goldstein A (1976) Effect of synaptic transmission blockade on morphine action in the guinea-pig myenteric plexus. J Pharmacol Exp Ther 196:97–106

Dingledine R, Goldstein A, Kendig J (1974) Effects of narcotic opiates and serotonin on the electrical behavior of neurons in the guinea pig myenteric plexus. Life Sci 14:2299–2309

Dolly JO (1988) Potassium channels – what can the protein chemistry contribute? Trends Neurosci 11:186–188

Domoto T, Gonda T, Oki M, Yanaihara N (1984) Coexistence of substance P- and methionine-enkephalin-like immunoreactivity in nerve cells of the myenteric ganglia in the cat ileum. Neurosci Lett 47:9–13

Donnerer J, Holzer P, Lembeck F (1984) Release of dynorphin, somatostatin and substance P from the vascularly perfused small intestine of the guinea-pig during peristasis. Br J Pharmacol 83:919–925

Fox JET, Daniel EE (1987) Activation of endogenous excitatory opiate pathways in canine small intestine by field stimulation and motilin. Am J Physiol 253:G189–G194

Furness JB, Costa M (1980) Types of nerves in the enteric nervous system. Neuroscience 5:1–20

Furness JB, Costa M, Miller RJ (1983) Distribution and projections of nerves with enkephalin-like immunoreactivity in the guinea-pig small intestine. Neuroscience 8:653–664

Garzón J, Höllt V, Herz A (1987) Cholecystokinin octapeptide activates an opioid mechanism in the guinea-pig ileum: a possible role for substance P. Eur J Pharmacol 136:361–370

Gintzler AR (1979) Serotonin participation in gut withdrawal from opiates. J Pharmacol Exp Ther 211:7–12

Gintzler AR (1980) Substance P involvement in the expression of gut dependence on opiates. Brain Res 182:224–228

Grider JR, Makhlouf GM (1987) Role of opioid neurons in the regulation of intestinal peristalsis. Am J Physiol 253:G226–G231

Harry J (1963) The action of drugs on the circular muscle strip from the guinea-pig isolated ileum. Brit J Pharmacol 20:399–417

Hirst GDS, McKirdy HC (1974) A nervous mechanism for descending inhibition in guinea-pig small intestine. J Physiol 238:129–143

Hodgkiss JP, Lees GM (1986) Transmission in enteric ganglia. In: Karczmar AG, Koketsu K, Nishi S (eds) Autonomic and enteric ganglia. Transmission and its pharmacology. Plenum, New York, pp 369–405

Johnson SM, Costa M, Humphreys CMS (1988) Opioid mu and kappa receptors on axons of cholinergic excitatory motor neurons supplying the circular muscle of guinea-pig ileum. Naunyn Schmiedeberg's Arch Pharmacol 338:397–400

Kosterlitz HW, Lees GM (1964) Pharmacological analysis of intrinsic intestinal reflexes. Pharmacol Rev 16:301–339

Kromer W (1988) Endogenous and exogenous opioids in the control of gastrointestinal motility and secretion. Pharmacol Rev 40:121–162

Kromer W (1989) The current status of opioid research on gastrointestinal motility. Minireview. Life Sci 44:579–589

Kromer W, Pretzlaff W (1979) In vitro evidence for the participation of intestinal opioids in the control of peristalsis in the guinea pig small intestine. Naunyn-Schmiedeberg's Arch Pharmacol 309:153–157

Kromer W, Schmidt H (1982) Opioids modulate intestinal peristalsis at a site of action additional to that modulating acetylcholine release. J Pharmacol Exp Ther 223:271–274

Kromer W, Pretzlaff W, Woinoff R (1980a) Opioids modulate periodicity rather than efficacy of peristaltic waves in the guinea pig ileum in vitro. Life Sci 26:1857–1865
Kromer W, Scheibelhuber E, Illes P (1980b) Functional antagonism by calcium of an intrinsic opioid mechanism in the guinea-pig isolated ileum. Neuropharmacology 19:839–843
Kromer W, Höllt V, Schmidt H, Herz A (1981a) Release of immunoreactive-dynorphin from the isolated guinea pig small intestine is reduced during peristaltic activity. Neurosci Lett 25:53–56
Kromer W, Pretzlaff W, Woïnoff R (1981b) Regional distribution of an opioid mechanism in the guinea-pig isolated intestine. J Pharm Pharmacol 33:98–101
Kromer W, Steigemann N, Shearman GT (1982) Differential effects of SKF 10,047 (N-allyl-normetazocine) on peristalsis and longitudinal muscle contractions of the isolated guinea-pig ileum. Naunyn-Schmiedeberg's Arch Pharmacol 321:218–222
Leslie FM, Chavkin C, Cox BM (1980) Opioid binding properties of brain and peripheral tissues: evidence for heterogeneity in opioid ligand binding sites. J Pharmacol Exp Ther 214:395–402
Lord JA, Waterfield AA, Hughes J, Kosterlitz HW (1977) Endogenous opioid peptides: multiple agonists and receptors. Nature (London) 267:495–499
Majeed NH, Lason W, Przewlocka B, Przewlocki R (1985) Serotonergic regulation of the brain and gut beta-endorphin and dynorphin content in the rat. Pol J Pharmacol Pharm 37:909–918
Makhlouf GM (1987) Isolated smooth muscle cells of the gut. In: Johnson LR (ed) Physiology of the gastrointestinal tract, 2nd edn. Raven, New York, pp 555–569
Mathias JR, Sninsky CA (1985) Motility of the small intestine: a look ahead. Am J Physiol 248:G495–G500
Mihara S, North RA (1986) Opioids increase potassium conductance in submucous neurons of guinea-pig caecum by activating δ-receptors. Br J Pharmacol 88:315–322
Morita K, North RA (1981) Opiates and enkephalin reduce the excitability of neuronal processes. Neuroscience 6:1943–1951
Morita K, North RA (1982) Opiate activation of potassium conductance in myenteric neurons: inhibition by calcium ion. Brain Res 242:145–150
North RA, Tonini M (1977) The mechanism of action of narcotic analgesics in the guinea-pig ileum. Br J Pharmacol 61:541–549
North RA, Williams JT, Surprenant A, Christie MJ (1987) μ and δ receptors belong to a family of receptors that are coupled to potassium channels. Neurobiology 84:5487–5491
Paton WDM (1957) The action of morphine and related substances on contraction and on acetylcholine output of coaxially stimulated guinea-pig ileum. Br J Pharmacol 11:119–127
Ramabadran K, Bansinath M, Turndorf H, Puig MM (1988) Stereospecific inhibition of gastrointestinal transit by κ opioid agonists in mice. Eur J Pharmacol 155:329–331
Ross DH, Medina MA, Cardenas HL (1974) Morphine and ethanol: selective depletion of regional brain calcium. Science 186:63–65
Sarna SK, Lang IM (1985) Dose- and time-dependent biphasic response to morphine on intestinal migrating myoelectric complex. J Pharmacol Exp Ther 234:814–820
Sato T, Takayanagi I, Takagi K (1973) Pharmacological properties of electrical activities obtained from neurons in Auerbach's plexus. Jpn J Pharmacol 23:665–671
Schaumann W (1957) Inhibition by morphine of the release of acetylcholine from the intestine of the guinea-pig. Br J Pharmacol 12:115–118
Schulz R, Wüster M, Rubini P, Herz A (1981) Functional opiate receptors in the guinea-pig ileum: their differentiation by means of selective tolerance development. J Pharmacol Exp Ther 219:547–550
Sekiya K, Funakoshi A, Nakano I, Nawata H, Kato K, Ibayashi H (1986) Effect of methionine-enkephalin analog (FK 33-824) on plasma motilin. Gastroenterol Jpn 21:344–348
Shook JE, Pelton JT, Hruby VJ, Burks TF (1987) Peptide opioid antagonist separates peripheral and central opioid antitransit effects. J Pharmacol Exp Ther 243:492–500
Smith TW, Hughes J, Kosterlitz HW, Sosa RP (1976) Enkephalins: isolation, distribution and function. In: Kosterlitz HW (ed) Opiates and endogenous opioid peptides. Elsevier/North-Holland Biomedical Press, Amsterdam, pp 57–62
Sosa RP, McKnight AT, Hughes J, Kosterlitz HW (1977) Incorporation of labelled amino acids into the enkephalins. FEBS Lett 84:195–198
Szerb JC (1980) Effect of low calcium and of oxotremorine on the kinetics of the evoked release of [^{3}H]acetylcholine from the guinea-pig myenteric plexus; comparison with morphine. Naunyn-Schmiedeberg's Arch Pharmacol 311:119–127

Szurszewski JH (1987) Electrical basis for gastrointestinal motility. In: Johnson LR (ed) Physiology of the gastrointestinal tract, 2nd edn. Raven, New York, pp 383–422

Telford GL, Caudill A, Condon RE, Szurszewski JH (1988) Ketocyclazocine, a κ-opioid receptor agonist, and control of intestinal myoelectric activity in dogs. Am J Physiol 255:G566–G570

Van Nueten JM, Janssen PAJ, Fontaine J (1976) Unexpected reversal effects of naloxone on the guinea pig ileum. Life Sci 18:803–810

Vaught JL, Cowan A, Jacoby HJ (1985) μ and δ, but not κ, opioid agonists induce contractions of the canine small intestine in vivo. Eur J Pharmacol 109:43–48

Wang FS, Tsou J (1989) Substance P and [leucine]enkephalin release in guinea pig ileum during naloxone-precipitated morphine withdrawal. Pharmacol Exp Ther 249:329–332

Ward SJ, Takemori AE (1982) Relative involvement of receptor subtypes in opioid-induced inhibition of intestinal motility in mice. Life Sci 31:1267–1270

Waterfield AA, Kosterlitz HW (1975) Stereospecific increase by narcotic antagonists of evoked acetylcholine output in guinea-pig ileum. Life Sci 16:1787–1792

Weisbrodt NW (1987) Motility of the small intestine. In: Johnson LR (ed) Physiology of the gastrointestinal tract, 2nd edn. Raven, New York, pp 631–663

Weisbrodt NW, Sussman SE, Stewart JJ, Burks TF (1980) Effect of morphine sulfate on intestinal transit and myoelectric activity of the small intestine of the rat. J Pharmacol Exp Ther 214:333–338

Wolter HJ (1985) Dynorphin-A(1–8) and gamma-melanotropin exist within different myenteric plexus neurons of rat duodenum. Biochem Biophys Res Commun 131:821–826

Wood JD (1987) Physiology of the enteric nervous system. In: Johnson LR (ed) Physiology of the gastrointestinal tract, 2nd edn. Raven, New York, pp 67–109

Yamamoto H, Harris RA, Loh HH, Way EL (1978) Effects of acute and chronic morphine treatments on calcium localization and binding in brain. J Pharmacol Exp Ther 205:255–264

CHAPTER 19

A Neuroendocrine Perspective of Sex and Drugs

O.F.X. Almeida and D.G. Pfeiffer

1 Introduction

The last 20 years have witnessed an alarming increase in the use of addictive drugs (including opiates), coincident with a more liberal view on sexuality. These contemporaneous events inevitably led to the popular association between drugs and sex which may well have contributed to the desire of many young people to "experience drugs". On the other hand, the attack on drug-taking by some sectors of society was probably only using this accidental timing of events to show a disdain for the "new sexuality". One wonders if the present "drug problem" would have been averted more effectively had it been appreciated that the body itself makes and uses opiate-like substances (endogenous opioid peptides, EOP) and perhaps opiates also, and that these substances, by and large, suppress many reproductive processes, of which sex is but one. Indeed, this chapter could well be retitled: "Opioidergic inhibition of reproductive functions".

This review will be confined to the role of opioids in the regulation of the reproductive-neuroendocrine axis, which traditionally includes the hypothalamus, pituitary, gonads, mammary glands and uterus, as well as the brain centres which coordinate reproductive behaviours (Fig. 1). Although recent work has identified opioids in peripheral reproductive tissues (e.g. gonads, placenta and mammaries; see Schafer et al. this Vol.), we will focus on the control of reproduction by hypothalamo-pituitary opioids only. Even within this narrow framework, our review offers limited coverage of the subject – a point better appreciated when one considers that a highly selective bibliography covering the period 1979–1984 lists 317 key references (Ellendorf 1985) and that a recent symposium on brain opioid systems in reproduction resulted in a volume of some 360 pages (Dyer and Bicknell 1989)!

Why emphasize hypothalamo-pituitary opioids? Emphasis on these is justified on at least three grounds. Firstly, peripheral reproductive tissue opioids have only recently been identified and therefore their functions (paracrine?) have been less well studied. Secondly, opioids originating in the hypothalamus are "better placed" to participate in the integration of reproductive processes (cf. Fig. 1). Thirdly, both the posterior pituitary (neurohypophysis, NP) and the anterior pituitary (AP; adenohypophysis) are now known to contain opioid receptors; while the former receives central opioid afferents, the latter is itself a major source of EOP but it is also known to receive hypothalmic opioids via the hypothalamo-hypophysial vasculature.

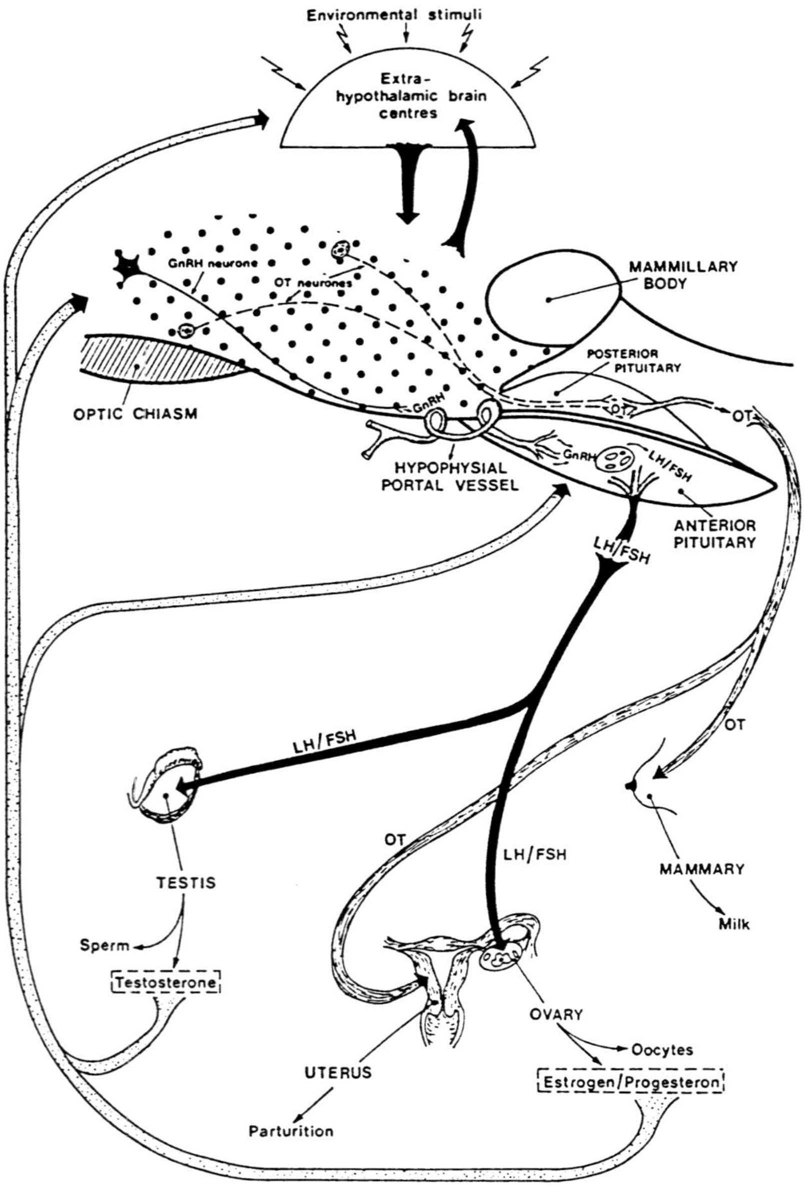

Fig. 1. Schematic representation of integration within the hypothalamo-pituitary-gonadal axis, and the hypothalamo-pituitary-mammary/uterine axis. The *stippled area* represents the hypothalamus. In the rat, gonadotropin-releasing hormone (GnRH) cell bodies are found mainly in the preoptic area; their axons terminate in the median eminence. GnRH, carried in the hypophysial-portal vessels, stimulates the release of luteinizing hormone (LH) and follicle stimulating hormone (FSH) from pituitary gonadotropes. The steroids (testosterone, estrogen and progesterone) feedback upon the hypothalamo-pituitary unit. Oxytocin (OT) cell bodies in the supraoptic and paraventricular nuclei release OT from their terminals in the posterior pituitary from which the hormone reaches its target organs, the mammary glands and uterus, via the general circulation

This chapter does not cover the control of the AP hormone prolactin (PRL) by EOP, although PRL seems to play an important part in the secretion of gonadotropins, the production of milk, and reproductive behaviours (see Knobil et al. 1988). It is sufficient to state that EOP stimulate PRL secretion, apparently by inhibiting the activity of tuberinfundibular dopaminergic (DA) neurons.

2 Opioidergic Inhibition of Gonadotropin and Gonadal Secretions

2.1 Background

Hypothalamic neurosecretory cells release gonadotropin releasing hormone (GnRH) from their terminals in the median eminence into the hypothalamo-hypophysial portal vessels, and thence into the AP. Here, GnRH stimulates the synthesis and secretion of the gonadotropins luteinizing hormone (LH) and follicle stimulating hormone (FSH) from the gonadotropes (Fig. 1); LH and FSH reaching the gonads via the general circulation promote steroid (androgen, estrogen and progestin) synthesis and release (LH) and maturation of ova and spermatozoa (FSH). The major gonadal steroids (tesosterone in the male; estradiol and progesterone in the female) influence germ cell maturation and also induce and maintain secondary sexual characteristics and sexual behaviours while, at the same time, feeding back on the hypothalamic-pituitary unit to maintain homeostasis in the system (Fig. 1).

Since GnRH is secreted from neurons, it is perhaps not surprising that it is released in distinct quanta or pulses (measured directly in the hypothalamo-hypophysical portal blood, or reflected in the pattern of LH appearance in the systemic blood). LH release seems to be subject to frequency and amplitude modulation, and the LH secretory profile is strongly influenced by the steroid milieu. This is particularly obvious during the progress of the ovarian cycle (see Knobil et al. 1988). Males of some species also display significant frequency/amplitude modulation of LH release. In contrast, male rats provide a simple model for studying steroid-gonadotropin interactions: LH release appears to occur at a more or less constant amplitude and low frequency and, although both parameters increase upon castration, single-point measures of LH still seem to adequately reflect the dynamic state of the hypothalamic-pituitary unit.

The discovery of discrete LH pulses led to the search for the "GnRH pulse generator". The "generator" obviously comprises the GnRH neurons, but these must themselves be driven by a complex neural circuitry. As mentioned, gonadal steroids modify GnRH and LH release, but GnRH neurons apparently lack steroid binding sites. Steroids must therefore act through other neurotransmitter/modulator systems in the brain. Classical neurotransmitters (e.g. DA, norepinephrine, serotonin) have been implicated as possible mediators (see Kalra and Kalra 1983) but, more interestingly, some hypothalamic opioid neurons have been found to bind steroid hormones (Morrell et al. 1985).

2.2 *Hypothalamic or Pituitary Site for Opioidergic Inhibition of GnRH/LH Secretion?*

Studies on the opiate regulation of reproduction began at least 50 years before the discovery of the EOP (see Barraclough and Sawyer 1955), using the prototypic opioid ligand morphine. The classical paper of Barraclough and Sawyer (1955), describing morphine inhibition of ovulation in rats, pinpointed the site of opiate action to the hypothalamo-pituitary unit (they had earlier postulated that ovulation occurs as a result of a surge discharge of LH which is controlled by a "neurogenic timing factor"). Barraclough and Sawyer (1955) also presented data which suggested the development of opiate tolerance within the neuroendocrine system: after an initial suppression, the formation of corpora lutea in the ovary was again observed in rats chronically exposed to a constant dose of morphine.

Later studies clearly showed that morphine exerts its antigonadotropic actions at a central site. For instance, only those doses of morphine which produce changes in the electrical activity of the brain have anti-ovulatory effects (Sawyer et al. 1955); the morphine blockade of ovulation is surmountable by electrical stimulation of the median eminence where GnRH neurons terminate (Sawyer 1963); and single, relatively low doses of GnRH overcome the anti-ovulatory and LH-suppressive actions of morphine (Cicero et al. 1977; Pang et al. 1977). Furthermore, morphine inhibits the efflux of GnRH both in vivo (Ching 1983) and in vitro (Mehmanesh et al. 1988).

Opioid receptor mediation of the above effects of morphine have been amply documented by the demonstration that they are reversible with the opiate receptor antagonist naloxone (NAL) (Bruni et al. 1977; Pang et al. 1977). Furthermore, NAL, which itself enhances GnRH and LH release under most conditions, fails to stimulate LH secretion if it is co-administered with a GnRH antagonist (Blank and Roberts 1982), or if the median eminence has been surgically ablated (Panerai et al. 1983); also, GnRH-deficient women fail to respond to NAL with an increase in LH levels (Blankstein et al. 1981). Data showing that natural and synthetic opioids can inhibit GnRH and LH secretion when applied either into the cerebral ventricle (i.c.v.), or onto isolated hypothalami in vitro, also support a central site of opioid action (see Mehmanesh et al. 1988). In addition, Leranth et al. (1988) demonstrated synaptic connections between EOPergic axons and GnRH cell bodies in the rat hypothalamus.

One problem relating to the central effects of opioids upon GnRH needs to be discussed. It concerns the failure to observe opioidergic suppression of FSH; the reason for this may lie in one of the following: (1) although GnRH also stimulates FSH secretion, the latter seems to depend upon the amplitude, rather than frequency, of GnRH pulses; note that opioids may modulate GnRH pulse frequency rather than amplitude (cf. Schanbacher 1985); (2) there is now evidence that there may be hypothalamic FSH-releasing factor neurons (other than GnRH neurons); these may be opioid-insensitive; (3) FSH has a longer biological half-life than LH, making its disappearance following opioid treatment difficult to detect; or (4) as mentioned below, some gonadotropes appear to bear opioid receptors which may respond to hypothalamic EOP; these may be LH-, rather than FSH-producing cells.

Until recently, the main arguments against an AP site of EOP inhibition of LH were based upon the inability to detect opioid receptors in this gland and the failure of opiate agonists and antagonists to alter LH release (basal and GnRH-stimulated) from pituitaries incubated in vitro (see Cicero et al. 1977). However, recent binding (Blank et al. 1986) and immunocytochemical (A.H.S. Hassan, A. Ableitner and O.F.X. Almeida, unpubl.) studies detected opioid receptors in the AP; there have also been new reports of NAL-reversible opiate and opioid suppresion of LH release from cultured gonadotropes (e.g. Cacicedo and Franco 1986). In addition, when the coincidence between increasing β-endorphin (β-END) concentrations in the hypothalamo-hypophysial blood and decreasing LH levels in the general circulation (Ferin et al. 1984) is considered, an AP site of action seems all the more plausible. Nevertheless, more work on the AP needs to be done to clarify the situation. Investigators in this area should be aware that effects on the hypothalamus may manifest themselves in the AP in a complex way, leading to erroneous interpretations; e.g. Barkan et al. (1985) found that opiate-induced changes in GnRH binding in the AP were actually due to alterations in GnRH secretion; an oversight of the hypothalamic aspect had previously led them to conclude that opioids may act directly on AP GnRH receptors.

2.3 Opioid Peptides, Receptors and Rhythms Mediating GnRH/LH Inhibition

All three major EOP found in the hypothalamus and AP (β-END; dynorphin, DYN; and methionine- and leucine-enkephalin, Met-ENK and Leu-ENK) suppress GnRH and/or LH in vitro and in vivo. These effects are apparently mediated by opioid receptors since they are NAL-reversible. As discussed (Simon this Vol.), β-END acts predominantly at μ-receptors (ε-receptors with which it also interacts have hitherto not been demonstrated in the AP or centrally), DYN at κ-receptors sites, and ENK at δ-receptor sites. The development of specific opioid antagonists made it possible to distinguish which receptors (and therefore peptides) are important regulators of GnRH/LH secretion; since each peptide may have a distinct physiological function, the results of such studies could potentially be applied to developing specific therapies for EOP-based disorders of GnRH/LH secretion. Utilization of specific μ (e.g. β-funaltrexamine, and low doses of NAL), κ (e.g. norbinaltorphimine) and δ (e.g. ICI 174684) opioid receptor antagonists revealed predominant μ-receptor regulation of LH secretion; κ, but not δ, receptors were also shown to play a role (Pfeiffer et al. 1987; Almeida et al. 1988d). Comparable data have been obtained by infusing antibodies specifically directed against each peptide (e.g. Schulz et al. 1981), although the penetrability, affinity and capacity of these molecules, as well as the possibility of delayed responses which might be missed, must be considered.

The observation that "opiate naive" subjects respond to opioid antagonists with elevations of GnRH/LH gave rise to the concept of "tonic opioid inhibition" of this neuroendocrine axis (see Almeida et al. 1986). However, the magnitude of the antagonist-induced LH response varies considerably from one experiment to another. Some of the variability might be attributed to fluctuations in the steroid background (see later), but at least some of it appears to be related to the time of day at which the tests are done. Blank and Mann (1981) made the first systematic study of this problem in immature female rats and suggested that diurnal variations in NAL-responsiveness might exist. Jacobson and Wilkinson (1986) later reported diurnal variations in opioid binding in the hypothalamus of similarly aged female rats. Our group found that adult male rats also display time-of-day related rhythms in their ability to release LH upon challenge with μ- and κ-receptor antagonists, and that the μ- and κ-responsive phases are asynchronous (Almeida et al. 1988d). While two daily periods of high μ-activity exist (approximately 12 h apart), there is only a single period (early dark phase) of high κ-activity (Fig. 2). Interestingly, these rhythms in opioid receptor function related to LH secretion showed some correlation with previously known rhythms in hypothalamic EOP content, but a cautionary note about the interpretation of simple correlations is necessary. A rhythm in the ability of μ-receptors to mediate the release of PRL was also observed in our experiments; however, only a single peak of high μ-activity, occurring out of phase with that for LH, was found (Fig. 2), indicating that separate EOP pathways with different temporal patterns of activity modulate the secretion of LH and PRL. These observations may be significant for understanding the role of opioids in the generation of behavioural and other endocrine circadian rhythms. Nevertheless, it is tempting to speculate that the high pulse frequency of LH secretion that occurs predominantly during the night in pubertal children and preovulatory women might represent "withdrawal from a periodically enhanced EOP tone"; this intriguing possibility is, to some extent, supported by data on the role of EOP in the organization of the preovulatory LH surge in rats (Lustig et al. 1989).

2.4 Opioids Mediate Corticotropin-Releasing Hormone Actions on GnRH/LH Secretion

Many types of stress inhibit GnRH/LH secretion. Stressors activate the hypothalamic (corticotropin-releasing hormone, CRH)-pituitary (proopiomelanocortin, POMC-derived peptides)-adrenal (glucocorticoids) axis, and it has long been suspected that the deleterious effects on reproduction result from the negative feedback actions of adrenal glucocorticoids upon the central pathways involved in GnRH release. While adrenal steroids may well suppress GnRH/LH secretion, it has recently become clear that the major part of the inhibition results from CRH-stimulation of hypothalamic EOP. The evidence summarized below is reviewed in detail elsewhere (Almeida et al. 1989).

LH secretion in rats is markedly reduced following the central (i.c.v.) application of CRH (Rivier and Vale 1984); however, this inhibition seems to occur primarily at a central site since CRH suppresses the efflux of GnRH from hypothalamic slices in vitro (Nikolarakis et al. 1986). In view of the fact that CRH is a potent secretagogue of POMC-derived peptides (e.g. β-END) from the pituitary (see Kley et al. this Vol.),

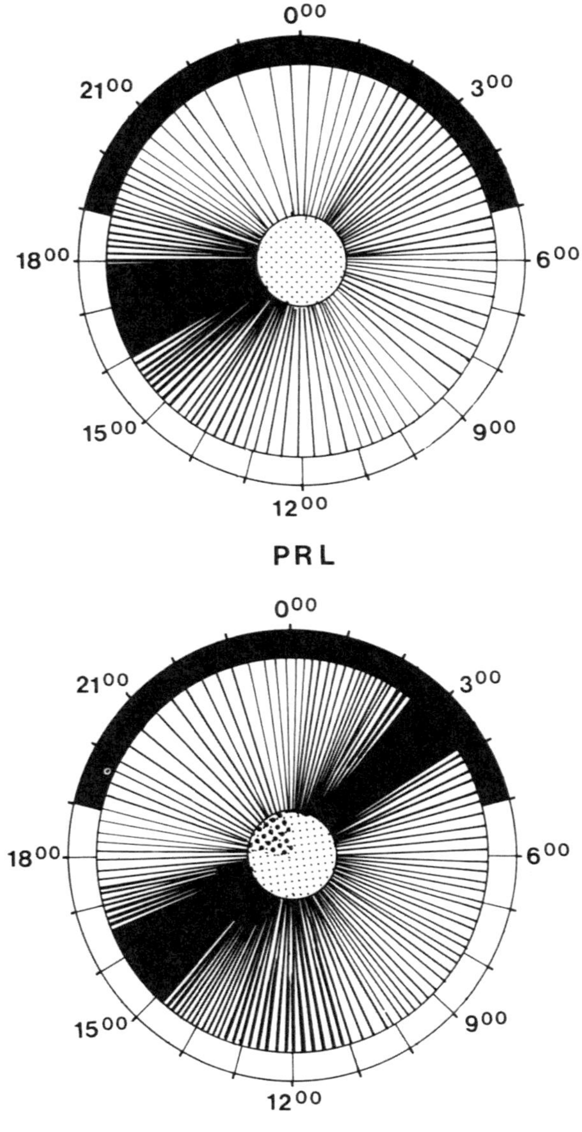

Fig. 2. Semi-quantitative representation of degree of EOP control of *PRL* and *LH* secretion throughout the 24 h light/dark cycle in male rats. Density of *radial lines*, and intensity of *stippling* in the centre are proportional to the magnitude of EOP (μ and κ, respectively) influences. Note that the periods of high and low EOP influences upon PRL and LH are not completely coincident. (Data drawn from Almeida et al. 1988d)

we hypothesized that CRH might also be responsible for the release of hypothalamic EOP which could, in turn, inhibit GnRH secretion. Indeed, CRH was found to release β-END, DYN and Met-ENK from the rat hypothalamus in vitro (see Burns and Nikolarakis this Vol.), and immunocytochemical studies also suggest CRH-EOP neuronal interactions in the hypothalamus (A.H.S. Hassan and O.F.X. Almeida, unpubl.). Studies from this and other laboratories later proved that EOP mediate the inhibitory actions of CRH upon GnRH/LH release by demonstrating an attenuation of the CRH (and stress) effects upon pretreatment with opioid antagonists and antibodies directed against opioid peptide sequences (Petraglia et al. 1986c; Almeida et al. 1988b). Experiments showing that CRH does not prevent LH secretion in animals exhibiting reduced sensitivity to opiates (long-term castrated male rats, see later; Almeida et al. 1988b) and in male rats made tolerant to morphine (Nikolarakis et al. 1988a) also supported the view that EOP mediate the anti-GnRH/LH actions of CRH. In most studies however, we observed that EOP account for only 70–80% of the actions of CRH, suggesting that CRH may act directly to inhibit GnRH release, or that other neural systems may be involved. In fact, recent immunocytochemical work has demonstrated direct CRH-GnRH neuronal connections (MacLusky et al. 1988).

Perhaps the most convincing proof of EOP mediation of the CRH inhibition of GnRH/LH secretion comes from in vivo (push-pull) and in vitro perfusions of the hypothalamus with a CRH antagonist, α-helical CRH_{9-41} (Nikolarakis et al. 1988b). These studies showed a concomitant decrease and increase in hypothalmic EOP and GnRH release, respectively. Furthermore, they extended the concept of "tonic inhibition" in suggesting that the inhibition may be secondary to a tonic stimulation of EOP neurons by CRH. In light of the rhythmic EOP control of LH discussed earlier, it would be interesting to know whether there are rhythmic fluctuations in the degree of this supposed CRH tone; such rhythms seem likely in view of the well-known diurnal rhythms in adrenal function.

2.5 *Opioid-Gonadal Steroid Interactions*

We previously referred to the importance of gonadal steroid negative feedback in the maintenance of GnRH/LH secretion. That EOP may mediate this feedback was first recognized by Cicero et al. (1979) who observed that NAL could counteract the LH-suppressive effects of testosterone in castrated male rats, most probably by acting at a central site. Later investigations showed that NAL could also "antagonize" the feedback actions of estrogen and progesterone upon LH secretion in females (van Vugt et al. 1982; Casper and Alapin-Rubillovitz 1985; Shoupe et al. 1985). Data obtained in a number of species showing that NAL-induced increases in serum LH levels are greatest when gonadal steroid secretion is maximal, and absent or markedly attenuated in the castrated or low steroid-secretory states, also support the notion of EOP mediation of steroid feedback (Reid et al. 1983; Melis et al. 1984; Almeida et al. 1988c).

The adult castrated male rat has proved an interesting model for studying opioid-steroid relationships since long-term castration (> 2 weeks) results in a hypersecretion of LH despite reduced hypothalamic GnRH gene expression, reserves and release (see Almeida et al. 1988a); this is accompanied by a subsensitivity to the LH-suppressive actions of testosterone (Gabriel et al. 1986) and of opiates such as morphine (Cicero et al. 1982; Almeida et al. 1987b). Treatment of castrated males with testosterone promotes GnRH production, storage and release, while morphine treatment raises hypothalamic GnRH levels. Likewise, testosterone replacement restores sensitivity to morphine, and castrated rats treated with morphine regain their responsiveness to testosterone (see Almeida et al. 1988a). It thus appears that morphine and testosterone can substitute for each other and that they may share common neural substrates.

There is now good evidence indicating that hypothalamic EOP systems are strongly influenced by gonadal steroids. Firstly, combined autoradiographic-immunocytochemical analysis has revealed that some EOP neurons bind gonadal steroids (Morrell et al. 1985). Secondly, sex steroid treatment results in altered (reduced) opioid (POMC) gene expression (Wilcox and Roberts 1985; Chowen-Breed et al. 1989). Thirdly, the content and release of hypothalamic EOP may be modulated by sex steroids (Almeida et al. 1987a; Nikolarakis et al. 1989) in a dose-dependent fashion; of particular significance are the findings that the in vitro basal release of β-END and DYN increase with increasing doses of testosterone replacement in castrated male rats (Almeida et al. 1987a). Recent (unpubl.) observations by us suggest that at least part of the gonadal steroid effects upon EOP neuronal activity may occur as a result of their modulation of CRH synthesis and release.

A number of groups have addressed the question of whether gonadal steroids alter opioid binding in the brain. While Hahn and Fishman (1985) reported increased binding in whole brain homogenates following castration, Wilkinson et al. (1985) noted that chronic estrogen (and, to a lesser extent testosterone) treatment increases opioid binding in the anterior hypothalamus. Other groups failed to detect any steroid-induced effects, thus leaving the question open to more rigorous examination of discrete hypothalamic nuclei.

There remain some aspects of the opioid mediation of steroid feedback that need clarification, e.g. (1) although most reports indicate that opiate agonists and antagonists are only effective in the presence of gonadal steroids, the reason(s) for some contrary findings (e.g. NAL-stimulation of LH secretion in castrated rats and hypogonadal men; Cicero et al. 1980; Veldhuis et al. 1982) are not completely clear. Factors that may need consideration include the period of steroid deprivation or replacement (in analogy to the way time of exposure to opiates determines the degree of tolerance and/or dependence developed; see Chaps. 22 and 23); the method of analyzing the LH secretory profile (e.g. castrated rams reportedly respond to morphine with reduced LH pulse frequency and amplitude, but to NAL with increased pulse frequency only; Schanbacher 1985); and the possibility that sex differences may exist in the steroid dependence of opiate responsiveness (see Almeida et al. 1988c). (2) Reports of castration-induced subsensitivity to opiate agonists exist alongside reports that NAL stimulates LH secretion in castrates (e.g. Cicero et al. 1980, 1982). The latter results suggest endogenous ligand occupation of opioid receptors, so it is surprising that agonists do not elicit a response (unless all the receptor sites are occupied by EOP – unlikely for reasons discussed earlier). (3) Both the EOP and morphine suppression of LH are attenuated during the preovulatory phase (period of elevated LH, estrogen and progesterone – "positive feedback of steroids") in the rat (Petraglia et al. 1986b). This case contrasts markedly with the usual one of simultaneous increased EOP activity and increased steroid secretion, and suggests that a switch in opioid activity occurs at this time, probably under the influence of rising sex steroid levels (see Jacobson and Kalra 1989; Lustig et al. 1989).

3 Opioids and Reproductive Development

3.1 EOPergic Activity and the Onset of Puberty

Literally, puberty in humans begins with the appearance of pubic hairs, a criterion that obviously cannot be applied to other species. Instead, the start of ovulatory cycles and sperm production are used as pubertal indices. However, the onset of puberty begins prior to these events; depending on the species, the process may start after only a few days of life or after several years. For the purposes of this chapter, the onset of puberty will be considered as the time when there is an increase in the activity of the GnRH/LH system. Our present understanding of the mechanisms leading to puberty has not advanced much beyond the "which comes first: the chicken or the egg?" stage. Some investigators maintain that the trigger for sexual development is of central origin. Others adhere to the so-called gonadostat theory, according to which the hypothalamo-pituitary unit is kept in check by low circulating levels of gonadal steroids until a dramatic reduction occurs in the sensitivity of the hypothalamus and/or pituitary to the negative feedback effects of gonadal steroids. While each of these views is supported by good arguments, there is, as yet, no concensus opinion. Could it be that EOP, by interacting with gonadal steroids, form part of the "gonadostat"? Or might the maturation of hypothalamic EOP systems determine the central control of the onset of puberty? Clues to the latter question await further studies on the ontogeny of hypothalamic opioid receptors (there is a dearth of information on this subject), but meanwhile, pharmacological experiments have cast some light on the first question.

If GnRH/LH secretion is restrained by EOP in the prepubertal period, NAL should cause a disinhibition of the GnRH/LH system. Studies in female rats from the perinatal period through to adulthood revealed fluctuations in the "EOPergic inhibitory tone", as judged by the LH response to NAL. Thus, NAL produced significant increases in serum LH concentrations between days 10–15 of life, at around age 25 days, and again, between the ages of 35–60 days. The last finding is perhaps the most significant since female rats commence their ovulatory cycles and show other signs of sexual maturity (e.g. vaginal opening) at around 40 days of age. In contrast, developing males show their first significant LH response to NAL from about days 30–40 of life (pubertal phase), and the magnitude of response steadily increases until sexual maturity is achieved. These data thus suggest that EOP activity may be increased before the onset of puberty, the timing of which differs between the sexes (Blank et al. 1979; Ieiri et al. 1979; Cicero et al. 1986).

Studies by Schulz et al. (1985) and Sylvester et al. (1985) showed that perinatal androgenization of females, and castration of males, results in "defeminization" of the LH response to NAL, i.e. genetic males now respond to NAL by releasing LH between days 10–15 of life. "Defeminization" results from androgens being aromatized to estrogens (see Almeida et al. 1988c); we recently showed that antagonism of estrogen receptors prevents this process insofar as the response to NAL is concerned (Almeida and Schulz 1988).

As in the adult rat, μ and κ opioid ligands appear to be responsible for suppressing GnRH/LH secretion in developing rats. The central administration of antisera raised against β-END and DYN, but not Met-ENK, stimulated LH secretion in immature female rats (Schulz et al. 1981), and only preferential μ and κ opioid antagonists increased serum LH levels in immature male rats (Cicero et al. 1988). These results are consistent with the limited data on the ontogeny of opioid receptors in rats (see Kornblum et al. 1987).

To test whether EOP may be part of the "central brake" on GnRH/LH release in the developing rat, Sirinathsinghji et al. (1985) injected female rat pups with NAL (a relatively short-lived antagonist), four times daily, during the first 10 days of life. This treatment resulted in significantly greater levels of LH and FSH, and also in an advance of the timing of vaginal opening and first displays of ovulation. Moreover, NAL treatment caused a greater number of ova to be released on the first ovulation. These observations suggest that EOP contribute to the regulation of the onset of puberty. However, several other groups (e.g. Meijs-Roelofs and Kramer 1988) failed to observe any advancement in the onset of puberty in female rats treated with either NAL or the longer-acting antagonist naltrexone (NLTRX). Likewise, NAL (several hours) and NLTRX (several weeks) infusions in boys with delayed puberty were ineffective (Veldhuis et al. 1982; Kulin et al. 1987). Before opioid antagonism is abandoned as a potential therapy, however, it may be worth trying prior or concomitant priming with gonadal steroids (in case the "gonadostat theory" applies, and because steroids seem to be conducive to opioid antagonist-induced LH secretion). It is pertinent to note that Petraglia et al. (1986a) found NAL to acutely stimulate LH secretion in boys and girls that were in their final stages of puberty, when gonadal steroids were already elevated. It is also interesting that Mauras et al. (1986) found chronic (1 month) NLTRX treatment to stimulate LH pulse frequency in late-pubertal boys, whereas the same treatment resulted in an inhibition of LH pulse frequency in sexually immature individuals.

There are inconsistent reports as to when the immature rat first shows opioid agonist-induced suppressions of LH secretion. Ieiri et al. (1979) found that morphine inhibits LH around day 15 of life in both sexes, whereas Cicero et al. (1986) found males to respond earlier (day 15) than females (days 30–35). As already discussed, NAL treatment elicits LH secretion in females earlier than in males. The result of Ieiri et al. (1979) may be simply explained in terms of a greater and earlier EOP influence in females; the latter may relate, in some way, to the different levels of sex steroids in developing males and females, or to the sex-dependent differences in the steroid feedback thresholds at a given age. The result of Cicero et al. (1986) is less easily understood: the sex difference in agonist- and antagonist-induced responses are apparently precisely out of phase; their data also suggest the occurrence of opiate insensitivity and/or dependence in one sex but not the other. The phenomenon in females appears to be unrelated to dependence since NAL elevates LH levels at a time when morphine has no effect.

We earlier highlighted the pivotal role of gonadal steroids in determining the adult LH response to opioids. Indeed, steroids also play an important role in the EOP control of LH in the developing rat, as shown by the extensive work of Bhanot and Wilkinson (1983, 1984). These authors found that the dose of agonist required to inhibit LH in prepubertal rats of both sexes increases with age (presumably as sex steroid levels increase), and that gonadectomy of 23-day-old male and female rats prevents the NAL enhancement and opiate suppression of LH secretion; the opiate responsiveness of these animals could be reinstated by estrogen and testosterone supplementation. Although the data from the NAL experiments are somewhat controversial (e.g. see Cicero et al. 1986), those with the opiate agonist in males have been confirmed (with morphine); the latter are considered to reflect tolerance in the opioid-responsive elements regulating the GnRH/LH system (Cicero et al. 1986).

3.2 Sex Differences in Gonadectomy-Induced Subsensitivity to Opiates

This subject is covered here because, in the rat at least, sexual differentiation of the brain is traditionally believed to occur during the first 10 days of post-natal life (see Almeida et al. 1988c). Some aspects of sex differences in the EOP control of GnRH/LH secretion prior to sexual maturity have already been discussed above.

Studies in our laboratory (Almeida et al. 1987b) have shown that ovariectomy of adult rats does not result in LH insensitivity to morphine (in contrast to prepubertal ovariectomy; see above); on the other hand, castration of adult males does. The latter result agrees with that previously obtained in adult male rats (Cicero et al. 1982). Whereas one might expect the prepubertal gonadectomy-induced opiate subsensitivity observed by Bhanot and Wilkinson (1983) to last through to adulthood, we found this to be only a transient effect: at around the normal age of sexual maturity, prepubertally castrated males regained their responsiveness to morphine (Almeida et al. 1987b). Morphine significantly inhibited LH secretion in prepubertally castrated rats given testosterone after the normal age of sexual maturity, but had no effect in prepubertal castrates supplemented with the steroid between the ages of 26 and 50 days (the latter period is one during which GnRH/LH secretion is normally accelerated). We, in fact, found that the "window" of testosterone sensitivity is much narrower (days 45–57 of age) (Almeida et al. 1987b). Apart from showing the importance of age and duration of castration in the determination of the GnRH/LH response to opiates, these observations shed new light on the idea of a "critical period" of sexual differentiation of the rat brain. They suggest that the differentiation that occurs in the perinatal period may remain plastic up to the time just preceding adulthood, at least insofar as EOP influences on GnRH/LH secretion are concerned.

4 Opioids in Seasonal Reproduction

Most mammals are seasonal breeders, the best experimental examples of which are the golden hamster and sheep which are stimulated into reproductive activity by long and short daily photoperiods, respectively. The seasonal fluctuations in the activity of the reproductive system are in some respects reminiscent of the transition into puberty, and have often been referred to as "recurring puberty". The ecological significance of the phenomenon is that the young are produced at a propitious time of the year. The mechanisms by which day length is measured so as to time the preparation for reproduction are beyond the scope of this review; briefly, they depend upon the circadian secretion of the pineal hormone melatonin (Tamarkin et al. 1985), although the neural transducer(s) of the information contained in the melatonin signal to the GnRH/LH system are, as yet, unknown.

Recent studies have focussed on the possible involvement of EOP in translating the photoperiodic cues regulating reproduction. Hamsters and sheep (like the non-seasonal rat) are most sensitive to agonist supression, and antagonist disinhibition, of LH secretion when the photoperiod dictates increased gonadal steroid secretion (Eskes et al. 1984; Ebling and Lincoln 1985; Roberts et al. 1985). Careful analysis of the LH secretory profile in sheep treated with morphine and NAL revealed that these substances act primarily to modulate LH pulse frequency rather than amplitude (Ebling and Lincoln 1985; Schanbacher 1985).

The interaction between photoperiod and steroids in determining the LH response to exogenous opiates is more subtle however. Surgical or functional pinealectomy (PINX) uncouples reproductive cycles from the photoperiod, i.e. the animals show reproductive cycles that are asynchronous with the prevailing photoperiod. Ebling and Lincoln (1985) and Roberts et al. (1985) found that photoperiod-desynchronized sheep and hamsters only responded to morphine and NAL treatments at times when their gonads were maximally active. Other studies in the hamster showed that castration results in a loss of opiate responsiveness (Eskes et al. 1984); more interesting, however, was the observation that castrates replaced with testosterone regained their opiate responsiveness only when exposed to stimulatory day lengths (Roberts et al. 1985; Swann and Turek 1983). This result implies that photoperiod regulates the sensitivity of the EOP pathways to gonadal steroids, and that the lack of opiate sensitivity of photoinhibited animals is not due to low steroid secretion per se, although an action of EOP in the initial phases of photo-induced gonadal involution cannot be excluded (see Hastings et al. 1985). On the other hand, hamsters which become photorefractory after extended exposure to inhibitory day lengths and show photoperiod-independent gonadal recrudescence, regain their sensitivity to NAL; in this case, steroids appear to determine the EOP control of LH secretion, independently of the photoresponsive state of the animal (Roberts et al. 1985).

Hypothalamic EOP concentrations have been shown to be subject to the influence of both photoperiod and gonadal steroids. In hamsters, the hypothalamic content of β-END and Met-ENK are greater during photoinhibition than photostimulation (Kumar et al. 1984; Roberts et al. 1985). Moreover, there is a marked nocturnal increase in hypothalamic β-END levels in reproductively quiescent (but not photostimulated) hamsters; this pattern is also seen when gonadal involution is induced by PINX and daily injections of melatonin (Roberts et al. 1985; Hastings et al. 1985). Castration of photostimulated hamsters also results in significant increases in the hypothalamic content of β-END, an effect which is testosterone-reversible (Roberts et al. 1987). In contrast, no significant changes are seen following castration (with or without testosterone replacement) of photoinhibited animals (Roberts et al. 1987).

The plasma and AP content of β-END also seem to be related to photoperiod and, therefore, perhaps to steroid levels. Photostimulation in the sheep is accompanied by reduced pituitary reserves, and increased plasma levels of β-END (Ebling and Lincoln 1987). However, plasma levels of β-END are elevated in photoinhibited hamsters (Chen et al. 1984; Kumar et al. 1984). These findings are confusing since short daily photoperiods are stimulatory for sheep and inhibitory for hamsters; it could well be that AP and/or plasma β-END concentrations are not related to GnRH/LH secretion, and that they do not reflect photoperiod and steroid modulation of central EOP activity. This seems all the more surprising since CRH is a secretagogue for both hypothalamic and AP β-END (see Burns and Nikolarakis this Vol.); the role of CRH in the control of seasonal reproduction has not been studied to date.

Seasonal breeders clearly show many similarities to the standard rat models in their opioidergic control of reproduction. However, their profound susceptibility to changes in day length adds a further interesting dimension to EOP-reproductive research. Apart from this added exoticism, studies on the reproductive effects of EOP in such animals are important because many species of agricultural importance breed seasonally. In this regard, it is noteworthy that Chen et al. (1984) partially prevented photo-induced testicular regression in hamsters given daily injections of NAL. Studies in the economicallly more important sheep succeeded in inducing increased LH pulse frequency by injecting NAL six times daily (Ebling and Lincoln 1985); although the LH response rapidly dwindled, the authors were apparently optimistic that this could be overcome by adjusting (increasing) the intervals between successive injections.

5 Opioid Control of Sexual Behaviour

Sexual behaviour is essential for reproduction in that it provides the opportunity for the fusion of gametes. It is a social behaviour involving the interaction between individuals of the same species which commences in the form of play much before sexual maturity. In the adult, sexual behaviour involves "mate choice" (sexual attraction and/or selection of a partner with certain physical or genetic attributes or with a particular deportment), "sex drive" (motivation for sexual interaction, possibly triggered by factors listed under "mate choice"), and "sexual performance" (the physical aspect of sexual interaction). Each of these components is estrogen- and androgen-dependent and sexually differentiated; after sexual maturity, the action of steroids is probably primarily "permissive", allowing brain centres to be able to "choose" and "drive", and spinal reflexes to "perform".

Social and even legal taboos (see Mendelson et al. 1978) have hampered research on the effects of opiates on sexual behaviour in humans, but from the little information available (mainly on males), one may conclude that opiates dampen sexual appetite and performance. For example, heroin and methadone users have a reduced sexual desire and ability to gain or maintain penile erection (see Serra et al. 1988) and there are documented (e.g. Buffum 1982) and anecdotal reports that heroin is used in some cultures to delay ejaculation. The spontaneous orgasms, erections and ejaculations accompanying opiate withdrawal also suggest that opioid receptor activation inhibits sexual behaviour (see Serra et al. 1988). An important observation with relevance to the management of erectile impotence in men is that opioid antagonists can, even in the absence of erotic stimuli, induce erections in opiate-naive subjects (Mendelson et al. 1978; Fabbri et al. 1989). This result implies that EOP may play a role in the pathophysiology of erection and possibly, sexual motivation; it is not at present possible to distinguish between brain and/or spinal sites of this action. Currently, there is no information available on the effects of opioid receptor activation upon sexual behaviour of women, other than sketchy reports that libido is suppressed (see Pfaus and Gorzalka 1987).

The laboratory rat is a useful model for studying opioid effects on sexual behaviour and, in some respects, closely resembles the human. For example, most investigators who have injected NAL into opiate-naive or -dependent male rats will have frequently observed that an early behavioural response to the antagonist is the development of penile erection and, in the absence of a sexually receptive female, penile licking (autosexuality?). Of particular relevance to the potential therapeutic use of opioid antagonists in sexual dysfunction is the observation of Gessa et al. (1979) that sexually inactive male rats show a rapid onset of full copulatory behaviour following the systemic injection of NAL. However, this finding is equivocal (see later).

Because of the cyclic nature of sexual proclivity in the female, experiments in female rats have generally employed ovariectomized animals that have been primed with estrogen and sometimes with progesterone prior to testing. Both the single and combined steroid treatments elicit a series of characteristic responses culminating in lordosis which allows intromission by the male. Although there is some disagreement in the literature, most authors report that lordosis behaviour is facilitated following the administration of opioid antagonists (e.g. Sirinathsinghji 1984; Allen et al. 1985). Interestingly, Allen et al. (1985) found that NLTRX-treated females primed with estrogen alone elicited lordosis with a latency similar to that found in females primed with estrogen and progesterone, indicating that NLTRX can substitute for the actions of the latter (see Lustig et al. 1989, for comparable data in relation to LH secretion).

EOP inhibition of lordosis probably occurs within the mesencephalic central grey (MCG) area of the brain. This region has a dense, high EOP innervation and opioid receptor population; in addition, it receives GnRH projections from the hypothalamus which facilitate lordosis. One mechanism by which NAL may induce lordosis is by disinhibition of GnRH release in the MCG since the effect of microinfusions of the antagonist therein can be counteracted with either anti-GnRH serum or a GnRH antagonist (Sirinathsinghji 1984). The EOP responsible for inhibiting lordosis appears to be β-END, since infusions of an antiserum against this peptide (but not DYN or Met-ENK) induces lordosis, a response blocked by the injection of an anti-GnRH serum. Sirinathsinghji (1984) also reported that CRH injections into the hypothalamus or MCG inhibit sexual receptivity in the female rat and subsequently (Sirinathsinghji 1985) demonstrated that CRH probably acts by stimulating β-END secretion, although a direct CRH-GnRH interaction was not excluded.

Other studies by Sirinathsinghji (1987) implied that EOP inhibition of copulatory behaviour in the male rat may be due to increased CRH neuronal activity, in analogy to the situation in the female, and drew attention to the fact that depressed patients with sexual dysfunctions have elevated cerebrospinal fluid levels of CRH. So far it is not clear whether CRH is responsible for the physiological control of sexual behaviour, or whether it only mimics stressful situations which disrupt sexual behaviour (cf. Sect. 2).

As might be expected from the foregoing, opiates and EOP reduce the expression of male and female sexual behaviours in a variety of species; μ-opioid receptors appear to mediate these inhibitory effects (Serra et al. 1988). In addition to the neurotransmitters already mentioned, DA has also been implicated in the neural circuitry through which EOP might alter sexual behaviour (see Clark et al. 1988; Serra et al. 1988).

It was previously mentioned that sexual behaviour is highly dependent on gonadal steroids. It is believed that steroids modify sexual motivation, but studies in male rats also imply a role for them in coital performance. Thus, castrated males show a gradual decline in copulatory behaviour; this is manifested as a reduction in the occurrence of penile erections, intromissions and ejaculations in the company of a receptive female. The loss of erectile, intromissive and ejaculatory capacity is testosterone-reversible. Chronic morphine treatment of testosterone-supplemented castrates overrides the effects of testosterone upon copulatory behaviour, but not upon penile reflexes. These data reveal a dissociation between the effects of morphine and testosterone upon the different components of masculine behaviour, and lead to the conclusion that the deleterious effects of opioids result from reduced sexual arousal rather than erectile ability (Clark et al. 1988). However, further studies of this subject are required since a large proportion of castrates continue to show sexual motivation in the form of mounting behaviour.

One aspect of the role of EOP in sexual arousal has been studied by Donchin et al. (1984). They showed that the central injection of a stable opioid peptide analogue, metkephamide, into lambs induces flehmen, during the breeding season. Flehmen, which involves a curling of the upper lip with rapid licking movements and mouthing is thought to waft sexually stimulating pheromones into the vomeronasal organs; it is a typical (mainly male) adult ungulate behaviour. The finding of Donchin et al. (1984) suggests that EOP may facilitate the perception of sexual stimulants.

While there are many reports that opioid antagonists stimulate male sexual behaviour, there are others which indicate the opposite (see Miller and Baum 1987). The work of the latter authors, using a place-preference paradigm shows that the castration-induced loss of sexual interest is similar to that induced by NAL in sexually satiated intact male rats, and that combined castration and NAL treatment accentuates the reduction of sexual motivation seen after each manipulation alone. Based on their own results, and those of others showing that opioid antagonists inhibit mating between dominant male monkeys and their preferred dominant females (in "harem species"), Miller and Baum (1987) concluded that opioids (like steroids) may contribute to the positive interpretation of incentive motivational stimuli (e.g. olfactory cues) leading to copulation. In support of their conclusion, they cite other behaviours (e.g. feeding, drinking and ingestion of sweet substances) which are suppressed by NAL, indicating that EOP may have generalized facilitatory actions on motivated behaviours (see also Shippenberg and Bals-Kubik this Vol.). Furthermore, they provided evidence that EOP may act by stimulating DAergic transmission in the ventral tegmental area to increase the reward value of a receptive female, a finding consistent with the results of studies on the motivational properties of opioids in other behavioural contexts (see Shippenberg and Bals-Kubik this Vol.). The studies of sexual behaviour may be criticized, however, in that they have apparently not considered that the inhibitory effects of NAL may be secondary to the stressful effects of the antagonist (see Prezłocki et al. this Vol.), including the production of hyperalgesia (see Millan et al. this Vol.).

Related (although somewhat contrarily) to the work of Miller and Baum (1987) are the studies of Hughes et al. (1988). The latter workers found that central β-END administration to male rats reduced their perception of the motivational properties of a receptive female, and that increasing doses of the peptide resulted in a diversion of interest away from the female (towards ingestion of a sweetened diet). Furthermore, they demonstrated that the degree of sexual behaviour displayed by such males depends upon the molar ratio between β-END and other POMC-derived peptides (e.g. α-MSH and N-Ac-β-END, both non-opioids). Since these ratios were altered in the hypothalami of castrated rats, Hughes et al. (1988) proposed that steroid-regulated post-translational processing of POMC might be an important factor in determining the EOP control of sexual behaviour.

To return to a favourite theme of this chapter, the interaction between steroids and opioids, we should mention that the interaction, insofar as sexual behaviour is concerned, appears to be a subtle one indeed. There seems no reason to doubt that steroids modulate the activity of EOP systems although the exact mechanism or extent of the interaction is unknown. The limited data available suggest that gonadal steroids are essential primers of the neural substrate upon which EOP act. It remains to be shown, however, whether there exist analogies between the EOP control of sexual behaviour and that of GnRH/LH release. This could be achieved by testing whether opioid antagonists can improve sexual function in women with disturbed ovarian cycles resulting from increased EOP inhibition (e.g. after excessive exercise or fasting) and in hypogonadal men.

6 Opioids in Parturition and Lactation

EOP apparently also play a role in the final processes of reproduction, parturition and lactation. Although this chapter does not cover the important preceding stage of pregnancy (because of the sparse and confusing literature), the reader should note that pregnancy is an important preparatory phase for delivery of a mature foetus and its ex utero maintenance. Thus, the high steroid levels found during pregnancy, among other changes, make an important contribution to the structure, biochemical activity and responsiveness of the uterus and mammary glands which are the principal tissues active in parturition and lactation, respectively. Also, in view of the steroid-EOP interactions described earlier, it is highly probable that the increased steroid concentrations accompanying pregnancy (see Bridges and Ronsheim 1987) may modulate EOP effects upon parturition and lactation.

Both parturition and lactation ultimately depend upon contractions induced by the nonapeptide oxytocin (OT) (Fig. 3). The source of the OT responsible for these events are the cell bodies in the magnocellular divisions of the hypothalamic supraoptic (SON) and paraventricular (PVN) nuclei. These "giant" cells synthesize pro-OT which is transported down the axon. The axons, which are relatively long, traverse the blood-brain barrier to terminate in the neurohypophysis (NP) from which OT is released into the blood stream upon demand.

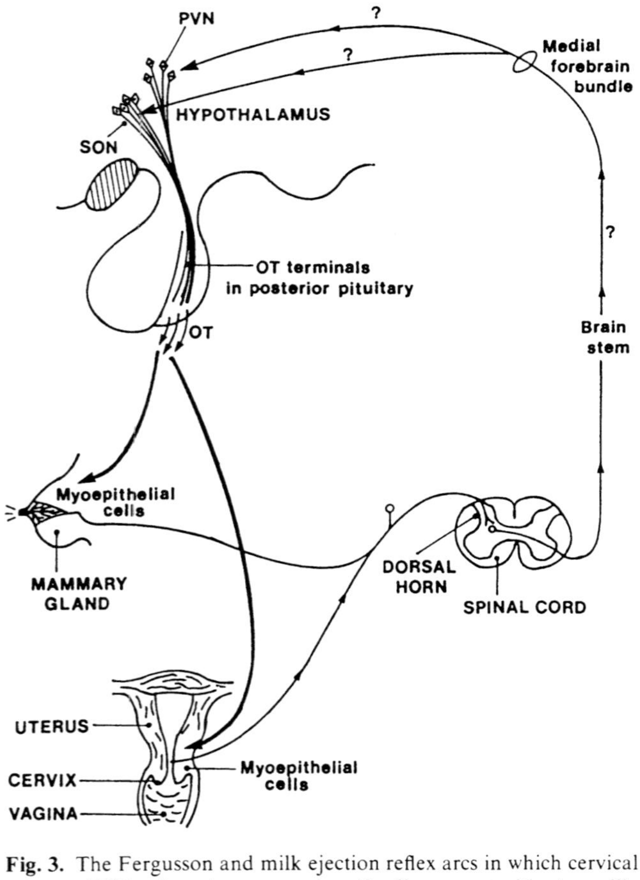

Fig. 3. The Fergusson and milk ejection reflex arcs in which cervical or nipple stimulation results in a release of OT and uterine contractions (leading to parturition) or milk-duct contractions (leading to milk ejection), respectively. (Adapted from Johnson and Everitt 1984)

6.1 Parturition

As seen with many other reproductive processes, "timing" is an important feature of parturition and lactation. This is achieved by complex interactions between central transmitter systems, but peripheral neural afferents also play an important role. The foetus signals its own maturity by hormonal secretions and by exerting pressure on the mother's cervix. This triggers a neuroendocrine reflex (Fergusson reflex) which then becomes centrally coordinated, resulting in OT release (Fig. 3) and a stimulation of uterine prostaglandin (PG) secretion. These local hormones (PG) cause the myoepithelial cells of the uterus (myometrium) to contract, and the process of birth (parturition) begins.

Two characteristics of OT neurons are their capacity to fire without fatigue, and their ability to release large amounts of hormone without significant depletion of the releasable pools. Consideration of these features is important in view of the relative ease with which the Fergusson reflex arc can be activated. It is imperative that "gating mechanisms" are incorporated so as to safeguard against the delivery of a premature foetus. The necessity for this becomes all the more obvious when one notes that OT is almost invaribly released by noxious stimuli. Signals from the foetus, an adequate number of estrogen-induced OT receptors, and sufficient amounts of the ovarian hormone relaxin are among the "fail-safe" mechanisms which ensure the correct timing of birth.

EOP have recently been suggested to participate in the timing of parturition. Although EOP produced by the mother, placenta and foetus may serve in the local coordination of this event, we will restrict our discussion to the modulation of the hypothalamo-neurohypophysial system (HNS) by central opioids.

Studies on the control of parturition by EOP were most probably stimulated by earlier observations that opiates, EOP and NAL modulate OT release both in vivo (Clarke et al. 1979) and in vitro (Bicknell and Leng 1982); briefly, NAL enhanced OT release, and opiates and EOP and were shown to inhibit OT secretion. Interestingly, the discovery of Clarke et al. (1979) was made in the course of lactational studies; the likelihood of common factors and mechanisms in lactation and parturition will become more evident as we proceed.

Experiments by Leng et al. (1985) and Leng et al. (1988) have clearly demonstrated that NAL reduces the intervals between successive births in rat litters, by enhancing OT secretion; this finding implies a "spacing" role for EOP in parturition. A particularly interesting observation was that NAL only potentiates OT release immediately before and during delivery, but not postpartum; furthermore, it did not alter plasma OT levels in non-pregnant female rats (Hartman et al. 1986; Leng et al. 1988).

Leng et al. (1988) also made a careful analysis of whether EOP may be responsible for the delay in parturition caused by a specific stress, namely, the transfer of the mother to a novel environment once delivery of the litter had started. Although NAL prevented the stress-induced disruption of parturition, comparison of the NAL responses of disturbed and undisturbed rats indicated that increased EOPergic activity does not necessarily account for the disruption; thus, the delay in pup delivery was primarily due to an inhibition of OT release, and NAL alleviated the situation by merely enhancing OT neuronal activity. These results illustrate the complexity of one physiological process in which opioids are implicated. They also raise the question of why OT is apparently not stimulated by stress during certain physiological states. Perhaps, as suggested earlier, EOP serve a "gating" role in preventing birth at suboptimal times, and it should be noted that brain EOP concentrations are elevated during late pregnancy, parturition and immediately postpartum (e.g. Wardlaw and Frantz 1983; Petraglia et al. 1985).

The association between NAL-stimulated OT secretion and reduced inter-birth intervals (Leng et al. 1988) suggests that EOP influence parturition by acting within the HNS; this is also supported by other evidence, e.g. Gosden et al. (1985) reported that morphine interrupted established parturition in rats. This action was considered to occur centrally (if the NP is considered a central site) since the opiate, in vitro, had no effect on either basal or OT-stimulated contractions of uteri from recently parturient mothers. In another study these authors showed that the morphine effect was due to a suppression of OT secretion (Cutting et al. 1986).

So far, the data described were derived from experiments in which morphine or NAL were administered once parturition had commenced. There is some evidence, however, that EOP might also help in timing the onset of birth by mediating the OT-suppressive effects of relaxin: NAL was found to reverse the prolongation of gestation and inhibition of OT secretion caused by infusions of relaxin (Jones and Summerlee 1986). Relaxin concentrations in the maternal blood fall markedly just prior to the onset of birth and it has been postulated that relaxin stimulates EOP pathways to inhibit OT secretion; a fall in relaxin levels would then "relieve" the EOPergic inhibition of OT neurons (Summerlee 1989). This schema may be considered an example of "endogenous dependence" (see Collier 1984). In view of what we know about dependence (see Schulz this Vol.), one may guess that the relaxin-EOP-OT interaction occurs at the OT cell bodies rather than terminals. Further discussion of the mechanisms underlying the EOPergic inhibition of OT secretion will be given after we describe the role of opioids in lactation. Suffice it to say here that opioids might also act directly upon OT terminals during parturition.

6.2 Lactation

As with parturition, lactation also involves the activation of a neuroendocrine reflex pathway; during lactation, however, the reflex occurs several times a day (each episode lasting minutes-hours, depending on the species) over extended periods (weeks-months). The process is highly OT-dependent, and this is well provided for by high levels of OT synthesis and storage. The milk-ejection reflex is triggered by the young sucking on the mother's nipple, and the neural pathway beyond this step is probably identical to that of the Fergusson reflex. The sucking reflex results in increased firing of, and secretion by, magnocellular OT neurons; this OT causes the myoepithelial cells surrounding the milk ducts to contract and eject milk (Fig. 3).

A role for EOP in the physiological regulation of lactation was first suggested by Haldar and Sawyer (1978). They observed that morphine reduced milk yield in mice, an action which was NAL-reversible. Since milk ejection could be fully reinstated by OT injections, they inferred that the opiate effect was due to a blockade of OT release, rather than to unresponsiveness of the mammary tissue to OT (Haldar and Sawyer 1978). Shortly after these initial observations, Clarke et al. (1979) reported that NAL-antagonized the inhibition of milk ejection in rats treated with morphine. They also observed that although morphine inhibited OT release, it did not affect the electrical activity of OT cell bodies in the SON. Furthermore, they found morphine to inhibit intramammary pressure and OT release induced by carbachol and electrical stimulation of the NP, respectively. They therefore concluded that EOP inhibit OT secretion and, thus, milk ejection, by actions on (or in the vicinity of) OT terminals. Since NAL, in the absence of exposure to exogenous opiates, markedly enhanced the intramammary pressure responses to sucking, carbachol, and electrical stimulation of the OT terminals, Clarke et al. (1979) proposed that EOP exert a tonic inhibition upon OT release in the lactating rat. This suggestion should be regarded cautiously however: Clarke and Wright (1989) recently reported that the same dose of NAL (10 mg/kg), while effectively antagonizing the inhibitory actions of opiate agonists, failed to alter milk yield or number of ejections in opiate-naive suckled dams. Moreover, Haldar and Bade (1981) observed that milk yield in mice was not significantly altered by 5 mg/kg of NLTRX, whereas 10 and 20 mg/kg of NLTRX resulted in marked reductions. Assuming that NLTRX was acting as an antagonist at all doses tested, one is left with two possible explanations for these discrepant findings: (1) NAL/NLTRX might have enhanced OT release by counteracting the efficacy of EOP released by the very stimulus (sucking) that stimulates OT release; and (2) the NAL data may be complicated by the use of anaesthetics such as urethane (Clarke et al. 1979, used urethane-treated mothers; Clarke and Wright 1989, used conscious mothers). The latter possibility seems to be an important complicating factor since urethane acts as an osmotic stimulus (Hartman et al. 1987), and activates HNS vasopressin (AVP) neurons that may be involved in the opioid inhibition of OT release (see later). The likelihood that urethane anaesthesia may be responsible for the NAL response in some studies is supported by recent experiments by Bicknell et al. (1988a) on urethane-anaesthesized, lactating (but unsuckled) rats in which NAL was again found to increase OT secretion.

Besides morphine, other opioid agonists have also been demonstrated to inhibit OT secretion and/or milk let-down, e.g. electrically stimulated secretion of OT in vitro has also been shown to be suppressed by opioid peptides, e.g. β-END and [D-Ala]-Met-ENK (Haldar et al. 1982) and DYN (Bondy et al. 1988; Zhao et al. 1988). Of greater interest perhaps are the observations of Wakerley et al. (1983) using hypothalamic slices in vitro: they found that morphine and the δ-agonist D-Ala-D-Leu-ENK (DADLE) suppressed the electrical activity of OT cell bodies in the SON, a result which differs from that of Clarke et al. (1979). Wakerley's studies were done on tissues from male rats, but this is unlikely to explain the difference because Bicknell et al. (1988c) observed that NAL excited OT perikarya in the SON of lactating dams treated chronically with morphine.

The last result (Bicknell et al. 1988c) indicates that OT neurons can be made dependent following chronic exposure to morphine, but it does not resolve the issues of whether EOP tonically inhibit OT neuronal activity, and whether OT secretion reflects withdrawal from a state of endogenous dependence (cf. Clarke et al. 1979). These are both unlikely events, however, since Bicknell et al. (1988c) did not observe any alterations in the electrical activity of OT cell bodies when NAL was given to lactating, but non-sucking, mothers (recall that it was proposed earlier that sucking-induced OT secretion might raise EOPergic influences on OT neurons); also, NAL only enhanced stimulated (not basal) OT release from the NP (Bicknell and Leng 1982).

Chronic treatment with morphine also produces tolerance in the mechanisms governing OT secretion and milk ejection. Thus, Rayner et al. (1988) found that, following an initial reduction in milk yield from rat mothers treated chronically with morphine, the milk delivered to the young rapidly returned to control levels. OT neurons thus provide a central model for studying the relative importance of neuronal soma and terminals in the development of dependence and tolerance, in analogy to studies in the peripheral nervous system (see Schulz this Vol.). The large size of OT neurons would facilitate such studies, providing an insight into the way these phenomena occur in other neuroendocrine systems (cf. Almeida et al. 1986; Mehmanesh et al. 1988).

Maternal behaviour has a major influence upon milk transfer from mother to young, and should be considered in the interpretation of experimental results (obviously, not an issue in studies on anaesthetized mothers). Opiates have been shown to disrupt the maternal response of rat mothers to their pups (see Pedersen and Prange 1987). Interestingly, Rayner et al. (1988) observed that chronic morphine treatment resulted in disturbed maternal behaviour in only a subpopulation of mothers, and inferred that either tolerance develops in the mechanisms underlying maternal behaviour, or that morphine only affects the initiation, but not the maintenance of, maternal behaviour.

Related to the above is the emotional state of the mother. Early studies in the rabbit demonstrated that OT secretion and milk ejection are interrupted when the mother is stressed (Cross 1955). More recently, Haldar and Bade (1981) showed that NLTRX reversed the OT-inhibiting effects of heat stress in lactating mice. Studies of conscious, lactating subjects are, thus, clearly complicated; it is critical to distinguish between physiological and pathophysiological effects, a caution which also applies to studies of parturition. Studies on the role of CRH in the EOP control of lactation have recently commenced, and the results should add to our knowledge of the central integration of this process.

6.3 More About Mechanisms

Some hints about the way in which EOP regulate OT secretion have been provided by experiments in lactating rats. Because of the common neural pathways involved in lactation and parturition, there is every reason to believe that similar mechanisms operate during parturition. So far we have seen that EOP may help ensure the appropriate timing of parturition and lactation by inhibiting OT secretion. Whether this control is exerted tonically and represents endogenous dependence is far from clear at present. There are data demonstrating EOP modulation of OT secretion via pre- and post-synaptic interactions, but a concensus view is still lacking. An intriguing possibility worth examining in the meantime is that EOP may also influence OT axonal transport and processing because magnocellular OT (and AVP) neurons lack a protective myelin sheath.

Opioid receptor antagonists such as NAL and NLTRX have been useful in demonstrating that opiates and opioids affect parturition and lactation by altering OT secretion. However, they provide no information about whether the EOP-OT interaction is a direct one, or if it is mediated by intervening neurons. In this respect, receptor mapping studies have provided some insight into the problem. Some years after Simantov and Snyder's (1977) autoradiographic demonstration of opioid receptors in the NP (which had a relatively weak affinity for μ- and δ-ligands), a third class of opioid receptor (κ) was discovered (Chavkin and Goldstein 1981). There is now convincing evidence that the NP is predominantly endowed with κ-receptors (Herkenham et al. 1986) which are matched by DYN (efferent) fibres originating in the SON and PVN (Palkovits 1986). Indeed, κ-agonists suppress OT release from the NP (e.g. Bondy et al. 1988; Zhao et al. 1988). The possibility that δ-receptors are also located in the vicinity of OT terminals cannot be dismissed since their putative endogenous ligands (Leu-/Met-ENK) are found in the NP (Rossier et al. 1979), and the δ-preferential agonist DADLE has been shown to inhibit electrically evoked NP secretion, albeit that of AVP (Iversen et al. 1980). Although morphine was assumed to presynaptically inhibit OT release via an interaction with μ-receptors (Clarke et al. 1979), it is possible that such effects reflected actions at δ- or κ-receptors since the selective μ-agonist DAGO was ineffective in altering terminal release of OT (see Bicknell et al. 1988b).

The picture regarding possible EOPergic inhibition of OT cell bodies in the hypothalamus is less clear. The SON and PVN contain cells and receive fibres, which stain immunopositively for both DYN and ENK; they also receive β-END afferents from the arcuate nucleus (Palkovits 1984). These nuclei may thus be expected to be subject to μ-, δ- and κ-opioid receptor modulation. Autoradiographic studies have, however, detected only κ-receptors in these sites (e.g. Mansour et al. 1987). Although a recent immunocytochemical study also indicated κ-receptors to be present in the SON, cross-reactivity of the receptor antibody with μ- and δ-sites could not be ruled out (Hassan et al. 1989). Nevertheless, at least one report indicates a role for non-κ opioid ligands in inhibiting the activity of OT cell bodies (Wakerley et al. 1983); for reasons discussed by those authors, it is unlikely that their observations with morphine and DADLE resulted from an inhibition of stimulatory afferents to OT cells.

The precise location of κ-receptors in the NP has not yet been discussed. Recent work by Zhao et al. (1988) has shown these to be located on the OT nerve terminals, in confirmation of several earlier studies. These receptors comprise a separate population from those found in association with the pituicytes (astrocytes) that surround the OT and AVP nerve terminals (Lightman et al. 1983). Although the exact mechanism by which the pituicytes convey the EOP signals is not known, it seems that encapsulation of the terminals by these cells, and formation of "synaptic-like" contacts, occurs at times when EOPergic influences are greatest (see Cross and Leng 1983). Moreover, there is now good morphological evidence indicating that OT terminals are "released" from their pituicyte enclosures during parturition (Tweedle and Hatton 1982) and lactation (Theodosis et al. 1988).

Even though the evidence for κ-ligand inhibition of OT release from the neurohypophysis is quite strong, and DYN peptides have been located in this tissue, the regulatory mechanisms are rather complex. Neurohypophysial DYN originates in hypothalamic AVP cell bodies (Sherman et al. 1986), and presumably affects OT secretion via κ-receptors located on the OTergic terminals themselves, or on the pituicytes. When it is recalled that OT and AVP neurons respond differentially to stimuli such as stress (almost all stressors stimulate OT secretion, whereas only dehydration stimulates that of AVP), and that DYN is colocalized with AVP, one may question whether DYN originating in AVP cell bodies can really regulate OTergic terminal activity. Part of the answer to this problem may lie in the recent observations that lactating rats are less responsive to restraint (Carter and Lightman 1987) and dehydration (Hartman et al. 1987) stresses compared to control animals.

The problem concerning ENK regulation of OT terminals is similar in that the stress-induced synthesis of ENK in the PVN is reduced in lactating rats (Lightman and Scott-Young 1989). Until recently, it was thought that the ENK in the NP was colocalized with OT (i.e. originated in the magnocellular part of the PVN), but this view has since been challenged (see Bicknell et al. 1988c). It now seems that ENK reaches the NP through fibres emanating from the parvocellular division of the PVN, where they are most likely to be under the control of CRH (Lightman and Scott-Young 1989). We thus arrive back at our starting point that EOP help optimize OT secretion to ensure that birth and lactation occur at advantageous times.

So far we have concentrated on EOPergic inhibition of parturition and lactation by actions within the HNS. However, recent data from Wright (1985) suggest that EOP might interfere with the transmission of the suckling stimulus within the spinal cord. Since it is likely that this spinal tract is common to the milk ejection and Fergusson reflex arcs, this finding may also apply to the EOPergic control of parturition.

6.4 Practical Considerations

Opiates (e.g. pentazocine, a partial κ-ligand) are commonly used in the clinic to relieve pain during labour (see Clarke and Wright 1984). From the foregoing experimental data, one would expect a reduction in the frequency or intensity of uterine contractions, but apparently, no systematic studies have been undertaken. This, as well as possible interference with lactation, deserves future attention. It is known that milk ejection is delayed in women undergoing Caesarean sections, a finding that has been commonly attributed to the trauma of surgery and dehydration resulting from anaesthesia. The possibility that these conditions activate EOP pathways that are inhibitory to OT secretion should be examined since opioid antagonist therapy postpartum might easily overcome the problem.

References

Allen DL, Renne KJ, Luine VN (1985) Naltrexone facilitation of sexual receptivity in the rat. Hormone Behav 19:98–103

Almeida OFX, Schulz R (1988) Sexual differentiation of the luteinizing hormone response of neonatal rats to the narcotic antagonist naloxone: critical role of estrogen receptors. Biol Reprod 39:1009–1012

Almeida OFX, Schulz R, Herz A (1986) Paradoxical LH and prolactin responses to naloxone after chronic treatment with morphine. J Endocrinol 108:181–189

Almeida OFX, Nikolarakis KE, Herz A (1987a) The significance of testosterone in regulating hypothalamic content and in vitro release of β-endorphin and dynorphin. J Neurochem 49:742–749

Almeida OFX, Nikolarakis KE, Schulz R, Herz A (1987b) A "window of time" during which testosterone determines the opiatergic control of LH release in the adult male rat. J Reprod Fertil 79:299–305

Almeida OFX, Nikolarakis KE, Herz A (1988a) Neuropharmacological analysis of the control of LH secretion in gonadectomized male and female rats: altered hypothalamic responses to inhibitory neurotransmitters in long-term castrated rats. J Endocrinol 119:15–21

Almeida OFX, Nikolarakis KE, Herz A (1988b) Evidence for the involvement of endogenous opioids in the inhibition of luteinizing hormone by corticotropin-releasing factor. Endocrinology 122:1034–1041

Almeida OFX, Nikolarakis KE, Schulz R, Herz A (1988c) Sex differences in the opioidergic control of LH secretion in rats. In: Genazzani AR, Montemagno U, Nappi C, Petraglia F (eds) The brain and female reproductive function. Parthenon, Lancs, pp 73–80

Almeida OFX, Nikolarakis KE, Webley GE, Herz A (1988d) Opioid components of the clockwork that governs luteinizing hormone and prolactin release in male rats. FASEB J 2:2874–2877

Almeida OFX, Nikolarakis KE, Sirinathsinghji DJS, Herz A (1989) Opioid-mediated inhibition of sexual behaviour and luteinizing hormone secretion by corticotropin-releasing hormone. In: Dyer RG, Bicknell RJ (eds) Brain opioid systems in reproduction. Oxford Science Publications, Oxford, pp 149–164

Barkan AL, Duncan JA, Schiff M, Papavasiliou S, Garcia-Rodriguez A, Kelch RP, Marshall JC (1985) Opioid and neurotransmitter regulation of pituitary gonadotropin-releasing hormone (GnRH) receptors in the ovariectomized estradiol-treated rat: role of altered GnRH secretion. Endocrinology 116:1003–1010

Barraclough CA, Sawyer CH (1955) Inhibition of the release of pituitary ovulatory hormone in the rat by morphine. Endocrinology 57:329–337

Bhanot R, Wilkinson M (1983) Opiatergic control of gonadotropin secretion during puberty in the rat: a neurochemical basis for the hypothalamic 'gonadostat?' Endocrinology 113:596–603

Bhanot R, Wilkinson M (1984) The inhibitory effect of opiates on gonadotropin secretion is dependent upon gonadal steroids. J Endocrinol 102:133–141

Bicknell RJ, Leng G (1982) Endogenous opiates regulate oxytocin but not vasopressin secretion from the neurohypophysis. Nature (London) 298:161–162

Bicknell RJ, Coombes JE, Fink G, Russell JA, Sheward WJ (1988a) Effects of naloxone and long-term morphine administration on oxytocin release into hypophysial portal blood in urethane-anaesthetized rats. J Physiol 401:35P

Bicknell RJ, Leng G, Russell JA, Dyer RG, Mansfield S, Zhao B-G (1988b) Hypothalamic opioid mechanisms controlling oxytocin neurons during parturition. Brain Res Bull 20:743–749

Bicknell RJ, Leng G, Lincoln DW, Russell JA (1988c) Naloxone excites oxytocin neurones in the supraoptic nucleus of lactating rats after chronic morphine treatment. J Physiol 396:297–317

Blank MS, Mann DR (1981) Diurnal influences on serum luteinizing hormone responses to opiate receptor blockade with naloxone or to luteinizing hormone-releasing hormone in the immature female rat. Proc Soc Exp Biol Med 168:338–343

Blank MS, Roberts DL (1982) Antagonist of gonadotropin-releasing hormone blocks naloxone-induced elevations in serum luteinizing hormone. Neuroendocrinology 35:309–312

Blank MS, Panerai AE, Friesen HG (1979) Opioid peptides modulate luteinizing hormone secretion during sexual maturation. Science 203:1129–1131

Blank MS, Fabbri A, Catt JK, Dufau ML (1986) Inhibition of luteinizing hormone release by morphine and endogenous opiates in cultured pituitary cells. Endocrinology 118:2097–2101

Blankstein J, Reyes FJ, Winter JSD, Faiman C (1981) Endorphins and the regulation of the human menstrual cycle. Clin Endocrinol 14:287–294

Bondy CA, Gainer H, Russell JT (1988) Dynorphin A inhibits and naloxone increases the electrically stimulated release of oxytocin but not of vasopressin from the terminals of the neural lobe. Endocrinology 122:1321-1327
Bridges RS, Ronsheim PM (1987) Immunoreactive beta-endorphin concentrations in brain and plasma during pregnancy in rats: possible modulation by progesterone and estradiol. Neuroendocrinology 45:381-388
Bruni JF, Van Vugt D, Mashall S, Meites J (1977) Effects of naloxone, morphine and methionine enkephalin on serum prolactin, luteinizing hormone, follicle stimulating hormone, thyroid stimulating hormone and growth hormone. Life Sci 21:461-466
Buffum J (1982) Pharmacosexology: the effects of drugs on sexual function. A review. J Psychoact Drugs 14:5-44
Cacicedo L, Franco SF (1986) Direct action of opioid peptides and naloxone on gonadotropin secretion by cultured rat anterior pituitary cells. Life Sci 38:617-625
Carter DA, Lightman SL (1987) Oxytocin responses to stress in lactating and hyperprolactinaemic rats. Neuroendocrinology 46:532-537
Casper RF, Alapin-Rubillovitz S (1985) Progestins increase endogenous opioid peptide activity in postmenopausal women. J Clin Endocrinol Metab 60:34-36
Chavkin C, Goldstein A (1981) Specific receptors for the opioid peptide dynorphin: structure-activity relationships. Proc Natl Acad Sci USA 78:6543-6547
Chen HJ, Targovnik J, McMillan L, Randall S (1984) Age difference in endogenous opiate modulation of short photoperiod-induced testicular regression in golden hamsters. J Endocrinol 101:1-6
Ching M (1983) Morphine suppresses the proestrous surge of GnRH in pituitary portal plasma of rats. Endocrinology 112:2209-2211
Chowen-Breed J, Fraser HM, Vician L, Damassa DA, Clifton DK, Steiner RA (1989) Testosterone regulation of proopiomelanocortin messenger ribonucleic acid in the arcuate nucleus of the male rat. Endocrinology 124:1697-1702
Cicero TJ, Badger TM, Wilcox CE, Bell RB, Meyer ER (1977) Morphine decreases luteinizing hormone by an action on the hypothalamic-pituitary axis. J Pharm Exp Ther 203:548-555
Cicero TJ, Schainker BA, Meyer ER (1979) Endogenous opioids participate in the regulation of the hypothalamic-pituitary-luteinizing hormone axis and testosterone's negative feedback control of luteinizing hormone. Endocrinology 104:1286-1291
Cicero TJ, Wilcox CE, Bell RD, Meyer ER (1980) Naloxone-induced increases in serum luteinizing hormone in the male: mechanisms of action. J Pharm Exp Ther 212:573-578
Cicero TJ, Meyer ER, Schmoeker PF (1982) Development of tolerance to the effects of morphine on luteinizing hormone secretion as a function of castration in the male rat. J Pharm Exp Ther 223:784-789
Cicero TJ, Schmoeker PF, Meyer ER, Miller BT, Bell RD, Cryton SM, Brown CC (1986) Ontogeny of the opioid-mediated control of reproductive endocrinology in the male and female rat. J Pharm Exp Ther 236:627-633
Cicero TJ, Meyer ER, Miller BT, Bell RD (1988) Age-related differences in the sensitivity of serum luteinizing hormone to prototypic mu, delta and kappa opiate agonists and antagonists. J Pharm Exp Ther 246:14-20
Clark JT, Gabriel SM, Simpkins JW, Kalra SP, Kalra PS (1988) Chronic morphine and testosterone treatment. Effects on sexual behaviour and dopamine metabolism in male rats. Neuroendocrinology 48:97-104
Clark G, Wright DM (1984) A comparison of analgesia and suppression of oxytocin release by opiates. Br J Pharmacol 83:799-806
Clarke G, Wright DM (1989) Opioid inhibition of milk-yield in conscious lactating rats. Adv Biosci 75:607-610
Clarke G, Wood P, Merrick L, Lincoln DW (1979) Opiate inhibition of peptide release from the neurohumoral terminals of hypothalamic neurons. Nature (London) 282:746-748
Collier HOJ (1984) Cellular aspects of opioid tolerance and development. In: Hughes J, Collier HOJ, Rance MJ, Tyers MB (eds) Opioids, past, present and future. Taylor & Francis, London Philadelphia, pp 109-125
Cross BA (1955) Neurohormonal mechanisms in emotional inhibition of milk ejection. J Endocrinol 12:29-37
Cross BA, Leng G (1983) The neurohypophysis: structure, function and control. Elsevier, Amsterdam, p 542

Cutting R, Fitzsimons N, Gosden RG, Humphreys EM, Russell JA, Scott S, Stirland JA (1986) Evidence that morphine interrupts parturition in rats by inhibiting oxytocin secretion. J Physiol 371:182P
Donchin Y, De Vane GW, Caton D (1984) Metkephamid-induced flehmen in lambs. Physiol Behav 33:335–337
Dyer RG, Bicknell RJ (eds) (1989) Brain opioid systems in reproduction. Oxford Science Publications, Oxford, 365 pp
Ebling FJP, Lincoln GA (1985) Endogenous opioids and the control of seasonal LH secretion in Soay rams. J Endocrinol 107:341–353
Ebling FJP, Lincoln GA (1987) β-endorphin secretion in rams related to season and photoperiod. Endocrinology 120:809–818
Ellendorff F (1985) Opiates in the central control of reproduction. Bibliogr Reprod 45:B1–B12
Eskes GA, Wilkinson M, Bhanot R (1984) Short-day exposure eliminates the LH response to naloxone in golden hamsters. Neuroendocrinology 39:281–283
Fabbri A, Jannini EA, Gnessi L, Moretti C, Ulisse S, Franzese A, Lazzari R, Fraioli F, Frajese G, Isidori A (1989) Endorphins in male impotence: evidence for naltrexone stimulation of erectile activity in patient therapy. Psychoneuroendocrinology 14:103–111
Ferin M, Van Vugt D, Wardlaw S (1984) The hypothalamic control of the menstrual cycle and the role of endogenous opioid peptides. Rec Prog Hormone Res 40:441–485
Gabriel SM, Berglund LA, Kalra SP, Kalra PS, Simpkins JW (1986) The influence of chronic morphine treatment on negative feedback regulation of gonadotropin secretion by gonadal steroids. Endocrinology 119:2762–2767
Gessa G, Paglietti E, Pellegrini-Quarantotti B (1979) Induction of copulatory behaviour in sexually inactive rats by naloxone. Science 204:203–205
Gosden RG, Humphreys EM, Johnston V, Liddle S, Russell JA (1985) Morphine acts centrally to interrupt established parturition in rats. J Physiol 364:59P
Hahn EF, Fishman J (1985) Castration affects male rat brain opiate receptor content. Neuroendocrinology 41:60–63
Haldar J, Bade V (1981) Involvement of opioid peptides in the inhibition of oxytocin release by heat stress in lactating mice. Proc Soc Exp Biol Med 168:10–14
Haldar J, Sawyer WH (1978) Inhibition of oxytocin release by morphine and its analogs. Proc Soc Exp Biol Med 157:476–480
Haldar J, Hoffman DL, Zimmerman EA (1982) Morphine, beta endorphin and D-Ala 2 met-enkephalin inhibit oxytocin release by acetylcholine and suckling. Peptides 3:663–668
Hartman R, Rosella-Dampman LM, Emmert SE, Summy-Long JY (1986) Inhibition of release of neurohypophysial hormones by endogenous opioid peptides in pregnant and parturient rats. Brain Res 382:352–359
Hartman RD, Rosella-Dampman LM, Summy-Long JY (1987) Endogenous opioid peptides inhibit oxytocin release in the lactating rat after dehydration and urethane. Endocrinology 121:536–543
Hassan AHS, Almeida OFX, Gramsch Ch, Herz A (1989) Immunocytochemical demonstration of opioid receptors in rat brain and neuroblastoma × glioma hybrid (NG108-15) cells, using a monoclonal antiidiotypic antibody. Neuroscience 32:269–278
Hastings MH, Herbert J, Martensz ND, Roberts AC (1985) Melatonin and the brain in photoperiodic mammals. In: Photoperiodism, melatonin and the pineal. Ciba Found Symp 117:57–77
Herkenham M, Rice KC, Jacobson AE, Rothman RB (1986) Opiate receptors in rat pituitary are confined to the neural lobe and are exclusively kappa. Brain Res 382:365–371
Hughes AM, Everitt BJ, Herbert J (1988) The effects of simultaneous or separate infusions of some proopiomelanocortin-derived peptides (β-endorphin, α-melanocyte stimulating hormone, and corticotrophin-like intermediate polypeptide) and their acetylated derivatives upon sexual and ingestive behaviour of male rats. Neuroscience 27:689–698
Ieiri T, Chen HT, Meites J (1979) Effects of morphine and naloxone on serum levels of luteinizing hormone and prolactin in prepubertal male and female rats. Neuroendocrinology 29:288–292
Iversen LL, Iversen SD, Bloom FE (1980) Opiate receptors influence vasopressin release from nerve terminals in rat neurohypophysis. Nature (London) 284:350–351
Jacobson W, Kalra SP (1989) Decreases in mediobasal hypothalamus and preoptic area opioid ([^{3}H]naloxone) binding are associated with the progesterone-induced luteinizing hormone surge. Endocrinology 124:199–206
Jacobson W, Wilkinson M (1986) Association of diurnal variations in hypothalamic but not cortical

opiate ([^{3}H]-naloxone)-binding sites with the ability of naloxone to induce LH release in the prepubertal female rat. Neuroendocrinology 44:132–135
Johnson MH, Everitt BJ (1984) Essential reproduction. Blackwell, Oxford, p 367
Jones SA, Summerlee AJS (1986) Relaxin acts centrally to inhibit oxytocin release during parturition: an effect that is reversed by naloxone. J Endocrinol 111:99–102
Kalra SP, Kalra PS (1983) Neural regulation of luteinizing hormone secretion in the rat. Endocrine Rev 4:311–351
Knobil E, Neill J, Ewing L, Greenwald G, Pfaff D (eds) (1988) The physiology of reproduction. Raven, New York
Kornblum HI, Hurlbut DE, Leslie FM (1987) Postnatal development of multiple opioid receptors in rat brain. Dev Brain Res 37:21–41
Kulin HE, Demers LM, Rogol AD, Veldhuis JD (1987) The effect of long-term opiate antagonist administration to pubertal boys. J Androl 8:374–377
Kumar MSA, Besch EL, Millard WJ, Sharp DC, Leadem CA (1984) Effect of short photoperiod on hypothalamic methionine-enkephalin and LHRH content and serum β-endorphin-like immunoreactivity (β-end LI) levels in golden hamsters. J Pineal Res 1:197–205
Leng G, Mansfield S, Bicknell RJ, Dean ADP, Ingram CD, Marsh MIC, Yates JO, Dyer RG (1985) Central opioids: a possible role in parturition? J Endocrinol 106:219–224
Leng G, Mansfield S, Bicknell RJ, Blackburn RE, Brown D, Chapman C, Dyer RG, Hollingsworth S, Shibuki K, Yates JO, Way S (1988) Endogenous opioid actions and effects of environmental disturbance on parturition and oxytocin secretion in rats. J Reprod Fertil 84:345–356
Leranth C, MacLusky NJ, Shanabrough M, Naftolin F (1988) Immunohistochemical evidence for synaptic connections between proopiomelanocortin-immunoreactive axons and LHRH neurons in the preoptic area of the rat brain. Brain Res 449:167–176
Lightman SL, Scott-Young III W (1989) Lactation inhibits stress-mediated secretion of corticosterone and oxytocin and hypothalamic accumulation of corticotropin-releasing factor and enkephalin messenger ribonucleic acids. Endocrinology 124:2358–2364
Lightman SL, Ninkovic M, Hunt SP, Ivesen LL (1983) Evidence for opiate receptors on pituicytes. Nature (London) 305:235–237
Lustig RH, Fishman J, Pfaff DW (1989) Ovarian steroids and endogenous opioid peptide action in control of the rat LH surge. In: Dyer RG, Bicknell RJ (eds) Brain opioid systems in reproduction. Oxford Science Publications, Oxford, pp 3–26
MacLusky NJ, Naftolin F, Leranth C (1988) Immunocytochemical evidence for direct synaptic connections between corticotropin-releasing factor (CRF) and gonadotropin-releasing hormone (GnRH)-containing neurons in the preoptic area of the rat. Brain Res 439:391–395
Mansour A, Khachaturian H, Lewis ME, Akil H, Watson SJ (1987) Autoradiographic differentiation of mu, delta, and kappa opioid receptors in the rat forebrain and midbrain. J Neurosci 7:2445–2464
Mauras N, Veldhuis JD, Rogol AD (1986) Role of endogenous opiates in prepubertal maturation; opposing actions of naltrexone in prepubertal and late pubertal boys. J Clin Endocrinol Metab 62:1256–1263
Mehmanesh H, Almeida OFX, Nikolarakis KE, Herz A (1988) Hypothalamic LH-RH release after acute and chronic treatment with morphine studied in a combined in vivo/in vitro model. Brain Res 451:69–76
Meijs-Roelofs HMA, Kramer P (1988) Effects of treatment with the opioid antagonists naloxone and naltrexone on LH secretion and on sexual maturation in immature female rats. J Endocrinol 117:237–243
Melis GB, Paoletti AM, Gambacciani M, Mais V, Fioretti P (1984) Evidence that estrogens inhibit LH secretion through opioids in postmenopausal women using naloxone. Neuroendocrinology 39:60–63
Mendelson JH, Mello NK, Ellingboe J (1978) Effects of alcohol on pituitary-gonadal hormones, sexual function, and aggression in human males. In: Lipton MA, Di Mascio A, Kiliam KF (eds) Psychopharmacology: a generation of progress. Raven, New York, pp 1677–1992
Miller RL, Baum MJ (1987) Naloxone inhibits mating and conditioned place preference for an estrous female in male rats soon after castration. Pharm Biochem Behav 26:781–789
Morrell JI, McGinty JF, Pfaff DW (1985) A subset of β-endorphin- or dynorphin-containing neurons in the medial basal hypothalamus accumulates estradiol. Neuroendocrinology 41:417–426
Nikolarakis KE, Almeida OFX, Herz A (1986) Corticotropin-releasing factor (CRF) inhibits gonado-

tropin-releasing hormone (GnRH) release from superfused rat hypothalami in vitro. Brain Res 377:388-390

Nikolarakis KE, Almeida OFX, Herz A (1988a) Hypothalamic opioid receptors mediate the inhibitory actions of corticotropin-releasing hormone on luteinizing hormone release: further evidence from a morphine-tolerant animal model. Brain Res 450:360-363

Nikolarakis KE, Almeida OFX, Sirinathsinghji DJS, Herz A (1988b) Concomitant changes in the in vitro and in vivo release of opioid peptides and luteinizing hormone releasing hormone from the hypothalamus following blockade of receptors for corticotropin releasing factor. Neuroendocrinology 47:545-550

Nikolarakis KE, Almeida OFX, Herz A (1989) Multiple factors influencing the in vitro release of [Met5]-enkephalin from rat hypothalamic slices. J Neurochem 52:428-432

Palkovits M (1984) Distribution of neuropeptides in the central nervous system: a review of biochemical mapping studies. Prog Neurobiol 23:157-189

Panerai AE, Martini A, Casanueva F, Petraglia F, DiGiulio AM, Mantegazza P (1983) Opiates and their antagonists modulate luteinizing hormone acting outside the blood brain barrier. Life Sci 32:1751-1756

Pang CN, Zimmermann E, Sawyer CH (1977) Morphine inhibition of the preovulatory surges of plasma luteinizing hormone and follicle stimulating hormone in the rat. Endocrinology 101:1726-1732

Pedersen CA, Prange AJ Jr (1987) Effects of drugs and neuropeptides on sexual and maternal behavior in mammals. In: Meltzer HY (ed) Psychopharmacology: the third generation of progress. Raven, New York, pp 1477-1483

Petraglia F, Baraldi M, Giarrè G, Facchinetti F, Santi M, Volpe A, Genazzani AR (1985) Opioid peptides of the pituitary and hypothalamus: changes in pregnant and lactating rats. J Endocrinol 105:239-245

Petraglia F, Bernasconi S, Iughetti L, Loche S, Romanini F, Facchinetti F, Marcellini C, Genazzani AR (1986a) Naloxone-induced luteinizing hormone secretion in normal, precocious, and delayed puberty. J Clin Endocrinol Metab 63:1112-1116

Petraglia F, Locatelli V, Facchinetti F, Bergamaschi M, Genazzani AR, Cocchi D (1986b) Oestrous cycle-related LH responsiveness to naloxone: effect of high oestrogen levels on the activity of opioid receptors. J Endocrinol 108:89-94

Petraglia F, Vale W, Rivier C (1986c) Opioids act centrally to modulate stress-induced decrease in luteinizing hormone in the rat. Endocrinology 119:2445-2450

Pfaus JG, Gorzalka BB (1987) Opioids and sexual behaviour. Neurosci Biobehav Rev 11:1-34

Pfeiffer DG, Pfeiffer A, Almeida OFX, Herz A (1987) Opiate suppression of LH secretion involves central receptors different from those mediating opiate effects on prolactin secretion. J Endocrinol 114:469-476

Rayner V, Robinson ICAF, Russell JA (1988) Chronic intracerebroventricular morphine and lactation in rats: dependence and tolerance in relation to oxytocin neurons. J Physiol 396:319-347

Reid RL, Quigley ME, Yen SSC (1983) The disappearance of opioidergic regulation of gonadotropin secretion in postmenopausal women. J Clin Endocrinol Metab 57:1107-1110

Rivier C, Vale W (1984) Influence of corticotropin-releasing factor on reproductive functions in the rat. Endocrinology 114:914-921

Roberts AC, Hastings MH, Martensz ND, Herbert J (1985) Naloxone-induced secretion of LH in the male Syrian hamster: modulation by photoperiod and gonadal steroids. J Endocrinol 106:243-248

Roberts AC, Martensz ND, Hastings MH, Herbert J (1987) The effects of castration, testosterone replacement and photoperiod upon hypothalamic β-endorphin levels in the male Syrian hamster. Neuroscience 23:1075-1082

Rossier J, Battenberg E, Pittman Q, Bayon A, Koda L, Miller R, Guillemin R, Bloom F (1979) Hypothalamic enkephalin neurons may regulate the neurohypophysis. Nature (London) 277:653-655

Sawyer CH, Critchlow BV, Barraclough CH (1955) Mechanism of blockade of pituitary activation in the rat by morphine, atropine and barbiturates. Endocrinology 57:345-354

Sawyer CH (1963) Neuroendocrine blocking agents and gonadotropin release. In: Nalbandov AV (ed) Advances in neuroendocrinology. Univ Ill Press, Urbana, pp 444-451

Schanbacher BD (1985) Endogenous opiates and the hypothalamic-pituitary-gonadal axis in male sheep. Domest Anim Endocrinol 2:67-75

Schulz R, Wilhelm A, Pirke KM, Gramsch Ch, Herz A (1981) β-endorphin and dynorphin control serum luteinizing hormone level in immature female rats. Nature (London) 294:757–759

Schulz R, Wilhelm A, Pirke KM, Herz A (1985) Sex-dependent endorphinergic and adrenergic control mechanisms of luteinizing hormone secretion in immature rats. Acta Endocrinol 109:198–203

Serra G, Collu M, Gessa GL (1988) Endorphins and sexual behaviour. In: Rodgers RJ, Cooper SJ (eds) Endorphins, opiates, and behavioural processes. John Wiley & Sons, New York, pp 237–249

Sherman TG, Civelli O, Douglass J, Herbert E, Watson SJ (1986) Coordinate expression of hypothalamic pro-dynorphin and pro-vasopressin mRNAs with osmotic stimulation. Neuroendocrinology 48:16–24

Shoupe D, Montz FJ, Lobo RA (1985) The effects of estrogen and progestin on endogenous opioid activity in oophorectomized women. J Clin Endocrinol Metab 60:178–183

Simantov R, Snyder SH (1977) Opiate receptor binding in the pituitary gland. Brain Res 124:178–184

Sirinathsinghji DJS (1984) Modulation of lordosis behavior of female rats by naloxone, β-endorphin and its antiserum in the mesencephalic central gray: possible mediation via GnRH. Neuroendocrinology 39:222–230

Sirinathsinghji DJS (1985) Modulation of lordosis behaviour in the female rat by corticotropin releasing factor, β-endorphin and gonadotropin releasing hormone in the mesencephalic central gray. Brain Res 336:45–55

Sirinathsinghji DJS (1987) Inhibitory action of corticotropin releasing factor on components of sexual behaviour in the male rat. Brain Res 407:185–190

Sirinathsinghji DJS, Motta M, Martini L (1985) Induction of precocious puberty in the female rat after chronic naloxone administration during the neonatal period: the opiate "brake" on prepubertal gonadotropin secretion. J Endocrinol 104:299–307

Summerlee AJS (1989) Relaxin, opioids, and the timing of birth in rats. In: Dyer RG, Bicknell RJ (eds) Brain opioid systems in reproduction. Oxford Science Publications, Oxford, pp 257–270

Swann J, Turek FW (1983) Effect of naloxone on serum LH levels in testosterone-treated castrated animals exposed to short or long days. Fed Proc 42:300

Sylvester PW, Sarkar DK, Briski KP, Meites J (1985) Relation of gonadal hormones to differential LH response to naloxone in prepubertal male and female rats. Neuroendocrinology 40:165–170

Tamarkin L, Baird CJ, Almeida OFX (1985) Melatonin: a coordinating signal for mammalian reproduction? Science 227:714–720

Theodosis DT, Poulain DA, Montagnese C, Vincent JD (1988) In: Pickering BT, Wakerley JB, Summerlee AJS (eds) Neurosecretion: cellular aspects of the production and release of neuropeptides. Plenum, New York London, pp 157–166

Tweedle CD, Hatton GI (1982) Magnocellular neuropeptidergic terminals in neurohypophysis: Rapid glial release of enclosed axons during parturition. Brain Res Bull 8:205–209

Van Vugt DA, Sylvester PW, Aylsworth CF, Meites J (1982) Counteraction of gonadal steroid inhibition of luteinizing hormone release by naloxone. Neuroendocrinology 34:274–278

Veldhuis JD, Kulin HE, Warner BA, Santner SJ (1982) Responsiveness of gonadotropin secretion to infusion of an opiate-receptor antagonist in hypogonadotropic individuals. J Clin Endocrinol Metab 55:649–653

Wakerley JB, Noble R, Clarke G (1983) Effects of morphine and D-ALA, D-LEU enkephalin on the electrical activity of supraoptic neurosecretory cells in vitro. Neuroscience 10:73–81

Wardlaw SL, Frantz AG (1983) Brain β-endorphin during pregnancy, parturition, and the postpartum period. Endocrinology 113:1664–1668

Wilcox JN, Roberts JL (1985) Estrogen decreases rat hypothalamic proopiomelanocortin messenger ribonucleic acid levels. Endocrinology 117:2392–2396

Wilkinson M, Brawer JR, Wilkinson DA (1985) Gonadal steroid-induced modification of opiate binding sites in anterior hypothalamus of female rats. Biol Reprod 32:501–506

Wright DM (1985) Evidence for a spinal site at which opioids may act to inhibit the milk-ejection reflex. J Endocrinol 106:401–407

Zhao B-G, Chapman C, Bicknell RJ (1988) Functional κ-opioid receptors on oxytocin and vasopressin nerve terminals isolated from the rat neurohypophysis. Brain Res 462:62–66

CHAPTER 20

Motivational Effects of Opioids: Neurochemical and Neuroanatomical Substrates

T. S. Shippenberg and R. Bals-Kubik

1 Introduction

All voluntary behavior, whether it be the approach and withdrawal responses of lower animals or the diverse behavioral repetoire of phylogenetically advanced species, is controlled by its consequences. Animals will work to obtain those stimuli, which by satisfying a particular need state, are positively reinforcing; they will avoid those which induce pain or discomfort. Reinforcement, its recognition and procurement, motivates an animal to react and, as such, ensures its continued survival.

Opioids and other drugs of abuse exert profound effects upon mood and motivation. For some individuals, these agents may, in fact, become the primary stimuli which drive behavior. The subjective and positive reinforcing effects of these drugs are well documented. μ-Opioid receptor agonists such as morphine and heroin produce euphoria in humans and compulsive drug-seeking behavior in a variety of species.

Until recently, attempts to explain the abuse liability of opioids and other drugs have focused upon their ability to produce tolerance, physical and/or psychic dependence. There is now, however, substantial evidence which indicates that it is the activation of reward pathways in the brain, which is responsible for the motivational properties of all drugs of abuse. Furthermore, endogenous opioidergic systems appear to be an integral component of these pathways, such that a disruption in their normal activity can markedly affect the motivational state of an individual.

This chapter reviews work relating to such ideas and examines the neural substrates which may mediate the reinforcing and aversive effects of opioids. Although a variety of behavioral methods have been used to examine the motivational effects of drugs, this review will focus upon those data obtained from self-administration (SA) and conditioned place preference (CPP) techniques.

2 Theoretical Perspectives

Within the perspective of learning theory, drug use, whether casual or compulsive, is regarded as a behavior which is maintained by its consequences. Consequences which increase behavior are referred to as reinforcers and may be positive or negative. Positive reinforcers are those stimuli whose presentation increases the frequency of a response. In humans, they typically produce euphoria or a state of well-being. Negative reinforcers are those stimuli whose removal increases the frequency of a behavior. A stimulus may also be punishing in which case those behaviors which lead to its presentation are suppressed.

Both operant and classical conditioning techniques can be used to evaluate the motivational effects of drugs. In operant techniques, such as SA or intracranial self-stimulation (ICSS), administration of a stimulus (e.g., a drug) is contingent upon the subject performing a specific behavioral task (e.g., lever pressing). The ability of a drug to directly control behavior is assessed by evaluating drug-induced changes in performance. Data so derived provide a measure of primary reinforcement processes. An alternate approach used to characterize drug-induced motivational effects is the CPP paradigm. This classical conditioning procedure examines the associations which develop when the administration of a drug is paired with a previously neutral stimulus (e.g., place). Evaluation of a subject's behavior (approach, avoidance) following presentation of the stimulus provides a measure of the secondary reinforcing effects of a drug: a conditioned preference for the drug-paired place indicates positive reinforcement, whereas a place aversion indicates aversive (e.g., punishing) effects of a drug.

For a detailed discussion of these methodologies, their theoretical basis, advantages and disadvantages, the reader is referred to two recent reviews (Bozarth 1987b; Carr et al. 1989). It is, however, important to note that the results obtained from these methods are not always in agreement. Whether the disparate findings sometimes obtained result from differences in the methodologies used (i.e., drug-experienced subjects in the case of SA vs drug-naive in CPP) or the motivational effects measured (primary vs secondary reinforcement; acquisition of reinforcement vs its maintenance) is unknown and remains a source of controversy.

3 Motivational Effects of Opioid Agonists

A variety of systemically administered opioid agonists and mixed agonist-antagonists function as positive reinforcers in the SA and CPP paradigms. Data from both procedures have shown that this effect is characteristic of agonists with high intrinsic activity at μ-receptors (Young et al. 1983; Mucha and Herz 1985). Indeed, a strong correlation exists between the potencies of such agonists in suppressing the morphine abstinence syndrome and their positive reinforcing effects (Young et al. 1983; Brady and Lukas 1984).

The intracerebral route of administration has been used to examine the effects of those opioids (e.g., peptides) that do not readily cross the blood-brain barrier after systemic application. These studies have shown that endogenous peptides such as β-endorphin (B-END) and Leu-enkephalin (Leu-ENK) as well as the ENK analog D-Ala-D-Leu-ENK (DADLE) function as positive reinforcers in both paradigms (Belluzzi and Stein 1977; Katz and Gormezano 1979). The effectiveness of Leu-ENK, a putative endogenous ligand for the δ-receptor, as well as DADLE in producing reinforcement suggested that the activation of this receptor type may induce reinforcing states. Since, however, these peptides bind to μ- and δ-receptors (Magnan et al. 1982) the role of δ-receptors in opioid reinforcement was not resolved. A recent study has, however, shown that the selective δ-agonist DPDPE (Mosberg et al. 1983) produces marked place preferences in rats (Shippenberg et al. 1987). These findings, and those from studies (see Herz and Shippenberg 1989, for review) examining the influence of selective μ- (CTOP, B-FNA) and δ- (ICI 174,864) antagonists (Takemori et al. 1981; Cotton et al. 1984; Pelton et al. 1986) indicate an involvement of both μ- and δ-receptors in opioid-induced reward and suggest that the activation of either receptor type produces positive reinforcing effects. Furthermore, the data show that δ- as well as μ-opioid receptor agonists may be positively reinforcing in humans and, as such, have a marked potential for abuse.

In subhuman primates, opioids with κ-agonist activity (ethylketazocine, N-furyl-substituted benzomorphans) do not maintain behavior leading to their administration (Woods et al. 1979) and may function as negative reinforcers (Kandel and Schuster 1977; Hoffmeister 1979). CPP studies (Mucha and Herz 1985; see Herz and Shippenberg 1989, for review) later provided the first indication that the administration of κ-agonists is, in fact, highly aversive. Thus, the selective κ-agonists U-69593 and U50,488H (von Voightlander et al. 1983; Lahti et al. 1985) as well as a series of mixed agonist-antagonists produced conditioned place aversions and the potencies of these compounds in this respect was correlated with differences in their binding affinity to κ-receptors (Mucha and Herz 1985; see Herz and Shippenberg 1989, for review). A subsequent clinical study (see Pfeiffer this Vol.) revealed that preferential κ-agonists produce aversive as well as dysphoric effects in human subjects. Thus, results from both clinical and experimental studies demonstrate that the activation of κ-opioid receptors induces aversive states.

In summary, data from both SA and CPP studies have shown that the motivational effects of opioid agonists differ depending upon the receptor type with which they interact. μ- and δ-receptor agonists function as positive reinforcers, whereas κ-agonists are negative reinforcers and induce aversive states. Similar results have been obtained in other animal models and in human subjects.

4 Motivational Effects of Opioid Antagonists: Existence of an Endogenous Opioid Reward Pathway

Opioid antagonists such as naloxone (NAL) or naltrexone (NLTRX) lack positive reinforcing effects in both humans and experimental animals (Grevert and Goldstein 1977; Hollister et al. 1981). Indeed, NAL serves as a negative reinforcer in nonopioid-dependent monkeys; maintaining those behaviors that terminate its administration (Downs and Woods 1976). Subsequent CPP studies (see Herz and Shippenberg 1989, for review) have demonstrated that these antagonists are not only negative reinforcers but are highly aversive, producing marked place aversions in nonopioid-dependent rats and mice. Negative results with the inactive stereoisomers of these antagonists have confirmed that such effects are opioid receptor-mediated.

The finding of such marked effects of opioid antagonists are particularly noteworthy in that they suggest the existence of a tonically active endogenous opioid reward pathway, the disruption of which induces aversive states. Indeed, the demonstration over three decades ago (Olds and Milner 1954) that animals would work to electrically stimulate certain brain regions suggested the existence of a specialized brain system which mediates reward. Furthermore, the findings that DADLE and naturally occurring opioid agonists such as Leu- or Met-ENK enhance, whereas opioid antagonists reduce, such stimulation indicated that endogenous opioids were a critical component of these reward pathways (Belluzzi and Stein 1977; Broekkamp et al. 1979; Stapleton et al. 1979; Olds and Williams 1980).

A recent study has provided insights as to the identity of one such opioid "reward" peptide. Mucha et al. (1985) showed that lesions of the mediobasal arcuate nucleus, the primary site of β-END synthesis in the CNS (Watson et al. 1978), attenuates the aversive effects of NAL, whereas the motivational effects of μ- or κ-agonists are not modified. Although such lesions may alter other neurotransmitter systems, the ability of such lesions to decrease CNS levels of β-END, but not ENK suggest that there is a tonically active β-END system which, when disrupted (either by an inhibition of β-END release or, in the case of NAL, by blockade of the receptor with which β-END interacts) results in aversive states.

An examination of the motivational properties of a series of opioid antagonists has shown that aversive effects are also produced by the selective μ-receptor antagonist CTOP (Bals-Kubik et al. 1989a). In contrast, the δ-receptor antagonist ICI 174,864 and the κ-antagonist nor-binaltorphimine (nor-BNI: Takemori and Porthoghese 1988) lack reinforcing or aversive effects. Such findings suggest that the aversive effects of both NAL and NLTRX result from the blockade of μ-opioid receptors. Furthermore, they indicate that the release of β-END and the tonic activation of μ-opioid receptors are required for the maintenance of neutral motivational states. The apparent ineffectiveness of nor-BNI would also suggest the absence of an opposing tonically active κ-opioidergic system.

Clinical studies examining the motivational effects of selective opioid antagonists are lacking. Although aversive effects of both NAL and NLTRX have been documented in nonopioid-dependent subjects, the effects reported were relatively weak (Grevert and Goldstein 1977; Janowsky et al. 1979; Hollister et al. 1981). Furthermore, other studies were unable to confirm such results (Grevert and Goldstein 1978). Although these data may indicate species differences in the response to opioid antagonists, the failure of the above clinical studies to obtain full dose-response curves limits data interpretation and additional studies are needed to clarify this issue.

5 Central vs Peripheral Site of Action of Opioids

Data from both SA and CPP studies have shown that the rewarding effects of opioids are centrally mediated. Thus, the intracerebroventricular (icv) administration of methylnaloxonium, a quarternary derivative of NAL, which does not readily cross the blood brain barrier, abolishes intravenous (iv) heroin SA (Vaccarino et al. 1985b), whereas the peripheral administration of quarternary NAL or NLTRX is without effect (Koob et al. 1984). Animals will also work to obtain icv injections of β-END, ENK, as well as a variety of μ- and δ-agonists (Belluzzi and

Stein 1977; van Ree et al. 1979). Furthermore, for those opioids tested, the doses which support icv SA are substantially lower than those observed in iv SA studies (Amit et al. 1976; Smith et al. 1976; Bozarth and Wise 1981b). Similarly, the icv administration of morphine produces marked conditioned place preferences and the doses producing this effect are ca. 60-fold lower than those effective systemically (Shippenberg et al. 1987; van der Kooy et al. 1982).

To date, only the CPP procedure has been used to evaluate the sites mediating the aversive effects of opioid ligands. The finding that the icv administration of quaternary opioid antagonists produces place aversions at doses which are ineffective following systemic application, together with data indicating a substantially higher potency of icv as compared to systemically administered NAL (Bals-Kubik et al. 1989a), strongly indicate a central site of action. There is, however, conflicting opinion regarding the effect of κ-agonists. Bechara and van der Kooy (1987), by analyzing the dose-response curves for subcutaneously and intraperitoneally applied U50,488H, susceptibility to antagonism by MR2266, and the effects of vagotomy, which destroy opiate receptors located on vagal sensory axons, concluded that κ-agonists produce aversive effects in the periphery and reinforcing effects which are centrally mediated. The effects of icv application were, however, not assessed. In contrast, a recent study (Bals-Kubik et al. 1989a) in which the dose-response curves for systemic and icv applied U50,488H were compared, as well as studies (Bals-Kubik et al. 1989b; Shippenberg et al. 1989) employing intracranial microinjection and lesion techniques (see below) indicate that the CNS is an important site of action for the aversive effects of κ-agonists. Furthermore, microinjection studies have been unable to demonstrate any reinforcing effects of synthetic κ-agonists. The recent availability of a selective κ-agonist with limited capacity to cross the blood-brain barrier (Tachibana et al. 1988) may help to resolve this issue.

6 Neural Substrates Mediating the Motivational Effects of Opioids

The fact that the icv administration of opioids can produce reinforcing or aversive effects provides little information regarding the neural substrates which mediate such actions. In studies examining this issue, two general approaches may be employed. In the first, agonists and/or antagonists are microinjected into discrete brain areas and a map of those sites which produce or block a specific motivational effect is generated. In the second approach, specific neurotransmitter systems, nuclei or cell bodies located in a particular brain region are lesioned, and the ability of such manipulations to modify the effects of a test drug is examined. Since the advantages and disadvantages of these approaches have been discussed elsewhere (see Katz 1989, for review), they will not be extensively elaborated upon. It must, however, be emphasized that in microinjection studies of any kind, the volume, pH, osmolality, rate and pressure of infusion are critical factors in determining both drug diffusion and the amount of tissue damage produced. Furthermore, most studies to date are limited by the fact that the extent of drug diffusion has not been directly assessed.

6.1 Neuroanatomical Substrates

Intracranial morphine SA was first demonstrated in the lateral hypothalamus (LH) of rats (Stein and Olds 1976). SA of D-Ala-D-Met-ENK was also observed in this region (Olds and Williams 1980). These studies, however, used animals that were previously trained to press levers for rewarding electrical brain stimulation and data from experimentally naive subjects was not presented. The use of animals with a prior history of responding for a reinforcing stimulus confounds data interpretation since reinforcement involves not only the maintenance, but also the acquisition of a response. Therefore, whether this structure plays a role in the acquisition of opioid reinforcement remains unclear.

More recent studies have pointed to the mesolimbic dopaminergic (DA) system, and the opioid receptors located therein, as a critical site for the initiation of opioid reinforcement. Bozarth and Wise (1981a) demonstrated that rats will rapidly learn to self-administer morphine (100 ng) into the ventral tegmental area (VTA), where the cell bodies of this DA system are localized (Ungerstedt 1971). In contrast, infusion of the same dose of morphine into other sites including the caudate, periventricular gray, substantia nigra, LH and the nucleus accumbens (NAC); the latter as a primary projection site of VTA neurons were ineffective. In view of (1) the ability of morphine to increase the single unit activity of VTA DA cells (Gysling and Wang 1983; Matthews and German 1984): (2) the apparent localization of opioid receptors either on, or proximal to, these DA cell bodies (Moskowitz and Goodman 1984; Mansour et al. 1988); and (3) evidence that μ-agonists increase DA release in this system (see Illes and Jackisch this Vol.), Bozarth and Wise (1981a) concluded that VTA DA neurons are critical for opioid reinforcement. Furthermore, the domonstrated involvement of this site in the reinforcing effects of psychostimulants such as cocaine and amphetamine led to the hypothesis that a common DA substrate may mediate the reinforcing properties of all drugs of abuse (Wise 1980; see Wise and Rompre 1989, for review).

It is particularly noteworthy that several studies have shown that neither the SA nor the continuous infusion of morphine into the VTA produce physical dependence (Bozarth and Wise 1984). Such findings are of critical importance since they indicate that the neural substrates underlying reinforcement are distinct from those underlying dependence and withdrawal, thus demonstrating that it is the reinforcing properties of these drugs rather than their capacity to induce dependence, which underlies their abuse potential.

More recent SA studies (Koob and Goeders 1989) also suggest an involvement of the VTA in opioid reinforcement. However, the relative importance of DA (see below) as well as the NAC have been questioned. Thus, it has been shown that rats will self-administer morphine into the NAC at a three fold higher dose than that tested by Bozarth and Wise and they will also self-administer the δ-agonist, Met-ENK (Olds 1982; Koob and Goeders 1989). The higher doses of morphine required for NAC SA may indicate that this structure is not of primary importance in opioid reinforcement. However, the predominance of δ-receptors (Goodman et al. 1980), as well as the ability of Met-ENK to support SA in this region, suggest that both sites are important for opioid reinforcement. With regard to this conclusion, however, it is interesting to note that whereas an abolition of iv heroin SA following VTA (but not NAC) microinjection of quarternary nalorphine has been reported (Britt and Wise 1983), another study using the antagonist methyl naloxonium demonstrated a greater abolition of heroin SA following NAC as compared to VTA injections (Vaccarino et al. 1985a). Although an explanation for these data is lacking, the different regional distributions of opioid receptor types and the differing affinities of various ligands for these sites suggest that dose-response curves are critical for all mapping studies.

Several studies have examined the effects of lesioning specific brain areas upon opioid SA. Microinjection of the neurotoxin kainic acid, which destroys cell bodies but not fibers of passage (Schwarcz et al. 1979), into the NAC was shown to markedly decrease iv SA of both opioids and psychostimulants (Zito et al. 1985) and a similar effect was observed following selective destruction of cell bodies in the ventral pallidum by ibotenic acid injection (Hubner and Koob 1987). Such findings strongly indicate that these areas and the receptors located therein are crucial for opioid reinforcement. Interestingly, Smith and Lane (1983) also reported a disruption of morphine SA following kainic acid injections into the NAC. However, an increase, rather than a decrease, in morphine SA was observed. The authors concluded that the increased intake represented an attempt to compensate for the loss of neurons which normally facilitate opioid reinforcement processes. Although the data just described appear contradictory, differences in the extent of the lesions in each study may underlie the different results obtained. Thus, depending on the efficacy of the lesion (e.g., the percentage of remaining cells), cessation of SA or a shift to the right in the dose-response curve may occur. As such, both findings are compatible with the hypothesis that intrinsic and efferent neurons of the NAC and ventral pallidum are necessary for opioid reinforcement.

Data regarding the effects of such lesions to other brain regions are limited. Microinjection of kainic acid into the LH failed to modify either heroin or cocaine SA (Britt and Wise 1981), suggesting that this region is not critical for opioid or psychostimulant reinforcement. To date, neurotoxic lesions have not been used to examine the contribution of neurons originating in the VTA. Such information is, however, crucial in view of the continued controversy regarding the neurochemical substrates mediating opioid reward.

CPP studies also suggest an important role for the mesolimbic system in opioid reinforcement – although, again, controversy exists regarding the relative importance of the NAC vs the VTA. Unilateral or bilateral injections of morphine into the VTA produce place preferences, whereas injections into sites around this nucleus are without effect (Bozarth 1987a). NAC injections of opioids also induce place preferences (Olds 1982). However, this effect is generally less than that in the VTA and is not observed by all laboratories.

Morphine-induced place preferences have also been observed following injection into the LH and periaqueductal gray, whereas microinjections into the caudate, amygdala or nucleus ambiguous were without effect (van der Kooy et al. 1982). However, confirmation of these data, as well as additional studies examining the influence of ibotenic or kainic acid lesions of these areas, are needed to determine whether they are facilitatory or necessary for opioid reward.

There are only a few reports regarding the sites mediating the aversive effects of opioids. Mapping studies in our laboratory indicate that unilateral microinjections of the opioid antagonist NAL or of the κ-agonist U-50,488H into either the VTA or NAC produce place aversions (Bals Kubik et al. 1989b; Shippenberg et al. 1989). Furthermore, the doses producing such effects are lower than those observed after icv or peripheral administration. Injections of κ-agonists, but not opioid antagonists, into the medial prefrontal cortex (MFC) also produce aversive effects. Although these data are preliminary, the failure to elicit conditioning from other areas suggest that the VTA, NAC and MFC are specific sites which mediate the aversive effects of opioids.

In summary, as shown in Figs. 1 and 2, mapping and lesions studies indicate important, and perhaps critical roles of the VTA and NAC in mediating the motivational effects of opioid agonists and antagonists. Furthermore, although the neural substrates mediating the reinforcing and aversive effects of opioids appear to overlap (e.g., the mesolimbic system), it is probable that multiple sites are involved in the expression of both these effects.

6.2 Neurochemical Substrates

Several lines of evidence suggest an involvement of central DAergic neurons and, in particular the mesolimbic DA system, in reinforcement processes. Administration of the DA receptor antagonists haloperidol, raclopride or SCH 23390 attenuate the reinforcing effects of "natural rewards" such as food and water, as well as the effects of ICSS (Spyraki et al. 1982; Nakajima and McKenzie 1986; Hoffman and Beninger 1986; Nakajima and Baker 1989). The positive reinforcing effects of psychostimulants also appear to be DA-dependent, since DA antagonists or manipulations which disrupt mesolimbic DA neurotransmission attenuate amphetamine as well as cocaine SA, and abolish the place preference produced by these drugs (Yokel and Wise 1975, 1976; Lyness et al. 1979; Pettit et al. 1984). Indeed, the critical role of DA and the apparent involvement of the mesolimbic system in these various types of reward have prompted the suggestion that the mesolimbic DA pathway may mediate the reinforcing properties of *all* drugs of abuse.

The ability of μ- and δ-opioid agonists to function as positive reinforcers following microinjection into either the VTA (the site of origin of this DA system) or the NAC (a terminal field of this nucleus) are also compatible with the involvement of the mesolimbic system in opioid reward. Furthermore, the effectiveness of these sites in producing aversive states suggest that this system may mediate the reinforcing as well as the aversive effects of opioids.

In order to establish whether the motivational effects of opioids are DA-dependent, it is necessary to show that (1) opioid agonists influence DA neuronal activity; (2) DA activity is differentially affected by μ- and δ- (reinforcing) as compared to κ-(aversive) agonists; (3) disruption of DA transmission, either by pretreatment with receptor antagonists or by lesioning DA-containing neurons, attenuates both motivational effects of opioids.

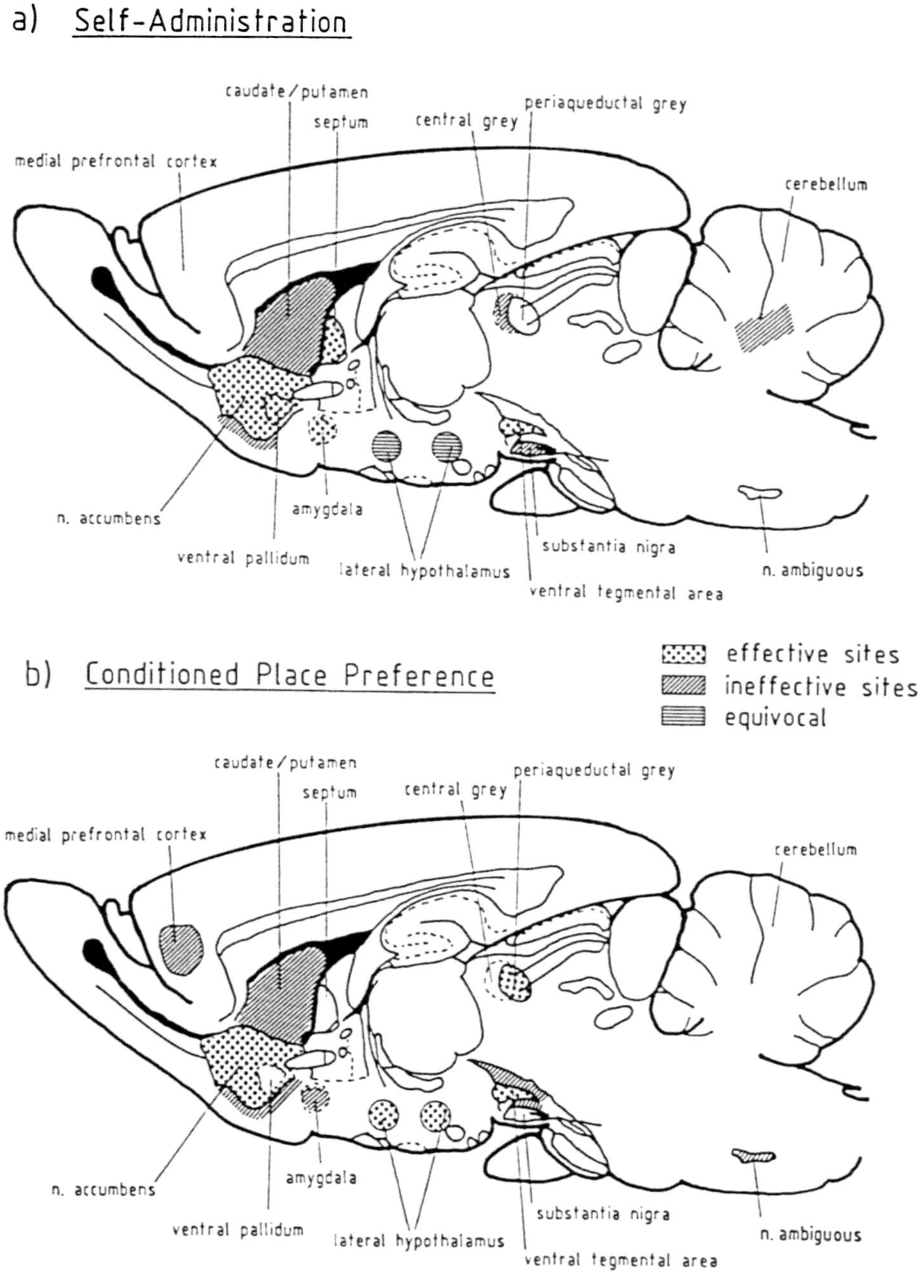

Fig. 1. Neuroanatomical substrates implicated in the mediation of the positive reinforcing effects of opioids. Data obtained from self-administration and conditioned place preference studies are shown in **a** and **b**, respectively. Regions outlined by *hatched lines*, although represented, would not actually be present in the section depicted. Note that both procedures suggest an involvement of the ventral tegmental area and the n. accumbens in opioid reinforcement, whereas data concerning other areas are conflicting

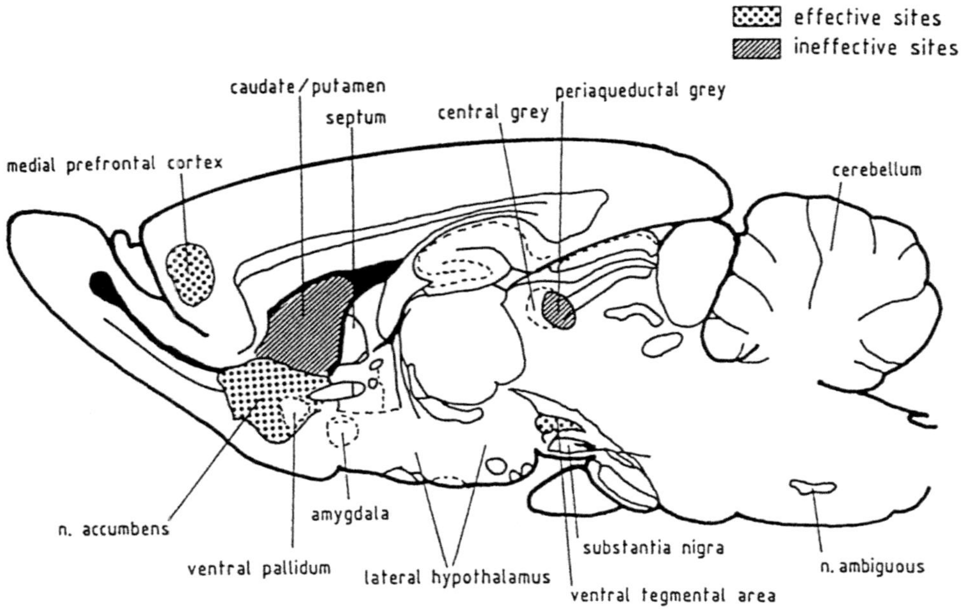

Fig. 2. Neuroanatomical substrates implicated in the mediation of the aversive effects of κ-opioid agonists. Data from conditioned place preference studies indicate an involvement of the ventral tegmental area, n. accumbens, and the medial prefrontal cortex in mediating the aversive effects of U-50488H and the dynorphin analog E-2078

Electrophysiological studies have shown that μ-opioid receptor agonists increase the single unit activity of mesolimbic DA neurons (Gysling and Wang 1983). In contrast, κ-agonists decrease this activity (Jeziorski and White 1989). Neurochemical data also suggest marked effects of opioids on DA function. Thus, recent microdialysis studies have shown that the systemic administration of μ-agonists increases the release and metabolism of DA in the NAC, whereas κ-agonists decrease these parameters (DiChiara and Imperato 1989). The icv administration of β-END or the selective δ-receptor agonist DPDPE were also shown to increase NAC DA release, whereas a metabolically stable analog of DYN decreased release from this structure (R. Spanagel, A. Herz and T. Shippenberg, unpubl. observ.). Furthermore, the findings that other drugs of abuse including psychostimulants, ethanol and nicotine selectively increase NAC DA release (Imperato and DiChiara 1986; Imperato et al. 1986; Carboni et al. 1989a) suggest that this action may underlie their reinforcing effects.

In view of these data, the ability of several drugs, which decrease DA activity in this area (e.g., U-65953, U-50488H, SCH 23390; see below), to produce aversive states has prompted the hypothesis that this decrease may underlie their aversive effects.

Interestingly, studies to date have been unable to demonstrate a decrease in DA release in response to opioid antagonists (DiChiara and Imperato 1989). If, as has been suggested, there is a tonically active reward pathway upon which opioid antagonists act to produce their aversive effects, it would be expected that these agents would decrease DA release. The failure to observe this appears to preclude an involvement of DA in the aversive effects of opioid antagonists. However, it must be noted that early studies evaluating the effects of opioid antagonists on striatal DA metabolism found no alterations in levels of DA or its metabolites, indicating the lack of a tonically active nigrostriatal opioidergic system. Subsequent studies (Wood et al. 1989), in which the effects of endogenous opioids were potentiated by use of the enkephalinase inhibitor thiorphan have, however, prompted a reappraisal of this hypothesis. Therefore, until similar experiments are conducted in the mesolimbic system, the role of DA in mediating the aversive effects of opioid antagonists remains unconfirmed.

Data regarding the influence of DA receptor blockade or DA denervation upon behavioral indices of opioid-induced reward and aversion are less than clear. Both pimozide and α-flupenthixol produce dose-related decreases in heroin SA (Ettenberg et al. 1982; Bozarth 1983). However, the pattern of responding observed after such pretreatment has generated considerable controversy. Typically, treatment with DA antagonists initially increases and then decreases psychostimulant SA. A similar pattern is also observed with opioids following NLX or NLTRX pretreatment and is thought to reflect an animal's initial attempt to obtain reinforcement and its subsequent frustration at being unable to obtain it. In contrast, pretreatment with D-2 or D-1/D-2 DA receptor antagonists only produces a cessation in responding. Thus, it has been argued that abolition of opioid SA results from sedative or toxic effects of these agents rather than from a true pharmacological antagonism (Ettenberg et al. 1982). However, as discussed by Wise and Rompre (1989), it must be noted that following pretreatment with DA antagonists, SA of the direct-acting DA agonist apomorphine also shows a simple decrease in response rate, an effect which is generally considered to reflect an attenuation of the reinforcing properties of this drug. Therefore, differences in the mechanisms by which drugs enhance DA function may determine whether there is an initial compensatory increase or only a cessation of intake.

Interestingly, although extensive discussion is beyond the scope of this chapter (see Wise and Rompre 1989, for review) recent electrophysiological studies have provided one explanation as to why the combination of DA antagonists and opioids may not produce compensatory increases in responding. Thus, several investigations have shown that overstimulation of DA systems results in depolarization-induced inactivation of DA neurons (Bunney and Grace 1978; White and Wang 1983; Grace and Bunney 1986). Since morphine and D-2 DA antagonists stimulate the firing of DA neurons and can each, at appropriate doses, cause depolarization inactivation (White and Wang 1983; Grace and Bunney 1986), administration of increasing doses of opioids with moderate doses of a DA antagonist could amplify the effects of DA antagonists causing complete shutdown of the DA system. Thus, although animals may initially respond to an opioid in the presence of low doses of DA antagonists, once the threshold for inactivation is achieved, failure of SA would occur.

CPP data regarding DA receptor antagonists are conflicting. Thus, whereas an attenuation of morphine or heroin reinforcement has been reported by some investigators (Bozarth and Wise 1981b; Spyraki et al. 1983), such findings have not been universal (Mackey and van der Kooy 1985).

Results of recent studies employing selective D-1 and D-2 receptor antagonists may provide an explanation for these data and shed new light on the role of DA in both opioid reinforcement and aversion. Thus, it has been shown that the selective D-1 antagonist SCH 23390 (Billiard et al. 1984) abolishes the place preferences produced by morphine, whereas selective D-2 antagonists are without effect (Shippenberg and Herz 1987, 1988). A similar antagonism is also observed after the acute microinjection of SCH 23390 into the NAC but not other brain regions (Shippenberg et al. 1989). Interestingly, systemic or NAC injections of this D-1 antagonist also abolish the place aversions produced by the κ-agonists U-69593 and U-50488H, whereas the D-2 antagonists spiperone or sulpiride are ineffective (Shippenberg et al. 1989). The motivational effects produced by a drug of a different pharmacological class, lithium, were not modified by this treatment, suggesting that the antagonism observed in the case of opioids did not result from the inability of animals to learn a conditioned response.

Doses of SCH 23390 higher than those used in the above studies produce conditioned place aversions (Shippenberg and Herz 1988; Shippenberg and Bals-Kubik, unpubl. observ.). In contrast, D-2 antagonists are devoid of reinforcing or aversive effects (Spyraki et al. 1982; Mackey and Van der Kooy 1985). Such findings are analogous to those obtained with opioid antagonists and have prompted the hypothesis that tonic activation of D-1 receptors is required for the maintenance of neutral motivational states (Fig. 3). Furthermore, in view of the effects of opioids and other drugs of abuse upon mesolimbic DA release, it would appear that an increase in the basal release of DA and a subsequent increase in the tonic activity of D-1 receptors in the NAC underlies, at least in part, the reinforcing effects of these agents. Results obtained in our laboratory regarding κ-agonists and SCH 23390 also suggest that a decrease in the tonic activation of NAC D-1 receptors either by the administration of a competitive antagonist or, in the case of κ-opioid agonists, by a decrease in DA release, results in aversive states.

Fig. 3. Postulated involvement of the mesolimbic system in mediating the motivational effects of opioids. β-ENDergic fibers projecting from the mediobasal hypothalamus terminate upon interneurons located in the ventral tegmental area (*VTA*). These interneurons impinge upon VTA DA neurons and are presumed to be inhibitory in nature. VTA DA neurons, in turn, project to the n. accumbens (*NAC*) and other limbic areas. **a** Neutral motivational states: The tonic release of β-END and the subsequent inhibition of VTA interneurons results in the tonic release of DA in the NAC and a tonic activation of the NAC D-1 receptor. **b** Reinforcing states: Increased release of β-END (or administration of μ- and δ-agonists) results in a disinhibition of VTA DA neuronal activity and a subsequent increase in DA release. Increased activation of the D-1 receptor leads to positive reinforcing states. **c** Aversive states: Inhibition of β-END release or the blockade of μ-opioid receptors in the VTA (**a**) increases the activity of VTA inhibitory interneurons. The subsequent inhibition of DA release results in a decrease in the tonic activation of the D-1 receptor. Alternatively, increased release of *DYN* (**b**) or the administration of a κ-agonist inhibits DA neuronal activity resulting in an inhibition of DA release and a subsequent decrease in the activity of the NAC D-1 receptor

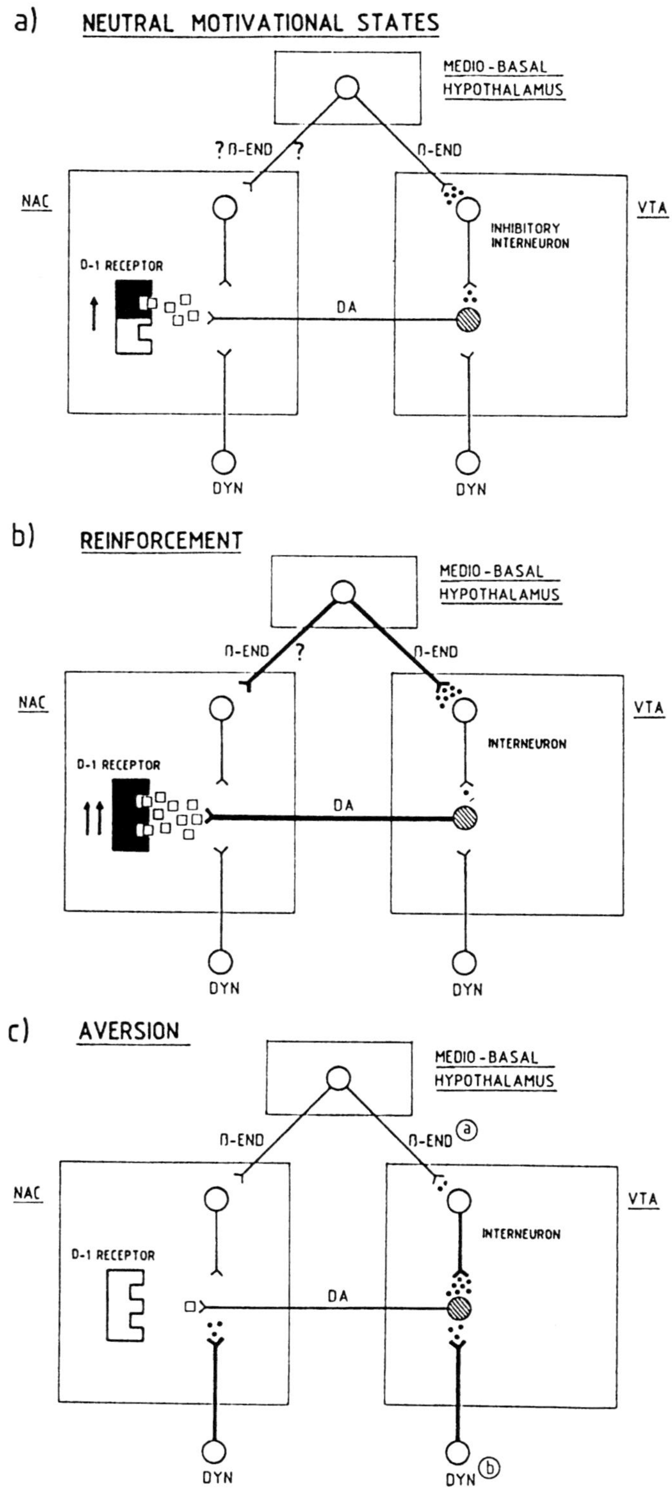
a) NEUTRAL MOTIVATIONAL STATES
MEDIO-BASAL HYPOTHALAMUS
? ß-END ?
ß-END
NAC
VTA
INHIBITORY INTERNEURON
D-1 RECEPTOR
DA
DYN
DYN
b) REINFORCEMENT
MEDIO-BASAL HYPOTHALAMUS
ß-END ?
ß-END
NAC
VTA
INTERNEURON
D-1 RECEPTOR
DA
DYN
DYN
c) AVERSION
MEDIO-BASAL HYPOTHALAMUS
ß-END
ß-END ⓐ
NAC
VTA
INTERNEURON
D-1 RECEPTOR
DA
DYN
DYN ⓑ

Interestingly, a recent study (Nakajima and Wise 1987) has shown that the systemic administration of SCH 23390 disrupts heroin SA. The pattern of responding produced consists of a compensatory increase in heroin SA followed by cessation. Similar effects are observed for psychostimulant SA (Nakajima and Wise 1987) as well as for ICSS (Nakajima and McKenzie 1986). Since SCH 23390 differs from D-2 antagonists in that it does not influence DA cell firing (Hand et al. 1987) and, as such, does not cause depolarization-induced inactivation, it would appear that this action may, as suggested by Wise and Rompre (1989), explain the different effects of DA antagonists upon opioid SA. Furthermore, the apparent importance of the D-1 receptor for opioid reinforcement as well as aversion is consistent with data indicating that D-1 receptor activation may enable the expression of D-2 receptor-mediated effects.

The influence of 6-OHDA lesions of the NAC has also been used to examine the role of the mesolimbic DA system in reinforcement and aversion. Microinjection of this neurotoxin, following pretreatment with a norepinephrine re-uptake inhibitor, permits the selective destruction of DA neurons. However, the extent of DA depletion is dependent on the dose and volume of toxin administered and may greatly influence the results obtained.

Pettit et al. (1984) trained rats to self-administer heroin and cocaine until stable intake for each drug was attained. The animals then received 6-OHDA lesions of the NAC which resulted in a greater than 90% depletion of DA. Cocaine SA decreased to 30% of prelesion values. In contrast, although heroin SA was initially decreased, it gradually recovered, reaching 76% of prelesion values. An initial compensatory increase in cocaine, but not heroin, SA was seen. Based on these results, the authors concluded that DA mediates the reinforcing effects of psychostimulants, but not of opioids.

Smith et al. (1985) examined the effects of 6-OHDA lesions of the NAC (14% depletion of DA) upon morphine SA in rats previously made physically dependent on morphine. In contrast to the above results, these investigators found such lesions to markedly increase opioid SA. Analysis of the dose-response curves indicated that twice the dose of morphine was necessary to maintain rates of intake similar to that of prelesion values. The authors concluded that the DA innervation of the NAC is necessary for the reinforcing effects of opioids such that its loss results in a compensatory increase in drug intake.

The reasons for the different results obtained in these studies are unclear. Procedural differences (e.g., degree of physical dependence or schedules of drug availability) as well as the extent and exact location of the lesions may be important factors. In addition, as discussed by Smith et al. (1985), the lack of effect observed by Pettit et al. (1984) may have been dose-related. Thus, the dose of cocaine used may have been a threshold dose for SA, whereas that used for heroin may have been further to the right on the dose-response curve. If such was the case, then cocaine, but not heroin SA might be affected. Unfortunately, however, no dose-response curves were obtained in that study and, at present, the role of DA awaits further clarification. Furthermore, it must be noted that in both studies the animals had had extensive drug experience. Therefore, maintenance of reinforcement, rather than its acquisition, may have been measured. Thus, SA studies examining the influence of such lesions upon the initiation of drug-taking in naive animals would appear critical for the resolution of this long-running controversy.

Several CPP studies have shown that DA denervation alters the reinforcing effects of opioids. In one of the first studies of its kind, Schwartz and Marchok (1974) found that the icv administration of 6-OHDA or treatment with the catecholamine synthesis inhibitor α-methylparatyrosine attenuated morphine-induced place preferences. In contrast, depletion of NE was without effect. Other studies (Spyraki et al. 1983; Shippenberg et al. 1989) have shown that 6-OHDA lesions of the NAC attenuate the place preferences produced by μ-agonists. Lesions to other brain areas or of noradrenergic neurons were without effect. Taken together, these data clearly indicate that, in the CPP procedure, the mesolimbic DA system is critical for opioid reinforcement.

Preliminary studies in our laboratory have also shown that 6-OHDA lesions of the NAC abolish the aversive effects of κ-agonists. In contrast, lesions of the MFC or striatum are ineffective. Although such findings must be replicated by other laboratories, they again suggest an involvement of DA in both motivational effects of opioids and are consistent with the schema shown in Fig. 3.

More recently, the role of serotonin (5-HT) in opioid reinforcement processes has been examined. Microdialysis studies (Carboni et al. 1989b; R. Spanagel and T.S. Shippenberg, unpubl. observ.) have provided evidence that morphine, β-END and other opioids that function as positive reinforcers stimulate 5-HT release in the NAC. Furthermore, 5,7 dihydroxytryptamine (5,7-DHT) lesions of the NAC, which reduced 5-HT levels in this nucleus to 50% of control values, were found to increase morphine SA in physically dependent rats (Smith et al. 1987). Interestingly, the lesions increased the intake of various doses of morphine, essentially abolishing the dose-response relationship for morphine SA. Although, studies in nondependent animals are lacking, electrolytic lesions of the raphe nuclei were also shown to shift the dose-response curve for morphine SA (Glick and Cox 1977), suggesting that 5-HT may play an important role in opioid reinforcement.

Results of recent CPP studies support this hypothesis. Thus, 5,7-DHT lesions of the NAC attenuate the place preferences produced by morphine (Nomikos et al. 1986). In addition, pretreatment with 5-HT3 receptor antagonists, such as ICS 205-930 and MDL 7222, or with the 5-HT2 receptor antagonist ritanserin, were also effective in attenuating the place preferences produced by morphine (Nomikos and Spyraki 1988; Carboni et al. 1989c). Interestingly, Carboni et al. (1989b) have recently shown that a 5-HT3 antagonist blocks the stimulatory effects of morphine upon DA release in the NAC, suggesting that opioid reinforcement may require functionally intact 5-HT and DA neurons.

In summary, there is increasing evidence suggesting the involvement of both DA and 5-HT in the motivational effects of opioids. Whether or not other neurotransmitter systems may modulate or mediate the motivational effects of opioids has not, as yet, been assessed. Such information, however, is critical for our understanding of the neurochemical and neuroanatomical substrates which underlie the reinforcing properties of opioids and other drugs of abuse.

7 Conclusions

Drug addiction is a major health problem, which affects either directly or indirectly, a large segment of society. The development of animal models which permit assessment of the motivational effects of a drug has contributed greatly to our understanding of the processes underlying the genesis of compulsive drug-seeking behavior. The existence of endogenous reward pathways in the brain, upon which various classes of abused drugs act to produce their positive reinforcing effects, appears well established. Furthermore, there is substantial evidence which indicates that the VTA and NAC are important sites for the expression of the motivational effects of opioids and other drugs of abuse. Fundamental questions, however, remain regarding the neurotransmitter systems mediating the reinforcing and aversive effects of various drugs. Additional studies, in which the behavioral and neurochemical correlates of these motivational effects are examined in parallel, will be needed to resolve these issues. Such data will substantially increase our understanding of the neuronal systems underlying reinforcement processes and may also have relevance for the treatment of disorders of mood and affect.

References

Amit Z, Brown ZW, Sklar LS (1976) Intraventricular self-administration of morphine in naive laboratory rats. Psychopharmacology 48:291

Bals-Kubik R, Herz A, Shippenberg TS (1989a) Evidence that the aversive effects of opioid antagonists and κ-agonists are centrally mediated. Psychopharmacology 98:203–206

Bals-Kubik R, Shippenberg TS, Herz A (1989b) Aversive effects of kappa-opioid agonists: involvement of the mesolimbic dopaminergic system. In: Abstr 12th Annu Meet Eur Neurosci Assoc, 3–7 Sept 1989. Oxford Univ Press, Turin, p 309

Bechara A, van der Kooy D (1987) Kappa receptors mediate the peripheral aversive effects of opiates. Pharmacol Biochem Behav 28:227–233

Belluzzi JD, Stein L (1977) Enkephalin may mediate euphoria and drive-reduction reward. Nature (London) 266:556–558

Billiard W, Ruperto V, Crosby G, Iorio L, Barnett A (1984) Characterization of the binding of ^{3}H-SCH 23390, a selective D-1 receptor antagonist ligand in rat striatum. Life Sci 35:1885–1893

Bozarth MA (1983) Opiate reward mechanisms mapped by intracranial self-administration. In: Smith JE, Lane JD (eds) The neurobiology of opiate reward processes. Elsevier/Northholland Biomedical Press, Amsterdam, pp 331–359

Bozarth MA (1987a) Neuroanatomical boundaries of the reward-relevant opiate-receptor field in the ventral tegmental area as mapped by the conditioned place preference method in rats. Brain Res 414:77–84

Bozarth MA (1987b) Intracranial self-administration procedures for the assessment of drug reinforcement. In: Bozarth MA (ed) Methods of assessing the reinforcing properties of abused drugs. Springer, Berlin Heidelberg New York, pp 173–188

Bozarth MA, Wise RA (1981a) Intracranial self-administration of morphine into the ventral tegmental area in rats. Life Sci 28:551–555

Bozarth MA, Wise RA (1981b) Heroin reward is dependent on a dopaminergic substrate. Life Sci 29:1881–1886

Bozarth MA, Wise RA (1984) Anatomically distinct opiate receptor fields mediate reward and physical dependence. Science 224:516

Brady JV, Lukas SE (1984) Testing drugs for physical dependence potential and abuse liability. NIDA Res Monogr 52. US Dep Health Human Serv, Washington DC

Britt MD, Wise RA (1981) Opiate rewarding action: independence of the cells of the lateral hypothalamus. Brain Res 222:213–217
Britt MD, Wise RA (1983) Ventral tegmental site of opiate reward: antagonism by a hydrophillic opiate receptor blockade. Brain Res 258:105–108
Broekkamp CL, Phillips AG, Cools AR (1979) Facilitation of self-stimulation behavior following intracerebral microinjection of opioids into the ventral tegmental area. Pharmacol Biochem Behav 11:289–295
Bunney BS, Grace AA (1978) Acute and chronic haloperidol treatment: comparison of effects on nigral dopaminergic cell activity. Life Sci 23:1715–1728
Carboni E, Imperato A, Perezzanil, DiChiara G (1989a) Amphetamine, cocaine, phencyclidine and noninfensine increase extracellular dopamine concentrations preferentially in the nucleus accumbens of free moving rats. Neurosci 28:653–661
Carboni E, Acquas E, Frau R, DiChiara G (1989b) Differential inhibitory effects of a 5-HT_3 antagonist on drug-induced stimulation of dopamine release. Pharmacology 164:515–519
Carboni E, Acquas E, Leone P, DiChiara G (1989c) 5-HT_3 receptor antagonists block morphine- and nicotine- but not amphetamine-induced reward. Psychopharmacology 97:175–178
Carr GD, Fibinger HC, Phillips AG (1989) Conditioned place preference as a measure of drug reward. In: Liebman JM, Cooper SJ (eds) The neuropharmacological basis of reward. Clarendon, Oxford, p 264
Cotton R, Giles MG, Miller L, Shaw JS, Timms D (1984) ICI 174,864: a highly selective antagonist for the opioid δ-receptor. Eur J Pharmacol 97:331–332
DiChiara G, Imperato A (1988) Opposite effects of μ- and κ-opioid agonists on dopamine release in the nucleus accumbens and in the dorsal caudate of freely moving rats. J Pharmacol Exp Ther 244:1067–1080
Downs DA, Woods JH (1976) Naloxone as a negative reinforcer in rhesus monkeys: effects of dose, schedule, and narcotic regimen. Pharmacol Rev 27:397–436
Ettenberg A, Pettit HO, Bloom FE, Koob GF (1982) Heroin and cocaine intravenous self-administration in rats: mediation by separate neural systems. Psychopharmacology 78:204–209
Glick SD, Cox RD (1977) Changes in morphine self-administration after brainstem lesions in rats. Psychopharmacology 52:151–156
Goodman RR, Snyder SH, Kuhar MJ, Young WS (1980) Differentiation of δ- and μ-opiate receptor localization by light microscopic autoradiography. Proc Natl Acad Sci USA 77:6239–6243
Grace AA, Bunney BS (1986) Induction of depolarization block in midbrain depomaine neurons by repeated administration of haloperidol: analysis using in vivo intracellular recording. J Pharmacol Exp Ther 238:1092–1100
Grevert P, Goldstein A (1977) Effects of naloxone on experimentally induced ischemic pain and on mood in human subjects. Proc Natl Acad Sci USA 74:1291–1294
Grevert P, Goldstein A (1978) Endorphins: naloxone fails to alter experimental pain or mood in humans. Science 199:1093–1095
Gysling K, Wang RY (1983) Morphine-induced activation of A10 dopamine neurons in the rat. Brain Res 277:119–127
Hand TH, Kasser RJ, Wang RY (1987) Effects of acute thioridazine, metoclopramide and SCH 23390 on the basal activity of A9 and A10 dopamine cells. Eur J Pharmacol 137:251–255
Herz A, Shippenberg TS (1989) Motivational effects of opioids: neurochemical and neuroanatomical substrates. In: Goldstein A (ed) Molecular and cellular aspects of the drug addictions. Springer, Berlin Heidelberg New York, pp.111–141
Hoffman DC, Beninger RJ (1986) Feeding behaviour in rats is differentially affected by pimozide treatment depending on prior experience. Pharmacol Biochem Behav 24:259–262
Hoffmeister F (1979) Preclinical evaluation of reinforcing and aversive properties of analgesics. In: Beers RF, Bassett EG (eds) Mechanisms of pain and analgesic compounds. Raven, New York, pp 447–466
Hollister LE, Johnson K, Boukhabza D, Gillepsie HK (1981) Aversive effects of naltrexone in subjects not dependent on opiates. Drug Alcohol Depend 8:37–41
Hubner CB, Koob GF (1987) Ventral pallidal lesions produce decreases in cocaine and heroin self-administration in the rat. Proc Soc Neurosci 13:1717
Imperato A, DiChiara G (1986) Preferential stimulation of dopamine release in the nucleus accumbens of freely moving rats by ethanol. J Pharmacol Exp Ther 239:219–228

Imperato A, Mulas A, DiChiara G (1986) Nicotine preferentially stimulates dopamine release in the limbic system of freely moving rats. Eur J Pharmacol 132:337–338

Janowsky DS, Judd LL, Huey L, Segal D (1979) Effects of naloxone in normal, manic and schizophrenic patients: evidence for alleviation of manic symptoms. In: Usdin E, Bunney WE Jr, Kline NS (eds) Endorphins in mental health research. Macmillan, London, pp 435–441

Jeziorski M, White FJ (1989) Electrophysiological effects of selective opioid receptor agonists on A10 dopamine (DA) neurons. Soc Neurosci Abstr 15:1001

Kandel DA, Schuster CR (1977) An investigation of nalorphine and phrephenazine as negative reinforcers in an escape paradigm. Pharmacol Biochem Behav 6:61–71

Katz JL (1989) Drugs as reinforcers: pharmacological and behavioural factors. In: Cooper SJ, Lieberman JM (eds) The neuropharmacological basis of reward. Clarendon, Oxford, pp 164-213

Katz RJ, Gormezano G (1979) A rapid and inexpensive technique for assessing the reinforcing effects of opiate drugs. Pharmacol Biochem Behav 11:231–233

Koob GF, Goeders NE (1989) Neuroanatomical substrates of drug self-administration. In: Cooper SJ, Lieberman JM (eds) The neuropharmacological basis of reward. Clarendon, Oxford, pp 214–263

Koob GF, Pettit HO, Ettenberg A, Bloom FE (1984) Effects of opiate antagonists and their quarternary derivatives on heroin self-administration in the rat. J Pharmacol Exp Ther 229:481–485

Lahti RA, Mickelson MM, McCall JM, von Voightlander PF (1985) [^{3}H]-U69593: a highly selective ligand for the κ-opioid receptor. Eur J Pharmacol 109:281–284

Lyness WH, Friedle NM, Moore KE (1979) Destruction of dopaminergic nerve terminals in nucleus accumbens: effects on d-amphetamine self-administration. Pharmacol Biochem Behav 11:553–556

Mackey WB, Van der Kooy D (1985) Neuroleptics block the positive reinforcing effects of amphetamine but not of morphine as measured by place conditioning. Pharmacol Biochem Behav 22:101–105

Magnan J, Paterson SJ, Tavani A, Kosterlitz HW (1982) The binding spectrum of narcotic analgesic drugs with different agonist and antagonist properties. Naunyn-Schmiedeberg's Arch Pharmacol 319:197–205

Mansour A, Khachaturian H, Lewis ME, Akil H, Watson SJ (1988) Anatomy of CNS opioid receptors. Trends Neurosci 11:308–314

Matthews RT, German DC (1984) Electrophysiological evidence for excitation of rat ventral tegmental area dopamine neurons by morphine. Neuroscience 11:617–625

Mosberg HI, Hurst R, Hruby VJ, Gee K, Akiyama K, Yamamura HI, Galligan JJ, Burks TF (1983) Bis-penicillamine enkephalins possess highly unproved specificity toward delta opioid receptors. Proc Natl Acad Sci USA 80:5871

Moskowitz AS, Goodman RR (1984) Light microscopic autoradiographic localization of opioid binding sites in the mouse central nervous system. J Neurosci 4:1331–1342

Mucha RF, Herz A (1985) Motivational properties of κ- and μ-opioid receptor agonists studied with place and taste preference conditioning procedure. Psychopharmacology 86:274–280

Mucha RF, Millan MJ, Herz A (1985) Aversive properties of naloxone in non-dependent (naive) rats may involve blockade of central β-endorphin. Psychopharmacology 86:281–285

Nakajima S, Baker JD (1989) Effects of D2 dopamine receptor blockade with raclopride on intracranial self-stimulation and food-reinforced operant behaviour. Psychopharmacology 98:330–333

Nakajima S, McKenzie GM (1986) Reduction of the rewarding effect of brain stimulation by blockade of dopamine D_1 receptor with SCH 23390. Pharmacol Biochem Behav 24:919–923

Nakajima S, Wise RA (1987) Heroin self-administration in the rat suppressed by SCH 23390. Soc Neurosci 13:1545 (Abstract)

Nomikos GG, Spyraki C (1988) Effects of ritanserin on the rewarding properties of α-amphetamine, morphine and diazepam revealed by conditioned place preference in rats. Pharmacol Biochem Behav 30:853–858

Nomikos GG, Spyraki C, Galanopoulou P, Papadopoulou Z (1986) Amphetamine and morphine induced place preference in rats with 5,7-dihydroxytryptamine lesions of the nucleus accumbens. Psychopharmacology 89:26

Olds ME (1982) Reinforcing effects of morphine in the nucleus accumbens. Brain Res 237:429–440

Olds J, Milner P (1954) Positive reinforcement produced by electrical stimulation of septal area and other regions. J Comp Physiol Psychol 47:419–427

Olds ME, Williams KN (1980) Self-administration of d-Ala-Met-Enkephalinamide at hypothalamic self-stimulation sites. Brain Res 194:155–170

Pelton JT, Kazmierski W, Gulya K, Amamura HI, Hruby VJ (1986) Design and synthesis of confor-

mationally constrained somatostatin analogues with high potency and specificity for μ-opioid receptors. J Med Chem 29:2370–2375
Pettit HO, Ettenberg A, Bloom FE, Koob GF (1984) Destruction of dopamine in the nucleus accumbens selectively attenuates cocaine but not heroin self-administration in rats. Psychopharmacology 84:167–173
Schwarcz R, Hokfelt T, Fuxe K, Jonsson G, Goldstein M, Terenius L (1979) Ibotenic acid-induced neuronal degeneration: a morphological and neurochemical study. Exp Brain Res 37:199–216
Schwartz AS, Marchok PL (1974) Depression of morphine-seeking behaviour by dopamine inhibition. Nature (London) 248:257–258
Shippenberg TS, Herz A (1987) Place preference conditioning reveals the involvement of D1-dopamine receptors in the motivational properties of μ- and κ-opioid agonists. Brain Res 436:169–172
Shippenberg TS, Herz A (1988) Motivational effects of opioids: influence of D-1 versus D-2 receptor antagonists. J Pharmacol 151:233–242
Shippenberg TS, Bals-Kubik R, Herz A (1987) Motivational properties of opioids: evidence that an activation of δ-receptors mediates reinforcement processes. Brain Res 436:234–239
Shippenberg TS, Bals-Kubik R, Spanagel R, Herz A (1989) Involvement of D1 dopamine receptors in the motivational effects of μ- and κ-opioid agonists. In: Int Narcotics Res Conf, 9–14 July 1989, Quebec, p 28
Smith JE, Lane JD (1983) Brain neurotransmitter turnover correlated with morphine self-administration. In: Smith JE, Lane JD (eds) The neurobiology of opiate reward processes. Elsevier, Amsterdam, pp 361–402
Smith JE, Guerin GF, Co C, Barr TS, Lane JD (1985) Effects of 6-OHDA lesions of the central medical nucleus accumbens on rat intravenous morphine self-adminstration. Pharmacol Biochem Behav 23:843–849
Smith JE, Shultz K, Co C, Goeders NE, Dworkin SJ (1987) Effects of 5,7-dihydroxytryptamine lesions of the nucleus accumbens on rat intravenous morphine self-administration. Pharmacol Biochem Behav 26:607–612
Smith SD, Werner TE, Davis WM (1976) Effect of unit dose and route of administration on self-administration of morphine. Psychopharmacology 50:103
Spyraki C, Fibinger HC, Phillips AG (1982) Attenuation by haloperidol of place preference conditioning using food reinforcement. Psychopharmacology 77:379–382
Spyraki C, Fibinger HC, Phillips AG (1983) Attenuation of heroin reward in rats by disruption of the mesolimbic depamine system. Psychopharmacology 79:278–283
Stapleton JM, Lind MD, Merriman VJ, Bozarth MA, Reid LD (1979) Affective consequences and subsequent effects on morphine self-administration of D-Ala^2-methionine enkephalin. Physiol Psychol 7:146–152
Stein EA, Olds J (1976) Direct intracerebral self-administration of opiates in the rat. Soc Neurosci Abstr 3:302
Tachibana S, Oshino H, Arakawa Y, Nakzawa T, Araki S, Kaneko T, Yamatsu K, Miyagawa H (1988) Design and synthesis of metabolically stable analogs of dynorphin-A and their analgesic characteristics. In: Naito Found Symp, 22–24 Oct 1987. Univ Press, Tokyo, pp 26–27
Takemori AE, Porthoghese PS (1988) Pharmacologic characteristics of nor-binaltorphimine (nor-BNI), a highly selective kappa opioid receptor antagonist. In: Proc INRC, 3–8 July, Albi, Fr, p 19
Takemori AE, Larson DL, Porthoghese PS (1981) The irreversible narcotic antagonistic properties of the fumarate methylester derivative of naltrexone. Eur J Pharmacol 70:445–451
Ungerstedt U (1971) Stereotaxic mapping of the monoamine pathways in the rat brain. Acta Physiol Scand Suppl 367:1–48
Vaccarino FJ, Floyd EB, Koob GF (1985a) Blockade of nucleus accumbens opiate receptors attenuates intravenous heroin reward in the rat. Psychopharmacology 86:37–42
Vaccarino FJ, Pettit HO, Bloom FE, Koob GF (1985b) Effects of intracerebroventricular administration of methyl-naloxonium chloride on heroin self-administration in the rat. Pharmacol Biochem Behav 23:495–498
Van der Kooy D, Mucha RF, O'Shaughnessy M, Buceneiks P (1982) Reinforcing effects of brain microinjections of morphine revealed by conditioned place preference. Brain Res 243: 107–117
Van Ree JM, Smyth DG, Colpaert FC (1979) Dependence creating properties of lipotropin C-fragment (β-endorphin): evidence for its internal control of behaviour. Life Sci 24:495–502

Von Voightlander PF, Lahti RA, Ludens JH (1983) U50,488H: a selective and structurally novel non-mu (kappa) opioid agonist. J Pharmacol Exp Ther 224:7–11

Watson SJ, Akil H, Richard CW, Barchas JD (1978) Evidence for two separate opiate peptide neuronal systems. Nature (London) 275:226–228

White FJ, Wang RY (1983) Comparison of the effects of chronic haloperidol treatment on A9 and A10 dopamine neurons in the rat. Life Sci 32:983–993

Wise RA (1980) Action of drugs of abuse on brain reward systems. Pharmacol Biochem Behav 13:213–223

Wise RA, Rompre PP (1989) Brain dopamine and reward. Annu Rev Psychol 40:191–225

Wood PL, Frederickson RCA, Jyengar (1989) A review of tonic opioid systems as revealed in perturbed systems and with studies of naloxone or enkephalinase inhibitors. In: Int Narcotics Res Conf, July 9–14, Quebec, p 25

Woods JH, Smith CB, Mezihradsky F, Swain HH (1979) Preclinical testing of new analgesic drugs. In: Beer RF, Bassett EG (eds) Mechanisms of pain and analgesic compounds. Raven, New York, pp 429–445

Yokel RA, Wise RA (1975) Increased lever pressing for amphetamine after pimozide in rats: implications for a dopamine theory of reward. Science 187:547–549

Yokel RA, Wise RA (1976) Attenuation of intravenous amphetamine reinforcement by central dopamine blockade in rats. Psychopharmacology 48:311–318

Young AM, Woods JH, Herling S, Hein DW (1983) Comparison of the reinforcing and discriminative stimulus properties of opioids and opioid peptides. In: Smith JE, Lane JD (eds) The neurobiology of opiate reward processes. Elsevier/Northholland Biomedical Press, Amsterdam, pp 147–174

Zito KA, Vickers G, Roberts DCS (1985) Disruption of cocaine and heroin self-administration following kainic acid lesions of the nucleus accumbens. Pharmacol Biochem Behav 23:1029–1036

CHAPTER 21

Psychotomimetic Effects of Opioids

A. Pfeiffer

1 Introduction

The attraction of opioids for over 2000 years and possibly the cause for their addictive properties are closely connected to their euphorigenic properties and their ability to suppress miserable feelings, hunger, and pain (Schmidbauer and von Scheidt 1977). The name morphine is derived from Morpheus, the greek god of dreams whose symbol is the opium poppy. The very name of this drug alludes to its trophic action on mental activity. Exogenous opioids act on receptors which are otherwise probably activated by endogenously produced opioid peptides. The use of exogenous opioids to mimic this endogenous system represents one of the first therapeutic exploitations of neuroendocrine circuits. Exogenous opioids may mirror (or represent an exaggeration of) the endogenus functions of opioids. It is therefore not surprising that endogenous opioid systems were ascribed a rewarding function. With this background, the discovery that some types of opioids elicit opposite, apparently unpleasant effects (Lasagna and Beecher 1954), was very exciting.

These opposing effects appear to be related to the particular type of opiate receptors activated. Morphine and its congeners act predominantly at μ-receptors. Other types of opiate receptors are named κ- and δ-receptors (Martin et al. 1976; Lord et al. 1977). The human brain contains approximately similar amounts of κ- and μ-receptors but fewer δ-receptors (Pfeiffer et al. 1981, 1982; Maurer et al. 1983) (Fig. 1). While the pharmacological actions of morphine-type μ-receptor agonists are quite well characterized, less is known about the actions of κ- or δ-agonists in humans.

The discovery of κ-receptors led to the search for nonaddictive, but strongly analgesic opioids. Early prototypes of these drugs like nalorphine or cyclazocine were not only noneuphorigenic but dysphorigenic and psychotomimetic (Lasagna and Beecher 1954; Keats and Telford 1964; Martin and Gorodetsky 1965; for review, see Jasinski 1977). The term "psychotomimetic" was used to describe a complex of variable, but mostly unpleasant, experiences which were observed after their administration. Much evidence gathered to date suggests that this activity is related to an interaction with κ-receptors. These receptors are present in high concentrations in various brain areas including those of the limbic system which is thought to play a role in emotionality (Fig. 1).

Many questions regarding the dysphorigenic actions of κ-agonists have not been resolved. One major problem is that the substances used in human studies are not highly selective for κ-receptors: they also interact with other opiate receptors, mostly

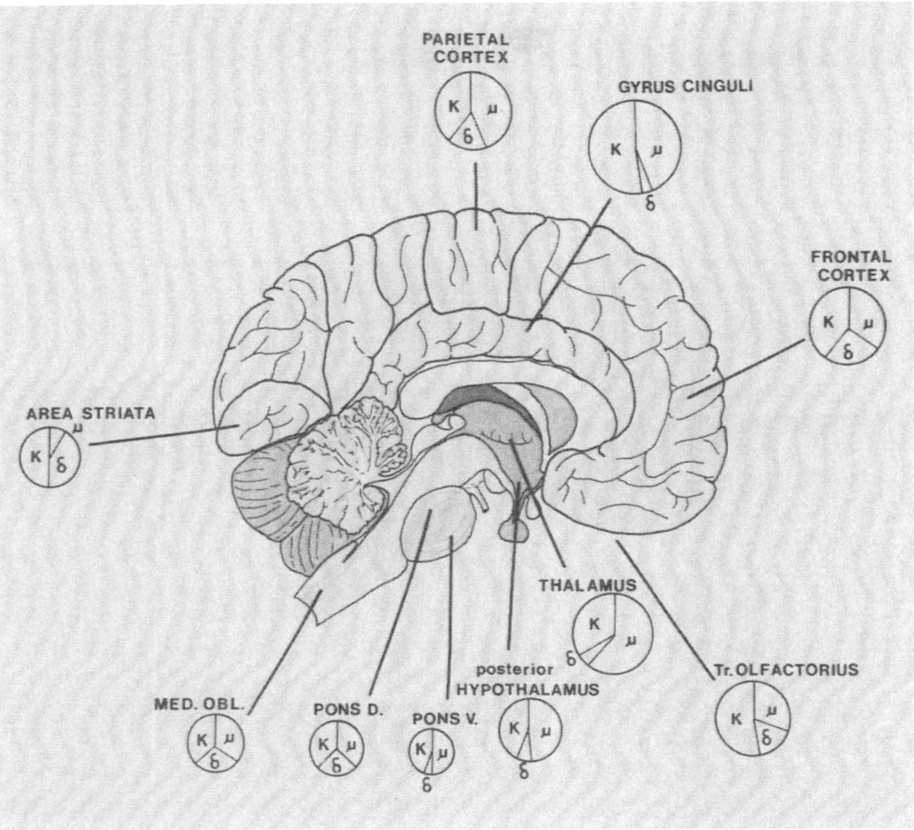

Fig. 1. Distribution of opiate receptor subtypes in various areas of the human brain. The *circles* indicate the relative amounts of opiate receptors in each area; the *fractions* within each circle indicate the relative amounts of opiate receptor subtypes. Data shown were calculated from Pfeiffer et al. (1982)

as antagonists or as weak agonists. Some compounds, including cyclazocine and nalorphine, additionally act at nonopiate σ/PCP-receptors (Manallack et al. 1986; Quirion et al. 1987). Such multiple interactions further complicate data interpretation as these receptors may themselves mediate psychotomimetic effects. Although selective κ-agonists have been used in human studies (Peters et al. 1987) data regarding mental effects have not been published.

This review attempts to summarize the mental effects ascribed to the activation of κ-receptors. The first part will focus on the nature of the mental disturbances. The second part will discuss the evidence for ascribing these mental effects to an activation of κ-receptors; this is necessary in view of the above-mentioned problems with the limited selectivity of the compounds employed. Finally, some of the clinical consequences regarding the possible therapeutic use of κ-agonists will be addressed. As alluded to above, the finding of euphorigenic and dysphorigenic effects mediated by two types of opiate receptors invites speculations as to their role in mental and emotional states and, eventually, in the development of abnormal mental states in man. This question is the subject of another review (see Emrich and Schmauss, this Vol.) and will therefore not be discussed in detail.

2 "Psychotomimetic Effects" of κ-Agonists?

The term "psychotomimetic" means that a drug provokes symptoms associated with major psychiatric illnesses such as schizophrenia or mania/depression. Important psychotic symptoms are hallucinations and/or delusions; additionally, incoherence of thought and speech, inappropriate affect, or catatonic behavior may occur (American Psychiatric Association 1980). "True" hallucinations have, to my knowledge, not been described following the administration of κ-agonists, i.e., hallucinations in which a subject underwent auditory or visual experiences which were totally implausible to others, including observers. However, hallucination-like syndromes or pseudohallucinations in which the subject knew that the hallucinated experience was not true, have been frequently observed. Martin and co-workers (1966) described visual imagery in one subject after cyclazocine who when he closed his eyes reached out to touch his visions. Martin and co-workers also (1965) described episodes of depression and "thoughts of violence that could not be controlled" after cyclazocine. Resnik and co-workers (1971) reported transient excitement and grandiose ideation in 1 of 60 subjects treated with cyclazocine, while 11 of the 60 patients had transient "perceptual illusions" like hearing their names being called, seeing vivid colors, or seeing visual images with closed eyes. Kumor and co-workers (1986) recently administered ketocyclazocine to drug experienced volunteers and compared its effects to those of cyclazocine and morphine. No psychotomimetic effects were observed after morphine (20 mg s.c.), while ketocyclazocine (which is more selective for κ-receptors than cyclazocine and has fewer antagonistic properties at μ-receptors) (Martin et al. 1976) elicited somewhat more pronounced psychotomimetic effects than cyclazocine. Hallucinatory-type experiences consisted of "seeing monsters or people in threatening situations", but all subjects realized that these experiences were not real. Our group (Pfeiffer et al. 1986a) observed loss of self-control, dream-like states associated with derealization and, occasionally, pseudohallucinations after the benzomorphan κ-agonist MR 2033 (Merz et al. 1973; Merz and Stockhaus 1979). MR 2033 has properties similar to those of the original prototype κ-agonist of Martin and co-workers (1976), ethylketozyclazocine. This compound apparently has few effects at nonopiate receptors and little antagonistic or agonistic activity at μ-receptors in vivo (Swain an Seevers 1974; Woods et al. 1979; Wood et al. 1980; Tyers 1980).

Delusions have not been described upon administration of κ-agonists. In one study we employed a questionnaire to test for disturbances of self-perception which was originally developed for evaluation of schizophrenic patients, but none of the subjects demonstrated marked deviations from the normal responses on this scale (Pfeiffer, Brantl, Herz and Emrich 1986, unpubl observ.).

Numerous reports describe weird and uncontrolled thoughts, increased dreaming, or difficulty to sleep despite tiredness upon treatment with κ-agonists. Thought disturbances, like incoherence or inability to communicate in a logical and ordered fashion, have not been described as isolated symptoms. However, loss of self-control and refusal to communicate have been reported (Pfeiffer et al. 1986a). Haertzen (1970) demonstrated that cognitive difficulties were frequently experienced after nalorphine or cyclazocine.

Another prominent set of symptoms reported upon treatment with all the above-named κ-agonists was related to alterations of sensory experiences: perception of space and color was frequently altered and there were feelings of body distortion and paresthesias (Lasagna and Beecher 1954; Martin and Gorodetzky 1965; Martin et al. 1965; Haertzen 1970; Kumor et al. 1986; Pfeiffer et al. 1986a). The Addiction Research Center Inventory (ARCI), is a 550-item questionnaire developed by members of the American National Institute of Mental Health to distinguish various drugs causing subjective disturbances, including morphine, LSD, alcohol, and barbiturates (Hill et al. 1963). After cyclazocine or nalorphine, positive responses were obtained on the LSD scale which contains statements like "I feel as if things were not real", "I feel as if my hands and feet are no longer part of me", "I have felt my body drift away from me", and "it seems as if someone else is answering these questions for me." Positive responses were also obtained on the Pentobarbital Chlorpromazine-Alcohol scale, which measures sedative effects (Haertzen 1970) and motor dyscoordination, but not on the Morphine-Benzedrine group scale which assesses euphoria. These responses have been reproduced (see Martin and Sloan 1977; Jasinski 1977; Kumor et al. 1986) with similar results with the exception of one study by Jasinski and colleagues (1968) in which cyclazocine caused a significant group response on the euphoria scale and no significant response on the LSD scale. Unfortunately, this was the only study which evaluated the naloxone (NAL) reversibility of cyclazocine effects in humans; however, because no psychotomimetic effects were observed, their NAL reversibility could not be demonstrated (see below).

In most subjects, κ-agonists were experienced as unpleasant, and subjects felt tense, shaky, moody, and impatient. Symptoms like tiredness, drunkeness, weakness, disorientation, and incoordination were frequent after cyclazocine and nalorphine (Haertzen 1970), but not after ketocyclazocine (Kumor et al. 1986). Bodily complaints like weakness, sweating, vertigo, and dizziness in response to MR 2033 were reported on specific scales. Sedation occurred with a lag after the peak incidence of psychotomimetic effects. However, although anxiety was increased, this was only marginally significant, and mood was not significantly deteriorated after the highest dose of the opiate-active (–)isomer of MR 2033 (Pfeiffer et al. 1986a), as indicated by self-rating scales.

In summary, the various κ-agonists tested to date produced a syndrome which is comprised of perceptual changes, cognitive alterations, and pseudohallucinations, a variable sedative component with decreased performance and an overall deterioration of well-being. The κ-agonists produced mental states which resembled psychotic states to some extent, but the major symptoms were not mimicked in the strict sense. Thus, disturbances of thought as well as disturbances of body image were similar to those seen in psychotic states, but the subjects were usually aware of the drug effect; the latter is incompatible with psychotic states. In contrast, PCP seems to cause frequent hallucinations which subjects cannot identify as drug effects

(Cohen et al. 1962; Allen and Young 1978). The term "psychotomimetic" is used to summarize the set of unusual experiences frequently described following treatment with κ-agonists. If one aims to distinguish between "true" versus "pseudo"-hallucination, the term "pseudopsychotomimetic" would describe the actions of κ-agonists more precisely. This would allow a terminological differentiation from psychotomimetic effects produced by drugs like PCP. However, the term psychotomimetic, in its broader sense, will be used in this review.

In the studies reported, the highest doses of the κ-agonists used were those which were still tolerable to the subjects. Dose-response curves for psychotomimetic effects of nalorphine and cyclazocine increased linearly over the dose ranges studied without evidence for reaching a plateau (Haertzen 1970; see Martin and Jasinski 1972). It, therefore, cannot be excluded that higher doses would have caused true hallucinations and other psychotomimetic effects.

The physicochemical and neuroanatomical basis for the occurrence of psychotic disturbances in man is unknown. Theoretically, the perception of the world around us is not a passive reception of sensory inputs. Rather, it involves an interpretation of sensory stimuli leading to purposeful reactions to events around us. The meaning of purposeful is determined by the ethics of a subject, which in turn is usually based on mostly implicit theories about the world and the needs of the subject living in it. This simple model requires that some activity of certain areas in the brain enables or creates the interpretation of sensory experiences. This activity must be restricted so as to avoid overinterpretation, on the one hand, and lack of activity or phantasy, on the other. Since most neurobiologists perceive the brain as being dependent on neurochemical principles, interference with such activity and thus, processes of interpretation would result in inappropriate reactions which are perceived by others as mental disturbances. As κ-agonists cause hallucinatory type experiences they would appear to enhance the creative aspect of mental activity. This might occur by inhibition of some control element which is normally required to avoid overinterpretation. The fact that we perceive pseudohallucinations as being untrue demonstrates the existence of further control elements which apparently check experiences for their probability of being compatible with an accepted (i.e., true) conception of the world. At present, however, we are ignorant about the neuroanatomical correlates of such phenomenologic descriptions of mental activity.

Regarding the neurochemical basis of opioid interference with neurotransmitter action, it is well known that opioids affect the release of serotonin (5-HT), noradrenaline (NA), dopamine (DA), GABA, acetylcholine (ACh), and probably other neurotransmitters in various experimental models. Disturbances of any of these neurotransmitters appear to cause mental disturbances as illustrated by LSD (5-HT), scopolamine (ACh), haloperidol (5-HT, DA), or cocaine (NA, DA). Heartzen (1970) compared the effects of κ-agonists with some of the above-named compounds in a very interesting discussion, which shows that both similarities and differences exist between them. Obviously, the most simple explanation for the psychotomimetic effects of κ-agonists appears to be an interference with the release of several neurotransmitters in various brain areas, resulting in a discoordination of central processes. The symptomatology therefore produced is rather nonspecific, and may be interpreted as the sum of such interferences, which need not be functionally related to each other in any regard. The fact that exogenously administered κ-agonists cause psychotomimetic symptoms therefore does not prove a similar function of endogenous κ-agonists.

It is presently impossible to elucidate the role of endogenous opioids in brain function. NAL has only minimal effects on cognitive functions at doses up to 280 mg/70 kg in humans (Cohen et al. 1981). This does not exclude a major function of endogenous opioids but may also indicate that other principles can apparently compensate for a lack of opioid activity. Although this probably shows that tonic opioid activity is unnecessary to create mood or emotional states, one may also conceive more complicated theories, assuming the existence of two opposed opioid systems. One of these systems, associated with μ-receptors, would participate in the expression of well-being while another, associated with κ-receptors, would mediate tense inner states and unpleasant emotional reactions. NAL then might not reveal any effect, as both systems are eliminated. The use of selective opioid antagonists will help to explore these possibilities. In this context, it is remarkable that NAL exerts clear-cut aversive effects in animal models; this suggests that endogenous opioids have rewarding properties (Mucha and Herz 1985; Mucha et al. 1985; see Shippenberg and Bals-Kubik, this Vol.).

3 Pharmacological Specificity of Psychotomimetic κ-Agonists

Psychotomimetic effects observed upon administration of κ-agonists were initially believed to be mediated by σ-receptors (Martin et al. 1976). The prototype agonist at this receptor was SKF 10,047 (N-allyl normetazocine) (Martin et al. 1976). Martin and colleagues (1976) demonstrated that this drug causes "canine delirium." These effects appeared to be reversible with high doses of naloxone and were therefore ascribed to σ-"opiate" receptors. Nalorphine and cyclazocine were thought to possess activity at these receptors, and tolerance to the stimulant actions of cyclazocine was associated with cross-tolerance to SKF 10,047 (Martin et al. 1976). Keats and Telford (1964) had described strong psychotomimetic effects in humans after SKF 10,047 and it was assumed that the psychotomimetic effects of cyclazocine and nalorphine were also mediated by an action at this σ-receptor. However, in further investigations the actions of SKF 10,047 were resistant to even high doses of NAL (Martin et al. 1980; Shearman and Herz 1982; Vaupel 1983; Vaupel et al. 1986). It was also noted that many actions of SKF 10,047 were similar to those of phencyclidine (PCP). This dissociative anaesthetic (Domino et al. 1965) is a powerful psychotomimetic agent which seems to closely mimic schizophrenia (Cohen et al. 1962; Allen and Young 1978). Its effects are not antagonized by NAL and therefore the receptors upon which PCP acts are classified as nonopiate (Vaupel 1983; Vaupel et al. 1986; Quirion et al. 1987). Although opiate receptors typically show a stereoselectivity for (−)-isomers, this receptor exhibited a greater preference for the (+)- as compared to the (−)-isomer of SKF-10,047. In binding assays cyclazocine showed activity at the σ/PCP receptor (Manallack et al. 1986). The psychotomimetic actions of cyclazocine could therefore have been mediated by σ/PCP receptors. The question as to whether cyclazocine elicits psychotomimetic effects by its interaction with opiate receptors or nonopiate receptors therefore concerns the NAL sensitivity of the actions of cyclazocine. This question was addressed by Jasinski and co-workers (1968) in one study. Unfortunately, the results from this study were unusual in that no psychotomimetic effects were observed after treatment with cyclazocine as compared to placebo. This was, however, due to a high psychotomimetic score on the LSD scale in the placebo group and was unlike other investigations (Haertzen 1970; Kumor et al. 1986). Therefore, this study did not demonstrate a significant NAL antagonism of the action of cyclazocine in humans as measured on the LSD/psychotomimetic scale of the ARCI questionnaire (Jasinski et al. 1968) and, to my knowledge, a similar study has not been reported. Nevertheless, although not statistically demonstrated, the study, in fact, reports reversal of the subjective effects of cyclazocine (Jasinski et al. 1968): NAL (15 mg s.c.) was administered 45 min after cyclazocine (1 mg s.c.) and antagonism of unpleasant subjective effects occurred in those subjects experiencing them. This report therefore would suggest that cyclazocine mediates psychotomimetic effects by interacting with κ-receptors. Resnik and co-workers (1971) treated 60 heroin addicts with 4 mg cyclazocine for prolonged periods. During the induction phase psychotomimetic effects occurred and were successfully treated with NAL (50–1000 mg p.o.). This high dose was necessary because of the rapid inactivation of NAL during first passage through the liver. Interestingly, cyclazocine was 16–24 times more potent than nalorphine in eliciting peak psychotomimetic effects (Haertzen 1970). This corresponds closely to the differing affinities of these drugs for κ-receptors in human brain membranes: 15-fold higher for cyclazocine (Kd = 0.46 nM) than for nalorphine (Kd = 6.9 nM) (Pfeiffer et al. 1981). Although such binding studies do not take into account possible pharmacokinetic differences between the drugs, the above data indicate a possible role of opiate receptors in mediating the psychotomimetic actions of cyclazocine.

Our group (Pfeiffer et al. 1986a,b,c) has investigated endocrine and psychotomimetic effects of the racemic benzomorphan κ-agonist MR 2033 (Merz et al. 1973; Merz and Stockhaus 1979), a compound with properties very similar to those of ethylketocyclazocine, the prototypic κ-agonist of Martin and colleagues (1976; Woods et al. 1979). This benzomorphan has no activity at σ/PCP receptors (Shearman and Herz 1982) and little, if any, μ-agonist activity at μ-receptors in humans. For example, the compound does not suppress LH-release at any of the doses tested, an effect which is typically observed with μ-opiate agonists and shows other endocrine actions, such as inhibition of TSH-secretion and a strong diuretic effect, which differ from those of μ-agonists (Pfeiffer et al. 1986b,c,). The psychotomimetic effects of MR 2033 were completely antagonized by NAL (10 mg i.v.) (Fig. 2). The effects were stereospecific and only observed after administration of the opiate active (−)-isomer, MR 2034; the opiate inactive (+)-isomer (MR 2035) was ineffective even at a dose fourfold higher than the maximally tolerated dose of the (−) isomer (3.8 μg/kg body weight, i.v.) (Fig. 3). The selective μ-agonist fentanyl did not cause psychotomimetic effects when administered in an analgesic dose (100 μg/subject i.v.) (Pfeiffer et al. 1986a). Therefore, in view of the observed NAL-sensitivity, opioid-like stereoselectivity, and effects clearly different from those mediated by μ-agonists, these data provide strong evidence for a role of opiate

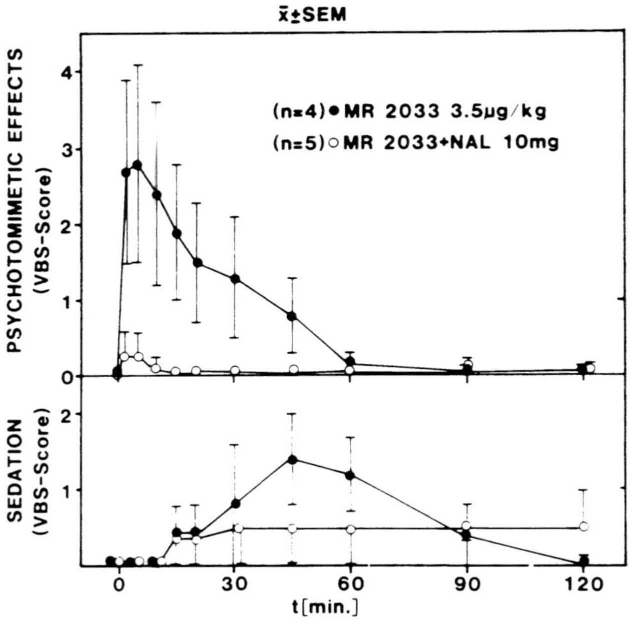

Fig. 2. Psychotomimetic effects of the κ-opiate agonist MR 2033 with or without naloxone (NAL). Psychotomimetic effects were assessed on the VBS scale (Verlaufs-Beobachtungs-Scala, course assessment scale), an eight-point rating scale that can be adapted to specific individual symptoms. Items evaluated were (i) disturbance in the perception of space; (ii) disturbance of the perception of time; (iii) abnormal visual experience; (iv) disturbance in body image perception; (v) depersonalization; (vi) derealization; and (vii) loss of self-control. For further details, see Pfeiffer et al. (1986a). Data are from Pfeiffer, Brantl, Herz, and Emrich (1986, unpubl. observ.)

receptors in mediating psychotomimetic effects in humans (Jasinski 1977; Kumor et al. 1986).

In view of the present lack of quantitative data concerning the antagonism of the effects of cyclazocine, ketocyclazocine or nalorphine it cannot be ascertained whether the psychotomimetic actions of these drugs are entirely mediated by opiate receptors. However, the antagonist effects of NAL in subjects receiving cyclazocine makes a mediation by opiate receptors very likely. Moreover, the close similarity of the syndromes provoked by the above drugs provides further support for a similar mechanism of action.

The highly selective κ-agonist U 62269 has been tested in humans with regard to diuretic effects (Peters et al. 1987). This report mentions "mild subjective effects" occurring in most volunteers. It would be interesting to obtain a more detailed description of the subjective effects of this compound and of similar drugs presently under development.

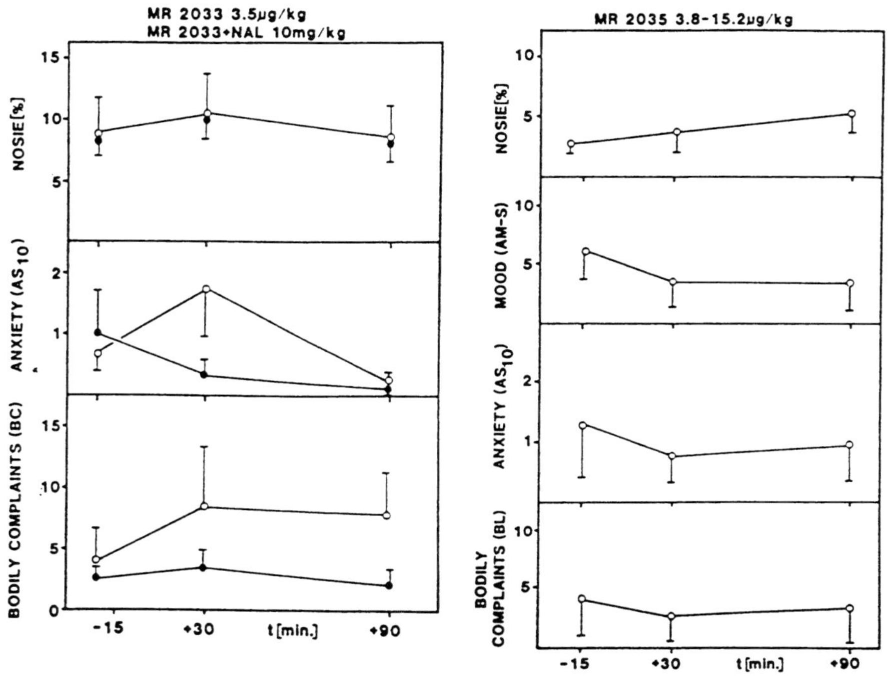

Fig. 3. Effects of MR 2033, with or without NAL, and of the opiate-inactive isomer MR 2035. Data are from a scale for abnormal behavior (*NOSIE*, nurses observation scale for in-patient evaluation) and from self-rating scales for mood, anxiety, or bodily complaints. An increase in the score indicates deterioration of psychic states. The scales were completed before, or 30 and 90 min after, i.v. injection of the drugs. For details and data for MR 2034, the opiate-active isomer, see Pfeiffer et al. (1986a). Data are from Pfeiffer, Brantl, Herz and Emrich (1986, unpubl. observ.)

4 Possible Clinical Use of κ-Agonists

The clinical interest in κ-agonists originates from attempts to develop a potent analgesic which is free from addiction liability and other unwanted side effects of morphine-like drugs (e.g., cardiovascular and respiratory depression, emesis, and constipation). Indeed, κ-agonists show little cardiovascular- or respiratory-depressant activity (Kumor et al. 1986; Pfeiffer et al. 1982, 1986c) and emetic effects have not been reported. To my knowledge, data on possible constipatory activity in humans have also not been reported. It is remarkable that there are only a few data on the analgesic actions of κ-agonists in humans, particularly with regard to the more recently developed compounds. The analgesic activity of κ-agonists differs markedly from that of μ-agonists, depending on the type of model employed (Tyers 1980).

The κ-agonists possess further properties which may be of interest clinically. Thus, they have been demonstrated to have potent anticonvulsant activity (Von-Voigtlander et al. 1988) to protect against ischemia-induced cerebral edema in the rat (Tang 1985), and to produce antidiuresis in experimental animals (Szligi and

Ludens 1982) and in humans (Nutt and Jasinski 1974; Pfeiffer et al. 1986c; Peters et al. 1987). The diuretic effect results from an increase in free water clearance which probably involves an inhibition of ADH-release from the posterior pituitary (Nutt and Jasinski 1974; Szligi and Ludens 1986), although a direct action on the kidney is also possible (Szligi and Ludens 1982, 1985; Pfeiffer et al. 1986c). The free water clearance caused by κ-agonists differs from that of the currently employed saluretics which cause a loss of either Na^+ or K^+. The saluretic action is undesirable in situations of hyponatremia such as in the syndrome of inappropriate ADH-secretion or, occasionally, in patients with decompensated cirrhosis of the liver. The diuretic activity of κ-agonists may be clinically useful if compounds with selectivity for this action and sufficient potency can be obtained.

Clearly, any use of κ-agonists for therapeutic purposes requires a separation of the psychotomimetic effects from their other effects. That this may be possible is illustrated by an anticonvulsant compound related to the benzeneacetamide κ-agonists in which many of the other κ-agonist effects been eliminated (Von-Voigtlander et al. 1988). The basis for such a separation may be found in the existence of subtypes of κ-receptors, associated with particular effects. Although the existence of subtypes of κ-receptors needs to be proven, there is some evidence for them from receptor binding studies in human brain (Pfeiffer et al. 1981) and in other species (Morris and Herz 1986). Moreover, the diuretic effect of κ-agonists is much more resistant to NAL-blockade than the psychotomimetic or endocrine actions of these compounds, providing further evidence for subclasses of κ-receptors (Pfeiffer et al. 1986a,b,c).

A major problem for the development of κ-agonists free of unwanted side effects is a lack of reliable animal tests for dysphoric effects in humans. The best choice at present may be drug discrimination or preference/avoidance paradigms (Mucha and Herz 1985; see Shippenberg and Bals-Kubik, this Vol.). Nonetheless, the promise of a new type of analgesic, and other therapeutic possibilities, should justify a further search within this group of compounds.

References

Allen R, Young S (1978) Phencyclidine induced psychosis. Am J Psychiat 135:1881–1883

American Psychiatric Association (ed) (1980) Diagnostic and statistical manual of mental disorders, 3rd edn. APA, Washington DC

Cohen BD, Rosenbaum G, Luby ED, Gottlieb JS (1962) Comparison of phencyclidine hydrochloride with other drugs. Arch Gen Psychiat 6:395–403

Cohen MR, Cohen RM, Pickar D, Weingartner H, Murphy DL, Bunney WE Jr (1981) Behavioural effects after high dose naloxone administration to normal volunteers. Lancet ii:1110

Domino EF, Chodoff P, Corrson G (1965) Pharmacologic effects of CI-581, a new dissociative anesthetic in man. Clin Pharm Ther 6:279–291

Haertzen CA (1970) Subjective effects of narcotic antagonists cyclazocine and nalorphine on the Addiction Research Center Inventory (ARCI). Psychopharmacologia 18:366–377

Hill E, Haertzen CA, Wolbach AB Jr, Miner EJ (1963) The Addiction Research Center Inventory: appendix. Psychopharmacology 4:184–205

Jasinski DR (1977) Assessment of abuse potentiality of morphine-like drugs. Methods used in man. In: Martin WR (ed) Drug addiction, vol 1. Springer, Berlin Heidelberg New York, pp 198–258

Jasinski DR (1979) Human pharmacology of narcotic antagonists. Br J Clin Pharm 7:287S–290S

Jasinski DR, Martin WR, Sapira JD (1968) Antagonism of subjective, behavioural, pupillary and respiratory depressant effects of cyclazocine by naloxone. Clin Pharm Ther 9:215–222

Keats AS, Telford J (1964) Narcotic antagonists as analgesics. In: Godd RF (ed) Drug design, advances in chemistry, Ser 45 Am Chem Soc, Washington, pp 170–176

Kumor KM, Haertzen CA, Johnsen RE, Kocher T, Jasinski D (1986) Human psychopharmacology of ketozyclazocine as compared with cyclazocine, morphine and placebo. J Pharmacol Exp Ther 238:960–968

Lahti RA, Collins RJ (1982) Opiate effects on plasma corticosteroids: relationship to dysphoria and self-administration. Pharmacol Biochem Behav 17:107–109

Lasagna L, Beecher HK (1954) The analgesic effectiveness of nalorphine and nalorphine-morphine combinations in man. J Pharmacol Exp Ther 112:356–363

Leander JD (1983) Further studies of kappa opioids on increased urination. J Pharmacol Exp Ther 227:35–41

Lord JAH, Waterfield AA, Hughes J, Kosterlitz HW (1977) Endogenous opioid peptides: multiple agonists and receptors. Nature (London) 267:495–499

Manallack DT, Beart PM, Gundlach AL (1986) Psychotomimetic sigma-opiates and PCP. Trends Pharmacol Sci 7:448–451

Martin WR (1984) Pharmacology of opioids. Pharmacol Res 35:283–345

Martin WR, Gorodetsky CW (1965) Demonstration of tolerance and physical dependence on N-allylnormorphine (nalorphine). J Pharmacol Exp Ther 150:437–442

Martin WR, Jasinski DR (1972) The mode of action and abuse potentiality of narcotic antagonists. In: Janzen R, Keidel WD, Herz A, Steichle C (eds) Pain. Thieme, Stuttgart, pp 225–234

Martin WR, Sloan JW (1977) Pharmacology and classification of LSD-like hallucinogens. In: Martin WR (ed) Drug addiction, vol 2. Springer, Berlin Heidelberg New York, pp 305–354

Martin WR, Fraser HF, Gorodetsky CW, Rosenberg DE (1965) Studies of the dependence producing potential of the narcotic antagonist cyclazocine. J Pharmacol Exp Ther 150:426–436

Martin WR, Gorodetsky CW, McClane TK (1966) An experimental study in the treatment of narcotic addicts with naloxone. Clin Pharm Ther 7:455–465

Martin WR, Eades CG, Thompson WO, Huppler RE, Gilbert PE (1976) The effect of morphine and nalorphine-like drugs in the non-dependent and morphine-dependent chronic spinal dog. J Pharm Exp Ther 197:517–532

Martin WR, Eades CG, Gilbert PE, Thompson JA (1980) Tolerance and physical dependence on N-allylnormetazocine in chronic spinal dogs. Subst Alcohol Actions Misuse 1:269–279

Maurer R, Cortes R, Probst A, Palacios JM (1983) Multiple opiate receptor in human brain: an autoradiographic study. Life Sci 33 (Suppl 1):231–234

Merz H, Stockhaus K (1979) N-(tetrahydrofuryl)alkyl and N-(alkoxyalkyl) derivatives of (−)normetazocine, compounds with differentiated opioid action profiles. J Med Chem 22:1475–1479

Merz H, Langbein A, Stockhaus K, Walther G, Wick H (1973) Structure activity relationships in narcotic antagonists with N-furylmethyl substituents. In: Braude MC (ed) Advances in biochemical pharmacology, vol 8: Narcotic antagonists. Raven, New York, pp 91–107

Morris BJ, Herz A (1986) Autoradiographic distribution in rat brain of a non-conventional binding site for the opiate 3H-diprenorphine. Brain Res 384:362–366

Mucha RF, Herz A (1985) Motivational properties of kappa and mu opioid receptor agonists studied with place and taste preference conditioning. Psychopharmacology 86:274–280

Mucha RF, Millan MJ, Herz A (1985) Aversive properties of naloxone in non-dependent rats may involve blockade of central β-endorphin. Psychopharmacology 86:281–285

Nutt JG, Jasinski DR (1974) Diuretic action of the narcotic antagonist oxilorphan. Clin Pharmacol Ther 15:361–367

Peters GR, Ward NJ, Antal EG, Lai PY, DeMaar EW (1987) Diuretic actions in man of a selective kappa-opioid agonist: U-62,066E. J Pharmacol Exp Ther 240:128–131

Pfeiffer A, Pasi A, Mehraein P, Herz A (1981) A subclassification of kappa-opiate sites in human brain by use of dynorphin 1–17. Neuropeptides 2:89–97

Pfeiffer A, Pasi A, Mehraein P, Herz A (1982) Opiate receptor binding sites in human brain. Brain Res 248:87–96

Pfeiffer A, Feuerstein G, Kopin IJ, Faden AI (1983) Cardiovascular and respiratory effects of mu-, delta- and kappa-opiate agonists microinjected into the anterior hypothalamic brain area of awake rats. J Pharm Exp Ther 225:735–741

Pfeiffer A, Brantl V, Herz A, Emrich HM (1986a) Psychotomimesis mediated by kappa-opiate receptors. Science 233:774–776
Pfeiffer A, Braun S, Mann K, Meyer H-D, Brantl V (1986b) Anterior pituitary hormone responses to a kappa-opioid agonist in man. J Clin Endocrinol Metab 62:181–185
Pfeiffer A, Knepel W, Braun S, Meyer HD, Lohmann H, Brantl V (1986c) Effects of a kappa-opioid agonist on adrenocorticotropic and diuretic function in man. Hormone Metab Res 18:842–846
Quirion R, Chiceportiche R, Contreras PC, Johnson KM, Lodge D, Tam WS, Woods JH, Zukin SR (1987) Classification and nomenclature of pencyclidine and sigma receptor sites. TINS 10:444–446
Resnik RB, Fink M, Freedman AM (1971) Cyclazocine treatment of opiate dependence: a progress report. Compreh Psychiat 12:491–502
Schmidbauer W, von Scheidt J (1977) Handbuch der Rauschdrogen. Fischer, Frankfurt, pp 294–306
Shearman GT, Herz A (1982) Generalization and antagonism studies with rats trained to discriminate an effect of the proposed sigma opiate receptor agonist N-allylnormetazocine (SKF 10,047). In: Colpaert FC, Slangen JL (Eds) Drug discrimination: Application in CNS pharmacology. Elsevier, Amsterdam, pp 37–47
Swain HH, Seevers MH (1974) Evaluation of new compounds for morphine-like physical dependence in the rhesus monkey. Bull Probl Drug Depend 36 Addendum 7:1168
Szligi GR, Ludens JH (1982) Studies on the nature and mechanism of the diuretic action of the opioid analgesic ethylketocyclazocine. J Pharmacol Exp Ther 220:585–591
Szligi GR, Ludens JH (1985) Displacement of (3)H-EKC binding by opioids in rat kidney: a correlate to diuretic activity. Life Sci 36:2189–2193
Szligi GR, Ludens JH (1986) Role of ADH in ethylketocyclazocine induced diuresis: studies in Brattleboro rats. Life Sci 38:2437–2443
Tang AH (1985) Protection from cerebral schemia by U-50488H, a specific opiate analgesic agent. Life Sci 37:1475–1482
Tyers MB (1980) A classification of opiate receptors that mediate antinociception in animals. Br J Pharmacol 69:503–508
Vaupel DB (1983) Phencyclidine does not interact with opioid receptors. Eur J Pharmacol 92:269–274
Vaupel DB, Risner ME, Shannon HE (1986) Pencyclidine and sigma-receptors. Drug Alcohol Depend 18:173–194
VonVoigtlander PF, Hall ED, Camacho-Ochoa M, Lewis RA, Triezenberg HJ (1987) U-55594A: a unique anticonvulsant related to kappa-opioid agonists. J Pharmacol Exp Ther 243:542–547
Wood PL, Stotland M, Richard JW, Rackham A (1980) Actions of mu, kappa, sigma, delta and agonist/antagonist opiates on striatal dopaminergic function. J Pharmacol Exp Ther 215:697–702
Woods JH, Smith CB, Medzihradsky F, Swain HH (1979) Preclinical testing of new analgesic drugs. In: Beers RF, Basset EG (Eds) Mechanisms of pain and analgesic compounds. Raven, New York, pp 429–445
Zukin RS, Zukin SR (1981) Binding sites of cyclazocine. Mol Pharmacol 20:246–254

CHAPTER 22

Psychiatric Aspects of Opioid Research

H.M. Emrich and C. Schmauss

1 Introduction

Functional psychoses, like endogenous depression and schizophrenia, have often been related pathogenetically to disturbances of serotoninergic, noradrenergic, and dopaminergic systems since classical antidepressants and neuroleptic compounds influence these classical neurotransmitters. On the other hand, these types of psychoses tend to spontaneously fluctuate during the time course of the illness, thereby suggesting the view that the processes regulating these neurotransmitters rather than the transmitters themselves, are involved in the pathogenesis of these disorders. Since endogenous opioids represent potent neurochemical regulators, modulating the activity of classical transmitters, these compounds have been the focus of many studies seeking to identify the causative factors in the initiation of these types of psychiatric diseases. The present chapter seeks to review the scientific approaches which have been used to examine the relationship between endogenous opioids and endogenous depression as well as schizophrenia. For the most part, information regarding the role of opioidergic systems in functional psychoses has been derived from psychopharmacological observations as well as from neuroendocrinological and ECT (electroconvulsive treatment) studies, whereas only little information stems from the neurochemistry of post-mortem brains.

2 Opioids in Affective Disorders

The observation that opioids have psychopharmacological effects in man especially on mood, more or less speculatively prompted the idea that endogenous opioids might play a pathogenetic role in affective types of psychoses. Opiates represent very powerful psychotropic drugs, exhibiting anxiolytic, tranquilizing, sedative, hypnotic, euphorogenic and possibly antidepressant effects. Since not only depressed mood but also anxiety represent integral parts of the psychopathology of endogenous depression and since opioids apparently have a spectrum of action which appears to be a "mirror image" of psychic symptoms of depressive disorders, an endorphin-deficiency hypothesis of depression appears plausible (for a summary, see Herz and Emrich 1983; cf. Byck 1976; Koob and Bloom 1983). Measurements of opioid activity in the CSF and plasma of depressed patients failed to demonstrate such a deficiency (cf. Schmauss and Emrich 1987, 1988). This finding, however, does not disprove the hypothesis since alterations in opioid activity may only occur in discrete brain areas, e.g. the limbic system, and, as such may not be of sufficient magnitude to be detected in the samples used for measurement.

More suggestive in the direction of a possible endorphin deficiency in endogenous depression were neuropsychoendocrinological findings, demonstrating a "blunted" PRL (prolactin) response to morphine and methadone by Extein et al. (1980) and Robertson et al. (1984). Our investigation regarding PRL response (stimulation) and LH (luteinizing hormone) response (inhibition) to 0.2 mg buprenorphine-challenge in 12 patients with major depressive disorders in comparison to 20 normal controls (Fig. 1) revealed no such abnormality, a finding which is in line with the observations by Judd et al. (1983) and Zis et al. (1985).

Another approach which can be used to evaluate the possible link between endogenous opioid activity and mood is to examine the effects produced by the application of opiate antagonists and to compare these to those of agonists. Although the opiate antagonist naloxone (NAL) does not exert effects on mood in non-opiate-dependent subjects (for a review, see Schmauss and Emrich 1988), it is apparent that the long-acting opioid antagonist naltrexone sometimes produces depression-like symptoms (Hollister et al. 1981). On the other hand, it is well known that opioids have euphorogenic effects in normal patients and can act therapeutically in depressive patients with affective disorders (Emrich et al. 1982; Weber and Emrich 1988). It should be mentioned in this regard that Belenky and Holaday

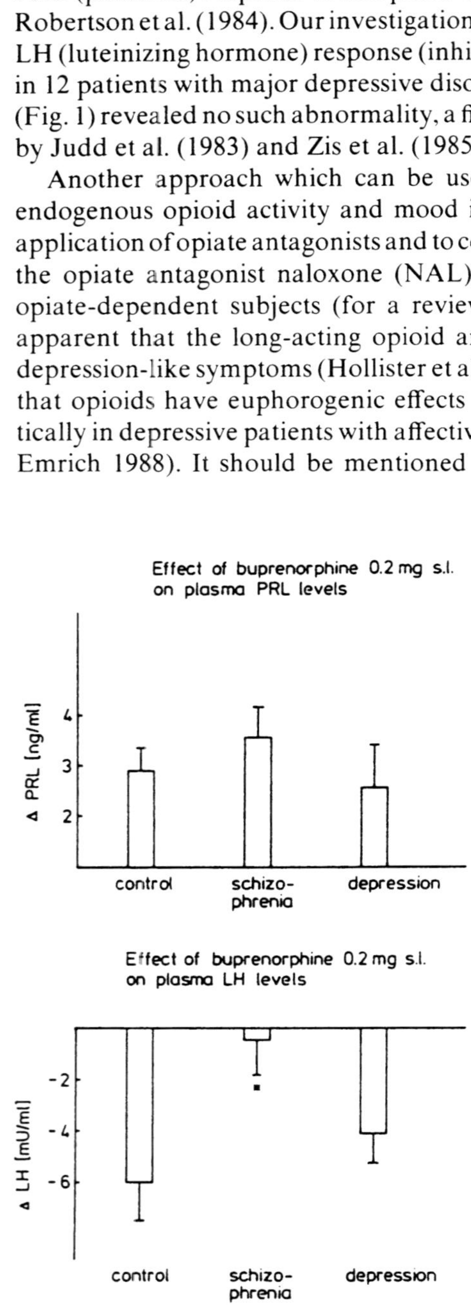

Fig. 1. Changes in plasma LH (ΔLH) and PRL (ΔPRL) levels 240 min following sublingual (s.l.) buprenorphine (0.2 mg) application. Data were obtained from drug-free patients suffering from either major depression or schizophrenia

(1979), applying electroconvulsive shocks (ECS) to animals, observed ECS-induced, NAL-reversible vegetative changes, indicating that ECS mobilizes endogenous opioids. Since ECS is an important therapy for resistant depression, Emrich et al. (1979) measured β-endorphin-(β-END) immunoreactivity in plasma of depressed patients after ECS and observed an elevation in this variable, a finding which has been reproduced by Alexopoulos et al. (1983). These findings gave rise to the hypothesis that ECS may act at least partially, via the release of β-END and led to the idea to treat therapy-resistant depressive patients with an opioid. Using the partial agonist buprenorphine, therapeutic effects could be demonstrated in about 50% of the patients (for a review, see Emrich 1984). From the data available as yet it is unclear whether opioids exert their therapeutic effects only on a symptomatic level or whether these actions are related to a pathogenetic opioid deficiency in depressive illness. However, in view of the neuroendocrine findings, shown in Fig. 1, a merely symptomatic mode of action appears probable.

3 Schizophrenia

The observation that certain partial opioid agonists, like cyclazocine and nalorphine (Jasinski et al. 1967), induce psychotomimetic effects in normal volunteers, in conjunction with the finding of Terenius et al. (1976) that CSF-opioid activity is elevated in patients with schizophrenia, gave rise to the hypothesis of a possible pathogenetic relationship between abnormally elevated opioids and schizophrenia. Also, the data of Pfeiffer et al. (1986) that the κ-agonist MR 2034 induced naloxone-reversible psychotogenic effects in normal volunteers, fit into this concept.

Consequently, several clinical studies have been performed examining the effect of the opiate-antagonist NAL on schizophrenic symptomatology. The negative findings of several groups, partially open studies, partially double-blind studies with very low dosages of NAL (for a review, see Schmauss and Emrich 1988) contrast with the positive findings under double-blind, placebo-controlled conditions of Emrich et al. (1977) and Berger et al. (1981), who observed significant reductions of schizophrenic productive symptoms 2–7 h after NAL administration (4 mg, i.v.). This delayed effect may be due to a compensatory release of opioids (Volavka et al. 1980) suggesting that not an antagonistic effect, but an agonistic mode of action might be responsible for the antipsychotic actions reported. In line with this is the observation of Jørgensen et al. (1979) that 1–3 mg of the opioid FK 33–824 induces a significant reduction of hallucinations in chronic schizophrenic patients. Also, Brizer et al. (1985) observed an improvement of severely ill, chronic schizophrenic patients when methadone was added to a neuroleptic therapy. Recently, Schmauss et al. (1987) used an acute treatment of opiate mixed agonist/antagonist buprenorphine (0.2 mg) in ten neuroleptic-free schizophrenic patients suffering from frequent hallucinations, delusions and severe formal thought disorders. Buprenorphine exerted a pronounced antipsychotic effect in patients with schizophreniform disorders and paranoid schizophrenia which lasted about 4 h, whereas in the patients with residual schizophrenia (according to DSM-III) no therapeutic effect could be observed.

Since opioids represent potent modulators of dopaminergic (DAergic) neurotransmission, these findings may indicate that an alteration in opioidergic regulation of DA function underlies the pathogenesis. To investigate this, prolactin- and LH-response upon buprenorphine challenge in schizophrenic patients, in comparison to controls, was recently measured in our laboratory (Fig. 1). These data (Schmauss and Emrich 1987, 1988; Schmauss et al. 1989) confirm that the release of prolactin (PRL) in humans can be stimulated dose-dependently by buprenorphine, whereas the release of LH is inhibited. While the stimulating effect of buprenorphine on PRL-release cannot be blocked by NAL, the same antagonist dose-dependently blocked the effect of buprenorphine on the inhibition of LH-release. NAL alone does not affect plasma levels of PRL, but dose-dependently stimulates the release of LH. Thus, the opioid receptors involved in mechanisms regulating LH- and PRL-release differ in their response to NAL. Interestingly, 18 patients suffering from nonresidual schizophrenia (according to DSM-III) show a blunted LH-response to buprenorphine when compared to a group of 20 matched healthy controls. The PRL-response to buprenorphine in these patients did not differ from that of the control response. Hypothetically, this blunted LH-response to buprenorphine may be interpreted as indicating an enhanced endorphinergic activity in these schizophrenic patients, which due to the increased inhibitory influence on DA neurons, would result in a reduction of basal DA activity (Schmauss and Emrich 1985).

As has been pointed out by Emrich (1982), pharmacological as well as other somatic therapeutic treatment of schizophrenia (neuroleptics, insulin coma, ECS, rotatory and cold-water stress) tend to increase central endorphinergic activity, which is in line with the hypothesis presented above. Thus, to conclude, in the functional psychoses, in manic depression as well as in schizophrenia the role of opioids appears to be more neuromodulatory than causal. However, the significance of the blunted LH-response to buprenorphine in schizophrenia should be further evaluated. It appears promising to estimate the state of activity of the endorphinergic activity of the LH/opioid system by applying different doses of NAL and of an opioid agonist and to relate these data to neuroendocrine responses as well as to clinical variables.

References

Alexopoulos GS, Inturrisi CE, Lipman R, Frances R, Haycox J, Doughert Jr JH, Rossier J (1983) Plasma immunoreactive beta-endorphin levels in depression. Arch Gen Psychiat 40:181–183

Belenky GL, Holaday JW (1979) The opiate antagonist naloxone modifies the effects of electroconvulsive shock (ECS) on respiration, blood pressure and heart rate. Brain Res 177:414–417

Berger PA, Watson SJ, Akil H, Barchas JD (1981) The effects of naloxone in chronic schizophrenia. Am J Psychiat 38:913–918

Brizer DA, Hartmann N, Sweeney J, Millman RB (1985) Effect of methadone plus neuroleptics on treatment-resistant chronic paranoid schizophrenia. Am J Psychiat 142:1106–1107

Byck R (1976) Peptide transmitters: a unifying hypothesis for euphoria, respiration, sleep, and the action of lithium. Lancet ii:72–73

Emrich HM (1982) A possible role of endorphinergic systems in schizophrenia. In: Namba M, Kaiya H (eds) Psychobiology of schizophrenia. Pergamon, Oxford, pp 291–297

Emrich HM (1984) Endorphins in psychiatry. Psychiat Dev 2:97–114
Emrich HM, Cording C, Pirée S, Kölling A, von Zerssen D, Herz A (1977) Indication of an antipsychotic action of the opiate antagonist naloxone. Pharmacopsychiatry 10:265–270
Emrich HM, Höllt V, Kissling W, Fischler M, Laspe H, Heinemann H, von Zerssen D, Herz A (1979) β-Endorphin-like immunoreactivity in cerebrospinal fluid and plasma of patients with schizophrenia and other neuropsychiatric disorders. Pharmacopsychiatry 12:269–276
Emrich HM, Vogt P, Herz A (1982) Possible antidepressive effects of opioids: action of buprenorphine. Ann NY Acad Sci 398:108–112
Extein I, Pottash ALC, Gold MS, Sweeney DR, Martin DM, Goodwin FK (1980) Deficient prolactin response to morphine in depressed patients. Am J Psychiat 137:845–846
Herz A, Emrich HM (1983) Opioid systems and the regulation of mood: possible significance in depression? In: Angst J (ed) The origins of depression: current concepts and approaches. Springer, Berlin Heidelberg New York, pp 221–234
Hollister LE, Johnson K, Boukhabza D, Gillespie HK (1981) Aversive effects of naltrexone in subjects not dependent on opiates. Drug Alcohol Depend 7:1–5
Jasinski DR, Martin WR, Haertzen CA (1967) The human pharmacology and abuse potential of N-allyl-noroxymorphone (naloxone). J Pharmacol Exp Ther 157:420–426
Jørgensen A, Fog R, Veilis B (1979) Synthetic enkephalin analogue in treatment of schizophrenia. Lancet i:935
Judd LL, Risch SG, Parker DC, Janowsky DS, Segal DS, Huey LY (1983) The effect of methadone challenge on the prolactin and growth hormone responses of psychiatric patients and normal controls. Psychopharmacol Bull 18:204–207
Koob GF, Bloom FE (1983) Behavioural effects of opioid peptides. Br Med Bull 39:89–94
Pfeiffer A, Brantl V, Herz A, Emrich HM (1986) Psychotomimesis mediated by κ-opiate receptors. Science 233:774–776
Robertson AG, Jackman H, Meltzer HY (1984) Prolactin response to morphine in depression. Psychiat Res 11:353–364
Schmauss C, Emrich HM (1985) Dopamine and the action of opiates: a reevaluation of the dopamine hypothesis of schizophrenia. With special consideration of the role of endogenous opioids in the pathogenesis of schizophrenia. Biol Psychiat 20:1211–1231
Schmauss C, Emrich HM (1987) The effects of opioids on behavior: possible role in psychotogenesis. In: Nemeroff CB, Loosen PT (eds) Handbook of psychoneuroendocrinology. Guilford, New York, pp 417–428
Schmauss C, Emrich HM (1988) Narcotic antagonist and opioid treatment in psychiatry. In: Rodgers RJ, Cooper SJ (eds) Endorphins, opiates and behavioural processes. John Wiley & Sons, New York London, pp 327–351
Schmauss C, Yassouridis A, Emrich HM (1987) Antipsychotic effect of buprenorphine in schizophrenia. Am J Psychiat 144:1340–1342
Schmauss C, Pirke KM, Bremer D, Weber M, Emrich HM (1989) Luteinizing hormone and prolactin response to buprenorphine in depression and schizophrenia. In: Lerer B, Gershon S (eds) New directions in affective disorders. Springer, Berlin Heidelberg New York, pp 276–279
Terenius L, Wahlström A, Lindström L, Widerlöv E (1976) Increased CSF levels of endorphins in chronic psychosis. Neurosci Lett 3:157–162
Volavka J, Bauman J, Pevnick J, Reker D, James B, Cho D (1980) Short-term hormonal effects of naloxone in man. Psychoneuroendocrinology 5:225–234
Weber MM, Emrich HM (1988) Current and historical concepts of opiate treatment in psychiatric disorders. Intern Clin Psychopharmacol 3:255–266
Zis AP, Haskett RF, Albala AA, Carroll BJ, Lohr NE (1985) Prolactin response to morphine in depression. Biol Psychiat 20:287–292

CHAPTER 23

Functional Correlates of Opioid Receptor Activation as Measured by Local Cerebral Glucose Utilization

A. Ableitner

1 Introduction

The 2-deoxyglucose (2-DG) autoradiographic method using [1-^{14}C]-2-DG (Sokoloff et al. 1977) has become a useful tool for delineating alterations of functional brain activity in laboratory animals subjected to various physiological and pharmacological stimuli. This method is based on the facts that under normal physiological conditions glucose is the sole substrate for energy metabolism of cerebral tissue (Siesjö 1978) and, further, that energy metabolism is closely correlated to functional activity (Kennedy et al. 1975; Hand et al. 1978; Miyaoka et al. 1979; Mata et al. 1980). Measurement of the rate of glucose utilization can thus provide a functional map of the central nervous system in response to different pharmacological or physiological manipulations.

In the present chapter, the functional alterations produced by the activation of particular opioid receptor types will be discussed with respect to opioid receptor distribution, the neuroanatomy of the brain regions affected, as well as their various roles in physiological processes.

2 The 2-Deoxyglucose Technique: Theory and Procedure

Descriptions of the theory and procedure as well as detailed reviews and reports on various aspects of this technique are available (Sokoloff et al. 1977, 1983; Sokoloff 1981, 1989; Gjedde 1987; Hawkins and Miller 1987; Nelson et al. 1987; Schmidt et al. 1989).

The method utilizes [1-^{14}C]-2-deoxyglucose as a tracer to quantify local cerebral glucose utilization (LCGU) and is based on a kinetic model of the biochemical properties of 2-deoxyglucose (2-DG) and glucose in the brain (Fig. 1): the 2-DG is transported across the blood-brain barrier by the same carrier mechanism as glucose and, in the tissue, 2-DG competes with glucose for hexokinase which phosphorylates both sugars (Sols and Crane 1954; Bidder 1968; Oldendorf 1971; Bachelard 1971).

Glucose-6-phosphate is subsequently either isomerized by phosphohexoseisomerase to fructose-6-phosphate and metabolized further via the glycolytic and tricarboxylic acid pathways to CO_2 and H_2O; or it can be oxidized by glucose-6-phosphatedehydrogenase. Alternatively, glucose-6-phosphate can be hydrolyzed back to free glucose by glucose-6-phosphate, although the activity of this enzyme has been reported to be very low in mammalian tissue (Hers 1957).

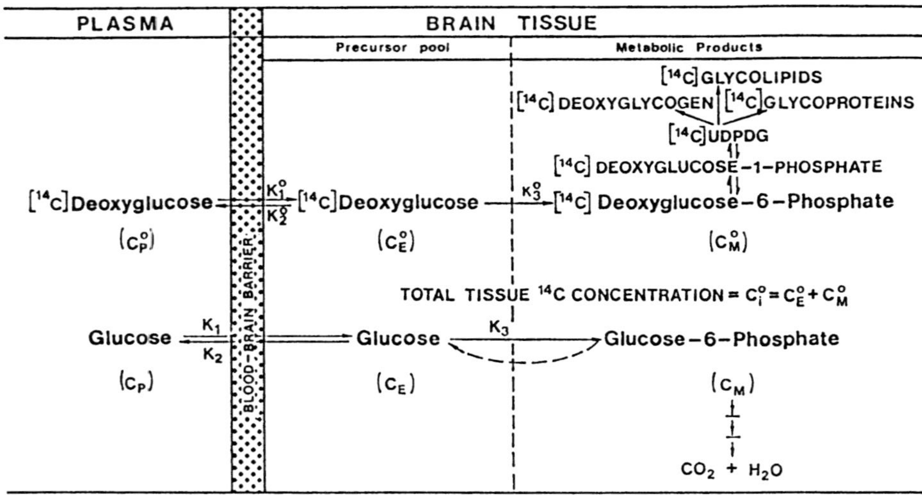

Fig. 1. Theoretical model of the deoxyglucose method. C_i^o represents the total ^{14}C concentration in a single homogeneous tissue of the brain. C_p^o and C_p represent the concentrations of [^{14}C]deoxyglucose and glucose in the arterial plasma, respectively; C_E^o and C_E represent their respective concentrations in the tissue pools that serve as substrates for hexokinase. C_M^o represents the combined concentration of [^{14}C]deoxyglucose-6-phosphate and its products in the tissue. The constants K_1^o, K_2^o, and K_3^o, represent the rate constants for carrier-mediated transport of [^{14}C]deoxyglucose from plasma to tissue, for carrier-mediated transport back from tissue to plasma, and for phosphorylation by hexokinase, respectively. The constants K_1, K_2, and K_3 are the equivalent rate constants for glucose. [^{14}C]-Deoxyglucose and glucose share and compete for the carrier that transports between both plasma and tissue and for hexokinase which phosphorylates them to their respective hexose-6-phosphates. The *dashed arrow* represents the possibility of glucose-6-phosphate hydrolysis by glucose-6-phosphatase activity, if any (Sokoloff 1989)

In contrast to glucose-6-phosphate, 2-DG-6-phosphate cannot be converted to fructose-6-phosphate, due to the lack of an OH-group on its second carbon atom. Furthermore, it is a poor substrate for glucose-6-phosphatedehydrogenase (Sols and Crane 1954) and the amount of 2-DG-6-phosphate hydrolyzed by phosphatases is extremely low (Sokoloff et al. 1977; Nelson et al. 1987). 2-DG-6-phosphate can be converted into 2-DG-1-phosphate, then into UDP-DG and subsequently into glycogen, glycoproteins, and glycolipids. However, in mammalian tissues only a small fraction of the 2-DG-6-phosphate formed is processed in this way (Nelson et al. 1984). Thus, 2-DG-6-phosphate and its relatively stable secondary products represent the products of deoxyglucose phosphorylation that are essentially trapped in the cerebral tissues.

If the time interval following administration of [1-^{14}C]-labeled 2-DG does not exceed 1 h, during which time there is a negligible loss of ^{14}C-DG-6-phosphate and its secondary products from the tissue, the amount of the accumulated, labeled products reflects the rate of 2-DG phosphorylation by hexokinase during this time interval. The rate of ^{14}C-2-DG phosphorylation, in turn, is quantitatively related to the rate of glucose phosphorylation and depends on the time course of the concentrations of glucose and ^{14}C-2-DG in the precursor pools and the Michaelis-

Menten kinetic constants of hexokinase for the two substrates. With cerebral glucose metabolism in a steady state, the net rate of glucose phosphorylation equals the rate of glucose utilization.

Mathematical analysis of these relationships has resulted in an "operational equation" which closely resembles the equation for the determination of the rate of biochemical reactions with radioactive tracers (Sokoloff et al. 1977; Fig. 2). Thus, local rates of cerebral glucose utilization (LCGU) can be calculated from (1) plasma levels of ^{14}C-2-DG and glucose during the experiment and (2) regional ^{14}C-concentrations. The values of the kinetic constants (for the carrier-mediated transport from plasma to tissue, for the carrier-mediated transport back from tissue to plasma, and for phosphorylation by hexokinase) and the lumped constant (an analog of a correction factor for an isotope effect) for the rat have previously been determined by Sokoloff and colleagues (Sokoloff et al. 1977).

Although the values of the constants determined by Sokoloff et al. (1977) can usually be used within a wide range of experimental conditions, it must be noted that

General Equation for Measurement of Reaction Rates with Tracers:

$$\text{Rate of Reaction} = \frac{\text{Labeled Product Formed in Interval of Time, O to T}}{\begin{bmatrix}\text{Isotope Effect}\\ \text{Correction Factor}\end{bmatrix}\begin{bmatrix}\text{Integrated Specific Activity}\\ \text{of Precursor}\end{bmatrix}}$$

Operational Equation of [^{14}C] Deoxyglucose Method:

$$R_i = \frac{\overbrace{\underbrace{C_i^{\circ}(T)}_{\text{Total }^{14}\text{C in Tissue at Time, T}} - \underbrace{k_1^{\circ} e^{-(k_2^{\circ}+k_3^{\circ})T}\int_0^T C_p^{\circ} e^{(k_2^{\circ}+k_3^{\circ})t}\,dt}_{^{14}\text{C in Precursor Remaining in Tissue at Time, T}}}^{\text{Labeled Product Formed in Interval of Time, O to T}}}{\underbrace{\left[\frac{\lambda V_m^{\circ} K_m}{\Phi V_m K_m^{\circ}}\right]}_{\text{Isotope Effect Correction factor}}\underbrace{\left[\underbrace{\int_0^T\left(\frac{C_p^{\circ}}{C_p}\right)dt}_{\text{Integrated Plasma Specific Activity}} - \underbrace{e^{-(k_2^{\circ}+k_3^{\circ})T}\int_0^T\left(\frac{C_p^{\circ}}{C_p}\right)e^{(k_2^{\circ}+k_3^{\circ})t}\,dt}_{\text{Correction for Lag in Tissue Equilibration with Plasma}}\right]}_{\text{Integrated Precursor Specific Activity in Tissue}}}$$

Fig. 2. Operational equation of the radioactive deoxyglucose method and its functional anatomy. T represents the time at the termination of the experimental period; and K_m^o and V_m^o equal the ratio of the distribution space of deoxyglucose in the tissue to that of glucose; Φ equals the fraction of glucose which, once phosphorylated, continues down the glycolytic pathway; and K_m^o and V_m^o represent the familiar Michaelis-Menten kinetic constants of hexokinase for deoxyglucose and glucose, respectively. The other symbols are the same as those defined in Fig. 1. Note the similarity in the structures of the operational and general equation (Sokoloff 1989)

under conditions such as severe hyperglycemia and ischemia (Hawkins et al. 1981; Schuier et al. 1981), hypoglycemia (Suda et al. 1981), or severe seizure activity (Diemer and Gjedde 1983) major changes occur in these constants. Under these particular conditions, therefore, it is necessary to redetermine their values. With regard to the requirement of a constant plasma glucose level during the experimental procedure (Sokoloff et al. 1977), a modified "operational equation" was derived which allows the use of the 2-DG-technique under conditions in which plasma glucose levels vary (Savaki et al. 1980).

The experimental procedure (see Sokoloff et al. 1977, 1983; Sokoloff 1989) involves an intravenous injection pulse of ^{14}C-labeled 2-DG at time zero, and the withdrawal of samples of blood via an arterial catheter at timed intervals for 45 min. These samples are used for the measurement of time courses of the appearance of ^{14}C in the plasma and plasma glucose concentration. The animals are usually decapitated 45 min after the injection of the tracer. The brains are removed, frozen, and sectioned (20 μm) in a cryostat maintained at -20°C. The sections are then mounted on coverslips and dried at 60°C. Autoradiograms are prepared by exposing the sections to an X-ray film together with a set of ^{14}C-plastic standards of increasing specific activity. Local tissue concentrations of ^{14}C are evaluated by densitometric analysis of the autoradiograms with reference to the standards. LCGU is then calculated from plasma and brain radioactivities and from the plasma glucose concentration using the operational equation (Sokoloff et al. 1977; Savaki et al. 1980).

3 Functional Alterations After Manipulating Central Opiodergic Systems

The 2-DG technique has been utilized in a number of experiments to characterize functional alterations associated with the activation of specific opioid receptor types. To interprete alterations of LCGU, however, it must be realized that systemic effects of a drug, e.g., changes in blood pressure, body temperature, and respiration, can markedly influence glucose utilization in the CNS. Thus, systemic hypotension is associated with marked increases in LCGU in a number of specific nuclei (Savaki et al. 1982). Furthermore, the catabolism of glucose is temperature-dependent (Siesjö 1978). As shown by McCulloch et al. (1982), LCGU decreases during hypothermia and increases during hyperthermia; furthermore, there are regional variations in the magnitude of these effects. There is also considerable evidence that hypercapnia, resulting from depressed respiration, induces a general reduction in the rate of glycolysis in the CNS (Miller et al. 1975; Borgström et al. 1976; DesRosiers et al. 1978). Since many opioids and especially μ-opioid receptor agonists exert a depressant action on respiration, this point has to be considered in the interpretation of opioid effects on LCGU.

3.1 Influences of μ-Opioid Agonists on LCGU

The influence of μ-opioid receptor activation on LCGU has proven difficult to evalute. Thus, several studies examining the effects of morphine and other μ-agonists on LCGU have produced conflicting data with regard to the regional changes which occur in response to these agents (see Table 1). On the one hand, this may possibly be due to the differences in doses, route, and time of drug application (Table 1). The respiratory depressant action of μ-agonists, which results in hypercapnia, however, has turned out to be a major obstacle in investigating μ-agonist effects on LCGU. Thus, the dose-dependent reductions in LCGU after acute morphine, which were found in preliminary studies by Sakurada et al. (1976), appear to result from hypercapnia rather than from a specific drug action within the CNS (Sakurada et al. 1976). $PaCO_2$ was also increased in the studies of Hiesinger et al. (1983). Hypercapnia and hypoxia, although weak, were also present in the studies of Fanelli et al. (1987), again making it difficult to differentiate specific drug effects and the effects of hypercapnia. It is important to note, however, that investigations on the effects of hypercapnia on LCGU have shown that uniform reductions occur throughout the brain (DesRosiers et al. 1978). Furthermore, there are particular brain regions, such as the superior colliculus, the hippocampus, and the white matter, which are more sensitive to hypoxia, as compared to other regions (Shimada 1981). Since these structures were not significantly affected, whereas, in contrast, significant changes were primarily restricted to the lateral habenula and/or thalamic nuclei (Hiesinger et al. 1983; Fanelli et al. 1987), these effects might in fact be specific, and the

Table 1. Regions showing alterations in LCGU after μ-agonists[a]

Morphine		Morphine	Oxymorphine
5 mg/kg i.v. 10 min prior to 2-DG[b]	5 mg/kg s.c. 60 min prior to 2-DG[c]	8 mg/kg s.c. 15 min prior 2-DG[d]	0.4 mg/kg 15 min prior 2-DG[d]
–	–	Paratenial thal. ncl. ↓	
Medial thal. ↓	–	Central medial thal. ncl. ↓	
–	–	Ventropost. thal. ncl. ↓ (medial + lat. part.)	
–	–	Ventrolat. thal. ncl. ↓	
–	–	Interanteromed. thal. ncl. ↓	
–	–	Gelatinosus thal. ncl. ↓	
Habenula ↓	Lat. habenular ncl. ↓ (medial part.)	–	
–	–	Median eminence ↓	
–	–	Dorsal tegmental ncl. ↓	
–	Medial mam. body ↑	–	
–	Subst. nigra, pars ret. ↑	–	

[a] No other region displayed significant changes.
[b] Data from Hiesinger et al. (1983).
[c] Data from Ito et al. (1983).
[d] Data from Fanelli et al. (1987).

functional implication of these changes with respect to specific pharmacological actions of μ-agonists may be considered. For example, in both studies glucose utilization was significantly decreased in the medial thalamic nuclei which contain a high density of μ-opioid binding sites (Goodman et al. 1980; Goodman and Pasternak 1985; Mansour et al. 1987). The intraperitoneal injection of morphine has previously been shown to inhibit the spontaneous and evoked unit electrical activity in nuclei of the medial thalamus (Dafny et al. 1979; Dafny and Gildenberg 1984). Moreover, since medial thalamic nuclei seem to participate in central analgesic mechanisms (Andy 1980, 1983), changes in LCGU observed in these areas may be related to such processes. In this regard it is interesting to note that microinjection of morphine into the medial thalamus produced no antinociceptive effect when measured in the tail-flick test. This morphine treatment did, however, increase the latency of response in terms of other parameters (e.g., vocalization) which are considered to reflect the affective response to pain (Yeung et al. 1978). The decrement in glucose utilization found in these thalamic nuclei after morphine treatment might therefore reflect an attenuation of these affective reactions.

Morphine treatment, however, did not induce any changes in glucose utilization in thalamic nuclei, but produced a selective decrease in LCGU in the lateral habenular nucleus and increases in LCGU in the mammillary body, and the substantia nigra (pars reticulata) in the studies of Ito et al. (1983). Although the lack of hypercapnia in this study, as compared to the others, represents an important point in the discussion of the different findings, the differences in the time of drug application must be considered. In the study of Ito et al. (1983), LCGU was evaluated 1 h following the application of the drug; thus, drug effects which reach a maximum within the first hour after administration may no longer be detectable, whereas others may become more prominent at this time. "Catatonia", evaluated with the bar test, was most prominent at 60 min and the altered glucose utilization in the substantia nigra (pars reticulata) most probably reflects this extrapyramidal sign. The decrease in the lateral habenular nucleus may also be interpreted in the context of this behavioral state. This nucleus receives, among others, afferent inputs from regions of the extrapyramidal motor system (e.g., the entopeduncular nucleus) and, in turn, projects to midbrain regions, e.g., the substantia nigra (pars compacta) and the raphe nuclei, two major sources of striatal efferents (Herkenham and Nauta 1977, 1979). In addition to the catatonia, a maximal effect in the response of the tail-flick to hot water at 1 h following drug administration was found (Ito et al. 1983). It seems, however, necessary to consider to what extent the motor deficiency might contribute to this test; moreover, no metabolic changes were found in structures which are involved in the perception and modulation of pain. Regarding the functional implication of the LCGU alteration observed in the mammillary body, additional information is needed on the possible involvement of this nucleus in opioid-induced behavioral states. In general, the mammillary body, as a part of the limbic system, has been implicated in the control of various neurobehavioral states, primarily affective processes (Kataoka et al. 1982; Shibata et al. 1986).

3.2 Influences of κ-Opioid Agonists on LCGU

The 2-DG method has been employed in various studies to delineate those regions and circuits in the CNS which are involved in the actions of κ-opioid agonists. Most of these studies, however, have been hampered by the use of nonselective ligands. Thus, Beck and Krieglstein (1986) evaluated the effects of tifluadom, a benzodiazepine derivative, and ketazocine on LCGU. Although tifluadom displays no affinity for benzodiazepine binding sites in vitro (Römer et al. 1982a,b), there is evidence that the in vivo effects of tifluadom are antagonized by the selective benzodiazepine antagonist Ro 15-1788 (Ruhland and Zeugner 1983). Further, ketazocine-like compounds display a high degree of cross-reactivity with μ- and, to a lesser extent, with δ-binding sites (Kosterlitz et al. 1981; Magnan et al. 1982). Also, nalbuphine, which was used in a study by Fanelli et al. (1987), is regarded as a partial κ-agonist (Jaffe and Martin 1985; Schmidt et al. 1985). The discrepancy in the effects of these drugs on LCGU (see Table 2) may therefore be related to their nonselective binding properties. However, they may also be due to their binding to different subtypes of κ-receptors, the presence of which have recently been suggested (Zukin et al. 1988). Moreover, in these studies the effects of only a single dose of each drug were evaluated. Increasing drug doses, however, might have revealed more widespread changes. In fact, this would be expected on the basis of the distribution of κ-binding sites. The involvement of the nucleus accumbens in the effects of ketazocine and tifluadom (Beck and Krieglstein 1986) represents an interesting observation in view of its high density of κ-opioid binding sites (Mansour et al. 1987), and the central role of this nucleus in mediating the aversive actions of κ-opioid agonists (see Shippenberg and Bals-Kubik, this Vol.). The increased LCGU in the globus pallidus (Fanelli et al. 1987) is consistent with the presence of κ-binding sites in this structure (Mansour et al. 1987); moreover, there are dynorphin (DYN)

Table 2. Regions, in which changes in glucose utilization were observed following systemic administration of κ-agonists[a]

Brain regions	Glucose utilization		
	Ketazocine[b]	Tifluadom[b]	Nalbuphine[c]
Frontal cortex	----	Decrease	----
Olfact.cortex	----	Decrease	----
Sens.mot.cort.	----	Decrease	----
Parietal cort.	----	Decrease	----
Caudate ncl.	----	Decrease	----
Globus pallidus	----	----	Increase
Accumbens ncl.	Increase	Increase	----
Spinal trigem. ncl. (oral/interpositus/caudal)	----	----	Increase

[a] No other region displayed significant changes.
[b] Data from Beck and Krieglstein (1986); drug treatment: ketazocine 5 mg/kg i.v., tifluadom 2.5 mg/kg i.v. 10 min prior to 2-DG.
[c] Data from Fanelli et al. (1987): drug treatment: nalbuphine 16 mg/kg i.v. 15 min. prior to 2-DG.

containing neurons in the globus pallidus (Vincent et al. 1982) and a striatonigral DYNergic projection has been demonstrated (Fallon et al. 1985).

In comparison to tifluadom, ketazocine, and nalbuphine (Beck and Krieglstein 1986; Fanelli et al. 1987), the selective κ-agonist U-50,488H (trans-3,4-dichloro-N-methyl-N-methyl-N-[2-(1-pyrrolidinyl)cyclohexenyl]-benzene-acetamide) induced a more complex pattern of changes (Ableitner and Herz 1989a). U-50,488H and a series of other benzeneacetamides have proven to be highly selective κ-agonists in various biochemical and pharmacological studies (Piercy et al. 1982; Szmuszkovicz and von Voigtlander 1982; Gillan et al. 1983; von Voigtlander et al. 1983; Lathi et al. 1985). The effects of U-50,488H on LCGU might therefore be considered to be due to specific κ-opioid receptor activation. The most pronounced changes in glucose utilization were observed in limbic forebrain structures and structures that lie within what Nauta (1985) has termed the "limbic midbrain area" (Tables 3 and 4). Functionally, these structures have been implicated in affective behaviors (Brady and Nauta 1955; Mogenson and Huang 1973; Graeff and Filho 1978; Goldstein and Siegel 1980; Sutherland 1982). Indeed, place conditioning studies which were conducted in a parallel group of animals revealed that doses that were effective in 2-DG experiments produced marked aversive effects. In

Table 3. Effects of U-50,488H, a kappa agonist, upon glucose utilization of limbic forebrain regions[a]

Brain regions	U-50,488H			
	0.5 mg/kg	1 mg/kg	2 mg/kg	5 mg/kg
Frontal cortex	----	-10%	-13%	----
Lat.sept.ncl.	----	----	----	+37%
Accumbens ncl.	----	----	----	+26%
Anterovent.thal.ncl.	----	----	----	+25%
Anteromed.thal.ncl.	----	+11%	+13%	+25%
Med.mammillary ncl.	----	----	+17%	+30%
Posterior mam.ncl.	----	----	+29%	+42%
Lat.habenular ncl.	+19%	+26%	+50%	+89%

[a] Data are from Ableitner and Herz (1989a).
Values shown are ± percent significant change from control; U-50,488H was injected prior to 2-DG.

Table 4. Effects of U-50,488H upon glucose utilization of brainstem regions[a]

Brain regions	U-50,488H			
	0.5 mg/kg	1 mg/kg	2 mg/kg	5 mg/kg
Median raphe ncl.	+13%	+21%	+25%	+26%
Dorsal raphe ncl.	----	----	----	+14%
Vent.tegment.ncl.	----	----	+22%	+23%
Dors.tegment.ncl.	----	----	----	+18%
Central gray, pons	----	+8%	+14%	+18%

[a] Data are from Ableitner and Herz (1989a).
Values shown are ± percent significant change from control; U-50,488H was injected 5 min prior to 2-DG.

view of the aversive effects of κ-agonists in experimental animals and humans (Pfeiffer et al. 1986), it is likely that the alterations in glucose utilization in these regions may reflect the aversive effects of U-50,488H. Increases in glucose utilization were also found in the dorsal raphe nucleus and the central gray pons. It is well established that electrical stimulation of these regions consistently elicits antinociception (analgesia) both in humans and experimental animals (Hosobuchi and Wenner 1977; Richardson and Akil 1977; Oliveras et al. 1979). The metabolic changes induced by U-50,488H in these regions, therefore, further support the finding of a supraspinal site of pain modulation in which κ-receptors have been implicated (Czlonkowski et al. 1987).

With regard to the relation of metabolic responses to the topography of κ-opioid binding sites, the alterations in LCGU were not confined to those regions which are known to contain specific κ-binding sites; this might be explained by the complexity of neuronal circuitries presumed to underlie drug-induced changes in LCGU. For example, the effects in the lateral habenular nucleus, and area apparently devoid of κ-binding sites (Morris and Herz 1986; Mansour et al. 1987), probably reflects an activation of its afferent inputs which originate in the nucleus accumbens and the median raphe nucleus (Herkenham and Nauta 1977). These nuclei displayed increases in glucose utilization and exhibit a dense labeling of κ-binding sites (Morris and Herz 1986; Mansour et al. 1987). A dose-dependent fall in blood pressure noted after the application of U-50,488H (Szligi et al. 1984) might further contribute to the increases in LCGU observed in the lateral habenular nucleus. Although systemic hypotension results in marked increases in LCGU in regions primarily involved in cardiovascular control, increases were additionally observed in regions (e.g., the lateral habenular nucleus) hitherto not considered to be involved in this physiological process (Savaki et al. 1982).

3.3 *Influences of δ-Opioid Agonists on LCGU*

The δ-agonists available are synthetically derived enkephalin peptide analogs. Due to the variation in the amount of peptides penetrating the blood-brain barrier, they are best applied intracerebroventricularly (i.c.v.), usually into the lateral ventricles. An important aspect in the interpretation of the effects on LCGU might therefore be the kinetics of peptide distribution within the ventricular system and brain tissue. Studies on the distribution of the ^{3}H-labeled enkephalin analog [D-ala^2-,D-leu^5]-enkephalin (^{3}H-DADLE) after injection into the lateral ventricle revealed that this peptide spreads, within 10 min, into all parts of the ventricular system including the spinal cord. The mean depth of penetration at this time was about 300 μm and increased with time (Haffmans et al. 1983). It can be expected that other peptides with chemical properties similar to DADLE also display such a distribution pattern and also diffuse within a reasonable period of time into the cerebral tissue. The amount of peptide within the various regions differs according to the concentration gradient from the ventricle system into the cerebral tissue, but nothing is known of the amount of peptide that is actually necessary to activate certain receptors within this region. Thus, the changes induced in LCGU after the i.c.v. application of an enkephalin-like peptide may not be due to the receptor activation of only a few specific regions in the vicinity of the lateral ventricles, but rather they may reflect the effects of widespread regional receptor activation.

Haffmans et al. (1984) used a modification of the 2-DG technique, as described by Meibach et al. (1980), to investigate the effects of the δ-receptor agonist DSTLE on local rates of metabolism in the CNS. In contrast to the original 2-DG technique (Sokoloff et al. 1977), plasma levels of radioactivity and glucose were not evaluated and the experimental period was reduced to 30 min. This methodological approach, however, has several intrinsic limitations. In general, if the radioisotope is applied intravenously and 45 min are allowed to elapse, the radioactivity present in the CNS almost exclusively represents DG-6-phosphate (reflecting local rates of metabolism) in regions of high glucose utilization in normoglycemic rats (Sokoloff et al. 1977). This is not valid for regions with lower metabolic rates. Furthermore, the fraction of total radioactivity in the CNS is increased by moderate hypergycemia or a reduction of the experimental period (Kelly and McCulloch 1981). The changes of radioactivity content of certain regions obtained after DSTLE (Haffmans et al. 1984) may therefore reflect alterations of local rates of glucose metabolism (represented by the amount of 2-DG-6-phosphate), as well as alterations in the fraction of unmetabolized 2-DG, a point that has to be considered in the interpretation of these data. After the i.c.v. administration of DSTLE (injected 0 and 2.5 min prior to 2-DG), the most pronounced and consistent increases in radioactivity content were observed in the frontal cortex, the subiculum – including the CA 1 area – and the cortical amygdala. In addition, in all these regions, epileptiform discharges were elicited after DSTLE, indicating that these areas play an important role in mediating enkephalin-induced seizures (Haffmans et al. 1984). Regarding the selectivity of the effects of DSTLE, this peptide not only displays a high affinity for δ-binding sites, but interacts, quite potently, with μl-binding sites (Clark et al. 1986). Thus, the effects of DSTLE might reflect the activation of either, or both, receptor types.

The conformationally restricted, cyclic, disulfide-containing enkephalin analog [D-Pen^2,D-Pen^5]-enkephalin (DPDPE) represents an enkephalin analog with high affinity for δ-sites which does not interact appreciably with μl sites. Therefore, this compound may be the most appropriate one in studies examining brain regions and circuits involved in the pharmacology of δ-opioid receptors (Mosberg et al. 1983; Clark et al. 1986). In preliminary studies, DPDPE has been found to induce selective changes in LCGU (Ableitner and Herz 1989b). Increases in glucose utilization were particularly prevalent in regions of the CNS involved in the control of motor activity, such as the caudate nucleus, globus pallidus, pars reticulata of the substantia nigra, red nucleus, and cerebellar structures, as well as the motor cortex and motor relay nuclei in the thalamus. This may well reflect the behavioral arousal and increased locomotion induced by DPDPE (Cowan et al. 1985). Further increases in LCGU were found in various anatomical components of the limbic system. The increase of LCGU in the hippocampal formation (Ableitner and Herz 1989b), confirming an involvement of δ-receptors in the regulation of the excitability of this structure, was in accordance with the study of Haffmans et al. (1984). Other components of the limbic system that displayed increases after DPDPE were the nucleus accumbens and the basolateral amygdaloid nucleus, two brain areas which contain the highest densities of δ-opioid binding sites (Gulya et al. 1986). The mammillary body apparently does not contain δ-opioid binding sites (Mansour et al. 1987); however, the DPDPE effects on LCGU found in this structure can easily be explained by the manifold interconnections of structures within the limbic system (Papez 1937; Cruce 1975; Sherlock and Raisman 1975; Holstege and Dekker 1979; Raisman et al. 1966).

3.4 Influences of β-Endorphin on LCGU

Sakurada et al. (1978) evaluated the effects of β-endorphin (4–8 μg) injected into the midbrain periaqueductal gray (PAG), 1 h prior 2-DG. This region has been demonstrated to play a major role in the analgesic actions of opiates (Cannon et al. 1982; Swajkowski et al. 1980). The generalized decrease in glucose utilization found after the microinjection of β-endorphin, however, provided little insight into the brain regions and circuitries underlying the actions of this opioid. Intracerebroventricular injection of β-endorphin (15–17 μg), in contrast, was reported to induce discrete increases in the hippocampal formation, the lateral septal nucleus, and the medial amygdala (Henriksen et al. 1979, 1982). Since epileptiform activity in the hippocampus was measured at the doses studied, a central role of these structures in the epileptogenic actions of β-endorphin was indicated. In fact, there is evidence that the epileptogenic action of β-endorphin may result from a disinhibitory effect within the microcircuitry of the hippocampus (Zieglgänsberger et al. 1979).

3.5 Influences of the Opioid Antagonist Naloxone on LCGU

Naloxone (NAL) has been proven effective in the treatment of human pathological conditions, e.g., schizophrenia, cerebral ischemia (Cohen et al. 1985; Hosobuchi et al. 1982); in animals, it has been shown to influence various behaviors. It improved spatial memory (Gallagher et al. 1983) and provoked aversive actions in the place preference test (Mucha and Iversen 1984). The data suggest that opioid systems involved in these functions are tonically active. Various studies have used the 2-DG technique to define brain regions and circuits that are possibly under tonic opiodergic control. However, in only one study (Hayashi and Nakamura 1984), were small but significant decreases in LCGU (in structures of the lower brain stem) found following NAL treatment. These occurred primarily in the circulatory and respiratory centers of the medulla oblongata. In the majority of studies, however, no effects were observed after NAL administration (Shigeno et al. 1983; Fanelli et al. 1987; Beck and Krieglstein 1986). This discrepancy may be explained by the use of different doses, times, and routes of drug administration. For example, Shigeno et al. 1983 and Fanelli et al. (1987) both applied NAL intraventricularly 5 or 10 min before the 2-DG, whereas Hayashi and Nakamura (1984) administered the 2-DG immediately after s.c. injection of NAL. Because of its short duration of action, the time of NAL administration would appear to be a critical issue.

4 Conclusion

The various studies on opioidergic influences on local cerebral glucose utilization have provided us with an appreciation of limitations of the particular approach and insights into functional processes which are induced by a particular drug. A *general* statement on the regions and circuits which respond to activation of particular opioid receptors, however, is difficult in view of the diversity of the results obtained to date. This will require further studies employing selective agonists and antagonists.

References

Ableitner A, Herz A (1989a) Limbic brain structures are important sites of κ-opioid receptor mediated actions in the rat. Brain Res 478:326–336

Ableitner A, Herz A (1989b) Functional alterations of activity in the rat brain in response to selective delta opioid receptor activation: a 2-deoxyglucose study. Proc Int Narcotics Res Conf, Quebec, Can, Abstr, p 115

Andy OJ (1980) Parafascicular-center median nuclei stimulation for intractable pain and dyskinesia (painful dyskinesia). Appl Neurophysiol 43:133–144

Andy OJ (1983) Thalamic stimulation for chronic pain. Appl Neurophysiol 46:116–123

Bachelard HS (1971) Specificity and kinetic properties of monosaccharide uptake into guinea pig cerebral cortex in vitro. J Neurochem 18:213–222

Beck T, Krieglstein J (1986) The effects of tifluadom and ketazocine on behaviour, dopamine turnover in the basal ganglia and local cerebral glucose utilization of rats. Brain Res 381:327–335

Bidder TG (1968) Hexose translocation across the blood-brain interface: configurational aspects. J Neurochem 15:867–874

Borgström L, Norberg K, Siesjö BK (1976) Glucose consumption in rat cerebral cortex in normoxia, hypoxia and hypercapnia. Acta Physiol Scand 96:569–574

Brady JV, Nauta WJH (1955) Subcortical mechanisms in emotional behaviour: the duration of affective changes following septal and habenular lesions in the albino rat. J Comp Physiol Psychol 48:412–420

Cannon JT, Prieto GJ, Lee A, Liebeskind JC (1982) Evidence for opioid and non-opioid forms of stimulation produced analgesia in the rat. Brain Res 243:315–321

Clark JA, Itzhak Y, Hruby VJ, Yamamura HI, Pasternak GW (1986) (D-Pen^2,D-Pen^5) enkephalin (DPDPE): a δ-selective enkephalin with low affinity for μ1 opiate binding sites. Eur J Pharmacol 128:303–304

Cohen MR, Pickar D, Cohen RM (1985) High-dose naloxone administration in chronic schizophrenia. Biol Psychiat 20:573–575

Cowan A, Zhu XZ, Porreca F (1985) Studies in vivo with ICI 174 864 and (D-Pen^2,D-Pen^5) enkephalin. Neuropeptides 5:311–314

Cruce JAF (1975) An autoradiographic study of the projections of the mammillothalamic tract in the rat. Brain Res 85:211–219

Czlonkowski A, Millan MJ, Herz A (1987) The selektive κ-agonist U-50, 488H, produces antinociception in the rat via supraspinal action. Eur J Pharmacol 142:183–184

Dafny N, Gildenberg P (1984) Morphine effects on spontaneous, nociceptive, antinociceptive and sensory evoked responses of parafasciculus thalami units in morphine naive and morphine dependant rats. Brain Res 323:11–20

Dafny N, Brown M, Burks TF, Rigor BM (1979) Pattern of unit responses to incremental doses of morphine in central gray, reticular formation, medial thalamus, caudate nucleus, hypothalamus, septum and hippocampus in unanaesthetized rats. Neuropharmacology 18:489–495

DesRosiers MH, Kennedy C, Sakurada O, Shinohara M, Sokoloff L (1978) Effects of hypercapnia on cerebral oxygen and glucose consumption in the conscious rat. Stroke 9:98

Diemer NH, Gjedde A (1983) Autoradiographic determination of brain glucose content and visualization of the regional lumped constant. J CBF Met 3 (Suppl 1):7

Fallon JH, Leslie FM, Cone RI (1985) Dynorphin containing pathways in the substantia nigra and ventral tegmentum: a double labeling study using combined immunofluorescence and retrograde tracing. Neuropeptides 5:457–460

Fanelli RJ, Szikszay M, Jasinsky DR, London ED (1987) Differential effects of μ and κ opioid analgesics on cerebral glucose utilization in the rat. Brain Res 422:257–266

Fanelli RJ, Walovitch RC, Jasinski DR, London ED (1988) Naloxone fails to alter local cerebral glucose utilization in the rat. Pharmacol Biochem Behav 31:481–485

Gallagher M, King RA, Young NB (1983) Opiate antagonists improve spatial memory. Science 221:975–976

Gillan MGC, Jan WQ, Kosterlitz HW, Paterson SJ (1983) A highly selective ligand for the κ-binding site (U-50,488H). Br J Pharmacol 79:275P

Gjedde A (1987) Does deoxyglucose uptake in the brain reflect energy metabolism. Biochem Pharmacol 36:1852–1861

Goldstein JM, Siegel J (1980) Suppression of attack behaviour in cats by stimulation of ventral tegmental area and nucleus accumbens. Brain Res 183:181–192

Goodman RR, Pasternak GW (1985) Visualization of δ-opiate receptors in rat brain by using a computerized autoradiograph substraction technique. Proc Natl Acad Sci USA 82:6667–6671

Goodman RR, Snyder SH, Kuhar MJ, Young III WS (1980) Differentiation of delta and mu opiate receptor localizations by light microscopic autoradiography. Proc Natl Acad Sci USA 77:6239–6243

Graeff FG, Filho NG (1978) Behavioural inhibition induced by electrical stimulation of the median raphe nucleus of the rat. Physiol Behav 21:477–484

Gulya K, Gehlert DR, Wamsley JK, Mosberg H, Hruby VJ, Yamamura HI (1986) Light microscopic autoradiographic localization of delta opioid receptors in the rat brain using a highly selective Bis-penicillamine cyclic enkephalin analog. J Pharmacol Exp Ther 238:720–726

Haffmans J, Blankwater YJ, Ukponmwan OE, Zijlstra JE, Vincent JE, Hespe W, Dzoljic MR (1983) Correlation between the distribution of ^{3}H-labeled enkephalin in rat brain and the anatomical regions involved in enkephalin induced seizures. Neuropharmacology 22:1021–1028

Haffmans J, Dekloef R, Dzoljic MR (1984) Metabolic rate in different rat brain areas during seizures induced by a specific delta opiate receptor agonist. Brain Res 302:111–115

Hand PJ, Greenberg JH, Miselis RR, Weller WL, Reivich M (1978) A normal and altered cortical column: a quantitative and qualitative ^{14}C-2-deoxyglucose (2-DG) mapping study. Soc Neurosci Abstr 4:553

Hawkins RA, Miller AL (1987) Deoxyglucose-6-phosphate stability in vivo and the deoxyglucose method. J Neurochem 49:1941–1949

Hawkins R, Phelps M, Huang SC, Kuhl D (1981) Effect of ischemia upon quantification of local cerebral metabolic rates for glucose with 2-(F-18) fluoro-deoxyglucose (FDG). J CBF Met 1 (Suppl 1):9

Hayashi T, Nakamura K (1984) Localized effects of naloxone on local cerebral glucose utilization in rat cerebral nuclei with met-enkephalinergic neurons. Jpn J Pharmacol 36:339–348

Henriksen SJ, Morrision F, Bloom FE (1979) β-Endorphin induced epileptiform activity increases local cerebral metabolism in hippocampus, amygdala and septum. Soc Neurosci Abstr 5:528

Henriksen SJ, Chouvet G, McGinty J, Bloom F (1982) Opioid peptides in the hippocampus: anatomical and physiological considerations. Ann NY Acad Sci 398:207–220

Herkenham M, Nauta WJH (1977) Afferent connections of the habenular nuclei in the rat. A horseradish perixodase study with a note on the first fiber of passage problem. J Comp Neurol 173:123–146

Herkenham M, Nauta WJH (1979) Efferent connections of the habenular nuclei in the rat. J Comp Neurol 187:19–48

Hers HG (1957) Le métabolisme du fructose. Arscia 102. Arscia, Bruxelles

Hiesinger EM, Voorhies RM, Lipschutz L, Basler G, Shapiro WR, Pasternak GW (1983) The effects of morphine on glucose metabolism (^{14}C-2-deoxyglucose) in rat brain as measured by quantitative autoradiography. Soc Neurosci Abstr 9:138

Holstege J, Dekker JJ (1979) Electron microscopic identification of mammillary terminals in the rats AV thalamic nucleus by means of anterograde transport of HRP. A quantitative comparison with the EM degeneration and EM autoradiographic techniques. Neurosci Lett 11:129–135

Hosobuchi Y, Wenner J (1977) Disulfiram inhibition of development of tolerance to analgesia induced by central gray stimulation in humans. Eur J Pharmacol 43:385–387

Hosobuchi Y, Baskin DS, Woo SK (1982) Reversal of neurological deficits by opiate antagonist naloxone after cerebral ischemi in animals and humans. J CBF Met 2:98–100

Ito M, Suda S, Namba H, Sokoloff L, Kennedy C (1983) Effects of acute morphine administration on local cerebral glucose utilization in rat. J CBF Met 3 (Suppl 1):574–575

Jaffe JH, Martin WR (1985) Opioid analgesics and antagonists. In: Gilman GA, Goodman LS, Rall TW, Murad R (eds) The pharmacological basis of therapeutics, 7th edn. Macmillan, New York, pp 491-531
Kataoka Y, Shibata K, Gomita Y, Ueki S (1982) The mammillary body is a potential site of antianxiety action of benzodiazepines. Brain Res 241:374-377
Kelly PAT, McCulloch J (1981) Errors associated with modifications of the quantitative 2-deoxyglucose technique. J CBF Met 1 (Suppl 1):60-61
Kennedy C, DesRosier MH, Jehle JW, Reivich M, Sharp F, Sokoloff L (1975) Mapping of functional neuronal pathways by autoradiographic survey of local metabolic rate with (^{14}C) deoxyglucose. Science 187:850-853
Kosterlitz HW, Paterson SJ, Robson LE (1981) Characterization of the opiate receptor in the guinea pig brain. Br J Pharmacol 73:939-949
Lahti RA, Mickelson MM, McCall JM, VonVoigtlander PF (1985) (^{3}H)U-69 593. A highly selective ligand for the opioid κ-receptor. Eur J Pharmacol 109:281-284
Magnan J, Paterson SJ, Tavani A, Kosterlitz W (1982) The binding spectrum of narcotic analgesic drugs with different agonist and antagonist properties. Naunyn-Schmiedeberg's Arch Pharmacol 319:197-205
Mansour A, Khachaturian H, Lewis ME, Akil H, Watson SJ (1987) Autoradiographic differentiation of mu, delta and kappa opioid receptors in the rat forebrain and midbrain. J Neurosci 7:2445-2464
Mata M, Fink DJ, Gainer H, Smith CB, Davidsen L, Savaki H, Schwartz W, Sokoloff L (1980) Activity dependent energy metabolism in rat posterior pituitary primarily reflects sodium pump activity. J Neurochem 34:213-215
McCulloch J, Savaki HE, Jehle J, Sokoloff L (1982) Local cerebral glucose utilization in hypothermic and hyperthermic rats. J Neurochem 39:255-258
Meibach RC, Glick SD, Ross DA, Cox RD, Maayani S (1980) Intraperitoneal administration and other modifications of the 2-deoxy-d-glucose technique. Brain Res 195:167-176
Miller AJ, Hawkins RA, Veech RL (1975) Decreased rate of glucose utilization by rat brain in vivo after exposure to atmospheres containing high concentrations of CO_2. J Neurochem 25:553-558
Miyaoka M, Shinohara M, Batipps M, Pettigrew KD, Kennedy C, Sokoloff L (1979) The relationship between the intensity of the stimulus and the metabolic response in the visual system of the rat. Acta Neurol Scand 60 (Suppl 72):16-17
Mogenson GJ, Huang YH (1973) The neurobiology of motivated behavior. Prog Neurobiol 1:53-83
Morris BJ, Herz A (1986) Autoradiographic localization in rat brain of κ-opiate binding sites labeled by ^{3}H-bremazocine. Neuroscience 19:839-846
Mosberg HI, Hurst R, Hruby VJ, Gee K, Yamamura HI, Galligan JJ, Burks TF (1983) Bis-penicillamine enkephalins possess highly improved specificity toward delta opioid receptors. Proc Natl Acad Sci USA 80:5871-5874
Mucha RF, Iversen SD (1984) Reinforcing properties of morphine and naloxone revealed by conditioned place preferences: a procedure examination. Psychopharmacology 82:241-247
Nauta WJH (1985) Hippocampal projections and related neural pathways to the midbrain of the cat. Brain 81:319-340
Nelson T, Kaufmann EE, Sokoloff L (1984) 2-Deoxyglucose incorporation into rat brain glycogen during measurement of local cerebral glucose utilization by the 2-deoxyglucose method. J Neurochem 43:949
Nelson T, Dienel GA, Mori K, Cruz NF, Sokoloff L (1987) Deoxyglucose-6-phosphate stability in vivo and the deoxyglucose method: response to comments of Hawkins and Miller. J Neurochem 49:1949-1960
Oldendorf WM (1971) Brain uptake of radiolabeled amino acids, amines and hexoses after arterial injection. Am J Physiol 221:1629-1638
Oliveras JL, Guilbaud G, Besson JM (1979) A map of serotoninergic structures involved in stimulation produced analgesia in unrestrained freely moving cats. Brain Res 164:317-322
Papez JW (1937) A proposed mechanism of emotion. Arch Neurol Psychiat 38:735-744
Pfeiffer A, Brantl V, Herz A, Emrich H (1986) Psychotomimesis mediated by κ-opiate receptors. Science 233:774-776
Piercy MF, Lahti RA, Schroeder LA, Einspar FJ, Bahrsun C (1982) U-50,488H. A pure kappa receptor antagonist with spinal analgesic loci in the mouse. Life Sci 31:1197-1200
Raisman G, Cowan MW, Powell TPS (1966) An experimental analysis of the efferent projection of the hippocampus. Brain 89:83-108

Richardson DE, Akil H (1977) Pain reduction by electrical brain stimulation in man. II. Chronic self-administration in the periventricular gray matter. J Neurosurg 47:184–194
Römer D, Büscher HH, Hill RC, Maurer R, Petcher TJ, Zeugner H, Benson W, Finner E, Milkowski W, Thies PW (1982a) An opioid benzodiazepine. Nature (London) 298:759–760
Römer D, Büscher HH, Hill RC, Maurer R, Petcher TJ, Zeugner H, Benson W, Finner E, Milkowski W, Thies PW (1982b) Unexpected opioid activity in a known class of drug. Life Sci 31 (Suppl 12/13):1217–1220
Ruhland M, Zeugner H (1983) Effects of the opioid benzodiazepine tifluadom and its optical isomeres on spontaneous locomotor activity of mice. Life Sci 33 (Suppl 1):631–634
Sakurada O, Shinohara M, Klee WA, Kennedy C, Sokoloff L (1976) Local cerebral glucose utilization following acute or chronic morphine administration and withdrawal. Soc Neurosci Abstr 2:613
Sakurada O, Sokoloff L, Jacquet YF (1978) Local cerebral glucose utilization following injection of β-endorphin into periaqueductal gray matter in the rat. Brain Res 153:403–407
Savaki HE, Davidsen L, Smith C, Sokoloff L (1980) Measurement of free glucose turnover in brain. J Neurochem 35:495–502
Savaki HE, MacPherson H, McCulloch J (1982) Alterations in local cerebral glucose utilization during haemorraghic hypotension. Circ Res 50:633–644
Schmidt K, Lucignani G, Mori K, Jay T, Palombo E, Nelson T, Pettigrew K, Holden E, Sokoloff L (1989) Refinement of the kinetic model of the 2-(^{14}C)deoxyglucose method to incorporate effects of intracellular compartmentation in brain. J CBF Met: 290–303
Schmidt WK, Tam SW, Shotzberger GS, Smith Jr DH, Clark R, Vernier VG (1985) Nalbuphine. Drug Alc Depend 14:339–362
Schuier F, Orzi F, Suda S, Kennedy C, Sokoloff L (1981) The lumped constant for the (^{14}C)deoxyglucose method in hyperglycemic rats. J CBF Met 1 (Suppl 1):63
Sherlock DA, Raisman G (1975) A comparison of anterograde and retrograde axonal transport of horseradish peroxidase in the connections of the mammillary nuclei in the rat. Brain Res 85:321–324
Shibata K, Kataoka Y, Yamashita K, Ueki S (1986) An important role of the central amygdaloid nucleus and mammillary body in the mediation of conflict behavior in rats. Brain Res 372:159–162
Shigeno T, Teasdale GM, Kirkham D, Mendelow, Graham DJ, McCulloch J (1983) Effect of naloxone on cerebral glucose metabolism in normal rats and rats with focal cerebral ischemia. J CBF Metab 3:528–529
Shimada M (1981) Glucose uptake in mouse brain regions under hypoxic hypoxia. Neurochem Res 6:993–1003
Siesjö BK (1978) Brain energy metabolism. John Wiley & Sons, New York
Szligi GR, Taylor CJ, Ludens JH (1984) Effects of the highly selective kappa opioid, U-50,488H, on renal function in the anesthetized dog. J Pharmacol Ther 230:641–645
Sols A, Crane RK (1954) Substrate specificity of brain hexokinase. J Biol Chem 210:581–595
Sokoloff L (1981) Localization of functional activity in the central nervous system by measurement of glucose utilization with radioactive deoxyglucose. J CBF Met 1:7–36
Sokoloff L (1989) Basic principles in the imaging of cerebral metabolism in vivo. In: Sharif NA, Lewis ME (eds) Brain imaging techniques and applications. Horwood, Chichester, pp 230–261
Sokoloff L, Reivich M, Kennedy C, DesRosier MH, Patlak CS, Pettigrew KD, Sakurada O, Shinohara M (1977) The ^{14}C-deoxyglucose method for the measurement of local cerebral glucose utilization: theory, procedure and normal values in the conscious and anesthetized albino rat. J Neurochem 28:897–916
Sokoloff L, Kennedy C, Smith CB (1983) Metabolic mapping of functional activity in the central nervous system by measurement of local glucose utilization with radioactive deoxyglucose. In: Björklund A, Hökfeld T (eds) Handbook of chemical neuroanatomy, vol 1: Methods in chemical neuroanatomy. Elsevier, Amsterdam, pp 416–441
Suda S, Shinohara M, Miyaoka M, Kennedy C, Sokoloff L (1981) Local cerebral glucose utilization in hypoglycemia. J CBF Met 1 (Suppl 1):62
Sutherland RJ (1982) The dorsal diencephalic conduction system: a review of the anatomy and functions of the habenular complex. Neurosci Behav Rev 6(1982):1–13
Swajkowski AR, Mayer DJ, Johnson JH (1980) Blockade by naltrexone of analgesia produced by stimulation of the dorsal raphe nucleus. Pharmacol Biochem Behav 15:419–423
Szmuszkovicz J, Von Voigtlander PF (1982) Benzeneacetamide amine: structurally novel non-mu opioids. J Med Chem 25:1125–1126

Vincent SR, Hökfeld T, Christensson I, Terenius L (1982) Dynorphin-immunoreactive neurons in the central nervous system of the rat. Neurosci Lett 33:185–190

Von Voigtlander PF, Lahti RA, Ludens JH (1983) U-50,488H: a selective and structurally novel non-mu (kappa) opioid agonist. J Pharmacol Exp Ther 224:7–12

Yeung JC, Yaksh TL, Rudy TA (1978) Effect on the nociceptive threshold and EEG activity in the rat of morphine injected into the medial thalamus and the periaqueductal gray. Neuropharmacology 17:525–532

Zieglgänsberger W, French ED, Siggins GR, Bloom FE (1979) Opioid peptides may excite hippocampal pyramidal neurons by inhibiting adjacent inhibitory interneurons. Science 205:415–417

Zukins RS, Eghbali M, Olive D, Unterwald EM, Tempel A (1988) Characterization and visualization of rat and guinea pig brain κ opioid receptors: evidence for κ1 and κ2 opioid receptors. Proc Natl Acad Sci USA 85:4061–4065

Section V

Opioid Tolerance and Dependence

CHAPTER 24

Tolerance and Physical Dependence: Physiological Manifestations of Chronic Exposure to Opioids

F.J. Ayesta

Acute administration of an opioid agonist produces a well-characterized constellation of effects (which have been reviewed in previous chapters of this book). When the opioid agonist is administered repeatedly, at a proper dose and frequency, two phenomena appear:

1. The intensity of these effects tends to decrease and an increase of the dose is necessary for them to be manifested again. This *acquired* change in the intensity of effects is called tolerance.
2. Several signs and symptoms appear when agonist concentrations in the organism decrease to a certain extent. This second phenomenon is called *withdrawal syndrome*. Its presence indicates the existence of physical (or *physiological*) dependence.

Chronic stimulation of opioid receptors through agonist administration may also lead to the appearance of psychological dependence, characterized by a subjective craving or a compulsive desire to reexperience the effects of the drug.

1 Tolerance

Tolerance can be described as an adaptation of a biological system to the continued or repeated effect(s) of a drug. In an operative sense, tolerance can be justly defined as a loss of potency of a drug after its repeated administration. In this broad sense, any decrease in the intensity of an effect, independently of how it is produced, will be considered tolerance. In the opioid field, the concept of tolerance is usually equated with the phenomenon termed "chronic tolerance", i.e., tolerance which (1) follows the repeated administration of an agonist, (2) is frequently associated with signs of physical dependence, and (3) in which an increase of the dose usually induces the original effects. Nevertheless, organisms have several mechanisms by which a decrease in the intensity of an effect can be obtained.

The development of chronic tolerance is characteristic of all opioids with agonist activity, regardless of the receptor type with which they interact. Tolerance to opioids first becomes evident as a shortening of the duration of drug action and is followed by a diminution of the peak effect. The rate at which tolerance develops depends on the pattern of use and on the characteristics of the drug used. Significant tolerance only develops when there is a more or less continuous drug action. Any prolonged drug-free interval results in a reduction of the degree of tolerance.

When chronic administration of one agonist confers tolerance to another, the term *cross-tolerance* is used. Cross-tolerance is usually observed between opioids acting upon the same receptor type. Given the lack of selectivity of most opioids, there is a considerable variability in the degree of cross-tolerance displayed by different agonists; this is especially true when high doses are tested. Nevertheless, the presence of cross-tolerance does not necessarily mean that both drugs act on the same receptor; rather, it suggests that one or more of the mechanisms which lead to the response measured are shared by both drugs. In this way, cross-tolerance to certain effects of opioids and other drugs may occur (see, e.g., Khanna et al. 1979; Hine 1985; McKearney 1985).

Tolerance develops at different rates, and to different degrees, for each opioid effect. As Cox (1978) has pointed out, it seems likely that the characteristics of tolerance resulting from prolonged exposure to opioid agonists are dependent, at least in part, on the properties of the neural substrate as well as on the subclass of opioid receptor which mediates a particular effect. Additionally, the effects of agonists (whether assessed in chronically treated or naive rats) are typically evaluated with respect to a particular function or response. Consequently, the development of tolerance is influenced by the anatomical, biochemical, and physiological substrates which underlie a given response. These substrates also determine the factors which may influence the development of tolerance.

The complexities of this phenomenon may best be understood by examining several characteristic effects of opioids. Opioids are used both for clinical and recreational purposes. Their main clinical use is in analgesia; their recreational use is linked to their subjective effects. In both cases, a limiting factor in their utility is their effects on respiratory function. Indeed, it is the respiratory-depressant effect of opioids which is responsible for most of the deaths which occur after morphine or heroin overdosage.

2 Analgesia: "Pharmacological" vs "Behavioral" Tolerance

Tolerance develops relatively rapidly to the analgesic effects of opioids. This tolerance is not due to pharmacokinetic or dispositional mechanisms (i.e., a decreased bioavailability of the agonist), since the concentrations of opioid agonists measured in cerebrospinal fluid or in plasma do not differ between tolerant and previously drug-naive animals (Goldstein et al. 1974).

The degree of tolerance developed (or displayed) to a particular agonist can be modified by behavioral factors. In fact, the repeated administration of a drug in a given stimulus context may induce the development of a conditioned response. This response occurs in anticipation of, and can interact with, the regular "unconditioned" effect of the drug. When the conditioned response counteracts, or is opposed to, the "unconditioned" effect, a decrease in the intensity of the latter (i.e., tolerance) is observed. Adams et al. (1969) were the first to note that the presence or absence of distinct environmental cues could markedly influence the development of tolerance to opioids. These investigators used the "hot-plate" procedure, in which pain sensitivity of rats was determined by measuring the latency to lick a paw when placed on a warm surface. Subjects were found to be more tolerant to the analgesic effects of morphine when they had been exposed to the test apparatus (i.e., hot plate) on each of the occasions that the drug was administered (even if the nociceptive stimulation was not applied until the last occasion) than if they were only introduced to the apparatus on the last occasion that the drug was administered. The authors also found that hot-plate exposure produced enhanced tolerance only when exposure was contiguous with drug delivery.

This phenomenon has been termed *environmentally specific* (or *associated*) *tolerance*. A source of complication in the interpretation of this phenomenon is that the acute effects of an agonist can also be associated with environmental stimuli and, consequently, can be elicited in the absence of the agonist (Grabowsky and Cherek 1983). For this reason, a challenge for any theory of tolerance is to specify when such conditioned responses should occur and, if occurring, whether they are compensatory to the unconditioned response or whether they mimic the measured unconditioned response (Paletta and Wagner 1986). Although there is still controversy about how this form of tolerance is most efficiently explained, and different mechanisms have been proposed (see, e.g., Siegel 1983; Baker and Tiffany 1985), the basic phenomenon is a reliable empirical finding which in some circumstances can play an important role in determining the pharmacological effects of a drug (Goudie and Griffiths 1986; Siegel 1988).

Conditioned responses usually occur as a result of the repeated injection of an agonist. An additional source of conditioned responses which comes into play in the laboratory assessment of drug-induced analgesia is the fact that repeated testing (and, consequently, repeated application of a painful stimulus) is necessary to characterize the time course of a particular drug or dose. This time course may be different in naive and in tolerant animals. The influence of repeated testing upon the development of tolerance was first shown by Kayan et al. (1969) using the hot-plate test. They divided rats into two groups: tested and nontested. The latter group consisted of animals which, although receiving drug injections, were not tested until the final day of the experiment. Morphine, given at different interdose intervals for varying periods of time, always produced a lesser effect in the tested group than in the nontested group. The difference between the groups was most striking when morphine was given once a week for 5 weeks. Indeed, in this case, tolerance was not observed in the nontested group, whereas it was seen in the tested group. This influence of repeated testing on opioid-induced analgesia has been also observed with other algesiometric tests (Advokat 1981).

The environmental context per se cannot account for all instances of morphine tolerance (Baker and Tiffany 1985). In fact, tolerance develops when chronic continuous administration systems (e.g., pellets or pumps) are employed independently of whether the animals have been previously tested (see, e.g., Way et al. 1969). This phenomenon is what most usually comes to mind when discussing concepts of tolerance or when defining "*pharmacodynamic tolerance*" and is referred to as *nonenvironmentally specific* (or *nonassociative*) *tolerance*. Sometimes (in opposition to behavioral tolerance) it is termed pharmacological tolerance, but the distinction of pharmacological vs behavioral is quite misleading, since animals may learn to utilize certain drug effects as conditioned stimuli for the acquisition of associative tolerance (Walter and Riccio 1983) and because the basic processes underlying learned behaviors and other acquired adaptations of the organism may have much in common (Balster 1985).

With an adequate experimental design, especially varying the interdose intervals, both components of tolerance (environmentally and nonenvironmentally specific) can be separated (Dafters et al. 1988; Tiffany and Maude-Griffin 1988). Similarly, different profiles of cross-tolerance can be found for one and the other, since environmental factors can sometimes influence the development of tolerance to drugs which do not directly interact with opioid receptors. Indeed, Jorgensen et al. (1986), analyzing the development of tolerance to the inhibitory effect of ethanol on the tail-flick reflex in the spinal rat, have reported that learned tolerance to ethanol (due to testing after agonist administration) caused cross-tolerance to morphine and clonidine, whereas pharmacologically induced tolerance (mere ethanol exposure without any testing after injections) did not have the same effect.

Although an adequate experimental design can allow for the independent assessment of either "behavioral" or "pharmacological" tolerance, in most cases, the procedures employed do not make this possible. Therefore, the possible influence of both learning and environmental factors must be considered when interpreting in vivo studies.

3 Acute and Functional Tolerance

3.1 Acute Tolerance

As applied to tolerance, the term *acute* has at least two possible meanings. One concept of acute tolerance is used to indicate changes in sensitivity to a drug *within* the duration of one continuous drug exposure and can be referred to as *intrasessional adaptation*. With central nervous system depressants, it is usually found, that at the same concentration of agonist, behavioral impairment is higher when drug concentrations are rising than when they are falling (Kalant et al. 1971). In the case of opioids, it is observed that the analgesic effect of a number of opioids declines rapidly in spite of the fact that a constant concentration was maintained by means of continuous drug infusion (Cox et al. 1968). A second concept of acute tolerance refers to that tolerance which occurs *after* one (or a few) exposure(s) to the drug and can be denoted as *intersessional adaptation*. The term *tachyphylaxis* is sometimes used as a synonym of acute tolerance but, although acute tolerance may sometimes be produced through a tachyphylactic mechanism (i.e., a mechanism by which the capacity to respond is abolished), acute tolerance does not always seem to be related to tachyphylaxis and different mechanisms may be responsible for the development of acute tolerance.

The relationship between acute and chronic opioid tolerance is not well understood. This is primarily because it appears that the adaptational processes that eventually produce chronic tolerance begin immediately after the first dose, and because the distinction between acute (short-term) and chronic (long-term) tolerance has often been based more on the experimental design, dose and schedule of administration, and latency of its development, rather than on any clear empirical separation (Hug 1972; Rosenfeld and Burks 1977). Consequently, it is still not clear if (or when) the mechanisms underlying acute tolerance are similar to (or the initial steps of) those underlying chronic tolerance (Hovav and Weinstock 1987).

Chronic tolerance develops to the respiratory actions of opioids (Martin and Sloan 1977). Acute tolerance also develops to these effects (Pentiah et al. 1966). Assessment of tolerance, as it occurs with regard to respiratory function, may offer some advantages to measuring other opioid actions because a spontaneous and on-going activity, rather than a response to an elicited stimulus, is measured. Consequently, repeated measures can be performed and the measured variable is not dependent upon learning or a conditioned behavioral response of the animal.

Martin and Sloan (1977) reported that when the onset of action of an opioid agonist is very rapid, a profound transient depression of respiration is seen; this reaches a peak in minutes and is followed (in minutes) by a partial recovery. According to the authors, the depression is associated with a rapid lowering of the set point of the respiratory homeostat, and the partial recovery or acute tolerance is associated with the accumulation of CO_2 to a level sufficient to once again drive the depressed respiratory center. Disappearance of the effects within minutes, despite continuous infusion, has also been reported for other opioid actions (see, e.g., Sander and Giles 1984).

Acute tolerance (in the sense of intersessional adaptation) to the respiratory-depressant effects of a μ-agonist was produced in decerebrate cats by injecting six similar doses of morphine at 1-h intervals (Flórez et al. 1972). In that experiment, the first dose of morphine decreased respiratory frequency and tidal volume. The second and third doses produced a further depression of tidal volume, but not of respiratory frequency, whereas the three subsequent doses of morphine did not induce any further depression of respiration. Due to the long half-life of morphine, basal values following drug injections did not return to preinjection values before the other doses were administered. These results are similar, in part, to those of Martin et al. (1968) regarding chronic tolerance. In the latter study, although basal values were lower than control values (which probably means that complete tolerance had not developed), large doses of morphine produced either no depression, or a modest depression of the sensitivity of the respiratory center to CO_2.

Those factors affecting the development of acute tolerance to the respiratory effects of opioids have been studied by Hovav and Weinstock (1987). These authors found that acute tolerance occurred to the elevation in the arterial partial pressure of CO_2 produced by a second dose of morphine given 4-h after the first, only when high doses were employed. The 4-h interval was chosen in order to obtain a return of basal respiratory values to control values. When the duration of action of the agonist was limited to 75 min by injection of naloxone or by the use of the short-acting opioid fentanyl, tolerance to a second dose of drug still developed, provided it was given at least 4-h later. The authors concluded that the development of acute tolerance to opioids depends on the degree of the initial drug effect and the time of receptor occupancy. The latter must be longer than that required to attain a peak response, since tolerance did not occur when naloxone was given at the peak of the morphine effect (30 min). According to the authors, these findings suggest that agonists must elicit a response of a certain magnitude (i.e., about 60–70% of the maximum) and for a certain time in order for the processes involved in the development of acute tolerance to be initiated, i.e., when a functional load is imposed upon the organism.

3.2 *Functional Tolerance*

The concept that the degree of tolerance which develops to an agonist may be proportional to (or dependent on) the functional impairment that it causes has long been proposed not only for acute, but also (and primarily) for chronic tolerance. Several results regarding cross-tolerance between agonists of different classes have been explained in this way (Kalant et al. 1971). This phenomenon has been most frequently reported with central nervous system depressants. Okamoto et al. (1978) support the view that only those systems that have been challenged or altered by a barbiturate display tolerance to its effects. For this reason, it is usually assumed that the response to a drug, and hence the development of tolerance, are profoundly affected by the functional state of the organism during the period of drug exposure (Edwards et al. 1981). It seems that a neuronal system which is required to function under the influence of an agonist adapts better than when it is exposed to this agonist without a requirement for functioning (Kalant 1985). Although these conclusions relate to central depressants, it is reasonable to assume that they may also apply to functional tolerance to opioids.

4 Tolerance to Opioid Effects with Biphasic Patterns

The development of tolerance to the actions of opioids which, in naive animals, occur in a biphasic pattern has distinctive features. For example, in the rat low to moderate doses of morphine produce an initial stimulation of activity, lasting 1 or 2 h, followed by a return to normal activity levels. Higher doses produce biphasic effects: an initial depression of activity is followed by a period of hyperexcitability. The duration of the initial depressant effect increases as a function of the dose of morphine. Chronic morphine treatment markedly alters the effects of morphine on locomotor activity. Tolerance develops rapidly to the depressant action of high morphine doses: the initial depression of activity is reduced and the later increase in activity is enhanced and shifted to the left in the time-effect curve. Tolerance develops more slowly to the stimulant effect of low doses of morphine and is evident only after chronic treatment with very large daily doses (Babini and Davis 1972; Vasko and Domino 1978). Despite these important differences in the characteristics of tolerance development, the stimulant and depressant effect of morphine on locomotor activity are readily antagonized by naloxone, suggesting that both actions are mediated by opioid receptors (Brady and Holtzman 1981).

Although influenced by a variety of factors (e.g., ambient temperature), the effects of acutely administered opioids on thermoregulation present a similar biphasic pattern of action. With repeated administration, tolerance readily develops to the hypothermic effect, which leads to an increase in the hyperthermic action

(Rosow et al. 1982). Nevertheless, when high doses of a drug are employed over sustained periods, tolerance to hyperthermia can also be observed (Mucha et al. 1987; Adler et al. 1988).

It is still not clear how an increase in the stimulatory effects of opioids, in the presence of tolerance to the depressant actions, should be interpreted. The enhancement of the stimulant action might merely be the consequence of a decrease in the depressant action and represent an overshoot, due to the development of tolerance to the opposite action, and the absence of a counteracting mechanism (Vasko and Domino 1978). An added source of difficulty in the interpretation of results regarding biphasic effects of opioids is the influence that pharmacokinetic factors may play in the later-appearing actions. Each opioid effect has a different temporal course and these temporal courses vary with the development of tolerance. Depending on the rapidity of the onset of these effects, which mainly depends on the lipophilicity of the agonist used and on the route of administration employed, not only quantitative but also apparent qualitative differences can be found between naive and tolerant animals (Adler et al. 1984, 1988).

5 Tolerance to Reinforcement and Subjective Effects of Opioids

In operant terms, reinforcement is said to occur when the presentation of a stimulus (e.g., drug) increases the probability or frequency of the behavior that presentation of the stimulus is contingent upon. Positive reinforcement refers to a situation where the presentation of a stimulus increases the frequency of a behavior, and negative reinforcement refers to the cases where the removal of some stimulus (usually aversive) results in the increase of some behavior. Punishment, on the other hand, represents a very different situation in which the presentation of a stimulus suppresses the behavior to which it is associated (Bozarth 1987). The reinforcing effects of a drug, therefore, determine its likelihood of being self-administered and are a result of its reinforcing (both positive and negative) and aversive properties. This potential is not an absolute value, and it changes under different environmental conditions or schedules of administration. Whether the reinforcing properties of an agonist (strictly speaking, the excess of reinforcing over aversive properties) are more a consequence of positive reinforcement (e.g., improvement of mood) or of negative reinforcement (e.g., relief of hypophoria or of withdrawal), is an empirical question and an issue that is not resolved by traditional methods used to assess reinforcement processes (Henningfield et al. 1987).

A variety of experimental paradigms can be used as measures of positive reinforcement in humans and laboratory animals. In some of these methodologies the development of tolerance has been a controversial issue. In this section, the evidence regarding the development of tolerance, according to the different methodologies, will be briefly discussed.

5.1 Self-Administration Studies

In these techniques administration of a drug is contingent upon the performance of a specific behavioral task (e.g., lever pressing), and the ability of the drug injection to directly reinforce behavior is determined. A body of evidence has shown that humans and animals allowed only limited access to opioids will maintain a stable level and pattern of opioid intake, which when interrupted, reveals neither tolerance nor physical dependence as measured by increases in drug intake or gross signs of physical abstinence. In contrast, chronic unlimited access to opioids invariably produces an increase of intake and a severe withdrawal syndrome (Koob and Bloom 1988).

Even though self-administration is the most direct and best measure of a drug's reinforcing properties, the question of whether tolerance develops to the opioid action(s) responsible for their reinforcing properties is difficult to assess directly because drug self-administration does not necessarily reflect a specific or a direct drug effect, but is rather a complex behavior that requires a risk-benefit assessment by the subject. An index, like the reinforcement/toxicity ratio (Brady and Griffiths 1983), which compares the relative potency of a drug as a reinforcer with its relative potency in eliciting disruptive sensory/motor effects, may be of use to further examine this issue. This ratio is different for each drug and may change with repeated administration when tolerance develops to some effects (that may be aversive), or when withdrawal symptomatology appears.

Tolerance seems to develop to the "impact" of the drug or to the effects responsible for their primary reinforcing potential. The increases observed in response rates support this view, although increments in the rate of responding should be interpreted cautiously (Roberts and Zito 1987; Wise 1987) and the presence of withdrawal symptomatology (a factor which will be discussed in the next section), as in the case of brain reward stimulation, may be the major factor responsible for such increases.

5.2 Brain Stimulation Reward

This method involves training animals to work for electrical brain stimulation and determining the effects of drugs on brain stimulation reward. Most addictive drugs enhance or facilitate brain stimulation reward; this facilitatory effect can be demonstrated by an increased rate of lever pressing for fixed intensity brain stimulation, or by a lowering of current thresholds for brain stimulation. Opioids produce biphasic alterations of intracranial self-stimulation responses (depression and facilitation of response, or increase and lowering of thresholds) of both a dose-dependent and time-dependent nature (Weibel and Wolf 1979).

After chronic administration of an opioid, there is a clear tolerance to the initial depression of response. Parallel to the decrease in the initial depression, the period of facilitation moves forward closer and closer to the time of injections. Under these conditions, facilitation does not diminish much, if at all, with repeated doses (Lorens and Mitchell 1973; Bush et al. 1976). In fact, peak effects may become larger with repeated doses. For these reasons, it is generally believed that tolerance does not develop to the facilitatory effects of opioid in brain stimulation reward (Esposito and Kornetsky 1978; Reid 1987). Nevertheless, it seems that tolerance (although difficult to attain as in the case of opioid hyperthermic actions) can, under some conditions, develop to this effect. In this regard, tolerance, measured by the responding rate, has been reported (Glick and Rapaport 1974); during the withdrawal state the reinforcement threshold for brain stimulation is elevated (Schaefer and Michael 1986) and lever pressing is diminished (Bush et al. 1976). Furthermore, if the dose-response curve for the threshold-lowering action of opioids were an U-shaped one (Kornetsky and Bain 1983), some of the results of Esposito and Kornetsky (1977) could be considered to reflect signs of tolerance.

5.3 Euphoric and Drug-Liking Effects

Different questionnaires have been designed to evaluate the subjective effects of drugs in humans. In reporting the subjective effects of addictive drugs in humans, special attention has been directed toward identifying response items (e.g., euphoria, drug liking) that correctly classify highly addictive drugs, since the euphorigenic (pleasure-producing) properties of these compounds are almost certainly related to the positive reinforcing action of self-administered drugs (Watson et al. 1989). Clinical evidence shows that tolerance develops to the euphoric effects of opioids. These effects, although of shorter duration, can, however, again be elicited by increasing the dose. Thus, McAuliffe and Gordon (1974) reported that physically dependent street users of heroin still report a sustained euphoria (or "high") associated with the intravenous injection of heroin and Mirin et al. (1976) showed that tolerance quickly develops to the euphorigenic effects as a function of chronic usage, although single injections remain capable of producing brief periods (30 to 60 min) of positive mood. Chronic administration of opioids not only leads to tolerance to their euphoric effects, but also to an exacerbation of hypophoric feelings (Martin and Jasinski 1977).

Using the place preference conditioning technique, Shippenberg et al. (1988) have shown that tolerance develops to the rewarding effects of opioids as well as to their aversive properties.

6 Tolerance Measured in the Presence or Absence of Withdrawal

When chronic tolerance is induced by means of repeated injections of agonists, it is assessed in the drug-free intervals in which a more or less intense withdrawal syndrome is present. With chronic infusion systems (like pellets or pumps) tolerance has usually also been analyzed in the presence of withdrawal symptomatology. Under such conditions, decreases in the intensity of opioid effects are usually found. However, it seems that tolerance presents different characteristics depending on whether the effects are analyzed in the absence or presence of withdrawal symptomatology (i.e., following pellet or pump removal). This has been reported at least for opioid-induced analgesia (Lange et al. 1980a,b, 1983; Paktor and Vaught 1984), respiratory depression (Roerig et al. 1987; Ayesta and Flórez 1989, 1990), and discriminative stimulus properties of opioids (Emmet-Oglesby et al. 1989). Withdrawal of the agonist also influences the results of in vitro assays of tolerance (Cox and Weinstock 1966).

All these latter cited studies indicate that when withdrawal is not allowed to occur (i.e., testing is done in the presence of the pellets or pumps), animals do not seem to be hyposensitive to opioid agonist effects. Specifically, when challenged with highly lipophilic drugs, or when routes of administration with a fast onset of action are employed, the potency of the agonists may not be decreased (they might even be increased). This happens despite the fact that tolerance has developed as indicated by: (1) the return to control levels despite the continuous presence of an agonist, of tail-flick latencies, basal respiratory values, or the capacity of stimulus detection; (2) the displacement to the right of the dose-response curves (also in the absence of withdrawal) when more hydrophilic agonists are challenged, or when routes of administration with a slower onset of action are employed; (3) the results obtained in withdrawal where, independently of the agonist and the route through which it is administered, rightward displacements of the dose-response curves can be found (Lange et al. 1980a,b, 1983; Paktor and Vaught 1984; Roerig et al. 1987; Emmet-Oglesby et al. 1989; Ayesta and Flórez 1989, 1990).

No definitive explanation has been found for these results. Nevertheless, it seems that, under some conditions, a system may display a lower degree of tolerance than the real degree of tolerance developed. The conditions which seem to be important include: (1) the avoidance of withdrawal (see above); (2) the measurement of the effects at their peak, which in tolerant animals occurs earlier than in naive animals; when these are measured in the recovery phase, a rightward shift and flattening of the dose-response curves are found (see, e.g., Bläsig et al. 1979; Mucha and Kalant 1980); and (3) the use of highly lipophilic drugs. In fact, it is not an unusual finding that the magnitude of the tolerance displayed to an agonist is inversely related to its potency (see, e.g., Porreca et al. 1982; Petersen and Fujimoto 1983; Sivam and Ho 1984; Roerig et al. 1985; Brase 1986; Stevens and Yaksh 1989), as well as to the type of receptors it occupies.

Testing during withdrawal may confound the phenomenon of tolerance with that of withdrawal-induced changes in sensitivity to opioids (Lange et al. 1983; Emmet-Oglesby et al. 1989). The above cited results leave the important question (with its methodological implications) of whether or not the mechanisms underlying the tolerance observed in the presence or absence of withdrawal are the same, and to what extent they are common.

7 Physical Dependence

Physical dependence is a state in which exposure to a drug requires the continued presence of that drug to maintain normal functions; i.e., discontinuation of the drug results in objective (patho)physiological signs of withdrawal or abstinence. The withdrawal syndrome, which can be precipitated by the application of an antagonist, can be terminated abruptly and dramatically by readministering the drug.

The ability of one drug to maintain the physically dependent state, and to suppress the manifestations of physical dependence produced by another, is referred to as cross-dependence. In vivo and in vitro methodologies have produced apparently contradictory results on cross-dependence between μ- and κ-agonists (see Gmerek 1988; Schulz 1988). In general, opioids show cross-dependence with other opioid agonists that act on the same receptor type, but they are not able to suppress the withdrawal syndrome caused by alcohol or other central nervous system depressants. Opioids also produce a *protracted* abstinence syndrome, which is not as well defined as the acute one, but which lasts for weeks or months. This syndrome can be relieved by agonist administration, a factor which is frequently responsible for relapses in addicts (Jaffe 1985).

The development of physical dependence is mainly influenced by the same factors that influence the development of chronic tolerance. The most important of these are: (1) the degree to which opioid receptors are activated, which depends mainly on the intrinsic activity of the agonist; and (2) the continuity of this activation, which depends mainly on the pharmacokinetic properties of the agonist employed. Thus, a suitable dose/schedule is needed to develop tolerance and/or physical dependence. The appearance of the withdrawal syndrome seems to be related to the rate of displacement of the agonist from the receptor: the longer the duration of action of an agonist, the less severe but the more protracted is the resulting withdrawal syndrome. The withdrawal syndrome is most severe when the agonist is displaced from its receptors by the application of an antagonist (*precipitated withdrawal*). Deprivation-induced and precipitated withdrawal differ primarily in their time courses and it seems that the mechanisms underlying the manifestation of both are similar (Gmerek 1988). The processes involved in the development of physical dependence seem to be activated from the time of the first injection of an opioid agonist. Indeed, withdrawal symptoms can be precipitated in humans within 45 min of a single injection of morphine (Heishmann et al. 1989). The appearance of withdrawal symptomatology after just one (or a few) injections of an agonist is termed *acute dependence*.

The withdrawal symptoms associated with opioids are partially characterized by rebound effects in those same physiological systems that were modified initially by the drug (e.g., hyperpnea, mydriasis, diarrhea, hyperalgesia, hyperreflexia). For this reason, it is referred to as a rebound hyperexcitability. However, not all the complex patterns of signs and symptoms seen during withdrawal appear to be rebound effects and the withdrawal syndrome also reflects several neurovegetative signs and symptoms, mainly of noradrenergic character (Jaffe 1985). An accurate and objective quantification of dependence, as manifested by withdrawal, is difficult, especially in higher species since the presence and intensity of a sign can be unrelated to the degree of dependence developed (the intensity of the "overshooting" seems to depend more on the characteristics of the system than on the degree of dependence) and because not all signs of physical dependence evolve or are manifested in parallel (they have different neuronal substrates and follow a characteristic temporal pattern) (Bläsig and Herz 1977). In humans, this is even more complicated because the assessment of severity, as decided by an observer using a rating scale of withdrawal signs, may not correlate with the patient's subjective opinion of severity (Turkington and Drummond 1989).

7.1 Relationship Between Tolerance and Physical Dependence

Tolerance is a very general phenomenon which involves, as we have seen, several independent mechanisms, some of which do not seem to be related to physical dependence. Nevertheless, as Goldstein (1989) has pointed out, by its own definition, physical dependence is always accompanied by some degree of tolerance. In fact, clinical and experimental evidence shows that both phenomena develop in a roughly parallel fashion and that similar factors influence their development and intensity. It therefore seems that chronic tolerance and physical dependence are somehow, and to some extent, related (Collier 1984), though this relationship is still far from being completely understood (Edwards et al. 1981). Whether the mechanisms underlying both phenomena are the same or not is still unresolved because, though chronic tolerance and physical dependence are usually linked, in some instances they appear to be separable processes with distinct spatial locations and with unique molecular mechanisms of action (Koob and Bloom 1988; see Schulz, this Vol.). At least a partial separability of tolerance and dependence can be expected since, even if the underlying mechanisms responsible for their initiation are the same, the neural substrates mediating their subsequent expression might be quite different. This is especially important in the case of some simplified responses which are considered signs of withdrawal since, physical dependence is only displayed by intact nervous circuits, and different intra- or extracellularly localized mechanisms may be involved in the signal transmission of an opioid-affected nervous pathway (Wüster et al. 1985).

7.2 Physical and Psychological Dependence

The distinction between physical (or physiological) and psychological dependence has proved useful in identifying the role of positive reinforcement processes in mediating the initiation and maintenance of drug-seeking behavior. However, the distinction between both causes confusion (Jaffe 1985) and not all authors agree with the employment of this terminology (see, e.g., Edwards et al. 1981). This is primarily due to the fact that psychological dependence as well as the concepts of "need" and "craving" to which it is related are imprecise terms which are difficult to quantitate (Halbach 1973), and to the fact that the physiological mechanisms responsible for the psychic symptoms of the withdrawal syndrome and of psychological dependence or drug-seeking behavior are not very well known.

The opioid withdrawal syndrome may act as a negative reinforcer and, consequently, maintain drug-seeking behavior (Yokel 1987). The aversive emotions triggered by the withdrawal syndrome apparently contribute more to negative reinforcement than do the physical symptoms (Watson et al. 1989). Nevertheless, the presence of some psychic withdrawal symptoms, which most authors (see, e.g., Haefely 1986; Wise 1988; Goldstein 1989) do not usually relate to psychological dependence, could play a role in initiating repetitive and compulsive drug use after the initial self-administration. This is based on the fact that a small but measurable degree of physical dependence is instituted even by the first drug dose, so that a withdrawal syndrome (however mild) follows as that initial dose wears off. Additional evidence in support of this hypothesis comes from the fact that both animals and humans respond, predictably, to a withdrawal syndrome with intense drug-seeking behavior (Goldstein 1989).

The question of whether the immediate and long-term effects of exogenous opioids on the brain, and the cellular and behavioral changes that result from withdrawal and that have motivational relevance to drug-seeking behavior, involve the same (Koob and Bloom 1988) or different (Wise 1988) neural circuits as those that participate in the positive-reinforcing effects associated with short-term use of opioids (prior to physical dependence) remains open. It will only be answered when the anatomical circuits and molecular mechanisms underlying each response are better identified.

8 General Physiological Mechanisms Underlying Tolerance and Physical Dependence

8.1 The Homeostatic Theory

It has not yet been possible to identify a primary site in the initiation of the adaptative changes that chronic administration of opioid agonists produces. Similarly, no satisfactory explanation has been found to account for the phenomena of tolerance and/or dependence. The only theory that is commonly accepted, at least as a general framework, is the homeostatic theory. Himmelsbach (1943) postulated that opioids disturb homeostasis and that the organism tries to regain this homeostasis by some adaptative processes: in the presence of an agonist, these processes counteract its actions. In the absence of an agonist, however, these homeostatic mechanisms give rise to the withdrawal syndrome. Most theories about the mechanisms of tolerance and/or dependence (see, e.g., Goldstein and Goldstein 1968; Martin 1968; Jaffe and Sharpless 1968), in fact, postulate some kind of counteradaptation. Indeed, Himmelsbach's theory is in agreement with much of the experimental data, particularly with respect to time course and dose relations, and it is probable that the body has several lines of defense against the perturbations induced by a drug (Goldstein 1979). As Jaffe (1985) has pointed out, in view of our knowledge of negative feedback control of the activities of regulatory molecules, it would be surprising if some form of central nervous system counteradaptation to the agonistic actions of the drugs did not occur. Furthermore, in view of the complex and long-lasting changes produced by the chronic administration of an opioid agonist, the existence of more than one homeostatic mechanism through which the organism adapts to chronic opioid administration would not be surprising and "appears necessary" (Herz and Bläsig 1978).

In the same way that functional alterations can lead to adaptive processes which produce a decrease in the intensity of effects, chronic stimulation of an opioid receptor can also lead to different adaptive mechanisms, e.g., at the level of adenylate cyclase (Sharma et al. 1975), or of receptor number (Blanchard and Chang 1988). Some, but not all, of these mechanisms may be related to the processes responsible for the development of chronic tolerance and/or physical dependence. The mechanisms activated by chronic opioid agonist administration do not necessarily act synergistically; in fact, not only different, but also opposite changes can be induced, depending on the different variables or physiological systems measured. It was proposed some years ago that a molecule with antagonist properties might be produced in the tolerant/dependent state; more recently it has been suggested that this substance might be an endogenous opioid ligand (Smith et al. 1988). Thus, upregulation of some opioid receptors might be expected and, consequently, despite the fact that chronic stimulation of a receptor (at least for some agonists) leads to downregulation, both down- and upregulation could be found.

8.2 Tolerance and Physical Dependence as Latent Hyperexcitability

An enhanced neuronal excitability has been implied as the primary change underlying the tolerant/dependent state (North and Karras 1978). Counteradaptation results in the development of a "latent hyperexcitability" in the neural systems affected by the drugs (Wikler 1972), which becomes manifest in the form of rebound or overshoot phenomena when drug administration is stopped, or when an antagonist is administered (Jaffe 1985). The hyperexcitability responsible for the withdrawal symptoms may be already present before the agonist is withdrawn and may contribute to the tolerance observed.

Despite the fact that in the tolerant state there is a decrease in the responses analyzed, this does not necessarily mean that there is a decreased responsiveness. Tolerance might, in fact, be an increased capacity to counteract a stimulus and, consequently, to reduce effects produced by the stimulus. The idea of tolerance as an increased compensatory mechanism underlies some of the theories which have tried to explain the environmentally specific tolerance (see, e.g., Wagner 1981; Siegel 1983); functional compensatory factors may play a role in the earlier appearance of the secondary effects in those functions upon which opioids have biphasic effects. Similarly, the flattening of the dose-response curves that can usually be observed in tolerant animals (see, e.g., Bläsig et al. 1979; Mucha and Kalant 1980, 1981), has been interpreted as indicating a role for functional compensation in the expression of tolerance (Ayesta and Flórez 1989).

According to the above, it is argued that although a state of hyposensitivity of receptors produces a rightward displacement of the dose-response curve, not every shift to the right necessarily indicates hyposensitivity. This is best illustrated in animals which have developed pharmacokinetic tolerance: in these, there is a decrease in effect in the absence of receptor hyposensitivity (Goldstein et al. 1974). Likewise, although a drug may continue to interact with its receptors in the usual way, the magnitude of its biological effects may be increasingly antagonized by homeostatic mechanisms; the latter may occur at the level of the receptor or intracellularly. The idea that, under some circumstances, tolerant animals may not be hyposensitive to opioid agonists, but normosensitive with an increased capacity to counteract agonist stimulation, was proposed by Colpaert (1978). It concords with the view of Martin (1984) that, in the opioid-tolerant and -dependent animal, opioids continue to exert their full agonistic activity, making it unnecessary to postulate an inactive or desensitized form of the opioid-receptor complex. Results from the previously cited articles (Lange et al. 1980a,b, 1983; Paktor and Vaught 1984; Roerig et al. 1987; Emmet-Oglesby et al. 1989; Ayesta and Flórez 1989, 1990), while referring to the characteristics of tolerance in the presence and absence of the withdrawal syndrome, support this interpretation; thus, in the tolerant state (at least when withdrawal is avoided) opioid agonists may still exert their full agonist activity, and the animals do not appear to be hyposensitive to opioids (see Ayesta and Flórez 1989, 1990). The reports of Rothman et al. (1986, 1989) that animals which have developed tolerance to an opioid agonist exhibit opioid receptor upregulation (a phenomenon usually associated with an increase in sensitivity to opioids: Zukin and Tempel 1986), also agree with this interpretation.

It should be emphasized that, despite the above arguments with regard to many in vivo experiments, tolerance may, in some instances, be based upon cellular hyposensitivity or desensitization. For example, hyposensitivity or desensitization may indeed account for many in vitro observations (see Schulz, this Vol.; Louie and Way, this Vol.), particularly when high doses of agonists have been used. This leaves open the questions as to the extent of similarity between the phenomena observed using the different models (whole animals, isolated organs or cells), and the validity of the general extrapolation of results obtained using different methodologies.

As in other fields of science, many unresolved questions remain in this field. Citations from two of the most recent reviews on this topic express this most clearly. Making reference to the chronic opioid effects in whole animals, Smith et al. (1988) have pointed out that "our ignorance in this area can hardly be overstated" and, writing about the mechanisms underlying tolerance and dependence, Gmerek (1988) has stated that, although some postulated mechanisms have been contested, "surprisingly little additional progress has been made" since Himmelsbach produced his theory. Hopefully, as new methodologies become available and are applied to analyzing the different levels at which adaptation may occur, new light will be shed on these problems.

References

Adams WJ, Yeh SY, Woods L, Mitchell CL (1969) Drug-test interaction as a factor in the development of tolerance to the analgesic effect of morphine. J Pharmacol Exp Ther 168:251–257

Adler MW, Rowan CH, Geller EB (1984) Intracerebroventricular vs. subcutaneous drug administration: apples and oranges? Neuropeptides 5:73–76

Adler MW, Geller EB, Rosow CE, Cochin J (1988) The opioid system and temperature regulation. Annu Rev Pharmacol Toxicol 28:429–449

Advokat C (1981) Analgesic tolerance produced by morphine pellets is facilitated by analgesic testing. Pharmacol Biochem Behav 14:133–137

Ayesta FJ, Flórez J (1989) Tolerance to the respiratory actions of sufentanil: functional tolerance and route-dependent differential tolerance. J Pharmacol Exp Ther 250:371–378

Ayesta FJ, Flórez J (1990) Tolerance to the respiratory actions of opioids: withdrawal tolerance and asymmetrical cross-tolerance. Eur J Pharmacol 175:1–12

Babbini M, Davis WM (1972) Time-dose relationships for locomotor activity effects of morphine after acute or repeated treatment. Br J Pharmacol 46:213–224

Baker TB, Tiffany ST (1985) Morphine tolerance as habituation. Psychol Rev 92:78–108

Balster RL (1985) Behavioral studies of tolerance and dependence. In: Seiden LS, Balster RL (eds) Behavioral pharmacology: the current status. Liss, New York, pp 403-418

Blanchard SG, Chang KJ (1988) Regulation of opioid receptors. In: Pasternak GW (ed) The opiate receptors. Humana, Clifton, pp 425-439

Bläsig J, Herz A (1977) Precipitated morphine withdrawal in rats as a tool in opiate research. In: Essman WB, Valzelli L (eds) Current developments in psychopharmacology, vol 4. Spectrum, New York, pp 129-149

Bläsig J, Meyer G, Höllt V, Hengstenberg J, Dum J, Herz A (1979) Noncompetitive nature of the antagonistic mechanism responsible for tolerance development to opiate-induced analgesia. Neuropharmacology 18:473-481

Bozarth MA (ed) (1987) An overview of assessing drug reinforcement. In: Methods of assessing the reinforcing properties of abused drugs. Springer, Berlin Heidelberg New York, pp 635-658

Brady JV, Griffiths RR (1983) Testing drugs for abuse liability and behavioral toxicity: progress report from the laboratories at the Johns Hopkins University School of Medicine. In: Harris LS (ed) Problems of drug dependence 1982. NIDA Res Monogr 43, US Gov Print Off, Washington DC, pp 99-124

Brady LS, Holtzman SG (1981) Effects of intraventricular morphine and enkephalins on locomotor activity in nondependent, morphine-dependent and postdependent rats. J Pharmacol Exp Ther 218:613-620

Brase DA (1986) Unequal opiate cross-tolerance to morphine in the locomotor-activation model in the mouse. Neuropharmacology 25:297-304

Bush ED, Bush MF, Miller MA, Reid LD (1976) Addictive agents and intracranial stimulation: daily morphine and lateral hypothalamic self-stimulation. Physiol Psychol 4:79-85

Collier HOJ (1984) Cellular aspects of opioid tolerance and dependence. In: Hughes J, Collier HOJ, Rance MJ, Tyers MB (eds) Opioids: past, present and future. Taylor & Francis, London, pp 109-125

Colpaert FC (1978) Narcotic cue, narcotic analgesia, and the tolerance problem: the regulation of sensitivity to drug cues and to pain by an internal cue processing model. In: Colpaert FC, Rosecrans JA (eds) Stimulus properties of drugs: ten years of progress. Elsevier/North-Holland Biomedical Press, Amsterdam, pp 301-321

Cox BM (1978) Multiple mechanism in opiate tolerance. In: van Ree JM, Terenius L (eds) Characteristic and function of opioids. Elsevier/North-Holland Biomedical Press, Amsterdam, pp 13-23

Cox BM, Weinstock M (1966) The effect of analgesic drugs on the release of acetylcholine from electrically stimulated guinea-pig ileum. Br J Pharmacol 27:81-92

Cox BM, Ginsburg M, Osman DH (1968) Acute tolerance to narcotic analgesic drugs in rats. Br J Pharmacol 33:245-256

Dafters RI, Obder J, Miller J (1988) Associative and non-associative tolerance to morphine: support for a dual-process habituation model. Life Sci 42:1897-1906

Edwards G, Arif A, Hodgson R (1981) Nomenclature and classification of drug- and alcohol-related problems. Bull WHO 59:225-242

Emmett-Oglesby MW, Shippenberg TS, Herz A (1989) Fentanyl and morphine discrimination in rats continuously infused with fentanyl. Behav Pharmacol 1:3-11

Esposito R, Kornetsky C (1977) Morphine lowering of self-stimulation thresholds: lack of tolerance with long-term administration. Science 195:189-191

Esposito R, Kornetsky C (1978) Opioids and rewarding brain stimulation. Neurosci Biobehav Rev 2:115-122

Flórez J, Armijo JA, Delgado G (1972) Characterization of the acute tolerance and dependence to morphine-induced respiratory depression in decerebrate cats. Rev Esp Fisiol 28:167-174

Glick SD, Rapaport G (1974) Tolerance to the facilitatory effect of morphine on self-stimulation of the medial forebrain bundle in rats. Res Commun Chem Pathol Pharmacol 9:647-652

Gmerek DE (1988) Physiological dependence on opioids. In: Rodgers RJ, Cooper J (eds) Endorphins, opiates and behavioral processes. John Wiley & Sons, New York, pp 25-52

Goldstein A (1989) Introduction. In: Goldstein A (ed) Molecular and cellular aspects of the drug addictions. Springer, Berlin Heidelberg New York, pp xiii-xviii

Goldstein A, Goldstein DB (1968) Enzyme expansion theory of drug tolerance and physical dependence. Proc Assoc Res Nerv Ment Dis 46:265-267

Goldstein A, Aronow L, Kalman SM (eds) (1974) Drug tolerance and physical dependence. In: Principles of drug action: the basis of pharmacology. John Wiley & Sons, New York, pp 569-621

Goldstein DB (1979) Physical dependence on ethanol: its relation to tolerance. Drug Alcohol Depend 4:33–42
Goudie AJ, Griffiths JW (1986) Behavioral factors in drug tolerance. Trends Pharmacol Sci 7:192–196
Grabowsky J, Cherek DR (1983) Conditioning factors in opiate dependence. In: Smith JE, Lane JD (eds) The neurobiology of opiate reward processes. Elsevier/North Holland Biomedical Press, Amsterdam, pp 175–210
Haefely W (1986) Biological basis of drug-induced tolerance, rebound and dependence. Contribution of recent research on benzodiazepines. Pharmacopsychiatry 19:353–361
Halbach H (1973) Defining drug dependence and abuse. In: Goldberg L, Hoffmeister F (eds) Psychic dependence. Springer, Berlin Heidelberg New York, pp 17–24
Heishmann SJ, Stitzer ML, Bigelow GE, Liebson IA (1989) Acute opioid physical dependence in humans: effects of varying the morphine-naloxone interval. I. J Pharmacol Exp Ther 250:485–491
Henningfield JE, Johnson RE, Jasinski DR (1987) Clinical procedures for the assessment of drug potential. In: Bozarth MA (ed) Methods of assessing the reinforcing properties of abused drugs. Springer, Berlin Heidelberg New York, pp 573–590
Herz A, Bläsig J (1978) Opiate tolerance and dependence: some concluding remarks. In: Herz A (ed) Developments in opiate research. Dekker, New York, pp 407–414
Himmelsbach CK (1943) Can the euphoric, analgetic and physical dependence effects of drugs be separated? IV. With reference to physical dependence. Fed Proc 2:201–203
Hine B (1985) Morphine and Δ^9-tetrahydrocannabinol: two-way cross tolerance for antinociceptive and heart-rate responses in the rat. Psychopharmacology 87:34–38
Hovav E, Weinstock M (1987) Temporal factors influencing the development of acute tolerance to opiates. J Pharmacol Exp Ther 242:251–256
Hug CC (1972) Characteristics and theories related to acute and chronic tolerance development. In: Mulé SJ, Brill H (eds) Chemical and biological aspects of drug dependence. CRC, Cleveland, pp 307–358
Jaffe JH (1985) Drug addiction and drug abuse. In: Gilman AG, Goodman LS, Rall TW, Murad F (eds) The pharmacological basis of therapeutics. Macmillan, New York, pp 532–581
Jaffe JH, Sharpless K (1968) Pharmacological denervation supersensitivity in the central nervous system: a theory of physical dependence. Proc Assoc Res Nerv Ment Dis 46:226–246
Jorgensen HA, Fasmer OB, Hole K (1986) Learned and pharmacologically-induced tolerance to ethanol and cross-tolerance to morphine and clonidine. Pharmacol Biochem Behav 24:1083–1088
Kalant H (1985) Tolerance, learning and neurochemical adaptation. Can J Physiol Pharmacol 63:1485–1494
Kalant H, Leblanc AE, Gibbins RJ (1971) Tolerance to, and dependence on, some non-opiate psychotropic drugs. Pharmacol Rev 23:135–191
Kayan S, Woods LA, Mitchell CL (1969) Experience as a factor in the development of tolerance to the analgesic effect of morphine. Eur J Pharmacol 6:333–339
Khanna JM, Le AD, Kalant H, Leblanc AE (1979) Cross-tolerance between ethanol and morphine with respect to their hypothermic effects. Eur J Pharmacol 59:145–149
Koob GF, Bloom FE (1988) Cellular and molecular mechanisms of drug dependence. Science 242:715–723
Kornetsky C, Bain G (1983) Effects of opiates on rewarding brain stimulation. In: Smith JE, Lane JD (eds) The neurobiology of opiate reward processes. Elsevier/North Holland Biomedical Press, Amsterdam, pp 237–256
Lange DG, Fujimoto JM, Fuhrman-Lane CL, Wang RIH (1980a) Unidirectional non-cross-tolerance to etorphine in morphine-tolerant mice and the role of blood-brain barrier. Toxicol Appl Pharmacol 54:177–186
Lange DG, Roerig SC, Fujimoto JM, Wong RIH (1980b) Absence of cross-tolerance to heroin in morphine-tolerant mice. Science 208:72–74
Lange DG, Roerig SC, Fujimoto JM, Busse LW (1983) Withdrawal tolerance and unidirectional non-cross-tolerance in narcotic pellet-implanted mice. J Pharmacol Exp Ther 224:13–20
Lorens SA, Mitchell CL (1973) Influence of morphine on lateral hypothalamic self-stimulation in the rat. Psychopharmacologia 32:271–277
Martin WR (1968) A homeostatic and redundancy theory of tolerance and dependence on narcotic analgesics. Proc Assoc Res Nerv Ment Dis 46:206–225
Martin WR (1984) Pharmacology of opioids. Pharmacol Rev 35:283–323

Martin WR, Jasinski DR (1977) Assessment of the abuse potential of narcotics analgesics in animals. In: Martin WR (ed) Drug addiction, vol 1. Springer, Berlin Heidelberg New York, pp 159–196
Martin WR, Sloan JW (1977) Neuropharmacology and neurochemistry of subjective effects, analgesia, tolerance and dependence produced by narcotic analgesics. In: Martin WR (ed) Drug addiction, vol 1. Springer, Berlin Heidelberg New York, pp 43–158
Martin WR, Jasinski DR, Sapira JD, Flanary HG, Kelly OA, Thompson AK, Logan CR (1968) The respiratory effects of morphine during a cycle of dependence. J Pharmacol Exp Ther 162:182–189
McAuliffe WE, Gordon RA (1974) A test of Lindesmith's theory of addiction: the frequency of euphoria among long-term addicts. Am J Sociol 79:795–840
McKearney JW (1985) Tolerance to behavioral effects of clonidine after chronic administration of morphine. Pharmacol Biochem Behav 22:573–576
Mirin SM, Meyer RE, McNamee HB (1976) Psychopathology and mood during heroin use. Arch Gen Psychiat 33:1503–1508
Mucha RF, Kalant H (1980) Log dose/response curve flattening in rats after daily injection of opiates. Psychopharmacology 71:51–61
Mucha RF, Kalant H (1981) Naloxone prevention of morphine LDR curve flattening associated with high-dose tolerance. Psychopharmacology 75:132–133
Mucha RF, Kalant H, Kim C (1987) Tolerance to hyperthermia produced by morphine in rat. Psychopharmacology 92:452–458
North RA, Karras PJ (1978) Tolerance and dependence in vitro. In: van Ree JM, Terenius L (eds) Characteristics and functions of opioids. Elsevier/North Holland Biomedical Press, Amsterdam, pp 25–36
Okamoto M, Boisse NR, Rosenberg HC, Rosen R (1978) Characteristics of functional tolerance during barbiturate physical dependency production. J Pharmacol Exp Ther 207:906–915
Paktor J, Vaught JL (1984) Differential analgesic cross-tolerance to morphine between lipophilic and hydrophilic narcotic agonists. Life Sci 34:13–21
Paletta MS, Wagner AR (1986) Development of context-specific tolerance to morphine: support for a dual-process interpretation. Behav Neurosci 100:611–623
Pentiah P, Reilly F, Borison HL (1966) Interactions of morphine sulphate and sodium salicylate on respiration in cats. J Pharmacol Exp Ther 154:110–118
Petersen DW, Fujimoto JM (1983) Differential tolerance to the intestinal inhibitory effects of opiates in mice. Eur J Pharmacol 95:225–230
Porreca F, Cowan A, Raffa RB, Tallarida RJ (1982) Tolerance and cross-tolerance studies with morphine and ethylketocyclazocine. J Pharm Pharmacol 34:666–667
Reid LD (1987) Tests involving pressing for intracranial stimulation as an early procedure for screening likelihood of addiction of opioids and other drugs. In: Bozarth MA (ed) Methods of assessing the reinforcing properties of abused drugs. Springer, Berlin Heidelberg New York, pp 391–420
Roberts DCS, Zito KA (1987) Interpretation of lesion effects on stimulant self-administration. In: Bozarth MA (ed) Methods of assessing the reinforcing properties of abused drugs. Springer, Berlin Heidelberg New York, pp 87–103
Roerig SC, Fujimoto JM, Franklin RB, Lange DG (1985) Unidirectional non-cross-tolerance (UNCT) in rats and an apparent dissociation between narcotic tolerance and physical dependence. Brain Res 327:91–96
Roerig SC, Fujimoto JM, Lange DG (1987) Development of tolerance to respiratory depression in morphine- and etorphine-pellet-implanted mice. Brain Res 400:278–284
Rosenfeld GC, Burks TF (1977) Single dose tolerance to morphine hypothermia in the rat: differentiation of acute from long-term tolerance. J Pharmacol Exp Ther 202:654–659
Rosow CE, Miller JM, Poulsen-Burke J, Cochin J (1982) Opiates and thermoregulation in mice. IV. Tolerance and cross-tolerance. J Pharmacol Exp Ther 223:702–708
Rothman RB, Danks JA, Jacobson AE, Burke TR, Rice KC, Tortella FC, Holaday JW (1986) Morphine tolerance increases μ-noncompetitive σ binding sites. Eur J Pharmacol 124:113–119
Rothman RB, Bykov V, Long JB, Brady LS, Jacobson AE, Rice KC, Holaday JW (1989) Chronic administration of morphine and naltrexone up-regulate μ-opioid binding sites labelled by [^{3}H][D-Ala^2, MePhe^4, Gly-ol^5]enkephalin: further evidence for two μ-binding sites. Eur J Pharmacol 160:71–82
Sander GE, Giles TD (1984) The influence of the inter-dose time interval on the cardiovascular response to methionine-enkephalin in the conscious dog. Peptides 5:797–800

Schaefer GJ, Michael RP (1986) Changes in response rates and reinforcement thresholds for intracranial self-stimulation during morphine withdrawal. Pharmacol Biochem Behav 25:1263–1269
Schulz R (1988) Dependence and cross-dependence in the guinea-pig myenteric plexus. Naunyn-Schmiedeberg's Arch Pharmacol 337:644–648
Sharma SK, Klee WA, Nirenberg M (1975) Dual regulation of adenylate cyclase accounts for narcotic dependence and tolerance. Proc Natl Acad Sci USA 72:3002–3006
Shippenberg TS, Emmet-Oglesby MW, Ayesta FJ, Herz A (1988) Tolerance and selective cross-tolerance to the motivational effects of opioids. Psychopharmacology 96:110–115
Siegel S (1983) Classical conditioning, drug tolerance, and drug dependence. In: Israel Y, Glaser FB, Kalant H, Popham RE, Schmidt W, Smart RG (eds) Research advances in alcohol and drug problems, vol 7. Plenum New York, pp 207–246
Siegel S (1988) Drug anticipation and drug tolerance. In: Lader M (ed) Psychopharmacology of addiction. Univ Press, Oxford, pp 73–96
Sivam SP, Ho IK (1984) Analgesic cross-tolerance between morphine and opioid peptides. Psychopharmacology 84:64–65
Smith AP, Law PY, Loh HH (1988) Role of opiate receptors in narcotic tolerance/dependence. In: Pasternak GW (ed) The opiate receptors. Humana, Clifton, pp 441–485
Stevens CW, Yaksh TL (1989) Potency of infused antinociceptive agents is inversely related to magnitude of tolerance after continuous infusion. J Pharmacol Exp Ther 250:1–8
Tiffany ST, Maude-Griffin PM (1988) Tolerance to morphine in the rat: associative and nonassociative effects. Behav Neurosci 102:534–543
Turkington D, Drummond DC (1989) How should opiate withdrawal be measured? Drug Alcohol Depend 24:151–153
Vasko MR, Domino EF (1978) Tolerance development to the biphasic effects of morphine on locomotor activity and brain acetylcholine in the rat. J Pharmacol Exp Ther 207:848–858
Wagner AR (1981) SOP: a model of automatic memory processing in animal behavior. In: Spear NE, Miller RR (eds) Information processing in animals: memory mechanisms. Erlbaum Hillsdale, pp 5–47
Walter TA, Riccio DC (1983) Overshadowing effects in the stimulus control of morphine analgesic tolerance. Behav Neurosci 97:658–662
Watson SJ, Trujillo KA, Herman JP, Akil H (1989) Neuroanatomical and neurochemical substrates of drug-seeking behavior: overview and future directions. In: Goldstein A (ed) Molecular and cellular aspects of the drug addictions. Springer, Berlin Heidelberg New York, pp 29–91
Way EL, Loh HH, Shen FH (1969) Simultaneous quantitative assessment of morphine tolerance and physical dependence. J Pharmacol Exp Ther 167:1–8
Weibel SL, Wolf HH (1979) Opiate modification of intracranial self-stimulation in the rat. Pharmacol Biochem Behav 10:71–78
Wikler A (1972) Theories related to physical dependence. In: Mulé SJ, Brill H (eds) Chemical and biological aspects of drug dependence. CRC, Cleveland, pp 359–377
Wise RA (1987) Intravenous self-administration: a special of positive reinforcement. In: Bozarth MA (ed) Methods of assessing the reinforcing properties of abused drugs. Springer, Berlin Heidelberg New York, pp 117–141
Wise RA (1988) The neurobiology of craving: implications for the understanding and treatment of addiction. J Abnorm Psychol 97:118–132
Wüster M, Schulz R, Herz A (1985) Opioid tolerance and dependence: reevaluating the unitary hypothesis. Trends Pharmacol Sci 6:64–67
Yokel RA (1987) Intravenous self-administration: response rates, the effects of pharmacological challenges, and drug preferences. In: Bozarth MA (ed) Methods of assessing the reinforcing properties of abused drugs. Springer, Berlin Heidelberg New York, pp 1–33
Zukin RS, Tempel A (1986) Neurochemical correlates of opiate receptor regulation. Biochem Pharmacol 35:1623–1627

CHAPTER 25

Aspects of Opioid Tolerance and Dependence in Peripheral Nerve Tissues

R. Schulz

1 Introduction

Adaptation to chronic opioid exposure is expressed in cells carrying opioid receptors and this phenomenon has been termed "cellular opiate dependence" (Collier 1980). Evidence in support of this notion stems from experiments on neuroblastoma × glioma hybrid cells, using the activity of adenylate cyclase as an index of chronic opioid actions (Sharma et al. 1975). The findings reported for these cells were, at that time, best explained by a "unitary theory" which hypothesized that the biochemical adaptations producing tolerance were also responsible for dependence (Shuster 1961; Goldstein and Goldstein 1961). However, chronic opioid effects are not confined to the adenylate cyclase system. Subsequent studies with hybridomas chronically exposed to opioids also revealed changes at the opioid receptor level, namely, a down-regulation of binding sites (Chang et al. 1982; Law et al. 1983), the clustering of receptors (Hazum et al. 1979), and an uncoupling of the receptor from its effector (Wüster et al. 1985). Although these effects may underlie desensitization and/or tolerance, they do not explain the development of tolerance as well as dependence, and thus contrast with the "unitary theory".

The changes in adenylate cyclase activity which occur during the development of tolerance and dependence to both opioids and nonopioids are well established in NG 108-15 cells (Hamprecht 1978). Therefore, studies were undertaken to investigate whether similar mechanisms also underlie the development of tolerance/dependence in the brain and peripheral nervous system, however, the outcome of those studies failed to confirm this notion. Moreover, other studies have largely failed to demonstrate changes in opioid receptor binding affinity or capacity during the tolerant/dependent state (Klee and Streaty 1974; Creese and Sibley 1981; Mocchetti and Costa 1987). In addition, although a few data indicate a down-regulation of receptors upon chronic opioid exposure (Dingledine et al. 1983; Tao and Law 1984), the changes in binding capacity and development of tolerance occur asynchronously.

The limited information derived from studies with whole animals has led to the use of isolated tissue preparations such as the mouse vas deferens (MVD) and the guinea-pig ileum (GPI) which have less complex neuronal arrangements as compared to the brain. Both tissues carry opioid receptors, as well as nonopioid receptors mediating the inhibitory actions of other transmitters (Schulz 1985). They have recently been used to study opioid tolerance and dependence, with particular emphasis upon receptor-associated membrane proteins, the guanine nucleotide binding proteins (G-proteins; Gilman 1987). G-proteins represent regulatory elements situated between receptors and their effectors (e.g., adenylate cyclase). The change in enzyme activity occurs in response to stimulatory and inhibitory agents mediated by different G-proteins, termed G_s and G_i (s, stimulator; i, inhibitory), respectively. These G-proteins are heterotrimeric and consist of subunits termed α, β, and γ. The α-subunits show a considerable heterogeneity with subtypes designated α_s, α_i, α_o, which themselves are further subdivided (e.g., α_{i1-3}). Bacterial toxins can affect the function of these G-proteins by permanently ribosylating $G_{\alpha s}$ (cholera toxin, CT) and $G_{\alpha i}$, $G_{\alpha o}$ (pertussis toxin, PTX), and have thus served as tools to examine possible G-protein adaptations upon chronic exposure to opioids and nonopioids.

2 Models of Tolerance and/or Dependence

2.1 *Mouse Vas Deferens (MVD)*

The MVD represents a relatively simple model with respect to its neuroanatomy. The smooth muscle receives its innervation from the lumbosacral-plexus. In preparation of the tissue for in vitro studies, the nerve somata are severed, leaving only axons and nerve terminals within the tissue. These terminals respond to electrical stimulation by releasing norepinephrine (NE) which, in turn, causes the muscle to contract. Opioid receptors are confined to the nerve terminals, and their activation causes an inhibition of electrically evoked twitches. Figure 1 schematically represents the innervation of the muscle and displays the inhibitory activity (depression of twitch tension) of the opioid agonist D-Ala-D-Leu-enkephalin (DADL). Chronic treatment of mice with DADL in vivo results in a reduced sensitivity (tolerance) of the MVD to DADL in vitro. However, the highly tolerant MVD fails to display any signs of dependence in response to the narcotic antagonist naloxone (NAL), regardless of whether NAL is applied during electrical stimulation (submaximal), or after stimulation has been turned off. This preparation, therefore, displays tolerance in the absence of detectable dependence, demonstrating the separability of these two phenomena, thus challenging the "unitary hypothesis" postulated earlier.

We employed PTX and CT to investigate the transduction mechanisms underlying opioid-controlled neurotransmission in the MVD. PTX was given intraperitoneally (up to 120 μg/kg) to opioid-naive mice; some MVD preparations were then exposed to an opioid (DADL) in vitro. Neither the in vitro electrically evoked twitches of preparations from opioid naive mice, nor the acute or chronic opioid effects, were affected by PTX (Lux and Schulz 1986). Analogous studies were conducted with CT in vitro. Again, neither the electrically evoked twitch tension nor acute or chronic opioid effects were found to be altered by CT (Lux and Schulz 1986). These experiments suggest that the biochemical systems linked to opioid receptors at nerve terminals of the MVD are not sensitive to PTX and CT and, thus, G-proteins are probably not involved in the development of tolerance in this preparation.

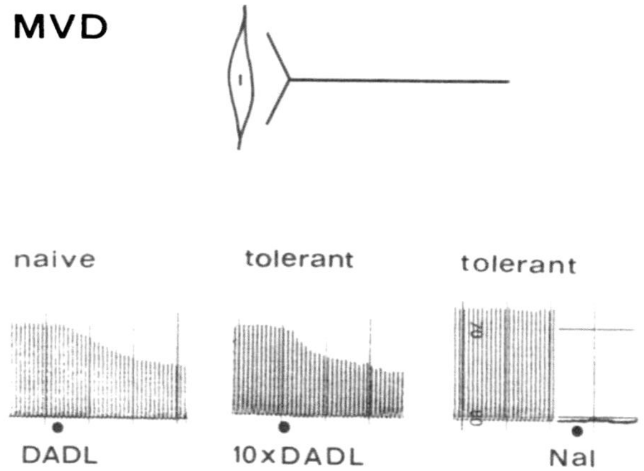

Fig. 1. Schematic representation of the isolated mouse vas deferens (MVD). Nerve terminals connected to an axon impinge on the smooth muscle. Nerve somata are eliminated during preparation of the tissue for in vitro studies. Electrical stimulation evokes twitches, and inhibition of twitch tension requires a certain concentration of a stable enkephalin derivative (*DADL*) to activate opioid receptors located at nerve terminals (naive: preparations from untreated mice). Chronic exposure of animals to a δ-opioid receptor agonist (*DADL*) renders the preparations tolerant, as is indicated by the 10-fold higher concentration of DADL required to cause a similar inhibition of twitch tension to that seen in naive tissues. Challenge of tolerant preparations (kept in vitro in the presence of an opioid) with the narcotic antagonist naloxone (*Nal*) fails to precipitate a withdrawal sign. The antagonist neither affected electrically induced twitches nor caused an increase of basic tension in the absence of electrical stimulation

2.2 Guinea-Pig Ileum (GPI, Longitudinal Muscle Myenteric Plexus Preparation)

The myenteric plexus, innervating the longitudinal muscles of the gut, consists of nerves which also have synaptic connections with each other (Gabella 1972; Fig. 2). The plexus contains opioid receptors of the μ- and κ-type (Wüster et al. 1983), located at both nerve terminals and somata (North and Williams 1983; Lux and Schulz 1986; North 1986). Like the isolated MVD, the GPI exhibits tolerance but, in contrast to the former tissue (MVD), the GPI also develops and displays dependence. This is indicated by NAL-precipitated withdrawal contractures in tissues chronically exposed to an opioid agonist (Fig. 2).

To investigate the role of G-proteins in the development of dependence, guinea pigs were treated with PTX (single injection, up to 120 μg/kg, i.p.), and subsequently rendered tolerant and dependent by infusion of an opioid. The isolated longitudinal muscle myenteric plexus preparations were then set up in vitro. These experiments showed that PTX fails to affect both electrically induced twitch tensions and the acute and chronic (tolerance) effects of opioids, clonidine, and chloroadenosine. PTX also fails to interfere with the excitatory actions of PGE_1 and neurotensin. Interestingly, the PTX-treated preparations did not exhibit any signs of dependence (withdrawal contracture), regardless of whether the chronically applied drug was an opioid or a nonopioid, e.g., clonidine (Lux and Schulz 1986). Other experiments

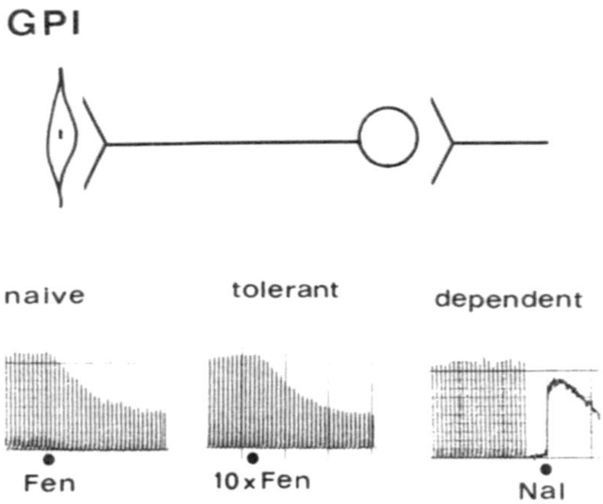

Fig. 2. Schematic representation of the isolated longitudinal muscle-myenteric plexus preparation of the guinea-pig ileum (*GPI*). The smooth muscle is innervated by a complete neuron which itself receives synaptic inputs from other nerves in the plexus. The opioid fentanyl (*Fen*) causes an inhibition of electrically evoked twitch tension of preparations from untreated (naive) guinea pigs. Chronic exposure of guinea pigs to fentanyl renders the preparations tolerant, since a tenfold higher concentration of the opioid is required to cause a similar effect to that observed in naive preparations. In contrast to the MVD, the GPI displays dependence since NAL precipitates a withdrawal contracture

were conducted with isolated GPI preparations using CT in vitro (up to 10^{-6} M). The toxin failed to affect the electrically induced twitch tension or the inhibitory actions of opioids and of clonidine and chloroadenosine. In contrast, CT dose-dependently inhibited precipitation of the withdrawal contracture and blocked the excitatory actions of serotonin, PGE_1, and neurotensin (Lux and Schulz 1986), demonstrating that the expression of withdrawal involves the transmission of excitatory signals within the myenteric plexus.

The above results demonstrate distinct differences between the MVD and GPI in terms of their sensitivity to PTX although, in both preparations, opioid receptors located at the neuromuscular junction (nerve terminals) do not appear to be affected by the toxin. These receptors control the release of excitatory neurotransmitters (GPI: acetylcholine; MVD: NE) and are thus responsible for the intensity of the twitch tension induced. It should be noted that the activity of opioid receptors not associated with the neuromuscular junction is undetectable when electrical stimulation is applied and when inhibition of twitch tension is used as a parameter for measuring opioid activity. In contrast to tolerance, which is displayed both by the MVD and the GPI, dependence is only exhibited by the GPI. Since the myenteric plexus contains complete neurons (including nerve somata), it is suggested that opioid receptors at the somata are linked to the phenomenon of dependence. Those receptors are sensitive to PTX, since the toxin prevents the development of dependence and the precipitation of withdrawal contractions by NAL. The scheme in Fig. 3 represents the location of PTX-sensitive opioid receptors at different neuroanatomical sites in the MVD and the GPI, and relates them to the capacity of a tissue to develop dependence.

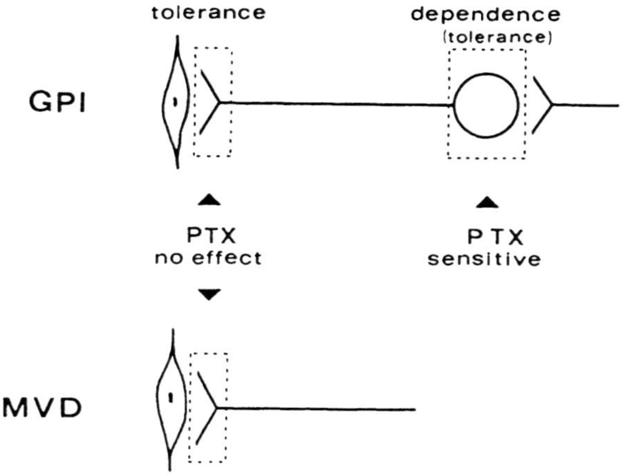

Fig. 3. The effect of pertussis toxin (PTX) on the guinea-pig ileum (GPI) and mouse vas deferens (*MVD*). Experimental data indicate that opioid receptors located at nerve terminals are not sensitive to the toxin, while the function of receptors at nerve somata is eliminated by PTX

3 Evidence for Alterations in G-Protein Levels During Dependence

The prevention of the withdrawal sign by PTX suggests an involvement of G-proteins in opioid dependence. In an attempt to study more closely the G-proteins linked to opioid receptors located at nerve somata, the GPI of animals chronically treated with fentanyl, so as to render them highly tolerant and dependent, were homogenized and centrifuged in a discontinuous sucrose gradient (Briggs and Cooper 1981). This latter procedure provided three fractions, containing mainly nerve somata (fraction I), somata and synaptosomes (fraction II), and synaptosomes only (fraction III). The ability of PTX to transfer ADP-ribose to the α-subunits of G-proteins was measured in these fractions. There was an increased incorporation of ^{32}P-NAD into a 40-kD band (which comigrates with the α-subunit of G_i) in preparations from animals chronically exposed to fentanyl (Fig. 4; Lang and Schulz 1989). This finding was more pronounced in fractions I and II, but was absent in fraction III (synaptosomes). Quantification of $G_{i\alpha}/G_{o\alpha}$ revealed a significant increase of these proteins in dependent preparations. When an antibody specific for $G_{\alpha o}$ was used this increase became even more obvious (Fig. 5; Lang and Schulz 1989).

4 Concluding Remarks

Work with the MVD and the GPI, prepared from animals chronically exposed to either an opioid or a nonopioid, suggests the following:

1. Detection of tolerance in both isolated preparations is confined to receptors located at nerve terminals, when the inhibition of electrically stimulated twitch tension is used as a parameter. Opioid receptors situated at nerve somata may also develop tolerance, but the current models do not permit detailed investigations.

   The mechanisms underlying opioid tolerance are not well understood at present. Data obtained from different tissues, including the GPI and the MVD,

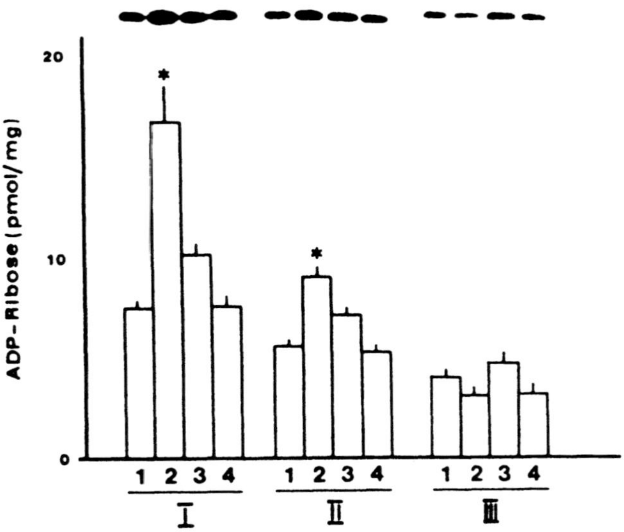

Fig. 4. ADP-ribosylation of GPI sucrose gradient fractions. Fraction **I** contains mainly somata, fraction **III** represents synaptosomes and fraction **II** contains a mixture of fraction **I** and **III**. The samples were ADP-ribosylated with either ^{32}P-NAD or PTX, and subsequently subjected to sodium dodecyl sulfate (SDS) polyacrylamide gel electrophoreses. *Lane 1* controls; *lane 2* chronic fentanyl; *lane 3* chronic NAL: *lane 4* fentanyl and NAL were given simultaneously for 6 days. Also shown (*top*) are the autoradiographic signals obtained in a representative experiment. Values represent means of triplicates ± SEM; $^{*}2p < 0.05$ (Data from Lang and Schulz 1989)

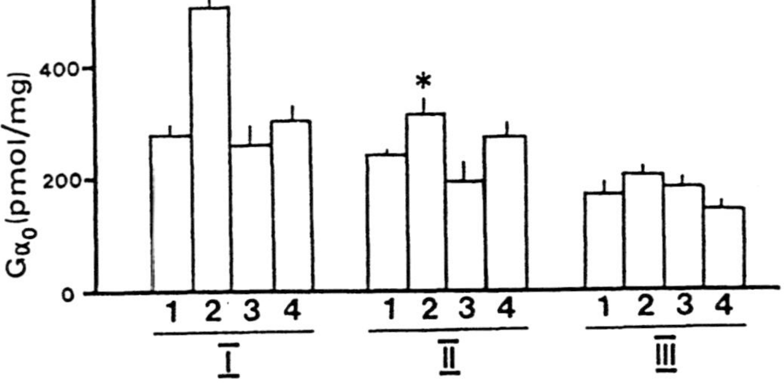

Fig. 5. Quantitative changes in G-protein (G_{α_0}) levels in the myenteric plexus after chronic treatment with fentanyl. For lane descriptions, see Fig. 4 (Data from Lang and Schulz 1989)

show no change in receptor capacity (Creese and Sibley 1981; Rubini et al. 1982) although they indicate a down-regulation (Chang et al. 1982; Law et al. 1983; Dingledine et al. 1983) and a reduction in opiate receptor reserve (Chavkin and Goldstein 1982). A further mechanism to be considered is the possible uncoupling of opioid receptors from G-proteins (Wüster et al. 1983a; Nestler et al. 1989), although the findings with PTX described above do not favor the notion that opioid receptors located at nerve terminals in the MVD and GPI are linked to these proteins. However, any of the mentioned changes in receptor characteristics could reflect a state of tolerance, even though they fail to explain dependence. The common biochemical mechanism once postulated to underly tolerance and dependence ("unitary theory") is therefore not supported by the experimental findings in the isolated MVD and GPI.

2. In contrast to the mechanisms which operate in the development of tolerance, the mechanisms underlying dependence require opioid receptors to be linked to their effector G-proteins. Indeed, precipitation of a withdrawal sign by NAL implicates the displacement of the narcotic agonist from its receptor. The withdrawal cascade must therefore require an intact signal transmission system (involving cell bodies, axons, and terminals and an interaction with further neurons).

   The fact that PTX prevents the development of dependence, regardless of whether an opioid or a nonopioid has been given chronically, suggests the involvement of G-proteins, most likely G_i- or G_o-proteins. This does not necessarily imply that the G-proteins or any of their subunits represent the components in signal transduction responsible for the phenomenon of dependence. Although a significant increase of $G_{\alpha s}$ has been observed 6 days after the start of fentanyl infusions, dependence (demonstrated by Nal-precipitated withdrawal) develops earlier than the measurable changes in G-protein levels (L. Nice and R. Schulz, unpubl. data). With respect to the association of dependence with the somata and PTX sensitivity, it is interesting that PTX has been shown to block the function of opioid receptors as well as α_2-adrenoceptors located on cell bodies of the locus coeruleus (Aghajanian and Wang 1986).

   The data obtained with the isolated MVD and GPI also suggest that tolerance, on the one hand, and dependence, on the other, may relate to distinct biochemical processes. As in other tissues (e.g., cat dorsal horn neurons), dependence can exist in the absence of tolerance (Zhao and Duggan 1987).

   A further indication for separate biochemical mechanisms underlying tolerance and dependence is the lack of cross-tolerance ("selective tolerance") between different opioid receptor types (Schulz et al. 1981), whereas cross-dependence develops simultaneously to different opioids and nonopioids (Schulz 1988). The data suggesting cross-dependence are of interest in light of reports that opioid receptors and α_2-adrenergic receptors control the same K^+-conductance via a G-protein (Andrade et al. 1983; Christie et al. 1987).

References

Aghajanian GK, Wang YY (1986) Pertussis toxin blocks the outward currents evoked by opiate and α_2-agonists in locus coeruleus neurons. Brain Res 371:390–394

Andrade R, Vandermalelen CP, Aghajanian GK (1983) Morphine tolerance and dependence in the locus coeruleus: single cell studies in brain slices. Eur J Pharmacol 91:161–169

Briggs CA, Cooper JR (1981) A synaptosomal preparation from the guinea pig ileum myenteric plexus. J Neurochem 36:1097–1108

Chang KJ, Eckel RW, Blanchard SG (1982) Opioid peptides induce reduction of enkephalin receptors in cultured neuroblastoma cells. Nature (London) 296:446–448

Chavkin C, Goldstein A (1982) Reduction in opiate receptor reserve in morphine tolerant guinea-pig ilea. Life Sci 31:1687–1690

Christie MJ, Williams JT, North RA (1987) Cellular mechanisms of opioid tolerance: studies in single brain neurons. Mol Pharmacol 32:633–638

Collier HOJ (1980) Cellular site of opiate dependence. Nature (London) 283:625–629

Cooper DMF, Londos C, Gill DL, Rodbell M (1982) Opiate receptor mediated inhibition of adenylate cyclase in rat striatal plasma membranes. J Neurochem 38:1164–1167

Creese J, Sibley DR (1981) Receptor adaptations to centrally acting drugs. Annu Rev Pharmacol Toxicol 21:357–391

Dingledine R, Valentino RJ, Bostock E, King ME, Chang KJ (1983) Down-regulation of delta but not mu opioid receptors in the hippocampal slice associated with loss of physiological response. Life Sci 33 (Suppl 1):333–336

Gabella G (1972) Fine structure of the myenteric plexus in the guinea pig ileum. J Anat 111:69–97

Gilman A (1987) G-proteins: transducers of receptor-generated signals. Annu Rev Biochem 56:615–649

Goldstein DB, Goldstein A (1961) Possible role of enzyme inhibition and repression in drug tolerance and addiction. Biochem Pharmacol 8:48

Hamprecht B (1977) Structural, electrophysiological, biochemical and pharmacological properties of neuroblastoma × glioma cell hybrids in cell culture. Int Rev Cytol 49:99–170

Hamprecht B (1978) Opioids and cyclic nucleotides. In: Herz A (ed) Developments in opiate research. Dekker, New York, pp 357–406

Hazum E, Chang KJ, Cuatrecasas P (1979) Opiate (enkephalin) receptors of neuroblastoma cells: occurrence in clusters on the cell surface. Science 206:1077–1079

Klee WA, Streaty RA (1974) Narcotic receptor sites in morphine dependent rats. Nature (London) 248:61–63

Lang J, Schulz R (1989) Chronic opiate receptor activation in vivo alters the level of G-protein subunits in guinea-pig myenteric plexus. Neuroscience 32:503–510

Law PY, Hom DS, Loh HH (1983) Opiate receptor down-regulation and desensitization in neuroblastoma × glioma NG 108-15 hybrid cells are two separate cellular adaptation processes. Mol Pharmacol 25:413–424

Lux B, Schulz R (1986) Effect of cholera toxin and pertussis toxin on opioid tolerance and dependence in the guinea-pig myenteric plexus. J Pharmacol Exp Ther 237:995–1000

Mocchetti J, Costa E (1987) In vivo studies of the regulation of neuropeptide stores in structures of the rat brain. Neuropharmacology 26:855–862

Nestler EJ, Erdos JJ, Terwilliger R, Duman RS, Tallman JF (1989) Regulation of G-proteins by chronic morphine in the rat coeruleus. Brain Res 476:230–239

North RA (1986) Opioid receptor types and membrane ion channels. Trends Neurosci 9:114–117

North RA, Williams JT (1983) How do opiates inhibit neurotransmitter release? Trends Neurosci 6:337–339

Rubini P, Schulz R, Wüster M, Herz A (1982) Opiate receptor binding studies in the mouse vas deferens exhibiting tolerance without dependence. Naunyn-Schmiedeberg's Arch Pharmacol 319:142–146

Schulz R (1978) The use of isolated organs to study the mechanism of action of narcotic analgesics. In: Herz A (ed) Developments in opiate research. Dekker, New York, pp 241–277

Schulz R (1985) Opioid receptors and peripheral autonomic mechanisms. In: Kalsner S (ed) Trends in autonomic pharmacology, vol 3. Taylor & Francis, London Philadelphia, pp 257–268

Schulz R (1988) Dependence and cross-dependence in the guinea-pig myenteric plexus. Naunyn-Schmiedeberg's Arch Pharmacol 337:644–648

Schulz R, Wüster M, Rubini P, Herz A (1981) Functional opiate receptors in the guinea-pig ileum: their differentiation by means of selective tolerance development. J Pharmacol Exp Ther 219:547–550
Sharma SK, Klee WA, Nirenberg M (1975) Dual regulation of adenylate cyclase accounts for narcotic dependence and tolerance. Proc Natl Acad Sci 72:3092–3096
Shuster L (1961) Repression and depression of enzyme synthesis as a possible explanation of some aspects of drug action. Nature (London) 189:314–315
Tao PL, Law PY (1984) Down-regulation of opiate receptors in rat brain after chronic etorphine treatment. Proc W Pharmacol Soc 127:557–560
Wüster M, Costa T, Gramsch C (1983a) Uncoupling of receptors is essential for opiate-induced desensitization (tolerance) in neuroblastoma × glioma hybrid cells NG 108–15. Life Sci 33 (Suppl 1):341–344
Wüster M, Schulz R, Herz A (1983b) A subclassification of multiple opiate receptors by means of selective tolerance development. J Receptor Res 3:199–214
Wüster M, Schulz R, Herz A (1985) Opioid tolerance and dependence: reevaluating the unitary hypothesis. Trends Pharmacol Sci 6:64–67
Zhao ZQ, Duggan AW (1987) Clonidine and the hyper-responsiveness of dorsal horn neurons following morphine withdrawal in the spinal cat. Neuropharmacology 26:1499–1502

CHAPTER 26

Overview of Opioid Tolerance and Physical Dependence

A.K. Louie, and E.L. Way

1 Introduction

The opioids have played a major role in promoting the understanding of the adaptive processes in the central nervous system. The homeostatic consequences provoked in the organism after chronic opiate administration are the well-known complex syndromes of tolerance and physical dependence. Both phenomena are dynamic and evolve over time. Tolerance to, and physical dependence on, an opioid may develop rapidly or slowly and to varying degrees. The evolution of these processes is exceedingly complex but general as well as specific explanations have been offered.

Tolerance can be defined as the decrease in effectiveness of a pharmacologic agent after prior administration. Thus, in order to maintain the same response, the dose administered must be increased. The term is encompassing and may cover the many appellations coined by investigators from various disciplines. The examples include acute tolerance, tachyphylaxis, protracted tolerance, cross-tolerance, behavioral tolerance, decreased sensitivity, increased resistance, desensitization, refractoriness, downregulation, immunotolerance, pharmacodynamic tolerance, adaptive tolerance, dispositional tolerance, etc. Some terms are synonymous; others imply that different modes of action are responsible for the phenomenon. A definition of some types will be provided in the sections relevant to the topic under discussion.

Physical dependence is defined as the need to maintain administration of a substance following prior exposure in order to prevent the appearance of an abstinence syndrome. A state of supersensitivity becomes manifest and with opiates, the signs and symptoms of a withdrawal reaction are generally opposite to those observed after their acute administration and seem to reflect an overshoot/rebound to the inhibited response.

1.1 Characteristics and General Theories of Opioid Tolerance and Physical Dependence

It is well established clinically that the development of tolerance after chronic administration of morphine, the prototypic opiate, is invariably accompanied by the development of physical dependence. When the subject is in the tolerant state and the administration of morphine is discontinued, highly characteristic abstinence signs and symptoms appear. This syndrome can be almost immediately terminated by readministration of morphine or one of its surrogates. In the particular case of morphine or heroin, various withdrawal signs and symptoms begin to appear 6–8 h after chronic administration is discontinued. They gradually attain a peak at about 36–48 h and largely subside within 1 week. In this detoxified state, the tolerance to, and the physical dependence on, the opiate are mostly abolished. Thus, in the clinic and on the street, the onset, offset, and degree of these two syndromes are nearly identical.

In the experimental laboratory, it is also possible to demonstrate a close coincidence between tolerance and physical dependence after sustained opiate administration. The concomitance of these two phenomena can be demonstrated in a variety of systems, ranging from cells in culture to laboratory animals, including rodents and nonhuman primates. It is not surprising, therefore, that many investigators believe that there is a close relationship between the two phenomena. In seeking to elucidate the basic mechanisms involved in opiate tolerance and physical dependence development, a common question is concerned with how they might be linked.

Nearly 50 years ago, Himmelsbach (1942) opined that the opiate tolerant/dependent state reflects a homeostatic response of the body to continued, frequent administration of the agent. The abstinence syndrome occurring after discontinuation of the opiate was seen as an overshoot or rebound phenomenon resulting from a compensatory response to the acute inhibitory effects of opioids. Subsequently, others have attempted to explain tolerance and physical dependence in physiological and biochemical terms but without identifying the specific mechanism involved. Thus, it was hypothesized that the adaptive processes were related to hypertrophy of alternate pathways (Martin 1968; Jaffe and Scharpless 1968), enzyme induction or expansion (Schuster 1961; Goldstein and Goldstein 1968), or to the induction of receptors (Collier 1965, 1980).

Demonstration of a biochemical basis for opiate tolerance and physical dependence development was reported nearly two decades ago. Early evidence to this effect was the blocking of opiate tolerance and physical dependence development by the protein synthesis inhibitor, cycloheximide. Thus, in chronically morphinized mice, opiate tolerance, as measured by the tail-flick test for analgesia, and physical dependence, as measured by antagonist-precipitated withdrawal jumping in mice, were both prevented by daily administration of cycloheximide (Loh et al. 1969). A variety of other agents which inhibit either RNA or protein synthesis were also shown to prevent the acquisition of tolerance to morphine (Cox and Osman 1970). More recent findings will be reserved for later discussion.

There are some investigators, who question that a relationship exists between tolerance and physical dependence and have submitted evidence to support this contention. However, most of these experiments miss the mark because, as will be pointed out, there are many indices for evaluating tolerance and physical dependence development which have different onset and offset times and intensities. Hence, it is no problem to derive data to dissociate some signs of tolerance and physical dependence and then argue that the two phenomena are unrelated. However, such experiments are not valid because they were not geared to measure common parameters with equal sensitivities in response. The study by Christie et al. (1987) may represent an exception. They reported that the development of tolerance to opioid inhibition of neuronal firing in the locus ceruleus was not accompanied by a rebound supersensitivity after naloxone (NAL) administration, but others have presented data to the contrary (Aghajanian 1978; Fry et al. 1980).

Rather than attempting to dissociate tolerance and physical dependence, it appears far more worthwhile to relate the experiences in the clinic, the street, and the experimental laboratory, and try to ascertain whether some related physiological process can be demonstrated to occur during opiate tolerance and physical dependence development.

In each instance, tolerance and physical development must follow agonist binding to the receptor. The dissociation between the two events after this may be sequentially related to a common process but involvement of different sites within the same or different neurons cannot be excluded. It is difficult to become enthused with the latter possibility in light of evidence to be presented later demonstrating that certain common responses can be linked to both tolerance and physical dependence development.

To derive explanations for tolerance and physical dependence in greater depth, it has become necessary not only to use in vitro systems such as excised organs or tissues but also isolated cells in culture. In applying such model systems, the basic principles involved in tolerance and dependence assessment should not be forgotten. In formulating testable hypotheses, besides having appropriate methodology for the quantitative assessment of tolerance and physical dependence, certain simplifying assumptions need to be made. These demands are no different from other experimental approaches that dictate the use of operational model systems for solving complicated biological questions.

1.2 Principles of Opioid Tolerance and Physical Dependence Assessment

It is well known that tolerance to any one opiate after repeated administration develops at a different rate and to varying degrees at various body sites. In theory, the processes related to the development of tolerance are initiated with the first dose but, to the subject on morphine in the clinic or on heroin in the street, tolerance does not ordinarily become apparent until about 2 or 3 weeks after frequent, daily administration. The patient senses a decrease in the effectiveness of the opiate to relieve pain, and the addict to produce a "rush" and euphoria. Both subjects also become more resistant to the respiratory-depressant and the antianxiety effects of their respective opiate but neither experiences much tolerance to their miotic or constipating effects. Thus, tolerance as a clinical entity may be discriminated qualitatively, but quantitative assessments are more complicated and the implementation requires that some degree of arbitrariness be imposed.

In designing meaningful laboratory experiments to test concepts that purport to explain tolerance mechanisms, the investigator encounters difficulties in attempting to evaluate the myriad of effects associated with clinical tolerance and hence must be parsimonious in approach. Certain assumptions are usually made that, in each instance, need to be validated later. In the assessment of tolerance in an intact preparation (animal or man) a single response is selected (usually analgesia) on the assumption or in the hope that the basic processes involved may be similar to those for euphoria, respiratory depression, and reducing anxiety, if not for miosis and constipation. Simple in vitro models are selected for similar reasons. However, in becoming reductionistic, one should not forget that integrative effects may nullify the information derived from a simple pathway or cellular system, especially since some responses seem to be mediated by groups of neurons acting in concert. For example, learning has been shown to take place not just at an individual nerve cell of the brain, but within groups of neurons acting in concert (Merzenich 1987). This may be of significance since analogies might be made to short- and long-term memory that could have relevance to acute and protracted opiate tolerance and physical dependence.

The caveats with respect to the assessment of tolerance hold equally well for the assessment of physical dependence. However, the latter phenomenon is even more complex. The behavioral signs that occur are peculiar for a given species. The parameters include a constellation of signs and symptoms of varying chronology and intensity that can interact and influence each other as well (Bläsig et al. 1973).

In the clinic, opiate physical dependence can be quantified using a point scoring system that associates the severity of the syndrome with selected signs and symptoms. Thus, lacrimation, rhinorrhea, yawning, hyperthermia, hyperventilation, diarrhea, etc. can be assigned numerical values, and the total scored over a fixed period provides a measure of the degree of physical dependence (Himmelsbach 1942; Quock et al. 1968). Such a global index, however utilitarian for clinical assessments, is difficult to apply for mechanistic evaluations.

In the experimental laboratory, although global assessments can be made, it becomes practical to select certain signs of the abstinence syndrome where a response is exaggerated and easily measured, such as, weight loss, hyperthermia, urination, defecation, jumping, etc. (Wei and Way 1975). Under such circumstances, the paramount consideration is whether the selected parameter truly reflects physical dependence. Even if this proves to be the case, the delayed and varying temporal effects of abrupt withdrawal make assessments inconvenient. To overcome this, antagonist-precipitated withdrawal has been widely employed. Unlike abrupt withdrawal, antagonist-precipitated abstinence is an explosive event in which virtually all of the abstinence signs occur with great intensity within a matter of minutes and then usually subside in less than 1 h (Wikler et al. 1953; Way et al. 1969).

The question that arises, however, is whether antagonist-precipitated withdrawal truly reflects the dependent state. By definition, agonist occupation of the receptors is required for the dependent state, and agonist removal for the abstinence syndrome. Since the receptors at many body sites might have different affinities for the agonist and if not, at least the clearance of agonist from these sites may vary due to differences in blood flow, it should be expected that the rate of agonist release should determine the onset and intensity of the abrupt withdrawal syndrome. On the other hand, the injection of an antagonist should displace the agonist almost immediately from all receptor sites so that the constellation of withdrawal signs and symptoms should appear almost simultaneously and with great intensity. We thought we had provided the evidence for this explanation (Shen and Way 1975) but, unfortunately, others did not agree (Dum et al. 1975; Catlin et al. 1977). However, I think both critics would prefer to believe that we are correct so let us, for the sake of argument, assume that we are. Even then there would still be some unresolved questions about the true nature of precipitated abstinence.

If antagonist-induced withdrawal is indeed indicative of the physically dependent state, we then encounter problems in semantics. Could antagonist reversal of an acute agonist effect then be a reflection of the incipient dependent state? Consider the following points. To have physical dependence there must be agonist occupation of the receptors and the related processes must have been initiated at the first agonist dose. On certain receptors, agonist displacement, i.e., withdrawal, can result in an overshoot rebound of the response that is depressed. By using an antagonist to effect rapid agonist removal, a supersensitivity can, in certain instances, be observed even after a single agonist dose. Furthermore, it is easily demonstrable that the NAL reversal of acute morphine overdosage is most dramatic with high doses of morphine. The higher the degree of respiratory depression attained, for example, the more pronounced is the hyperventilatory response observed after antidoting with an antagonist. This phenomenon appears to bear a close relationship to the well-established fact that an inverse relationship exists between the degree of physical dependence and the dose of antagonist needed to precipitate abstinence. In contrast to tolerance, where there is a decrease in sensitivity to agonists as tolerance increases, there is an increasing sensitivity to antagonists as physical dependence increases (Way et al. 1969). Taken together, these facts indicate that the line between antagonist-precipitated abstinence and antagonist reversal of acute agonist overdosage must be very fuzzy and difficult to define.

Antagonist-precipitated withdrawal might also be used to assess protracted physical dependence. In the experimental laboratory, it has been difficult to demonstrate protracted dependence on opioid agonists. Protracted abstinence in the rat was reported to last for as long as 4–6 months after withdrawal of morphine (Martin et al. 1963), but the time requirements and differences in signs between the addicted and control groups are not of the magnitude to encourage studies of the syndrome. However, we obtained evidence that the use of an antagonist might facilitate experiments in this area.

We accidentally noted that postmorphine-dependent mice which, over time, had lost their precipitated withdrawal responsivity to NAL, regained their sensitivity to NAL when primed with morphine. For pretreatment, a relatively small dose of morphine, which ordinarily would not be expected to induce physical dependence, was used (Brase et al. 1976). The mice were rendered dependent on morphine by pellet implantation (Way et al. 1969; Wei and Way 1975), a technique which is now widely used. At fixed intervals between 1 and 43 days after removal of the pellet, the capacity of NAL ED_{50} to precipitate withdrawal jumping in the postaddict mice was determined after 1-h pretreatment with morphine sulfate (30 mg/kg, s.c.) or saline. Placebo pellet-implanted mice pretreated with morphine served as the abstinent control group (Brase et al. 1976).

Marked development of physical dependence after a 3-day morphine pellet implantation period was demonstrable by the small NAL ED_{50} required to produce jumping. After the morphine pellet was removed, the mice lost most of their sensitivity to NAL within 1 day. Thus, even a near-convulsant dose of NAL (80 mg/kg) elicited only minimal withdrawal responses. Compatible data were reported by others (Cheney and Goldstein 1971; Wei and Loh 1972). If, however, abstinent mice were treated 1 h earlier with morphine, sensitivity to NAL was restored and could repeatedly be demonstrated over an extended period of time (Brase et al. 1976). This response to priming with morphine decayed so slowly that, even after 43 days, the capacity of NAL ED_{50} to induce jumping could be determined. Other NAL-precipitated withdrawal signs were also in evidence, including rearing, paw tremor, salivation, and body shakes. Under similar conditions, placebo-implanted animals primed with morphine showed no increase in sensitivity to NAL. Thus, prior dependence on morphine markedly increases sensitivity to NAL-precipitated abstinence after priming with morphine. Furthermore, that the effect reflected a true opiate response was confirmed by the usual criteria.

The sensitivity to NAL in morphine postaddict mice was found to be stereospecific. It could be demonstrated with the active isomers of morphine, levorphanol, and methadone, but not with dextrorphan, the inactive analgetic enantiomorph of levorphanol.

The ease of reinducing NAL-induced jumping after a single morphine injection in abstinent mice can be attributed to a remanifestation of an existing state of physical dependence. The induction of physical dependence evidently causes a homeostatic adaptative process which persists in a latent state but which can be unmasked by the administration of a single dose of an agonist. Apparently, agonist priming in postaddict mice induces the responding receptors to return to a conformational state that again becomes sensitive to antagonist challenge.

In summary, a precise measure of tolerance and physical dependence is extremely difficult because both processes are continually increasing as they are exposed to agonist which, in turn, by its presence complicates quantitative estimation. In theory, the determination of tolerance and physical dependence should be made at an agonist level that produces no acute agonist effects but suffices to suppress withdrawal hypersensitivity. When agonist potency is used to assess tolerance and the estimation of the agonist ED_{50} is made prior to this critical point in time, any agonist still present should contribute to acute actions and to lower the ED_{50} so that the degree of tolerance would be underestimated. On the other hand, if the level of agonist is not high enough to prevent abstinence, more agonist would be required to produce the desired acute agonist effect and, the resulting elevated ED_{50} would provide an overestimate of the degree of tolerance. Likewise, if the antagonist ED_{50} is used as an index for assessing physical dependence, the presence of agonist is required for manifesting dependence; but if the level of agonist is higher than that needed for preventing abstinence, an underestimate of the degree of physical dependence would result because the excess agonist would oppose the antagonist's action and the antagonist ED_{50} would tend to be elevated. Thus, in estimating tolerance and physical dependence it is important to remember that the acute and chronic actions of an agonist are in opposition, and that the two processes related to the latter effects are dynamic and continuously changing.

2 In Vitro Methodology

In earlier in vitro attempts to elucidate the basic processes involved in opiate actions, experiments were performed utilizing brain preparations such as homogenates and enriched nerve endings (synaptosomes). Although considerable data were generated, it was not always possible to temporally relate such findings to electrophysiological events. In recent years, efforts have been directed more toward using excised tissue preparations or cell cultures that exhibit selective responses to opioids. In particular, the guinea pig ileum, the mouse vas deferens, and the hybrid neuroblastoma × glioma cell have been used in our laboratory. Others have used sensory neurons, generally the dorsal root ganglion of mouse or chicken embryos in culture. The development of the transverse hippocampal slice, in which the local circuitry is essentially preserved for intracellular recording, has facilitated electrophysiological analysis of drug action (Nicoll 1988) but the technique has had limited use for studying opioid tolerance and physical dependence.

2.1 Guinea Pig Ileum (GPI)

This popular, isolated longitudinal muscle preparation for studying acute opioid effects and characterizing multiple opioid receptors by cross-tolerance experiments has also been useful for investigating opioid tolerance and dependence. In earlier studies, tolerance in the excised ileum was induced by prior chronic administration of an opioid to the intact animal (Mattila 1962; Haycock and Rees 1972; Goldstein and Schulz 1973; Ward and Takemori 1976; Huidobro-Toro et al. 1978). Later, it was found that tolerance could also be induced directly in vitro, by incubating the ileum in a bath containing an opioid (Hammond et al. 1976; Opmeer and van Ree 1978; North and Karras 1978; Collier et al. 1981). The procedure has been standardized for making quantitative assessments and this technique has facilitated the work. As much as threefold tolerance can be demonstrated after incubation of an IC_{50} concentration of certain opioids for 1–2 h at 37°C (Rezvani et al. 1983).

Physical dependence on opioids has also been demonstrated in the GPI and may be induced in vivo or in vitro. The criterion for the development of physical dependence is the ability of NAL to induce contraction of the longitudinal muscle (Ehrenpreis et al. 1972; Schulz et al. 1982), or an exaggerated response to electrical stimulation after rapid agonist removal (Rezvani and Way 1983). Studies of the nerve impulses from opioid-tolerant neurons of the myenteric plexus by means of suction electrodes indicated increased firing in these neurons during precipitated withdrawal (North and Karras 1978).

The validity of the GPI system for the study of tolerance and physical dependence was further established by the demonstration that the development of both phenomena is stereospecific, NAL reversible, and homologous in that sensitivity to other neurotransmitters, such as norepinephrine (NE) or adenosine, is not altered. Additionally, tachyphylaxis and other effects similar to those seen in the whole animal have been noted in vitro (Shohan and Weinstock 1974; Cox et al. 1975; Gillan et al. 1979). Another example is the decrease in the concentration of NAL required to induce contracture with increasing physical dependence (Goldstein and Schulz 1973; Collier 1980).

2.2 *Mouse Vas Deferens (MVD)*

The isolated MVD contains multiple opioid receptors and is highly sensitive to opioids. Like the GPI, field stimulation of the preparation evokes a twitch response that is inhibited by opioids. The preparation can be used for studying acute and chronic opioid effects (Gillan et al. 1979). Excised tissue preparations from mice chronically exposed to an opioid via osmotic minipumps develop an extremely high degree of tolerance (Schulz et al. 1980). Although, it has been claimed that under such conditions no physical dependence could be demonstrated and, therefore, tolerance and physical dependence are unrelated (Wüster et al. 1985; see Schulz, this Vol.), data from our laboratory document that tolerance and physical dependence can be demonstrated by recording a common response in the same preparation.

Tolerance and physical dependence were induced in vitro by incubation of the MVD with a predetermined concentration of D-alanine-D-leucine-enkephalin (DADLE) that completely abolished the electrically induced twitch response. Dependence was evidenced by the fact that, in the presence of NAL, a higher twitch response than that of the control preparation could be elicited. The extent of the supersensitivity was noted to be directly related to the agonist concentration and the degree of tolerance development. Furthermore, the development of supersensitivity was found to be stereospecific, NAL reversible, and preferentially μ-receptor mediated. Moreover, following the development of tolerance, the NAL-induced supersensitivity to electrical stimulation was concentration-dependent. Finally, on assessing the level of NE in the MVD, it was noted that DADLE decreased NE levels in the naive preparation and that the preparation could be rendered tolerant to this effect by in vitro incubation with DADLE. Subsequently, upon addition of NAL, supersensitivity occurred, as evidenced by an overshoot in NE release (Song et al. 1987). Thus, it appears that when tolerance and physical dependence are assessed by a common response in specific model systems, it is possible to demonstrate an association between the two phenomena.

2.3 *Neuroblastoma × Glioma NG 108-15 Hybrid Cells*

Study of the molecular mechanisms of opiate tolerance and physical dependence has been facilitated by the availability of this hybrid culture. This simple model lends itself easily to biochemical manipulation and analysis. It retains many neuronal characteristics including excitable membranes, dense core vesicles and the ability to form synapses after differentiation (Hamprecht 1977). Interest in this system was aroused after it was discovered that inhibition of adenylate cyclase by opiate agonists in a stereospecific and NAL-reversible manner could be demonstrated (Sharma et al. 1975). Subsequent studies revealed the presence of specific opiate binding sites of the δ-subtype which couple to adenylate cyclase to inhibit prostaglandin E_1 stimulated production of cAMP (Klee and Nirenberg 1974; Chang et al. 1978; Law et al. 1983a).

Tolerance to the inhibitory effect on adenylate cyclase could be demonstrated in less than 24 h after exposure of the cultured cells to opiates. Moreover, physical dependence development appeared to be present in the same preparation, as evidenced by a rebound increase in intracellular cyclic AMP levels, relative to controls, after addition of NAL (Sharma et al. 1975, 1977). This preparation has been used extensively to elucidate the biochemical events involved in opioid-adenylate cyclase interactions and this topic will be discussed later.

2.4 Dorsal Root Ganglion Cultures

Perikarya of embryonic chick and rodent dorsal root ganglia in culture, which have been used to study the acute effects of opioids, have been applied more recently to study the chronic effects of opiates. Intracellular recordings from explants of these sensory neurons indicate that tolerance can develop to produce opioid-inhibiting effects on evoked sensory responses, Ca^{2+} influx, and transmitter release. The mouse preparation has been used extensively to implicate the cAMP system in the development of tolerance to opioids and these series of studies have been summarized elsewhere (Crain et al. 1988).

3 Mechanisms Relating to Tolerance and Physical Dependence

In the 1960s and 1970s, numerous attempts were made to implicate certain neurotransmitters in the genesis of opiate tolerance and physical dependence. A tremendous amount of literature was generated and the subject has been repeatedly reviewed. Among the native ligands subjected to intensive scrutiny were acetylcholine, NE, 5-hydroxytryptamine, dopamine and, γ-aminobutyric acid (GABA). Although changes in the disposition of these neurotransmitters were noted to occur during tolerance and dependence development, none could be readily identified as the sole or primary causative agent associated with the two phenomena. It became increasingly apparent, therefore, that the changes after chronic exposure to an opiate were more apt to occur at receptor sites involving second messengers associated with effects on adenylate cyclase or ion channels. Such views were given impetus by prior behavioral and biochemical evidence.

An alteration in the characteristics of the receptor with tolerance/dependence development had been shown earlier in that, with increasing dependence, less antagonist is required to precipitate abstinence signs (Wikler et al. 1953; Way et al. 1969; Kosersky et al. 1974; Jacob et al. 1974). It has also been proposed that, with continuous morphine administration, a qualitative change of the opiate receptor occurs as tolerance increases and that this renders the receptor increasingly sensitive to NAL (Takemori et al. 1973; Tulunay and Takemori 1974; Tang and Collins 1978).

The finding in mice that the decrease in sensitivity to an agonist following repeated exposure to an opioid agonist is accompanied concomitantly by enhanced sensitivity to an antagonist suggests that the two phenomena might result from alterations at common receptor sites (Way et al. 1969). Indeed, Collier (1965, 1980) had earlier proposed that the manifestations of tolerance after sustained opiate administration could be explained by a decrease in efficiency or number of opiate receptors. Using more recent molecular parlance, the former effect may be attributed to a decrease in opiate receptor affinity or "desensitization" and, the latter, to receptor "downregulation" or "internalization".

Although changes in receptor affinity or number appear to be a logical approach toward studying tolerance and physical dependence, proof of the notion has not been easy, as evidenced by many binding studies with contradictory findings (Höllt et al. 1975). However, more recent studies suggest that under proper conditions, changes in affinity and number of opiate receptors may indeed occur during tolerance and dependence development, and that certain processes may be related to events associated with cyclic AMP or Ca^{2+}.

Early pharmacological evidence pointed to the possibility that cAMP might be involved in opiate action. Prior to the availability of methodology for measuring cAMP in opiate action, it was noted that the administration of cAMP could not only antagonize the analgetic effects of morphine (Ho et al. 1972), but that cAMP could also accelerate the development of tolerance and physical dependence (Ho et al. 1973). Based on these findings, it was reasoned that the opiate abstinence syndrome might be mimicked by elevated levels of cAMP and, indeed, it was found that opiate abstinence signs could be intensified in the opiate-dependent rat by giving IBMX, a cAMP derivative. Moreover, a quasi-withdrawal syndrome induced in naive rats by administering theophylline, a phosphodieserase inhibitor, could be augmented by NAL or attenuated by heroin (Collier and Roy 1974; Collier and Francis 1975). These early results provided compelling circumstantial evidence to suggest that cAMP might be involved in the acute and chronic effects of opiates.

Subsequently, cAMP was implicated in opioid action in neuroblastoma × glioma NG108-15 hybrid cells, mentioned earlier, by the observation that adenylate cyclase was inhibited by opiate agonists which bind to δ-receptors (Klee and Nirenberg 1974; Chang et al. 1978; Law et al. 1983a). Following the demonstration that tolerance develops to this inhibitory effect of morphine on adenylate cyclase after chronic exposure, and that physical dependence was present in the same preparation, as evidenced by a rebound increase in cyclic AMP levels, the adaptive mechanism was proposed to occur as follows. The opiate receptor was postulated to couple with, and inhibit, adenylate cyclase; however, homeostatic changes within other regulatory systems tended to counteract this opiate action. Removal of the opiate results in a compensatory increase in adenylate cyclase activity that becomes manifest as a rebound elevation in cAMP levels (Sharma et al. 1977). However, in light of more recent evidence, the proposed mode of opiate interaction with adenylate cyclase has undergone modification.

It is now well established that receptor effects on adenylate cyclase are mediated by guanine nucleotide regulatory proteins or G-proteins and that signal transduction pathways use G-proteins to couple neurotransmitter receptors to effector molecules (Rodbell 1980; Gilman 1984; Lang, this Vol.). More recent work on opioids has exploited these findings, and there is now considerable evidence showing that the effects of opioids involve an inhibitory protein, G_i (Koski and Klee 1981; Costa et al. 1983; Wüster and Costa 1984). The main evidence rests on experiments with pertussis toxin (PTX) which is recognized to inhibit G_i activity by catalyzing ADP ribosylation and reducing GDP hydrolysis (Murayama and Ui 1983). More recently, a stimulatory protein, G_s, has also been implicated in opioid action. In our laboratory, Law and associates demonstrated that the receptor alterations occurring after chronic opioid administration can be dissociated into two processes with different rates of appearance, desensitization (a relatively rapid event, discernible within 2 h) and downregulation (a slower process, becoming apparent after 24 h).

In a series of studies on the NG108–15 cell, Law and co-workers observed that, following chronic opioid administration, a reduction of agonist affinity and an uncoupling of receptor from adenylate cyclase occurred; this was attributed to a desensitization process (Law et al. 1983b; Louie et al. 1986). In other studies, it was reported that under certain conditions, opioid receptors on NG108–15 cells could exist in multiple affinity states and that these affinity states are related to the coupling of the opiate receptor to the inhibitory transducing protein, G_i, which regulates adenylate cyclase activity. It was proposed that the equilibrium between a high-affinity and low-affinity state of the opiate receptor after coupling bears a relationship to opiate tolerance. At any given time, a percentage of opiate receptors can be coupled to G_i and, in this coupled state, a higher affinity for opiate agonist is manifested. An agonist, by inducing coupling, can therefore increase the number of receptors in the high-affinity state. After prolonged agonist exposure, however, this distribution changes and the receptors in the high-affinity state become reduced while those of low affinity are increased. As this occurs, opiate inhibition of adenylate cyclase decreases. This tolerant state can be mimicked by PTX which, by covalent modification of the G_i protein, promotes uncoupling and alters the distribution of opiate receptors between the high- and low-affinity states. As a consequence, opiate inhibition of adenylate cyclase becomes blocked. The G_i protein, however, remains functional after exposure to opioids because, despite the presence of opiate tolerance, inhibition of adenylate cyclase could still be effected by ligands active on other types of receptor. Thus, it was reasoned that a specific lesion of the opiate receptor must be causing the tolerance (Law et al. 1985a). These findings on cultured cells have been extended to the brain. Opioid inhibition of adenylate cyclase activity in the striatum, but not in other brain regions, has been noted and the effect was found to be dose-dependently attenuated by pretreatment with PTX (Abood et al. 1985).

More recent studies reveal that the manner in which the opiate receptor is altered by desensitization during tolerance development may be related to effects on dephosphorylation. Using NG108–15 cell membranes, it was found that AppNHp, a nonhydrolyzable ATP analog which favors dephosphorylation, enhanced the development of tolerance, whereas ATP, which favors phosphorylation, did not (Louie et al. 1988). Additionally, direct treatment of NG108–15 membranes with a phosphatase resulted in a decrease in opiate inhibition of adenylate cyclase (Louie et al. 1988), and a purified opioid receptor from rat brain has been shown to exhibit phosphatase activity (Roy et al. 1986). Compatible with our findings, experiments by others with synaptic membranes from rat striatum also suggest an association between a dephosphorylated state and opiate tolerance, in that chronic morphine treatment reduced phosphorylation of certain proteins (O'Callaghan et al. 1979).

Opioid receptor downregulation is also a stage in the development of tolerance. A decrease in the amount of binding sites was shown by radioactive ligand-binding assays in neuroblastoma × glioma NG108–15 hybrid cells. After chronic treatment of the hybrid cells with [^{3}H]DADLE, accumulation of the [^{3}H]DADLE in the lysosome-enriched subcellular fraction was demonstrable when chloroquine was used to prevent receptor lysosomal degradation. The tritiated enkephalin (ENK) was shown to be associated with macromolecules in the lysosomes, indicating that the ligand-receptor complex is delivered, after internalization, to the lysosomes where it is degraded. Not only is opiate receptor downregulation in NG108–15 cells homologous and agonist concentration-dependent, but it is also time- and temperature-dependent, and blocked by metabolic inhibitors such as sodium azide and 2,4-dinitrophenol. These characteristics are compatible with the possibility that internalization of the ligand-receptor complex occurs during chronic opiate treatment (Law et al. 1984).

Consistent with the above findings and conclusions, Blanchard et al. (1983) noted accumulation of [^{3}H]DADLE in neuroblastoma N4TG1 cells after chronic treatment with the compound. Because the time course of the [^{3}H]-DADLE accumulation paralleled that of receptor downregulation and because both accumulation and receptor downregulation could be blocked by metabolic inhibitors, it was concluded that the findings reflect internalization of the opiate receptor. Furthermore, since, by the use of a fluorescent ENK analog, clustering of the opiate receptor in the neuroblastoma cell lines was observed, it was postulated that opiate receptor internalization involves coated pits and vesicles (Hazum et al. 1979). Morphine, a partial agonist in this system (Simantov and Amir 1983; Law et al. 1983b) could induce clustering of receptors but failed to effect downregulation. Thus, whenever an opioid agonist was demonstrated to downregulate the receptor, a full δ-opioid agonist was required to decrease the number of opiate receptors. It is notable, moreover, that the receptor subtype in the N4TG1 and the neuroblastoma × glioma NG108–15 hybrid cell appears to be exclusively of the δ-type. Also, compatible with these data, downregulation of δ-, but not μ-, opioid receptors was shown to occur in hippocampal slices (Dingledine et al. 1983).

Downregulation represents a later phase in opiate tolerance development since tolerance to opiate inhibition of adenylate cyclase in the NG108–15 cell may occur prior to evidence for downregulation. Also, chronic exposure to morphine, levorphanol, or cyclazocine, which all act as partial agonists in these cells, results in loss of inhibition of adenylate cyclase but not in downregulation of the receptor. Finally, even with full agonists, the downregulation lags behind the loss of opiate inhibition of cyclase and a significant amount of tolerance develops prior to the demonstration of receptor internalization (Law et al. 1983b).

The above findings in the NG108-15 cell have been incorporated into an explanation for physical dependence. The mechanism of the rebound increase in adenylate cyclase activity after chronic exposure to opiates and its removal, suggests involvement of at least one other G-protein besides G_i. Presumably, the receptor must still be coupled to some effector for the displacement of agonist to cause such a withdrawal reaction. It was reasoned, therefore, that when the opiate receptor becomes uncoupled from G_i it may still remain coupled to another transducer which manifests physical dependence. Evidence was provided for the involvement of both inhibitory and stimulatory guanine nucleotide binding proteins in the expression of chronic opiate regulation of adenylate cyclase activity in NG108-15 cells (Griffin et al. 1985). Inactivation of G_i by PTX resulted in elevated adenylate cyclase activity, comparable to that following chronic opiate treatment. However, the enzymatic activity of cells exposed to opiate previously could not be further increased by PTX treatment. In addition, procedures that prevented receptor-mediated activation of G_s, such as treatment with NaF or desensitization of the stimulatory receptors (prostaglandin E_1, adenosine), eliminated the increase in adenylate cyclase activity induced by NAL following chronic opiate exposure. As a consequence, it was proposed that the catalytic subunit of adenylate cyclase in NG108-15 cells is under tonic regulation by both G_i and G_s. Hence, the increase in enzymatic activity observed following chronic opiate treatment may be due to a loss of tonic inhibitory regulation of adenylate cyclase mediated through G_i, resulting in the unimpeded expression of G_s activity.

The search for the pathway expressing physical dependence in NG108-15 cells also led to attempts to isolate a cytosolic factor which might accumulate during chronic opiate exposure and stimulate adenylate cyclase. Two factors were found that stimulate adenylate cyclase, individually and synergistically. One of these factors was adenosine (Griffin et al. 1983) and the desensitization of NG108-15 cells by chronic exposure to adenosine decreased the NAL-induced stimulation of adenylate cyclase. The other factor from cytosol of physically dependent cells appears to be calmodulin, which has been shown to act directly upon the catalytic subunit of adenylate cyclase (Salter et al. 1981). Several studies note an interaction between opioids and calmodulin. This factor stimulates adenylate cyclase in a Ca^{2+}-dependent manner, which is consistent with findings that the rebound increase in cAMP levels during opiate withdrawal could be blocked by Ca^{2+} chelators (Law et al. 1984). Also, in NG108-15 cells, opioids have been reported to decrease the amount of calmodulin bound to plasma membrane (Nehmed et al. 1982), and to control calmodulin distribution in neuroblastoma glioma cells (Baram and Simantov 1983). In addition, morphine and β-endorphin were found to inhibit the high Ca^{2+} affinity/Ca^{2+}-ATPase of synaptic plasma membranes which is believed to be involved in the regulation of intracellular Ca^{2+} that triggers the release of neurotransmitters. The inhibition was blocked by NAL and reversed by calmodulin. Kinetic analyses revealed that opioids decreased the affinity of Ca^{2+}-ATPase for Ca^{2+}, and Ca^{2+} bound to the enzyme decreased the affinity of the enzyme for opioids (Lin and Way 1986).

Studies by other laboratories yielded data which implicate G-proteins in opioid tolerance and physical dependence processes. In the GPI, PTX was reported to inhibit opioid dependence without effecting acute potency (Collier and Tucker 1984; Lux and Schulz 1985). It was found that the toxin did not materially alter opioid inhibition of electrically stimulated contractions, but attenuated NAL-induced contracture after chronic exposure to agonist. That cAMP may be involved in the chronic and acute responses to opioids is indicated by studies on explant cultures of fetal mouse spinal cord with attached dorsal root ganglia (DRG). Acute exposure of the culture to opioids results in stereospecific, NAL-reversible, dose-dependent depression of sensory-evoked dorsal-horn synaptic-network responses. Chronic exposure of the explants to morphine and other opioids leads to a marked attenuation of these depressant effects. Brief treatment of explants with cAMP analogs or with forskolin, a selective activator of adenylate cyclase, also markedly attenuates the depressant effects of opioids. Moreover, exposure of explants to PTX blocks these depressant effects, indicating that the guanine nucleotide protein G_i is required for opioid receptor-mediated inhibition of adenylate cyclase. However, NAL did not elicit a rebound overshoot (Crain et al. 1986).

Intracellular recordings from DRG neurons in cord explants yielded compatible data. Opioid shortening of the action potential duration (APD) of DRG neuron perikarya has generally been accepted as an indication of inhibition of Ca^{2+} influx and transmitter release at presynaptic DRG terminals. Acute exposure of DRG neurons to the opioid peptide DADLE attenuated Ca^{2+} influx, as evidenced by the marked shortening of the duration of the Ca^{2+} component of the APD. In contrast, after chronic exposure the DRG perikarya became tolerant to this DADLE-inhibiting effect. Furthermore, these resistant neurons showed APD prolongation when tested acutely with higher concentrations of DADLE, similar to the effects of PTX or forskolin. In addition, opioid responsitivity tests in the presence of multiple K^+-channel blockers revealed (1): that DADLE may induce APD prologation of DRG neurons via novel opioid receptor subtypes which decrease a voltage-sensitive K^+ conductance; and (2) that conventional inhibitory opioid receptors may function within the same DRG cells by reducing a voltage-sensitive Ca^{2+} conductance. It was postulated that the opioid-induced APD prolongation provides evidence that opioids can also evoke direct excitatory effects on neurons. cAMP-enhanced excitatory, versus inhibitory, opioid receptor function in central DRG terminals may modulate presynaptic DRG inputs to the CNS during the development of tolerance (Crain 1988; Crain et al. 1988).

Reconstitution of purified μ-opioid receptors with purified guanine nucleotide binding regulatory proteins. G_i and G_o, has been claimed (Ueda et al. 1988) and later, also with δ- and κ-receptors (Ueda and Satoh 1988b); however, these exciting results remain to be confirmed. The method used for the purification of the μ-receptor appears to be virtually identical to that used to isolate a purified opiate receptor, but this purified protein shows a lack of membrane-spanning domains and may not be a G-protein-coupled opioid receptor (Schofield et al. 1989). Additional experiments are necessary to resolve this apparent inconsistency.

In summary, there is substantial evidence indicating a role for cAMP in tolerance and physical dependence development: However, definitive as the data on cultured cells may be, their immediate significance rests largely on their relevance to clinical and field situations. Even though some of the findings have been extended to the striatum and cortex, these are not the brain sites that are ordinarily examined for the commonly recognized opiate effects. Nor is the δ-receptor considered to be the prime site for expression of the classic signs of opiate withdrawal. Also, the fact that morphine acts as a partial agonist on hybrid culture cells raises further issues concerning the meaningfulness of the data. Despite these doubts, it is likely that it will be possible to incorporate considerable portions of the data derived from cultured cells into the body of evidence that will help promote an understanding of the basic processes involved in tolerance and physical dependence. In any event, the techniques and concepts that have been applied will certainly help guide the way for future approaches.

The story that unfolds with respect to cAMP effects on opioid action will almost certainly be meshed with those of Ca^{2+}. The involvement of G-proteins in the gating of Ca^{2+} by receptors support this notion (Gomperts 1983). Although the molecular processes with respect to Ca^{2+} and opioid interactions have not been delineated in detail as with cAMP, the data derived from whole animals and brain tissues have been woven into a plausible hypothesis.

A role for Ca^{2+} in opiate action has long been advocated based initially on pharmacological evidence. More recent findings may provide an explanation for some of these effects and implicate Ca^{2+} more definitively in opiate action and link its effects to the cAMP system. In particular, the new knowledge concerning the role of the polyphosphoinositide (PI) system in Ca^{2+} regulation and the isolation of what appears to be an opioid receptor segment may provide a gateway for understanding opioid action (see Smith and Loh, this Vol.). The purified opioid glycoprotein binds ligands when reconstituted with acidic lipids and possesses sequence homologies with certain other isolated proteins suggesting that it could be involved in cell recognition and adhesion, peptidergic ligand binding, or both. Located extracellularly, the purified opioid protein appears to be anchored to the membrane by lipids and it is likely that the C-terminal sequence of the protein serves as a signal for PI attachment to the cellular membrane (Schofield et al. 1989).

A role for PI-linked proteins in signal transduction is derived from findings that the release of the protein component of PI-linked proteins simultaneously releases phosphatidylinositol, diacylglycerol, or phosphatidic acid (Abdel-Latif 1986) which concomitantly affect cellular responses, in particular the mobilization of Ca^{2+} (Berridge and Irvine 1984) and the activation of protein kinase C (Nishizuka 1988). Possibly, cell contact or ligand binding of some PI-linked molecules, such as the characterized opioid receptor protein, may activate membrane-bound phospholipases which could cleave the PI-linkage molecules and initiate the PI-mediated release of free cellular Ca^{2+} and activation of protein kinase C. Hence, the isolated material has been designated OBCAM (opioid binding protein-cell adhesion molecule; Schofield et al. 1989). Thus, the earlier data implicating Ca^{2+} in opioid effects would not rule out a role for cAMP; indeed, the involvement of inositol phospholipids in opioid action may be the mechanism by which opioid effects on Ca^{2+} and cAMP are linked.

Two acute effects of opioids commonly observed are a reduction in neurotransmitter release and inhibition of neuronal electrical activity. The electrophysiological evidence indicates that opiates produce a decrease in Ca^{2+} influx by depression of voltage-sensitive Ca^{2+} channels (Mudge et al. 1979) and membrane hyperpolarization of cell bodies by increased K^+ conductance (North and Williams 1983). Another electrical finding that may be very important is that opiates also prolong Ca^{2+} dependence after hyperpolarization. This topic has been extensively reviewed (Duggan and North 1983; North, this Vol.). All three electrical findings may be related to Ca^{2+} disposition and may be explained by an opiate-induced decrease in intracellular Ca^{2+} binding. For example, it has been shown, in hippocampal slices, that elevated cystolic Ca^{2+} can cause increased K^+ conductance (Nicoll 1988). However, it is likely that opioids can alter Ca^{2+} disposition by modes that are tissue-, site-, receptor-, and time-dependent.

Early pharmacological evidence pointed the way, and provided strong circumstantial evidence, to implicate Ca^{2+} in opioid action. For instance, many of the observed effects of opiates can be altered by manipulating cellular Ca^{2+}. Invariably, maneuvers that tend to elevate neuronal Ca^{2+}, either with Ca^{2+} itself or with Ca^{2+} ionophores, reduce opiate action. On the other hand, procedures that lower cytosolic Ca^{2+}, such as reductions in the extracellular Ca^{2+} concentration, removal of extracellular Ca^{2+} with chelating agents, or blockage of Ca^{2+} entry with La^{3+}, all enhance opiate action (Chapman and Way 1980; Schmidt and Way 1980).

A hypothesis to explain the development of tolerance to, and physical dependence on, opiates was formulated on the basis of evidence that (1) Ca^{2+} and its ionophores antagonize opiate action; (2) Ca^{2+} antagonists (La^{3+} or EGTA) enhance opiate action; (3) cross-tolerance to La^{3+} and EGTA is exhibited during the morphine-tolerant state; (4) opiate abstinence can be alleviated by reducing neuronal Ca^{2+}; (5) acute opiate administration lower Ca^{2+} at synaptic plasma membranes and synaptic vesicles; and (6) chronic opiate administration elevates neuronal Ca^{2+} at these sites (Chapman and Way 1980).

It was suggested that opiates affect concomitantly, or sequentially, more than one Ca^{2+} compartment: the Ca^{2+} channel itself and intracellular Ca^{2+} binding sites, such as those on the inner membrane and/or synaptic vesicles. Depending on the loci involved, one may be more important than the other for the manifestation of the acute response. Thus, opiates could cause a transient displacement of Ca^{2+} from its intracellular binding sites to effect increased K^+ conductance, membrane hyperpolarization, decreased neuronal firing, and transmitter release. The Ca^{2+} released by opioids might act to reduce further Ca^{2+} entry, or the Ca^{2+} channel itself may be blocked by opiates, resulting in a fall in neuronal Ca^{2+}. However, this lowering sets into motion other homeostatic processes within the cell which become increasingly manifest as opiate administration continues. In particular, the counteradaptive effect occurs at the intracellular binding sites. The displacement of Ca^{2+} from its binding sites in the presence of the opiates becomes more difficult so that a higher dose of the opiate is required before an acute response can be elicited (tolerance). However, the counteradaptive effect to retain Ca^{2+} within the cell also increases, but requires the presence of the opiate (physical dependence). Removal of the opiate by discontinuing its administration or by an antagonist results in a rise in intracellular Ca^{2+}, higher excitability, and increased neurotransmitter release (abstinence syndrome; Chapman and Way 1980).

The increase in Ca^{2+} accumulation requires the continual presence of the opiate and this may explain physical dependence development. Thus, when opiate discontinuance or antagonist treatment removes the agonist, the high synaptosomal Ca^{2+} content in the absence of the agonist produces greatly increased neurotransmitter release. This neuronal hyperexcitability then gives rise to withdrawal signs and symptoms. Hyperalgesia, for example, may occur after dissipation of morphine at receptor sites below a critical level. Indeed, the hyperalgesia can be further enhanced by intracerebroventricular injection of Ca^{2+}. Also, according to this model, the abstinence syndrome can be attenuated by reducing intracellular Ca^{2+}. Compatible with this notion, La^{3+} administration reduces abrupt or NAL-induced withdrawal jumping in mice.

This model further suggests that Ca^{2+} administration, by opposing opiate effects on intracellular binding, should reduce tolerance development, whereas EGTA, by decreasing Ca^{2+} availability to the same site during opiate administration, should enhance tolerance development. Thus, the hypothesis suggests the Ca^{2+} site for mediating acute pharmacological effects may be distinct from that for producing tolerance and physical dependence. The acute effects may relate to gating at ion channels, whereas the chronic effects would involve slower events related to the cAMP system, adaptive changes in Ca^{2+} disposition, downregulation, and nuclear chromatin synthesis.

Another explanation that is compatible with the Ca^{2+} hypothesis has been proposed to account for acute and chronic opiate effects and is based on other changes in lipid metabolism. Acute opiate effects are considered to arise from alterations in membrane lipid structure owing to agonist-receptor binding. Since phospholipids are known to stimulate gene expression, it can be argued that prolonged drug treatment, and the consequent changes in lipid metabolism, may alter the synthesis of protein(s) necessary for tolerance and physical dependent development.

The specific Ca^{2+} pool(s) interacting with opiates have not been conclusively identified, but several possibilities have been considered. ATPase enzymes represent possible sites of opiate interference with Ca^{2+} flux because these enzymes have an important function in active ion transport. A number of investigators have reported positive effects of opiates on ATPase activity after in vitro, acute, and chronic treatment, but no clear pattern emerges from these studies that can adequately explain tolerance development. The data obtained with Mg^{2+}-dependent ATPase in synaptic vesicles perhaps have some relevance because the activity of this enzyme in synaptic vesicles was significantly increased after tolerance development, whereas the activities of the Mg^{2+}-dependent ATPase and Na^+, K^+ activated ATPase from synaptic plasma membrane fractions were not altered. Since synaptic vesicles are known to contain neurotransmitters, and Ca^{2+} is important for neurotransmitter release, the changes in synaptic vesicle Ca^{2+} content after opiate treatment may represent an interference with Mg^{2+}-dependent ATPase release mechanisms of synaptic vesicles since this enzyme has been implicated in the regulation of neurotransmitter release and Ca^{2+} accumulation.

Another possible binding site for opioids may be calmodulin. Calmodulin regulates the activity of many enzymes including phosphodiesterase and adenylate cyclase via the formation of complexes in response to Ca^{2+} fluxes. Calmodulin appears, therefore, to be a Ca^{2+} receptor site and represents a link between different types of cell messengers, namely, Ca^{2+} and cAMP, and may, thus, represent an important site for Ca^{2+}-opiate interactions. The interactions between opioids and calmodulin were mentioned earlier.

Another possible form of Ca^{2+}-opiate interaction is drug-induced inhibition of Ca^{2+} binding at synaptic membrane sites, but the evidence for such an effect needs considerable reinforcement. Opiates might displace Ca^{2+} from anionic binding sites on phospholipid molecules in neuronal membranes. The displacement of Ca^{2+} from these phospholipid opiate receptors would thus result in changes in membrane permeability to other ions and produce changes in neuronal excitability. Based on data that the phospholipid base-exchange reaction in nervous tissue is stimulated by Ca^{2+}, it is plausible that morphine alters the turnover and/or composition of membrane phospholipids by a direct effect on the Ca^{2+}-dependent base-exchange reaction. Marked alterations in exchange observed after chronic treatment possibly reflect a homeostatic adaptive change to overcome acute effects. The original studies supporting these conclusions are contained in a review (Chapman and Way 1980).

In summary, the development of physical dependence appears to involve several stages, some of which have commonalities with tolerance development, the most obvious step being receptor occupation. The ensuing gradual loss in sensitivity to agonist, accompanied by increasing sensitivity to antagonist with increasing exposure to agonist, may then follow identical or different pathways. If the route is the same, the answer might reside in how a common G-protein might be affected. Since the function of the G-protein itself is not affected by chronic opioid administration, and if the responsivity to agonist is dependent on the affinity state between ligand and receptor which affects receptor G-protein complex activity, it would appear that agonist modification of the conformational state of the receptor occurs. However, under such circumstances, it would be difficult to visualize how a given response becomes sensitized to an antagonist. One might well examine, in addition to adenylate cyclase, other G-proteins that interact with ion channels, phosphokinases, and their phosphoprotein substrates. The postulate discussed earlier that two G-proteins with opposing actions can be modified by agonist offers an attractive explanation. An unanswered question is why the presence of an agonist is required for an antagonist to elicit the amplified withdrawal response. Perhaps the main change is that the ligand-receptor-G-protein complex may become involved in controlling effector activity which is ordinarily regulated by the receptor-activated-G-protein complex. This would mean that the receptor-agonist complex becomes essential for physiological function. Finally, in making conceptual approaches, one might also well draw analogies with respect to the sequential events of learning and memory where acquisition, consolidation, storage, and retrieval are involved.

References

Abdel-Latif AA (1986) Calcium-mobilizing receptors, polyphosphoinositides, and the generation of second messengers. Pharmacol Rev 38:227-272

Abood ME, Law PY, Loh HH (1985) Pertussis toxin treatment modifies opiate action in the rat brain striatum. Biochem Biophys Res Commun 127:477-483

Aghajanian G (1978) Tolerance of the locus coeruleus to morphine and suppression of withdrawal response by clonidine. Nature (London) 276:186-188

Baram D, Simantov R (1983) Enkephalins and opiate agonists control calmodulin distribution in neuroblastoma-glioma cells. J Neurochem 40:55-63

Berridge MJ, Irvine RF (1984) Inositol triphosphate, a novel second messenger in cellular signal transduction. Nature (London) 312:315-321

Blanchard SG, Chang K-J, Cuatrecasas P (1983) Characterization of the association of tritiated enkephalin with neuroblastoma cells under conditions optimal for receptor down-regulation. J Biol Chem 258:1092-1097

Bläsig J, Herz A, Reinhold K, Zieglgänsberger S (1973) Development of physical dependence on morphine in respect to time and dosage and quantification of the precipitated withdrawal syndrome in rats. Psychopharmacologia 33:19-38

Brase DA, Iwamoto ET, Loh HH, Way EL (1976) Reinitiation of sensitivity to naloxone by a single narcotic injection in postaddict mice. J Pharmacol Exp Ther 197:317-325

Catlin DH, Liewen MB, Schaeffer JC (1977) Brain levels of morphine in mice following removal of a morphine pellet and naloxone challenge: no evidence for displacement. Life Sci 20:133-139

Chang KJ, Millen RJ, Cuatrecasas P (1978) Interaction of enkephalin with opiate receptors in intact cultured cells. Mol Pharmacol 14:961-971

Chapman DB, Way EL (1980) Metal ion interactions with opiates Annu Rev Pharmacol Toxicol 20:553-579

Cheney DL, Goldstein A (1971) Tolerance to opioid narcotics: time course and reversibility of physical dependence in mice. Nature (London) 232:477-478

Christie MJ, Williams JT, North RA (1987) Cellular mechanisms of opioid tolerance: studies in single brain neurons. Mol Pharmacol 32:633-638

Collier HOJ (1965) A general theory of the genesis of drug dependence by induction of receptors. Nature (London) 205:181-182

Collier HOJ (1980) Cellular site of opiate dependence. Nature (London) 283:625-629

Collier HOJ, Francis DL (1975) Morphine abstinence is associated with increased brain cyclic AMP. Nature (London) 255:159-162

Collier HOJ, Roy AC (1974) Morphine-like drugs inhibit the stimulation by E prostaglandins of cyclic AMP formation by rat brain homogenate. Nature (London) 248:24-27

Collier HOJ, Tucker JF (1984) Sites and mechanisms of dependence in the myenteric plexus of the guinea pig ileum. In: Sharo C (ed) Mechanisms of tolerance and dependence. US Gov Print Off ADM84-1330, Washington DC, pp 81-94

Collier HOJ, Cuthbert NJ, Francis DL (1981) Model of opiate dependence in the guinea-pig isolated ileum. Br J Pharmacol 73:921-932

Costa T, Aktories G, Schultz G, Wüster M (1983) Pertussis toxin decreases opiate receptor binding and adenylate cyclase inhibition in a neuroblastoma × glioma hybrid cell line. Life Sci (Suppl 1) 33:219-222

Cox BM, Osman OH (1970) Inhibition of the development of tolerance to morphine in rats by drugs which inhibit ribonucleic acid or protein synthesis. Br J Pharmacol 38:157-170

Cox BM, Ginsburg M, Willis J (1975) The offset of morphine tolerance in rats and mice. Br J Pharmacol 53:383-391

Crain SM (1988) Regulation of excitatory opioid responsitivity of dorsal root ganglion neurosn. In: Jlles P, Farsang C (eds) Regulatory roles of opioid peptides. VCH, Weinheim, pp 186-201

Crain S, Crain B, Peterson E (1986) Cyclic AMP or forskolin rapidly attenuates the depressant effect of opioids on sensory-evoked dorsal-horn responses in mouse spinal cordganglion explants. Brain Res 370:61-72

Crain S, Shen KF, Chalazonitis A (1988) Opioids excite rather than inhibit sensory neurons after chronic opioid exposure of spinal cord ganglion cultures. Brain Res 455:61-72

Dingledine R, Valentino RJ, Bostock E, King ME, Chang K-J (1983) Down-regulation of delta but not mu opioid receptors in the hippocampal slice associated with loss of physiological response. Life Sci 33 (Suppl 1):333–336

Duggan AW, North RA (1983) Electrophysiology of opioids. Pharmacol Rev 35:219–322

Dum J, Höllt V, Bläsig J, Herz A (1975) The influence of morphine tolerance/dependence on the interaction between morphine agonists and antagonists in vivo and in vitro. Naunyn-Schmiedeberg's Arch Pharmacol 287:R17

Ehrenpreis S, Light I, Schonbuch GH (1972) Use of the electrically stimulated guinea pig ileum to study potent analgesics. In: Singh JM, Millar LH, Lal H (eds) Drug addiction experimental pharmacology. Futura, New York, pp 319–342

Fry JP, Herz A, Zieglgänsberger W (1980) A Demonstration of naloxone-precipitated withdrawal on single neurons in the morphine-tolerant/dependent rat brain. Br J Pharmacol 68:585–592

Gillan MGC, Kosterlitz HW, Robson LE, Waterfield AA (1979) The inhibitory effects of presynaptic alpha-adrenoceptor agonists on contractions of guinea pig ileum and mouse vas deferens in the morphine-dependent and withdrawn states produced in vitro. Br J Pharmacol 66:601–608

Gilman AG (1984) Guanine nucleotide-binding regulatory proteins and dual control of adenylate cyclase. J Clin Invest 73:1–4

Goldstein A, Goldstein DB (1968) Enzyme expansion theory of drug tolerance and physical dependence. Res Publ Assoc Ment Dis 46:265–267

Goldstein A, Schulz R (1973) Morphine tolerant longitudinal muscle strip from guinea pig ileum. Br J Pharmacol 48:655–666

Gomperts BD (1983) Involvement of guanine nucleotide binding proteins in the gating of calcium by receptors. Nature (London) 306:64

Griffin MT, Law PY, Loh HH (1983) Modulation of adenylate cyclase activity by a cytosolic factor following chronic opiate exposure in neuroblastoma × glioma NG108-15 hybrid cells. Life Sci 33:365–369

Griffin MT, Law PY, Loh HH (1985) Involvement of both inhibitory and stimulatory guanine nucleotide binding proteins in the expression of chronic opiate regulation of adenylate cyclase activity in NG108-15 cells. J Neurochem 45:1585–1589

Griffin MT, Law PY, Loh HH (1986) Effects of phospholipases on chronic opiate action in neuroblastoma × glioma NG108-15 hybrid cells. J Neurochem 47:1098–1105

Hammond MD, Schneider C, Collier HOJ (1976) Induction of opiate tolerance in guinea pig ileum and its modification by drugs. In: Kosterlitz HW (ed) Opiates and endogenous opioid peptides. Elsevier/North Holland Biomedical Press, Amsterdam, pp 169–176

Hamprecht B (1977) Structural, electrophysiological, biochemical and pharmacological properties of neuroblastoma and glioma cell hybrids in cell culture. Int Rev Cytol 49:99–170

Harris J, Kazmierowski DT (1975) Morphine tolerance and naloxone receptor binding. Life Sci 16:1831–1836

Haycock VK, Rees JMH (1972) The effect of morphine pretreatment on the sensitivity of mouse and guinea-pig ileum to acetylcholine and to morphine. In: Kosterlitz HW, Collier HOJ, Villareal JE (eds) Agonist and antagonist action of narcotic analgesic drugs. Macmillan, New York, pp 234–239

Hazum E, Chang K-J, Cuatrecasas P (1979) Opiate (enkephalin) receptors of neuroblastoma cells: occurrence in clusters on the cell surface. Science 206:1077–1079

Himmelsbach CK (1942) Clinical studies of drug addiction: physical dependence, withdrawal and recovery. Arch Intern Med 69:766–772

Ho IK, Loh HH (1972) Cyclic 3′5′-adenosine monophosphate antagonism of morphine analgesia. J Pharmacol Exp Ther 185:336–346

Ho IK, Loh HH, Way EL (1973) Effect of cyclic 3′5′-adenosine monophosphate on morphine tolerance and physical dependence. J Pharmacol Exp Ther 185:347–357

Höllt V, Dum J, Blasig J, Schubert P, Herz A (1975) Comparison of in vivo and in vitro parameters of opiate receptor binding in naive and tolerant/dependent rats. Life Sci 16:1823–1828

Huidobro-Toro JP, Foree B, Way EL (1978) Single dose tolerance and cross tolerance studies with the endorphins in isolated guinea pig ileum. Proc W Pharmacol Soc 21:381–386

Jacob JJC, Barthelemy CD, Tremblay EC, Colombel ML (1974) Potential usefulness of single-dose actute physical dependence on and tolerance to morphine for the evaluation of narcotic antagonists. Adv Biochem Psychopharmacol 8:299–318

Jaffe JH, Sharpless SK (1968) Pharmacological denervation supersensitivity in the central nervous system: a theory of physical dependence. In: Wilker A (ed) The addiction states. Williams & Wilkins, Baltimore, pp 226–246

Klee WA, Nirenberg M (1974) A neuroblastoma and glioma hybrid cell line with morphine receptors. Proc Natl Acad Sci USA 71:3474–3477

Kosersky DA, Harris RA, Harris LS (1974) Naloxone-precipitated jumping activity in mice following the acute administration of morphine. Eur J Pharmacol 26:122–124

Koski G, Klee WA (1981) Opiates inhibit adenylate cyclase by stimulating GTP hydrolysis. Proc Natl Acad Sci USA 78:4158–4189

Law PY, Hom DS, Loh HH (1983a) Opiate regulation of adenosine 3′5′-cyclic monophosphate level in neuroblastoma × glioma NG108-15 hybrid cells. Mol Pharmacol 23:25–35

Law PY, Hom DS, Loh HH (1983b) Opiate receptor downregulation and desensitization in neuroblastoma × glioma NG108-15 hybrid cells are two separate cellular adaptation processes. Mol Pharmacol 25:413–424

Law PY, Hom DS, Loh HH (1984) Downregulation of opiate receptor in neuroblastoma × glioma NG108-15 hybrid cells: chloroquine promotes accumulation of tritiated enkephalin in the lysosomes. J Biol Chem 259:4096–4104

Law PY, Hom DS, Loh HH (1985a) Multiple affinity states of opiate receptors in neuroblastoma × glioma NG108-15 hybrid cells. J Biol Chem 260:3561–3569

Law PY, Louie AK, Loh HH (1985b) Effect of pertussis toxin treatment on down-regulation of opiate receptors in neuroblastoma × glioma NG108-15 hybrid cells. J Biol Chem 260:14818–14823

Law PY, Griffin MT, Loh HH (1984) Mechanisms of multiple cellular adaptation processes in clonal cell lines during chronic opiate treatment. NIDA Res Monogr 54:119–135

Lin SC, Way EL (1986) Effects of morphine and β-endorphin on Ca^{2+}-ATPase activity of synaptic plasma membranes. NIDA Res Monogr 75:117–120

Loh HH, Shen FH, Way EL (1969) Inhibition of morphine tolerance and physical dependence development and brain serotonin synthesis by cyclohexinide. Biochem Pharmacol 18:2711–2721

Louie AK, Law PY, Loh HH (1986) Cell free desensitization of opiate inhibition of adenylate cyclase in neuroblastoma × glioma NG108-15 hybrid cell membranes. J Nuerochem 47:733–737

Louie AK, Zhan J, Law PY, Loh HH (1988) Modification of opioid receptor activity by acid phosphatase in neuroblastoma × glioma NG108-15 hybrid cells. Biochem Biophys Res Commun 152:1369–1375

Lux B, Schulz R (1983) Cholera toxin selectively affects the expression of opioid dependence in the tolerant myenteric plexus of the guinea-pig. Eur J Pharmacol 96:175–176

Lux B, Schulz R (1985) Opioid dependence prevents the action of pertussis toxin in the guinea pig myenteric plexus. Naunyn-Schmiedeberg's Arch Pharmacol 330:184–186

Martin WR (1968) A homeostatic and redundancy theory of tolerance to and physical dependence on narcotic analgesics. In: Wikler A (ed) The addictive states. Williams & Wilkins, Baltimore, pp 206–225

Martin WR, Wikler A, Eades CG, Pescor FT (1963) Tolerance to and physical dependence on morphine in rats. Psychopharmacologia 4:247–260

Mattilla M (1962) The effect of morphine and nalorphine on the small intestine of normal and morphine tolerant rat and guinea pig. Acta Pharmacol Toxicol 19:47–52

Merzenich MM (1987) Dynamic neocortico processes and the origin of higher brain function. In: Changeux JP, Konishi HI (eds) The neurologic and molecular basis of learning. Dahlem Konferenzen. John Wiley & Sons, New York Chichester, pp 337–358

Mudge AW, Leeman SE, Fischbach GD (1979) Enkephalin inhibits release of substance P from sensory neurons in culture and decreases action potential duration. Proc Natl Acad Sci USA 76:526–530

Murayama T, Ui M (1983) Loss of the inhibitory function of the guanine nucleotide regulatory components of adenylate cyclase due to its ADP ribosylation by islet-activating protein pertussis toxin in adipocyte membranes. J Biol Chem 258:319–3326

Nehmed R, Nadlee U, Simantov R (1982) Effects of acute and chronic morphine treatment on calmodulin activity of rat brain. Mol Pharmacol 22:389–394

Nicoll R (1988) The coupling of neurotransmitter receptors to ion channels in the brain. Science 241:540–551

Nishizuka Y (1988) The molecular heterogeneity of protein kinase C and its implication for cellular regulation. Nature (London) 334:661–665

North RA, Karras PJ (1978) Opiate tolerance and dependence induced in vitro in single myenteric neurons. Nature (London) 272:73–75

North RA, Vitek L (1979) The effect of chronic morphine treatment on excitatory junction potentials in the mouse vas deferens. Br J Pharmacol 68:399-406
North R, Williams JT (1983) Opiate activation of potassium conductance inhibitory calcium action potentials in rat locus coeruleus neurons. Br J Pharmacol 80:225-228
O'Callaghan JP, Williams N, Clouet D (1979) The effect of morphine on the endogenous phosphorylation of synaptic plasma membrane proteins of rat striatum. J Pharmacol Exp Ther 208:96-105
Opmeer FA, van Ree JM (1978) Tolerance and dependence in vitro. In: van Ree JM, Terenius L (eds) Characteristics and functions of opioids. Elsevier/North Holland Biomedical Press, Amsterdam, pp 63-64
Quock C, Cheng J, Chan SC, Way EL (1968) The abstinence syndrome in long-term high dosage narcotic addiction. Br J Addict 63:261-270
Resvani A, Way EL (1983) Hypersensitivity of the opioid-tolerant guinea pig ileum to electrical stimulation after abrupt agonist withdrawal. Life Sci 33:349-352
Resvani A, Huidobro-Toro J, Hu J, Way EL (1983) A rapid and simple method for quantitative determination of tolerance development in the guinea pig ileum in vitro. J Pharmacol Exp Ther 225:251-255
Rodbell M (1980) The role of hormone receptors and GTP regulatory proteins in membrane transduction. Nature (London) 284:17-22
Roy S, Lee NM, Loh HH (1986) Mu opioid receptor is associated with phosphatase activity. Biochem Biophys Res Commun 140:660-666
Salter RS, Krinks MU, Klee CB, Neer BS (1981) Calmodulin activates the isolated catalytic unit of brain adenylate cyclase. J Biol Chem 256:9830-9833
Satoh M, Zieglgansberger W, Herz A (1976) Actions of opiates upon single unit activity in the cortex of naive and tolerant rats. Brain Res 115:99-110
Schmidt WK, Way EL (1980) Effect of a calcium chelator on morphine tolerance development. Eur J Pharmacol 63:243-250
Schofield PR, McFarland KC, Hayflick JS, Wilcox JN, Cho TM, Roy S, Lee NM, Loh HH, Seeburg PH (1989) Molecular characterization of a new immunoglobulin superfamily protein with potential roles in opioid binding and cell contact. EMBO J 8:489-495
Schulz R, Wüster M, Herz A (1979) Supersensitivity to opioids following chronic blockade of endorphin activity by naloxone. Naunyn-Schmiedeberg's Arch Pharmacol 306:93-96
Schulz R, Wüster M, Krens H, Herz A (1980) Selective development of tolerance without dependence in multiple opiate receptors of mouse vas deferens. Nature (London) 285:242-248
Schulz R, Seidl E, Wüster M, Herz A (1982) Opioid dependence and cross-dependence in the isolated guinea pig ileum. Eur J Pharmacol 84:33-40
Schuster L (1961) Repression and de-repression of enzyme synthesis as a possible explanation of some aspects of drug action. Nature (London) 189:314-315
Sharma SK, Klee WA, Nirenberg M (1975) Dual regulation of adenylate cyclase accounts for narcotic dependence and tolerance. Proc Natl Acad Sci USA 72:3092-3096
Sharma SK, Klee WA, Nirenberg M (1977) Opiate-dependent modulation of adenylate cyclase. Proc Nat Acad Sci USA 74:3365-3369
Shen J, Way EL (1975) Antagonist displacement of brain morphine during precipitated abstinence. Life Sci 16:1829-1830
Shohan S, Weinstock M (1974) The role of supersensitivity to acetylcholine in the production of tolerance to morphine in stimulated guinea-pig ileum. Br J Pharmacol 52:597-603
Simantov R, Amir S (1983) Regulation of opiate receptors in mouse brain: arcuate nucleus lesion induces receptor up-regulation and supersensitivity to opiates. Brain Res 262:168-171
Smits SE (1975) Quantitation of physical dependence in mice by naloxone-precipitated jumping after a single dose of morphine. Res Commun Chem Pathol Pharmacol 10:651-661
Song ZH, Rezvani A, Way EL (1987) Association of opiate tolerance and dependence development in the mouse vas deferens. Pharmacology 29:108
Takemori AE, Oka T, Nishiyama N (1973) Alteration of analgesic-antagonist interaction induced by morphihe. J Pharmacol Exp Ther 186:261-265
Tang A, Collins R (1978) Enhanced analgesic effects of morphine after chronic administration of naloxone in the rat. Eur J Pharmacol 47:473-474
Tempel A, Gardner EL, Zukin RS (1985) Neurochemical and functional correlates of naltrexone-induced opiate receptor up-regulation. J Pharmacol Exp Ther 232:439-444

Tulunay FC, Takemori AE (1974) The increased efficacy of narcotic antagonists induced by various narcotic analgesics. J Pharmacol Exp Ther 190:395–400

Ueda H, Satoh M (1988) Reconstitution of opioid receptors and GTP-binding proteins – signal transduction and its regulation of μ-, δ-, and κ-opioid receptor systems. In: Imura H, Shizume K, Yoshida S (eds) Progress in endocrinology. Elsevier, Amsterdam, pp 1137–1142

Ueda H, Harada H, Nozaki M, Katada T, Ui M, Satoh M, Takagi H (1988) Reconstitution of rat brain μ-opioid receptors with purified guanine nucleotide-binding regulatory proteins, G_i and G_o. Proc Natl Acad Sci USA 85:7013–7017

Ward A, Takemori AE (1976) Studies on the narcotic receptor in the guinea-pig ileum. J Pharmacol Exp Ther 199:117–123

Way EL, Loh HH, Shen FH (1969) Simultaneous quantitative assessment of morphine tolerance and physical dependence. J Pharmacol Exp Ther 167:1–8

Wei E, Loh H (1972) Morphine physical dependence unaltered by previous dependence on morphine. Nature (London) 238:396–397

Wei E, Way EL (1975) Application of the pellet implantation technique for the assessment of tolerance and physical dependence in the rodent. In: Ehrenpreis S, Neidle A (eds) Methods in narcotics research, vol 5. Dekker, New York, pp 241–259

Wikler A, Frazer HF, Isbell H (1953) N-allylnormorphine: effects of single doses and precipitation of acute "abstinence syndromes" during addiction to morphine, methadone or heroin in man (postaddicts). J Pharmacol Exp Ther 109:8–20

Wüster M, Costa T (1984) The opioid-induced desensitization (tolerance) in neuroblastoma × glioma NG108-15 hybrid cells: results from receptor uncoupling. In: Sharo C (ed) Mechanisms of tolerance and dependence. US Gov Print Off ADM84-1330, Washington DC, pp 136–145

Wüster M, Schulz R, Herz (1985) Opioid tolerance and dependence: re-evaluating the unitary hypothesis. Trends Pharmacol Sci 6:64–67

Section VI

Epilogue

CHAPTER 27

Some Thoughts on the Present Status and Future of Opioid Research

A. Herz

1 Introduction

In 1978, a few years after the identification of opioid receptors and the subsequent isolation of their endogenous ligands, our group published a monograph in which the state of opioid research at that time was represented. In the present work, published some 12 years later, more recent developments in this field are reviewed. The importance of opioids as biologically active peptides is probably reflected in their apparently long evolutionary history: opioids have been identified in a wide variety of species, ranging from primitive multicellular organisms to humans, and in both neural and nonneural tissue systems. At present, there is hardly an area of neurobiology in which opioid actions have not been implicated, and their involvement in nonnervous modulation of physiological processes is also receiving increasing recognition. Nevertheless, despite an immense amount of research, the real functional significance of these peptides, and the mechanisms underlying their actions, remains elusive.

The "new era" of opioid research depended largely upon the availability of opioid alkaloids which could easily penetrate the brain, and antagonists which could interact with the supposed receptors. In contrast to the situation for other neuropeptidergic systems, in which the detection of a new peptide initiated the search for related receptors and functions, opioid receptors were believed to exist well before their endogenous ligands were discovered; in this regard, opioid research benefited greatly from the availability of exogenous opioid-like (opiate) substances, and was thus able to (and still does) fulfill a "pace-setter" role in neuropeptide research.

2 The Elusive Receptor

In view of the impact of opioid research and vigorous attempts of competent groups all over the world, it is most surprising that elucidation of the molecular structure of the opioid receptor(s) has not yet been accomplished. [A specific opioid receptor binding protein has been cloned recently (Smith and Loh, this Vol.), its functional significance, however, has still to be evaluated.] This is indeed disappointing since the last few years have seen the description of the molecular structures of a number of other membrane receptors, including receptors of which very little is known in terms of their pharmacology, biochemistry, and physiology. The eventual deduction of the molecular structure of the opioid receptor(s) will not only represent a landmark of 15 years of opioid receptor research, but will open new avenues for the evaluation of the molecular aspects of opioid receptor function, e.g., signal transduction, receptor up- and downregulation, internalization, desensitization, and the development of tolerance and dependence. It will also clarify the question as to whether opioid receptor types and subtypes are generated from different genes, or whether they represent variations in the posttranslational processing of a common gene product.

3 More Than Three Opioid Genes?

The generally accepted view of three families of opioid peptides, deriving from three different genes, now seems to be open to reexamination. The detection of dermorphin, containing a D-amino acid in frog skin, signaled the possible existence of further opioid peptide systems in addition to proopiomelanocortin (POMC), proenkephalin (PENK), and prodynorphin (PDYN) some years ago. We have to wait for the identification of other D-amino-acid containing opioid peptides and their precursors, and to evaluate their phylogenetic relationships and functions before this issue can be settled.

4 Peptide and Receptor Multiplicity and Colocalization

The multiplicity of opioid ligands and receptors is not unique; it is also becoming apparent in other neuropeptide systems. While μ- and δ-opioid receptors seem to be, in some respects, related to each other (although there is no doubt that they represent different entities) κ-opioid receptors and κ-related phenomena often have quite different characteristics. For example, both μ- and δ-receptors trigger reward mechanisms, whereas κ-receptor activation produces aversion. It is also difficult to interpret the significance of the differences in the distribution of various opioid receptor types in various species, e.g., the prevalence of δ-receptors in the vas deferens of the mouse, of κ-receptors in the rabbit vas deferens, and of ϵ-receptors in the rat vas deferens. While there is good indication that the enkephalins represent physiological ligands of δ-receptors, and the dynorphins of κ-receptors, the problem of the natural ligand(s) of μ-receptors is still an open question. This is surprising since these receptors are widely distributed in the central nervous system and the effects mediated by this receptor type seem to be particularly important, as may be seen from the effects of the prototype opioid, morphine. There is some indication that β-endorphin (β-END) fulfills this role, at least with respect to some functions. But what about the spinal cord where β-END is largely missing (and μ-receptors are abundant) in the adult? The fascinating finding that mammals can synthesize morphinans raises the question of whether such alkaloids might be the physiological ligand of μ-receptors. In view of the very low concentration of such alkaloids present in the brain, this question is still open.

Another aspect of the multiplicity of opioid systems is represented by the site-specific processing of the various peptide precursors, well known for POMC in the anterior and intermediate pituitary lobes. This raises the question as to why a single precursor yields several peptides, each exhibiting very similar relative affinities for a single receptor type. The present scarceness of knowledge on the distribution and substrate specificity of enzymes processing and catabolizing opioid peptides may be one reason for our relative ignorance in the understanding of such problems.

A challenging phenomenon of current opioid research is the colocalization, not only of various opioid peptides in the same nerve terminal (and even synaptic vesicles), but also with nonopioid peptides and the classical neurotransmitters. The rapidly increasing data on this issue shows a bewildering complexity which is presently difficult to understand. The fact that so many different neuropeptides are present in circumscribed areas (e.g., more than one dozen in the spinal dorsal horn) is itself surprising. So what of the localization of opioids and other neuromodulators? Could it be that some of these compounds are simply phylogenetic relics, and that not all of them are of functional significance? At this stage, it seems rather hopeless to approach such questions in the intact mammalian nervous system. Simpler models may be more suitable to elaborate more on the nature of putative interactions between (opioid) peptides and neurotransmitters.

5 Nonneural Functions: Immunology

Recent developments indicate opioid mechanisms to be involved in immunological processes. Various opioid peptides, as well as opioid receptors, have been detected in several types of immune and inflammatory cells, and opioids were found to influence some functional parameters of these cells. The meaning of such (in detail often rather controversial) data, however, remains rather obscure. Some insight into the functional significance of these observations has been obtained from recent findings in our laboratory. We found that opioids exert peripherally mediated antinociceptive effects in inflamed (but not normal) tissue. Furthermore, we found that immune and inflammatory cells in the affected peripheral tissue to be rich in opioid peptides, and that the number of opioid receptors in inflamed tissue was somewhat increased. These findings suggest that immunological processes activate opioid mechanisms in inflammation.

6 New Horizons for Old Problems

Addiction, tolerance, and dependence remain an essential field of current opioid research. The finding that μ- and δ-opioid receptor-mediated mechanisms induce reward, whereas κ-receptor-mediated mechanisms not only lack such an effect but even induce the opposite (aversion), indicates opposing roles of various endogenous opioid systems in motivational processes. This also opens new aspects for the therapeutic use of various opioid receptor ligands. A topical question is whether the aversive effects of κ-ligands, as demonstrated in animal experiments and also in pilot experiments in humans, preclude the medical use of such ligands, e.g., as analgesics lacking abuse potential. In this regard, the demonstration of various κ-receptor subtypes becomes relevant, as it may be that certain κ-receptor subtypes do not mediate aversive effects.

The opposite motivational effects induced by the activation of different opioid systems suggests that, under "normal" conditions, a balance exists between reward and aversion. (Interestingly, both opioid effects involve dopamine, reward – stimulation, and aversion – inhibition of dopamine release.) From this, it may be suggested that other states of "pathological" imbalance between both mechanisms may also occur. It is, at present, difficult to validate such theories in humans, but it does not seem improbable that in the future, molecular biology may offer techniques to analyze the question of whether there is a hereditary component which renders certain individuals to be particularly susceptible to addiction, e.g., by a weakening of endogenous opioid reward mechanisms, or through an overactivity of endogenous opioid aversive mechanisms. This would prove a concept published a long time ago by Avram Goldstein.

7 Predicting the Future

Only a few fragmentary and very subjective aspects of opioid research, mainly arising from the personal interests of the author, and some of the work of his department over the years, have been discussed here. There are many other promising topics for research in the future, as is also reflected in many of the other chapters of this monograph. There is little doubt that opioids will continue to represent a most important area in neurobiological research. We eagerly await the cloning of the opioid receptor(s), which will open a new era for investigations of the broad spectrum of opioid functions. Our research into opioids led to the formulation of many new concepts, and also provided much firm data; many of these were surprises that would never have been predicted. No doubt, the future will be equally productive and full of surprises.

Subject Index

abuse potential 245, 255
acetylcholin (see ACh)
ACh (acetylcholine) 287, 288, 291, 292, 425
- intestinal motility 287 - 291
- opioid modulation 288 - 292
- release of 288
acupuncture 251
- antinociception, elicited by 249
acupuncture-induced antinociception 248
adaption of opioid systems to stress 229ff.
addictive drugs 394
- brain stimulation reward 394
- drug-liking effects 395
- euphoric effects 395
adenohypophysis 13, 14, 27, 38
adenosine 424
- A_1-receptor 220
adenylate cyclase 122, 128, 203, 204, 425
- NG 108-15 hybrid cells 426
- opioid gene expression 151, 155, 157
- role on regulation of opioid release 203 - 204
ADP-ribosylation, G-proteins 126
adrenal chromaffin cells
- adrenalectomy 13, 14, 18
- insulin stress 30
- neurotransmitters 156
- opioid gene expression 28 - 30, 156 - 160
- role of nicotine receptor 28, 156 - 158
affective disorders, opioids in 363 - 365
afferent neurons, primary 280, 283
agonist(s) 4, 273, 278, 279
- binding affinity 4
- treatment, chronic 188
alkaloids, quaternary 273
α2-adrenoceptors 220
4-aminopyridine (4-AP) 295
AMP, cyclic 426
amygdala and cortex 193
analgesia
- peripheral sites 256, 273
- stimulation-produced 246
analgesic(s) 214, 254, 255
- actions, peripheral 283
- effects 205, 273
- properties 245
antagonist(s) 4, 189, 201, 205, 207, 273, 278, 279
- administration, chronic 190
- binding affinity 4
- CRH receptor, effects 200 - 202
- ED_{50} 422
- negative intrinsic activity 110 - 112
- quaternary 277
antagonist-precipitated withdrawal 420, 421
anti-idiotypic monoclonal antibody 78
antinociception 254 - 256
- acupuncture-induced 248
- elicited by acupuncture 249
- environmentally-induced 247
- stimulation-produced (SPA) 247
antinociceptive
- actions 245
- effects 279, 280
-- peripheral 273, 274, 278 - 280
antisera generated against G-proteins 127
arcuate nucleus 17
arginine vasopressin (see AVP) 202, 205 - 207
ascending disinhibition, intestinal motility 287
AtT-20 cells 14, 15, 18, 20, 33
autoradiography 185
aversive/dysphoric effects 255, 256
- k agonists 333
- neuroanatomical substrates 335, 340
- opioid antagonists 334
AVP (arginine vasopressin) 202
- effects on opioid release 202, 205 - 207

basal ganglia
- anatomy 170, 171
- opioid receptor distribution 171 - 177
behavioral
- studies 278
- tolerance 388
β-CNA 77
β-END (β-endorphin) 199 - 208, 247, 291
- control of reproduction 305, 317, 321, 323
- immunocytochemical distribution in CNS 200
- release regulation in pituitary 203
β-endorphin/POMC, immunreactivity 55
- in basal ganglia 177
- CNS 55, 56
- CNS projections 57, 58
- peripheral tissues 55, 56
β-FNA 77
binding
- affinity 5
-- dynorphin A 6
-- endogenous peptides 6
-- morphine 5
- assays 4
- sites, separation and isolation 81
bioassays, mu-, d- and k-receptors 6
biochemical studies 80
bremazocine treatment 189

Ca^{2+}
- channels 15, 16, 28, 79, 141
- conductance 10, 130, 295
-- decrease 144
- entry 295
- influx 425
- opioid gene expression 15, 16, 28, 154 - 158
calcitonin gene-related peptide 261
calmodulin 433
cAMP (cyclic AMP) 14, 16, 28, 30, 31, 434
catecholamine release
- in CNS 213
- modulation 214
caudate putamen (see basal ganglia) 170ff.
CCK (cholecystokinin) 293, 294
CCK-8 293
cDNA 95
cellular opiate dependence 407
central opiodergic system 372
CGRP/SP, interactions 265
cholecystokinin (CCK) 293, 294
clinical

- reports 273
- studies 282, 283
cloning of opioid binding proteins 95
convulsive stiumuli 187
cortex 219
- and amygdala 193
- PENK 26
corticotropin-releasing hormone (see CRH)
CRH (corticotropin-releasing hormone) 13, 199 - 208
- autonomic and behavioral effects, association with stress 199
- control of opioid release 200 - 204, 308
- effects on reproduction mediation by opioids 306, 308, 316, 324
- receptor antagonist 200 - 202
- role in stress 199
CRH/opioid interaction 199
CRH/sex steroid, interaction 309
cross-tolerance 248, 252, 388
CT and PTX 408
CTX (cholera toxin), G-proteins 122ff.
cyclic AMP (see cAMP) 14, 16, 28, 30, 31

DA (dopamine) 217, 336, 425
- D-1 receptor 24, 35, 153, 338, 341, 342
- D-2 receptor 24, 35, 153, 341, 342
- mesolimbic system 336
- opioid gene expression 16, 24, 35, 153, 154
- receptor antagonsts 341
- reward pathways 338 - 342
- striatum 217
- terminals 191, 192
DA/opioid interactions 201, 303, 316, 336 - 344
DADLE 214 - 217
DAergic (dopaminergic)
- inhibition to hypothalamic release 201
- neurons 338
DAGO 77, 214 - 217
δ-agonists 279
δ-antagonists 279
δ-binding sites 295
δ-ligands 279
δ-receptors 254, 256, 277, 295, 296, 332, 426
δ-type 425, 428
2-deoxyglucose utilization
- β-endorphin 379
- δ-receptor activation 377 - 379
- κ-receptor activation 375 - 377
- μ-receptor activation 373, 374
- naloxone 379
- principles 369 - 372
- technique 369 - 372
dependence 147, 408, 409
- cellular opiate 407
- G-protein levels during 409
- mechanisms 399
- physical (physiological) 309, 312, 320 - 322, 336, 387, 396, 398, 417, 423, 424
- tolerance/physical, relationship 398
- withdrawal 306
depression, opioids in 363 - 365
desensitization 426, 427
distribution of opioids in brain and peripheral tissues 53ff.
dopamine (see DA)
dopaminergic system 341
dorsal
- horn 262
- root ganglion cultures 425
downregulation 426, 427
DPDPE 77, 214, 215
drug(s)
- discrimination 78
- not acting on opioid receptors 190ff.
DSBuLET 77
DSLET 77
DYN (dynorphin) 200 - 202, 205 - 207, 250 - 254, 261, 288, 291, 292 - 294
- (as for β-endorphin) 305, 321 - 323
- immunreactivity
-- in basal ganglia 177, 178
-- in CNS 61 - 63
- release 293
DYN/ENK (enkephalin), relationship 288
DYN/Met-ENK, interaction 202
DYN/SP, interaction 265
DYN_{1-8} 199
dynorphin (see DYN)

electrical stimulation 247
elusive receptor 443
endocrine state 186
endogenous opioid peptides 75
ENK (enkephalin) 76, 261, 288, 291 - 293, 428
- (as for β-endorphin) 305, 321 - 323
- (DADLE) 214
- (DAGO) 214
- immunreactivity
-- in basal ganglia 177, 178
-- in CNS 59
enkephalin (see ENK)
environmentally-induced antinociception 247
etorphine treatment 189

feeding behaviour 79
follicle stimulating hormone (see gonadotropins) 302 - 317
food and water deprivation 187
foot shock 246

G-protein(s) 121, 141, 203, 204, 208, 294, 408, 411
- ADP-ribosylation 126
- allosteric effects 106
- α-subunit 124
- antisera generated against 127
- cholera toxin (CTX) 122ff.
- classification 126
- distribution 128
- effector molecules 128
- functional states 107 - 109, 122, 123
- levels during dependence 409
- PTX (pertussis toxin) 122ff., 131
- PTX-sensitive 203
- quantification 126 - 128
- receptor coupling 103 - 106, 132 - 135

- role in control of opioid release 203, 204
- signal transduction 128 - 130
- structure 124
- subunit dissociation 102, 103, 106, 122, 123
- subunit heterogeneity 109, 121, 124 - 128
G-protein-coupled receptors 101
G-protein/receptor, interaction 102, 103131
GABA (gamma-aminobutyric acid) 201, 425
- opioid gene expression 17, 24, 153, 154
- role in hypothalamic opioid release 201
gamma-aminobutyric acid (see GABA)
gastrointestinal transit 296
globus pallidus (see basal ganglia) 170ff.
glucocorticoids 13, 15, 32
glucose utilization, local cerebral (see LCGU) 369ff.
GnRH (gonadotropin-releasing hormone)
- opioid regulation 18, 19, 302 - 317
GnRH/opioid interactions 303, 304
gonadotropin(s), control by opioids 302 - 317
GPI (guinea-pig ileum) 287, 288, 292, 293, 295, 296, 407, 409, 423, 429
- opioid tolerance 423
- physical dependence 423
GRH (gonadotropin-releasing hormone), opioid regulation 18, 19
guanine nucleotide regulatory proteins 101
guanine nucleotide-binding proteins 121
guanosine 5'-triphosphate 141
guinea pig 296
- ileum (see GPI)
- intestine 288

hallucinations 355
haloperidol 15, 24
heat and pressure stimuli 255
hippocampus 26, 36, 190
historical overview, opiate drugs 75
homeostatic theory, tolerance and physical dependence 399
5-HT (see serotonin) 201, 291, 292, 294, 345, 425
hyperexcitability, latent, tolerance and physical dependence 400
hypothalamic opioid peptide release 199ff.
hypothalamus 187, 199 - 201, 219
- opioid(s) 301, 303
-- gene expression 17, 25, 35
-- steroid receptors 303
- opioid/corticotropin-releasing hormone 306, 308, 316, 324

ICI 174,864 (selective δ-receptor ligand) 77
ileum, guinea-pig (see GPI) 287, 288, 292, 293, 295, 296, 407, 409, 423, 429
immune
- cells 280
- function, opioid involvement 234
- system 280, 283
-- PENK 31
-- POMC-derived peptides 20
immunoglobulin (Ig) superfamily 95
immunology, nonneural functions 445
inflammation 27, 37, 253, 278, 279, 280
- opioid involvement 234
- painful 252, 283
inflammatory
- agents and time of peak effect 275, 277
- conditions 279, 280
- hyperalgesic conditions 283
intestinal motility 287 - 296
- ascending disinhibition 287
- gastrointestinal transit 296
- mechanisms 287 - 295
- motility patterns 292, 293
- neurotransmitter effects 287 - 294
- opioid effects 291 - 296
- opioid receptors controlling 295, 296
- peristalsis 291, 292
- spasmogenic effects 291, 292
intestinal opioid action 294
intrasessional adaptation 390
ion channel 141
ir-β-END 248
isolated tissue 407
isolating opioid receptors, molecular approaches 89

K^+ channels 141, 294, 295
- G-protein coupling 294
- opioids 294, 295
- properties 294
K^+conductance 142
κ-agonists 279, 333
- possible clinical use 359
- psychotomimetic effects 353
κ-antagonists 279
- selective 79
κ-ligands 279
κ-receptors 245, 250, 251, 253 - 256, 277, 295, 296
kindling, cortical 36

LCGU (local cerebral glucose utilization) 369ff.
- β-endorphin 379
- δ-opioid agonists 377
- influences of opioid antagonists naloxone 379
- κ-opioid agonists 375
- μ-opioid agonists 373
Leydig cells 19, 38, 39
ligands, selective 77
locus coeruleus 142, 216
low-abuse potential 255
lutenizing hormone (see gonadotropins) 302 - 317

mammalian brain, opioid receptors 89
membrane-bound opioid receptors 80
mesolimbic dopaminergic system 336
Met-ENK (met-enkephalin) 199 - 202, 205 - 207, 249, 250, 262
migrating motor complex (see MMC) 291, 296
milk ejection, role of opioids 317 - 324
MMC (migrating motor complex) 291, 296
molecular
- approaches to isolating opioid receptors 89
- weight 91, 93, 97
monoclonal antibodies 94
- anti-idiotypic 78
mood 255, 256
morphine 245

- in animal tissue 4
- β-END 77
- binding affinity 5
- selective m-opioid ligand 7
- tolerance to 275
- treatment 188
motilin 293
motility pattern 292
motivation(al)
- effects of opioid
-- agonists 332
-- antagonists 333, 334
- neural substrates 335
- properties 316, 317
mouse vas deferens (MVD) 407, 408
μ-agonists 279
μ-antagonists 279
μ-binding properties 91
μ-ligands 279
- receptors 428
μ-receptors 245, 248, 253, 295, 332, 338, 428
- agonists 338
- antagonist 206
multiple opioid receptors 75ff.
MVD (mouse vas deferens) 407, 408
- opioid tolerance 424
- physical dependence 424
- tolerance of 408
myenteric
- neurons 296
- plexus 288, 291, 292, 295
-- effects 288 - 296
myogenic contractile activity 28

Na$^+$
- GTPase 112, 113
- opioid receptor activity 112
NAC (nucleus accumbens) 214, 219, 336, 337, 344, 424, 425
- D-1 receptors 342
- DA increase 340
- DA release 340
- opioid/DA interactions 338 - 344
- opioid/serotonin interactions 345
- role in opioid reinforcement 336, 337
NAL (naloxone) 213, 419, 421 - 426
- 2-deoxyglucose utilization 379
- influences of opioid antagonists naloxone 379
- treatment 190
NAL-antagonizable 216
naloxone (see NAL)
naltrindole 7, 77
NE (norepinephrine) 214, 425
- release 424
nerves, peripheral 283
neural substrates 335, 336
neuroanatomical substrates 340
neuroblastoma X glioma NG 108-15 hybrid cells 424
neuroblastoma-glioma hybrid cell 80
neurons
- DAergic 338
- myenteric 296
- primary afferent 280, 283
neurotransmitter release 79, 295, 429, 432
- acetylcholine (ACh) 288
- dopamine (DA) 340
- intestine 291 - 294
- serotonin 291, 292, 345
NG108-15 hybrid cells 424, 427, 428
- cholera toxin 111, 112
- GTPase 103, 104, 108 - 113, 133
- guanine nucleotides 103
- heavy/light membrane fractions 109
- Na$^+$ effect 112, 113
- opioid receptor down-regulation 107, 108, 133
- PENK gene expression 32
- pertussis toxin 108 - 128
- receptor/G protein interactions 103 - 105
nicotinic receptor, adrenal chromaffin cells 156 - 158
nociception 186, 261, 262
- behavioral measurement 262
non-neural functions, immunology 445
non-thermal stimuli 255
norbinaltorphimine 7
norepinephrine (see NE) 214, 424, 425
noxious
- pressure 252, 253
- stimulation, acute and chronic 246, 251
- stimulus 254
nucleus accumbens (see NAC) 219, 336, 337, 344, 424, 425

olfactory bulb 193
ontogenic
- development 14
- structure 78
opiate
- drugs, historical overview 75
- receptor 428
opioid(s)
- agonists 205, 207
-- motivational effects 332
- analgesia at peripheral sites 273
- antagonists 205, 206
-- motivational effects 333
- behavioral and physiological adaptation to stress 237
- control of reproduction 301 - 324
- effects 213
- gene expression 14, 16, 22, 24, 28 - 34, 39
-- adenylate cyclase 155, 157
-- Ca^{2+} 151, 154 - 159, 161
-- calmodulin 152, 155, 159
-- intracellular messengers 28, 29, 153, 155, 158, 159
-- kinase C 155, 158, 159, 161
-- protein kinases 152
-- regulation 153 - 158
- genes 444
- interactions with substance P 261
- neurotransmitter release 292, 293, 295
- peptide(s)
-- distribution 177 - 179
-- genes 11ff.

-- release 204
-- in stress, biochemical alteration 230
- peripheral tissues 63
- peristalsis effect 291
- potential clinical use 324
- psychotomimetic effects 351ff., 365
- receptor(s) 295
-- activity, Na^+ 112
-- antagonists 205, 206
-- in basal ganglia 169 - 181
-- central, modulation of 185ff.
-- characterization 94
-- cloning and sequencing 85
-- δ 295, 296, 305, 311
-- desensitization 133
-- distribution, basal ganglia
--- δ 171 - 173
--- κ 174 - 177
--- μ 173, 174
--- relation to opioid peptides 180, 181
--- species differences 171 - 181
-- down regulation 133
--- NG108-15 cells 107, 108, 133
-- drugs not acting on 190ff.
-- functional correlates, physiological 321 - 323
-- isolating, molecular approaches 89
-- κ 295, 296, 305, 311
-- ligands 76
-- localization
--- CNS 303, 309
--- gastrointestinal system 289, 295
--- periphery 273
--- periphery, including PNS 305, 323
--- smooth muscle cells 292
-- mammalian brain 89
-- modulation in stress 236, 237
-- μ, involvement in reproduction 305, 311
-- μ, in reward/motivation 332
-- μ-distribution 295, 296
-- multiple 75ff.
-- physiological modulation, sex steroid feedback 308- 312
-- presynaptically located 204
-- purification 90, 91
--- of covalently labeled 93
--- preparation of antibodies 94
--- of retaining ligand binding 91
-- response to stress 236
-- rhythms of sensitivity 306 - 308
-- selective agonists 214
-- selectivity 6
-- subtypes 3, 254
-- types 295
--- differentiation 80
- receptor-mediated effects 278
- receptor-specific antinociceptive effects, peripheral 280
- reinforcement 336
- in reproductive tissues 66
- research 443
- rewarding functions 316, 317
- sexually differentiated responses to 308 - 312
- spasmogenic effects 291
- tolerance 417, 423, 424
opioid-induced analgesia 205
opioids/steroid hormones, interactions 301 - 317
oxytocin 302
- role of opioids 317 - 320, 323, 324

PA (polyarthritis) 246, 250, 251, 253, 255
PAG (periaqueductal grey) 186, 190, 246, 247, 254
- pools 249
pain 245, 280, 282
- acute 280 - 283
- chronic 245, 252 - 254, 256, 280 - 283
-- management of 256
-- opioid involvement 234
- inflammatory 283
painful inflammation 252
pallidum (see basal ganglia 170ff.
paw pressure 274, 276
PCP 355
PDYN (prodynorphin) 34ff.
- in adrenal 64
- distribution of opioids in brain and peripheral tissues 59 - 62
- gene-products 58ff.
-- adrenal 64, 65
-- co-expression with AVP 233
-- distribution 233 - 235
-- gene expression 234
-- neuro-destructive/ -protective properties239
-post-translational processing 58, 233, 235
-- reproductive tissues 66, 67
-- stress 233 - 235
- gene(s) 22 - 39
-- post-translational processing mRNA in situ 23
-- regulation 35 - 39
-- regulatory elements 38
-- structure 23, 34, 35
PENK (proenkephalin) 22ff., 58ff.
- cortex 26
- distribution of opioids in brain and peripheral tissues 58, 59
- expression in glial cells 62
- gene(s)
-- regulation of expression 24 - 34
--- in chromaffin cells 156 - 160
--- intracellular mechanisms 158 - 160
--- by neurotransmitters 156, 157
--- by neurotransmitters glucocorticoids 24 - 34
-- regulatory elements 32
- gene/products 58ff.
-- CNS expression 59
-- glial expression 62, 63
-- in heart 64
-- post-translational processing 58
-- in reproductive tissues 66, 67
- heart 64
- stress 234
peptide, multiplicity and colocalization 444
periaqueductal grey (PAG) 186, 190, 246, 247, 254
peripheral
- analgesic actions 283
- antinociceptive 273, 274, 278
- components 277

- effects 282
- endings, substance P 280
- mechanism 275
- nerve terminals 283
- nerves 283
- opioid
-- antinociceptive effects 282, 283
-- receptor-specific antinociceptive effects 280
- receptors 283
- sites 279
-- opioid analgesia 273
- terminals 283
peristalsis effect, opioids 291
pertussis toxin (see PTX) 122ff., 408, 409, 427
pharmacological modulation 188
pheochromocytoma 12, 20
phospholipase A_2 128, 131
physical (physiological) dependence 336, 387, 396, 417, 423, 424
- and psychological 398
- and tolerance 387
pituitary (see also adenohypophysis) 194
- opioid(s) 301
-- gene expression 15 - 15, 27, 38, 152, 153
-- receptors 305, 323
pituitary-adrenal axis 283
place aversions 333
plexus
- myenteric 288, 291, 292, 295
- submucous 142
polyarthritic
- animals 253
-- spinal cord of 250
- rats 248, 251
polyarthritis (see PA) 246, 250, 251, 253, 255
polyclonal antibodies 94
POMC (pro-opiomelanocortin) 11ff., 20, 57, 152ff., 231, 232
- gene(s)
-- mRNA, in situ-hybridisation 13,14,18
-- post-translational processing 152
-- products 55ff.
--- in hypothalamus 55, 56
--- in pituitary 55, 56
--- post-translational processing 11, 55
--- release by CRH 231
--- in reproductive tissues 66, 67
--- in stress 230 - 232
-- regulation of expression 13 - 22, 152
--- chromaffin cells 19
--- corticotropin releasing hormone 153
--- CRH 14 - 16
--- intracellular mechanisms 154 - 156
--- neurotransmitters 153
--- pituitary 152 - 155
--- steroid hormones 18, 21
-- structure 11 - 22
-- synthesis and processing 247
-- transcription rate 13
- projections in the CNS 57
- regulatory elements 20
- synthesis and processing 247
positive reinforcement
- δ agonists 332
- effects 333
- μ agonists 332, 333, 339, 340
- neural substrates 336 - 345
- opioid reward pathway 333, 334
- role of DA 316, 317, 336
positive reward 255
potassium channels 79
pressure 278
- and heat stimuli 255
- paw 274, 276
presynaptic receptors 191, 192, 220
primary afferent neurons 280, 283
pro-dynorphin (see PDYN)
pro-enkephalin (s. PENK) 22ff., 58ff.
pro-opiomelanocortin (see POMC) 11ff., 20, 57, 152ff., 231, 232
prolactin 303, 306, 307
psychiatric aspects of opioid research 363
psychological dependence 387, 398
psychotomimetic effects
- of κ-agonists 353
- of opioids 351ff., 365
PTX (pertussis toxin) 122ff., 408, 409, 427
- and CT 408
- G-proteins 122ff.
PTX-sensitive G-protein 203
puberty, role of opioids 310 - 312
purification 89
- of covalently labeled opioid receptors 93
- of opioid receptors 90, 91
-- preparation of antibodies 94
-- retaining ligand binding 91

quaternary
- alkaloids 273
- antagonist 277
- compounds 277

raclopride 338
receptor(s)
- adenosine A_1-receptor 220
- affinity 273, 279
- antagonist
-- AVP 202
-- CRH 200 - 202
--- α-helical CRH 202
-- effect on opioid release 200 - 202
--- μ 206
-- opioid 205, 206
- bioassays, μ-, δ- and κ-receptors 6
- coupling, G-proteins 103 - 106, 132 - 135
cross-linking 81
- DA (dopamine)
-- D-1 and D-2 153, 342
-- PDYN 24
-- PENK 24
- δ 254, 256, 277, 332, 426
- G-protein-coupled 101
- G-protein interactions 102, 103, 131
-- intact membranes 103
-- NG108-15 cells 103 - 105
- κ 245, 250, 251, 253 - 256, 277, 295

- location of 280
- μ 245, 248, 253 - 255, 277, 295, 332
- multiplicity of 280
-- and colocalization 444
- NAC D-1 342
- nicotine, adrenal chromaffin cells 28, 156 - 158
- opioid 295
-- in basal ganglia 169 - 181
-- central, modulation of 185ff.
-- charactarization 94
-- cloning and sequencing 85
-- δ 305, 311
-- desensitization 133
-- distribution
--- basal ganglia, δ 171 - 173
--- basal ganglia, κ 174 - 177
--- basal ganglia, μ 173, 174
-- down regulation 133
--- NG108-15 cells 107, 108, 133
-- drugs not acting on 190ff.
-- functional correlates, physiological 321 - 323
-- isolating, molecular approaches 89
-- κ 305, 311
-- ligands 76
-- located peripherally 273
-- mammalian brain 89
-- membrane-bound 80
-- modulation in stress 236, 237
-- μ, involvement in reproduction 305, 311
-- μ-distribution 205, 296
-- multiple 75ff.
-- Na^+ 112
-- physiological modulation, sex steroid feedback 308 - 312
-- purification 90, 91
--- covalently labeled 93
--- preparation of antibodies 94
--- retaining ligand binding 91
-- receptor types, differentiation 80
-- relation to opioid peptides 180, 181
-- response to stress 236
-- rhythms of sensitivity 306 - 308
-- selective agonists 214
-- selectivity 6
-- species differences 171 - 181
-- subtypes 3, 254
- peripheral 283
- presynaptic 191, 192, 220
- σ 295
- σ/PCP 357
- types
-- multiplicity of 278
-- at peripheral sites 279
reinforcement 336, 337, 393
- processes 338
reinforcing effects 334
relaxin, opioid regulation 320
research, opioid 443
retina 220
rewarding properties 256
rhythms
- in opioid sensitivity 306 - 308
- reproductive rhythms 313, 314
SCH 23390 (DA receptor antagonist) 338, 342
schizophrenia
- opioid/dopamine, interactions in 366
- opioids in 365, 366
seasonal/photoperiodic reproduction, involvement of opioids 313, 314
segmenting mode 293
serotonin (5-HT) 201, 288, 345
- intestinal motility 288 - 294
- opioid release 201
sex steroid/opioid interactions 302 - 318
sexual behaviour, role of opioids 314 - 317
shock 13, 16
short chain alcohol 80
SIA (stress-induced antinociception) 247, 249, 251, 254
σ-receptors 295
σ/PCP receptor 357
signal transduction 121
- specificity 128
SKF 10,047 295
smooth muscle cell 292
somatostatin (SS) 144, 261
SP (substance P) 261, 280, 291, 293
SP/CGRP, interations 265
SP/DYN, interactions 265
SP/Met-ENK, interactions 262
SP/SS, interactions 266
SPA (stimulation-produced antinociception) 247
spasmogenic effect 292
- of opioids 291
spinal cord 186, 194, 251 - 255
- of polyarthritic animals 250
spinal dorsal horn
- encephalin gene expression 27
- PDYN 37
splicing, POMC 12
SS (somatostatin) 144, 261
SS/SP, interactions 266
stimulation-produced
- analgesia 246
- antinociception (SPA) 247
stimulus quality 255
stress 13, 16, 26, 30, 188, 247
- adaptation of opioid systems 229ff.
- opioid(s)
-- behavioral and physiological adaptation 237
-- involvement 229 - 239
-- peptides, biochemical alteration 230
-- receptor response 236
-- receptors, modulation in stress 236, 237
- PDYN (prodynorphin) system 233
- PENK (proenkephalin) system 234
- POMC (pro-opiomelanocortin) system 230
- pro-dynorphin gene/products in stress 233 - 235
- reproduction 306, 308, 319, 320, 322, 324
- visceral responses 238
stress-induced
- analgesia, opioid involvement 237 - 239
- antinociception (SIA) 247
striatum (see basal ganglia) 170ff., 191, 217, 428
- PDYN gene 35
- PENK gene expression 24

subsensitivity to opioids, steroid-induced 308 - 312
substance P (see SP) 261, 280, 291, 293
substantia nigra (see basal ganglia) 170ff., 193, 218

tachyphylaxis 390
testis 19, 31
tolerance 147, 248, 278, 387, 397, 408, 426
- acute 390
- analgesic effects 388
- behavioral 388
- environment specific 389
- functional 392
- hyperexcitability 400
- intracellular 14
- locomotor activity 392
- measured in the presence or absence of withdrawal 395
- morphine 275
- MVD 408
- opioid 417, 423, 424
-- effects with biphasic patterns 392
- physical/physiological 304, 309, 312, 322
-- dependence 387, 398
-- mechanisms 399
- reinforcement and subjective effects of opioids 393
- respiratory function 391
- selective 78
tonic
- inhibition in reproduction 305, 308, 310, 321
- stimulation of opioid release 308
transgenic mice 21
transmembrane sequences 85
transmitter release 295, 425
treatment
- bremazocine 189
- chronic agonist 188
- etorphine 189
- morphine 188
- NAL 190
tumors, POMC 19

UI (unilateral inflammation) 246, 247, 250 - 253, 256

vasoactive intestinal peptide (see VIP) 287 - 290, 293
vasopressin 15
- in reproduction 321
ventral tegmental nucleus (see basal ganglia 170ff.
VIP (vasoactive intestinal peptide) 287 - 290, 293
- release 293
vocalization 255, 256
- test 256

water and food depriviation 187
withdrawal 395 - 397, 409
- syndrome 387
writhing 274 - 276, 278

Xenopus 11, 12, 17

SPRINGER NATURE

GPSR Compliance

The European Union's (EU) General Product Safety Regulation (GPSR) is a set of rules that requires consumer products to be safe and our obligations to ensure this.

If you have any concerns about our products, you can contact us on ProductSafety@springernature.com

In case Publisher is established outside the EU, the EU authorized representative is:

Springer Nature Customer Service Center GmbH
Europaplatz 3
69115 Heidelberg, Germany

FSC
www.fsc.org
MIX
Papier | Fördert gute Waldnutzung
FSC® C083411

Kolibri 360 GmbH
Ferdinand-Jühlke-Straße 7
99095 Erfurt, Deutschland
produktsicherheit@kolibri360.de